100 YEARS
OF
SUPERSTARS!

from the pages of Squills Year Book

ie Racing Pigeon Publishing Co Ltd, 13 Guilford Street,
London, WC1N 1DX, United Kingdom

100 Years of Superstars!

Published by The Racing Pigeon Publishing
Co Ltd, 13 Guilford St, London, WC1N 1DX
United Kingdom

ISBN 0-85390-045-0

Text set in 10/10½ Times by RP Typesetters,
London.

Editor Rick Osman

Printed and bound in England
by Hillman Printers (Frome) Ltd, Somerset

British Library Cataloguing-in-Publication
Data. A catalogue record for this book is
available from the British Library.

Contents

CONTENTS

Introduction

A hundred years ago the first Squills Year Book was published as a small diary with a few stud pages, even in this modest form it was unique, being the earliest known example of a loft record book for fanciers ever produced. Three years later the next edition, still called Squills Diary, was printed in 1901 and this set the tone for future editions as it contained a few articles and a few advertisements from well-known fanciers of the time as well as stud pages and other loft record pages. Since then Squills Year Book has been published every year and, although the actual title has varied from Squills Diary to Squills Almanack to Squills Annual to Squills Year Book with a variety of combinations on the theme, the one constant has been the name of 'Squills'.

As no articles appeared in the early years, 1898, 1899 and 1900, we have used suitable reports from The Racing Pigeon to maintain continuity. In the 1906 edition Squills introduced the concept of the Stud Register which was hoped would clear up much of the confusion over pedigrees, especially as at that time not all birds carried rings. The idea lasted for a few years until ring records were established. It was also in the 1906 edition that Squills welcomed fanciers to "the first edition of his Annual", acknowledging one of the name changes in the book but providing some confusion for collectors of pigeon memorabilia in later years.

Certainly the original and first Squills, Alfred Henry Osman, could have had little idea that his concept of an annual publication for pigeon fanciers would grow to become the world famous book that it is today. Indeed with the vast number of books and videotapes available today it is hard to realise just how important an effect Alfred Henry Osman, writing as Squills, had on pigeon racing. His writings in those formative years of our sport were essential in giving novice and experienced fanciers alike an important grounding in the facts of pigeon racing. Some of his writings, as is proper, appear in this volume and they are good examples of how he was the first writer to recognise and explain much of what is taken for granted in pigeon racing today. His importance to pigeon racing can never be underestimated.

In selecting the best article from each Squills Year Book over the years two things became apparent, one is that 'there is always something new to learn' and the other is that 'there is nothing new under the sun'. Whereas every edition contained some gem or nugget of information that had been recorded for the first time, as often as not many of the fanciers who contributed bemoaned the same things that are complained of today, it sometimes seems that every year is the worst ever for losses! Interestingly the early breeding of young birds was being advocated some 80 years ago and at that same time some fanciers were extolling the advantages of Widowhood flying. Since then the arguments against professional pigeon breeders, mob flying, expensive corn and so forth have carried on in much the same manner for year after year. What has undoubtedly changed over the years has been the mixtures fed to our birds, those changes are clearly shown as you read through the articles year by year. The improvements in diet for our pigeons accurately reflect the increasing knowledge and thus improvements in diet amongst society in general.

As the editor of Squills Year Book I am in a privileged position, it is a job which is steeped in the history of the Year Book and thus it is steeped in the history of pigeon racing. In selecting the articles which appear in '100 Years of Superstars!' it was essential that one article be chosen from each year, inevitably this meant that, for some years, good articles have been passed over as only one could be chosen. This inevitably meant that some very fine fanciers who have appeared in Squills Year Book do not appear in this volume, that is the onerous side of being an editor. However one of the joys of editing Squills Year Book is that so many of the successful fanciers who have written for the annual have understood and acted upon the suggestion that they should write with the novice in mind. This has meant that many fanciers have been willing to impart their knowledge and experiences freely so as to benefit those starting in pigeon racing, in doing so they have often given other experienced, successful fanciers plenty of food for thought. Indeed I am certain that every fancier, no matter how successful, will find something new in an edition of Squills Year Book. I will not comment on individual articles except to point out a single instance which was said in 1975, "They are just a hobby and a relaxation to me", the writer was a double Up North Combine winner.

I finish by thanking those who have contributed to the production of '100 Years of Superstars!', Tony Cowan whose belief in the goldmine of information in the archives of The RP became this book, Dave 'Clun' Shelbourne who filled some of the gaps in our collection of Squills Year Books and, most importantly, the hundreds of fanciers who have contributed articles to Squills Year Books over the past 100 years. Only a proportion of them appear in '100 Years of Superstars!' but every one of them has contributed to the greater understanding of pigeon racing. On behalf of pigeon fanciers everywhere I thank them.

Rick Osman

The Home of 'Old 86'

When poor 'Carrier' wrote an account of Mr Logan's loft in 1883, he said that he felt like a dog in a tripe shop, and hardly knew where to begin. I cannot find words to more fittingly describe my own predicament, with the same task to accomplish, fifteen years later.

East Langton Grange stands but a stone's throw from East Langton Station on the Midland Railway, the centre of a noted hunting district. Mr Logan has the reputation of being one of the finest riders in the neighbourhood, and the first-class style in which he won the House of Commons' point-to-point race last year is, no doubt, fresh in the minds of my readers. The fine range of stabling in which the hunters are kept is shown in the full view of the outside of the loft and stables.

Although the information may not be new to all there must be many in the fancy today unacquainted with the details of Mr Logan's early connection with the sport of pigeon racing.

In 1867, Mr Logan lived at Wybersley Hall, near Stockport, and strolling out one day with his gun, a navvy, who worked for him, said "Don't shoot my birds, governor". He replied, "What are they? Is there anything remarkable about them?" The man showed him his pigeon house, and explained that his birds were homers, and pointed out to him one that had done London. From that moment, Mr Logan became a homing fancier, and determined to do more than other men had done with their birds. At this time, it was thought a great feat to get a bird to fly London, but in his first year Mr Logan had four out of five do that; and he then tried Dover with a nine months old bird, but lost it. After two years, he removed to Rotherham, but his business did not allow him to keep homers for two years, when he went to live at Sheffield in 1872, and started them again. here he so far improved on his former practice that he did what was then probably the longest journey that had up to that time been accomplished in England, viz, from Calais to Sheffield, 218 miles; and between 1872 and 1875 he twice got his birds to do Ostend to Sheffield 238 miles.

In 1875, he removed to his present residence, and in July, 1877, won 1st prize in the UCFC race from Paris, 291 miles, with a bird he bred at Sheffield. The following account of the race is extracted from the *Pigeon* of

'Old 86'.

**Mr J W Logan MP
For the Harboro' Division of
Leicestershire**

July 13, 1877: – "The birds were liberated at 4.45am, on the 4th inst., in front of the Northern Railway in Paris, and went straight away without rounding; but up to the time of the close of the race (two days), only one bird, a red chequer belonging to Mr Logan, had returned".

But, in spite of his success on the Paris road – for Mr Logan again won first prize from there in 1879 – he was not satisfied he had the class of birds he was striving for; therefore, he cleared out all his birds, went over to Belgium, and there bought the first bird back from the Rome race. it came home three days before the actual winner in the race, but belonging to a man outside the limits of the race, could not win, although it was the first home. This bird was afterwards known as 'Rome I', and was the foundation of the best and gamest birds ever bred. He also bought the entire loft (every bird) of N Barker, when he was at the very top of his success. Among these were 'Montauban', 'Orleanist', 'Young Gladstone', 'Lord Derby', 'Captain Webb', 'Anglaise', 'Montauban's' sister, and his (Barker's) old stock cock, and the two hens that nicked so well with him, 'Mareka', the mother of 'Montauban', and the little blue chequer hen, the mother of 'Young Gladstone'.

He also bought 'Debue' of Uccle's old cock, that had been a winner for years in all the big races in Belgium, and also a son of the old cock.

He also bought the 'Mausta' mealy cock, and a daughter of Hansenne's little red chequer cock, that won so much money for Hansenne.

He also got a good hen of Gits', which was imported for him by J O Allen.

In 1877, he won 1st prize, Paris.

In 1878, he won 1st prize, Amiens.

In 1879, he won 1st prize, Paris.

In 1880, he won 1st prize, Rheims.

Also 1st prize, Rheims, open to all England for over 250 miles, with a named bird.

In 1881, he won 6th prize, Penzance.

In 1882, he won 2nd prize, Arras.

In 1883, he won 1st, 2nd, and 3rd Cherbourg; 3rd Rennes; and also special highest velocity prize for a named bird.

In 1884, 1st, 2nd, and 3rd Rennes; and 1st, 2nd, and 3rd La Rochelle.

Outside view of lofts, showing the range of hunting stables.

In 1885, 3rd Rennes; 1st La Rochelle.

In 1886, as it was the last season of the UCFC, he determined to put all the strength of his loft on the road, and started sixteen birds. All went to the preliminary stages, including Ventnor, where he won 1st, 2nd, and 3rd for singles, and 1st, 2nd, and 3rd for a series of two birds.

With the above sixteen birds, he sent ten young birds of 1885 as far as Ventnor, and two of them were lost.

To Cherbourg, ten birds were sent, and all homed; but it was no race, as the birds were tossed on Sunday.

To Granville, eight birds were sent, and all homed. These eight birds did not go any further with the UCFC, but four afterwards went to Granville with the MFC, and returned home.

To Rennes, seven birds out of the ten sent to Cherbourg were sent, and he won 1st and 2nd prizes, 'Old 86' winning 1st.

To La Rochelle, seven birds were sent; six homed, and he won 1st, 3rd, and 4th prizes, day of toss, 'Old 86' winning 1st by nearly two hours.

Out of the sixteen birds originally started, only one bird was lost, and that was No 262 at La Rochelle, 444 miles.

The ages of the sixteen birds were:

1 ....7 years old
1 ....6 years old
3 ....5 years old
4 ....4 years old

Five favourites – No 538 and his two sons and two daughters.

$$\begin{array}{r} 4 \dots\dots\dots\dots\dots\dots\dots\dots\dots\dots\dots\dots\dots 3 \text{ years old} \\ \underline{3 \dots\dots\dots\dots\dots\dots\dots\dots\dots\dots\dots\dots\dots 2 \text{ years old}} \\ 16 \end{array}$$

The seven birds sent to La Rochelle only had three tosses across the water, viz, at Cherbourg, Rennes, and La Rochelle.

The following are the particulars of their numbers, ages, and previous work:

86 (better known as the famous 'Old 86', of which a life-like reproduction is depicted on the title page of *The Racing Pigeon*). – 7 years old, 1st La Rochelle 1884; 1st Rennes 1884; 3rd Rennes 1883; 1st Ventnor 1886; 1st Rennes 1886; 1st La Rochelle 1886.

116 – 6 years old, Rennes in 1883-4-5, each time a winner, and 4th prize La Rochelle 1886.

170 – 5 years old, 2nd Granville 1884; 3rd La Rochelle 1886.

152 – 5 years old, 1st La Rochelle 1885; went to and returned from La Rochelle in 1886.

252 – 4 years old, La Rochelle in 1885 and 1886.

258 – 4 years old, Nantes 1884; La Rochelle 1886.

262 – 4 years old, Granville 1884; lost at La Rochelle 1886.

The following are the distances from East Langton: Amiens 220 miles; Arras 222 miles; Paris 291 miles; Rheims 311 miles, Penzance 258 miles; Cherbourg 205 miles; Granville 269 miles; Rennes 309 miles; La Rochelle 444 miles.

The number of times in which the above seven birds failed to reach home on the day they were tossed, not in 1886 only, but during the whole years of their training, did not amount to fourteen times; for instance, Nos 170 and 262 never spent a single night out.

This brings us down to the date of the sensational sale at the Royal Aquarium. Never shall I forget the excitement which took place when this grand stud of birds was put up by auction. A few sensational sales have taken place since, but ~I venture to think nothing to equal that at the Aquarium. I remember poor old Barker, of Reading, coming to me after the sale and saying, "he had bought several lots at a good figure, but they were grand birds, and were worth all the money he had paid for them".

To descendants of birds sold at this sale can be traced many of the finest birds at present in this country, more particularly in the north, and it was a good thing for the English fancy that Mr Logan decided to sell off as he did, and, moreover, it was a good thing, too, that the reserve he had placed on 'Old 86' and one or two other particular favourites, was never reached, for whilst these birds were still in his loft, there was every hope that his retirement from the fancy would be but temporary. Fortunately this proved to be the case, for a few years later Mr Logan was again taking an active interest in the sport, and from that day to the present has continued to rank as the foremost fancier in the land.

It would be impossible for any pen to do justice to Mr Logan's many good offices to the fancy. His great ambition has always been to elevate the sport and raise it to a higher pinnacle. The fact that members of the Royal Family are now not ashamed to associte their names with racing pigeons is, I take it, conclusive evidence that the work of elevation commenced many years ago, when the old United Counties FC was formed, and continued to the present day, has been fruitful.

The outside view of Mr Logan's loft will hardly give the reader a proper conception of the thoroughness with which it has been constructed. Mr N Barker, who, I suppose, has been over as many lofts as any man living, describes it as the champion loft of the world.

The lower gable with an alighting board on the left-hand side shows the entrance to the old bird loft. The entrance to the young bird loft is at the next gable to the right. Mr Logan keeps his old and young birds separate, as he believes they do better if so treated.

The inside of Mr Logan's loft is, I suppose, as fine a piece of workmanship as one could possibly find, and I dare hardly estimate what it must have cost to have constructed these fittings, but money and patience are never taken into account by Mr Logan when he sets his mind on accomplishing any worthy object he has in view. The illustration of a section of the interior of Mr

Logan's old bird loft will give some idea of how it is fitted up. On the left it will be seen there are two nest boxes showing the entrance for the birds, and the ventilation holes above and below, also the lath door with which they can be closed if necessary. Each nest box is fitted with a small brass frame, which has a place for a card to slide in, in order that particulars of laying, date of hatching, &c, can be entered on the spot. Over the nest boxes are severla perches, under which it will be seen is a platform composed of the top of the nest boxes, on which the cock bird delights to play about and coo, but is warned against going on to the next box by a slight lathwork partition down the centre.

There are fourteen of these divisions, each one of which is occupied by a couple of pair of birds. The divisions can be closed at will, and when closed there is a pssage down the centre of the loft through which the birds can be seen from either side without in anyway disturbing them.

What a paradise for a pigeon to live in, but rest assured that the pigeons that find a home there have to be very choice indeed. As I entered the loft my eye caught sight of a grand chequer pied, a bird of wonderful quality, just my ideal of what a worker should be, and his pedigree stamped him as good as he looked, for he was a grandson of the 1894 National cock, and a great grandson of 'Old 86', with a cross off Dardenne's old pied. By the way, I fancy Mr Logan has the only living daughter of Dardenne's old pied bird in his loft, which he purchased of M Dardenne before he died. Mr Logan was after a sister at the time that had done a lot of work, but he got this bird and she has proved a very useful one.

Bird after bird in the loft I handled, but could not find a faulty one. There is no room for such here, for they would soon find a short shrift.

The illustration gives portraits of five of the grandest blue chequers I have ever seen in my life, and consists of:

No 538 – Blue chequer cock, bred in 1889, on the extreme right of the illustration. He is son of 170 and 199. 170 was son of 105 by 20, and flew La Rochelle 444 miles, to Langton in the day in 1886; 199 was by 156, a son of 'Rome I', that flew La Rochelle twice, out of 163, a daughter of D2 by B16, that flew Rochelle, 538 flew Nantes, 6th prize, in 1892, La Rochelle in Grand National in 1894, and Arbroath in 1896.

No 942 (the bird facing to the front of illustration). – Blue chequer cock, bred in 1894, by 538 out of 509, 509 being by 86 out of 163. 942 has flown Arbroath.

No 954 – Blue chequer cock, on the extreme left, was bred in 1895 by 538 out of Hansenne hen. 954 has flown Newcastle, 170 miles.

No 953 – Blue chequer hen, bred in 1895, by 538 out of Hansenne hen. 953 has flown Arbroath, 290 miles.

No 9533 – Blue chequer hen, next to No 538 in illustration, was bred in 1895 by 538 out of Hansenne hen. 9533 has flown Thurso, 430 miles.

Having been thoroughly through all the champions in the loft, amongst which I noticed several very old favourites, including:

509 – Blue chequer hen, bred in 1888, by 86 out of 163; flown Cherbourg.

The prisoners' aviary, showing East Langton Grange in the distance.

577 – Blue chequer hen, bred in 1891, by 86 out of 259. 259 by 577 flown Cherbourg.

609 – Blue chequer hen, bred in 1892, by 258 out of 531; 258, son of 'Rome I', that flew La Rochelle twice, and 531, a daughter of 86; I was next shown over the prisoners' aviary on the most elaborate scale, and with a view to the birds' living in the most natural state possible. It consists of two divisions, as shown in the illustration, which also shows East Langton Grange in the background.

At present the prisoners' aviary contains the Wegge birds Mr Logan bought at the famous Wegge sale. No 2, the red chequer Wegge cock is mated to a blue Wegge hen which Mr Logan went over especially to buy after the sale, as she was not included in the lots put up. I liked this hen very much. She is a very handsome bird, with rare shoulders, and I expect will produce some good workers. If she does not do so, no matter what she cost, her lodgings won't be for long at Langton, for the proof of the pudding is in the eating here. No 5, the blue Wegge cock, which was thought at the time of the sale to be the best bird put up, is mated to a bird of Mr Logan's old strain, descended from 'Old 86' and 'RI' and I should not be surprised if these birds hit well. No 15 Wegge, the mealy, is mated to a black hen purchased at Dardenne's sale. He is just the right shape to put with this hen. This hen was No 10 on Dardenne's sale list, and is from M Lucien Bastien, of Vervier loft, and is own sister to his 1st prize Biarritz, a winner of 7,050 francs in 1896.

Section of interior, showing nest boxes.

There was one pair of birds in the prison that I must not forget to mention; that is, a little short-faced Hansenne cock, which is mated to a hen of the same strain. It will be seen that there is some of the blood of this cock in the two daugthers of 538, depicted in the picture. He is one of the real old-fashioned Hansenne type, and full of intelligence in every look.

Just one other pair of prisoners I must mention; then I won't bother you any more about the birds, for I could write sufficient about the twenty-five pairs in the flying loft and the prisoners' to fill a book. The last pair I handled were pure Coopman's, both blues, and as nice a pair of birds as any fancier would wish to have. For years Mr Logan tried to get M Coopman's to let him have a pair of birds, and it was only after sticking to him with that dogged determination of an ardent lover of the sport that he got what he wanted, and they are prized all the more owing to the trouble they were to obtain.

Mr Logan being just about to put his birds in training, I thought I might be able to glean a few tips for the benefit of my young friends. He informed me that the tosses he usually gives his old birds are ten, twenty, thirty, fifty, and seventy miles, when they generally pick up the race birds. But sooner than risk a really bad day in training, he prefers to give his birds a longer life. Young birds he gives two or three tosses of a mile round home in different directions, if time, before making a start; he then sends them two, four, six, ten, twenty, thirty, forty, fifty-five, seventy, and 100 miles, which he does not think will hurt good youngsters. If he has anything he particularly fancies, he stops it at 100 miles, but if the weather is good he has come to the conclusion that 130 miles or more will not hurt youngsters a bit; in fact, he thinks the extra work does them good, providing they do not meet a very bad day.

To make a good start with long-distance homers, Mr Logan says a beginner should make sure he gets the right blood. Three hundred milers are plentiful enough, but it is birds that will cover 400 and 500 miles in this country that are had to find.

Beans are the favourite grain for young and old birds in the Langton loft, and they are spread out in useful bins with ventilation running underneath, and turned over with a wooden rake regularly; but Mr Logan is not a believer in keeping them until they are very old as long as they are in good, dry, sound condition. In winter the birds get a change, sometimes of one sort and sometimes another, but as soon as breeding commences then beans are the staple food, with a little rice and canary to entice them in on race days.

There is one novel feature about the loft I must not forget, and that is the electric fittings. On a bird's arrival on the platform of the trap an electric bell in the house is rung, and a disc indicator dropped, so that there can be no finding a bird in the loft here, and to guard against accidents, Mr Logan has now rigged up an indicator which drops in front of the window, just above the entrance, directly anything enters, and can be seen from the garden; so that the possibility of a bird slipping in by any manner of means is guarded against.

That the birds in the home of 'Old 86' are as good as they were in the old days I am quite convinced. Mr Logan's keen interest in politics keeps him a good deal in London, and therefore they do not get that amount of personal supervision that is necessary, but the following list of prizes won during recent years by the Langton loft shows that there are still some chips of the old block to be found at East Langton Grange: In 1895, 1st and special Berwick. In 1896, 1st and special Berwick; 2nd Arbroath; 5th Thurso. In

1897, 4th and 5th Berwick' 1st and 2nd Banff.

Mr Logan is president of the National Homing Union, the National FC and the Central Counties FC, and I feel sure that I am only echoing the sentiments of the entire fancy in hoping that he may live for many years to preside over the destinies of the sport which he, more than any man living, has raised to its present heavy state.

One of London's Coming Champions

Mr R Mattock

New Mills has been described as a hot-bed of the fancy up North. I think the same thing may be said of Wood Green in the South, for I do not know any district in the metropolitan area where there are so many good fanciers at such close quarters.

It was in 1892 when I was first introduced to Mr R Mattock, the autumn previous to which he started flying, which was early in 1893.

His introduction to the sport was through the instrumentality of the Mercer Bros, and from the first his career has been a most successful one, but hardly so successful as I believe it will be in the near future, as he has handled his birds carefully, with the result that he has now a fine reserve of old birds to fall back upon.

I am firmly convinced that success in pigeon racing only comes to those who strive for it. Mr Mattock has worked hard for the success he has attained, has waited patiently, and deserves the reward of his efforts. Not a little of his success, however, is due to the services of 'Dick' who looks after the birds in Mr Mattock's absence, for Mr Mattock who is in business with his father, is ever attentive to business before pleasure, and therefore Master Dick frequently has to attend to the birds for days at a time. That they get proper attention results prove.

Quite apart from a nice, smart team of youngsters which have been trained to Claypole this year, Mr Mattock has twenty-four trained old birds in his loft. He had eleven youngsters left last year, and the whole eleven have flown Newcastle this year, and are on the shelf to see what they are made of in 1900.

One of the best birds Mr Mattock ever owned was old 'Wingfield Pride'. I remember this good old pigeon well. He flew most consistently for several years, but like many another good one, eventually went down from some unknown cause. As 'Wingfield Pride' is answerable for many of the best birds in Mr Mattock's loft, I append the details of his work. Bred in 1892, he was untrained. In 1893 he flew Newark; 1894, Retford, York, Newcastle, 6th Berwick, 9th Arbroath; 1895, Retford, York, Newcastle, Berwick, Banff; 1896, Retford, York, Newcastle, 1st Arbroath, 5th Federation, being the only bird home in Palace Club before the close of the race, and flew Thurso, 500 miles, 7th prize, 1296 velocity, winning Mr G P Pointer's cup for best average velocity in AP old bird races. 'Wingfield Pride' was bred from Mercer's old strain.

Fortunately before losing this good game old bird, Mr Mattock was able to rear many a good youngster from him, amongst others the two chequer hens illustrated; but first of all it will be as well if I have a word or two to say about the game little Lerwick hen with which Mr Mattock won 1st County of Middlesex HS, 3rd Alexandra Palace HS, 3rd London North Road Federation, and 6th National Flying Club, flying 593 miles 910 yards, a record for the longest distance flown during the day of liberation in England.

With this preface by way of introduction, I think I might now give short details of the five birds illustrated.

The Lerwick hen is well depicted in her photo. She is not a big bird, but is very well made with good shoulders and flights; she is in her third season, having been bred in 1897, when she flew Essendine, and was 1st FPFC and 1st Federation in any age bird race, and also flew Claypole in APS open race; 1898 she flew Retford, York, and Newcastle; 1899 Retford, York, Newcastle, Perth, 1st FPFC 6th CMHS, and 3rd Produce Stake, finishing the season by winning 6th position in the Lerwick race, to which I have referred, besides the £3 3s barometer in the County for longest old bird race. Her sire is a blue cock bred by Mr Pointer from 'Old Brighton' and the blue hen. Dam, blue hen, bred by Mr G Page, from a red chequer cock from Turner's 'Old Thurso' and blue hen, bred by Cove, of Gloucester, 1st and 4th Banff winner.

Mr Mattock advised another arrival from Lerwick in the shape of a very handsome dark blue chequer cock. He was bred in 1896 and flew Essendine that year. 1897, Retford, York, Newcastle, and Arbroath; 3rd prize FPFC. 1898, Retford, York, Newcastle, Berwick, 4th APHS, and also flew Arbroath. 1899, Retford, York, Newcastle, Perth, and Lerwick. Sire, blue chequer, bred by Mr Kershaw from Evangelisti cock and Pointer hen. Dam, blue chequer, bred by G Tucker from a pair of birds, Evangelisti and Barker.

I need say little about the four Banff birds, Mr Hedges has caught them in a happy mood especially the blue cock. Bred in 1896; the following are his performances: 1896 Essendine, Claypole, Retford, and Doncaster; 1897 Retford, York, and also York again, 1st prize; 1898 Retford, York, and also York again, 3rd prize (APS open race); 1899 Retford, 2nd FPFC, 4th CMHS, 7th APHS, 11th LNRF, York, Newcastle, Perth, and Banff, 3rd CMHS, 4th APHS homing on the day of liberation. His sire is a blue cock, Pointer and Harding. His dam, blue hen, winner of 1st Arbroath, APHS, and 1st LNRF, 1897, bred from a pair of birds lent Mr Mattock by Mr Kershaw.

The dark blue chequer hen is another very handsome bird, with good shoulders and flights, the following being her performances: 1896, Essendine as a young bird. 1897 Retford, York. 1898 Retford, York, Newcastle, and Berwick, 5th prize APHS. 1899 Retford, York, Newcastle, 1st FPFC, 3rd APHS, 3rd CMHS, and 4th LNR Fed, Perth and Banff, 5th APHS and 4th CMHS, home on the day of liberation. Her sire was 'Wingfield Pride'. Her dam a pure Moore hen bred by Mercer Bros.

Mr R Mattock's Shetland hen.

The light blue chequer illustrated is own sister to the last-mentioned hen – in fact, her fellow-nestling – and is equally good for stock and racing, having flown: 1896 Essendine, 1st APHS, and 3rd LNRF. 1897 Retford and York. 1898 Retford, 3rd FPFC. 1898 York, 3rd FPFC. 1898 Newcastle and Arbroath, 1st FPFC, and 3rd LNR Fed. 1899 Retford, York, Newcastle, Perth, and Banff, 6th APHS, and 6th CMHS. Two youngsters from this hen have worked wonderfully well, and have flown Newcastle twice this year.

The mealy hen is hardly so taking as the two blue chequers, but is good in feather; she was bred in 1897 and flew Essendine and Claypole as a young bird. 1898 Retford, York, and Newcastle, 1st FPFC. 1899 Retford, York, Newcastle, Perth, and Banff, 9th CMHS. Sire red chequer, bred by G Page from his old stock pair, 'Stanhope' and 'Calf'. Dam blue hen, bred by F

Kershaw from his Price hen.

Another useful bird is a powerful red chequer cock, bred in 1897, and flown Essendine, 2nd APHS, and 3rd LNR Fed, also Claypole and Retford 1898, York 1899, Retford, 1st CMHs, 1st FPFC, 2nd APHS, and 2nd LNR Fed, York, Newcastle, Perth, and was sent to Banff, but was caught in a barn at Rugby early next morning. Both the sire and dam were bred by Mr G Page from his old stock pair.

The loft in which these birds are kept is simply an open aviary with a wire front, so that they are exposed to all weathers, and they are certainly the better for it, there seldom, if ever, being a sickly bird in the loft; but mind you there is no overcrowding here.

Mr Mattock is convinced that much of his success is due to patience. He seldom, if ever, mates his birds until well into March, and even then he only allows them to rear one round of youngsters. Most of the young birds he has been flying this year were reared by his prisoners, and nearly all were hatched after March.

There is nothing wonderful about the feeding or management. Everything is simple but systematic. Mr Mattock's business, that of a builder and contractor, necessitates early rising, and the birds are always out for the first fly soon after daybreak, and are allowed a good deal of liberty and are usually exercised about 3 to 4 times a day, but never driven.

Their food consists of good maple peas and beans about equal parts, which is distributed 3 times a day, early in the morning, about midday, and again at 5 o'clock, when the old food is taken up and a fresh supply given, as they eat freshly put down food more readily.

I have often heard it said that it requires a big loft to achieve success in pigeon racing, but I am quite convinced that this is a fallacy, and Mr Mattock's case is an example of what can be done with a mere handful of good birds properly handled. As I stated, the loft was only started in 1893, and I append a list of the prizes won:

1893 Essendine 1st and 8th; Newark 3rd.

1894 Doncaster 11th and 12th; Berwick 6th; Arbroath 9th.

1895 Retford 2nd and 11th, and 11th and 15th; York 8th; Newcastle 8th; Newark 7th, 8th, 9th.

1896 Essendine 1st and 2nd (and 3rd and 4th Fed); Claypole 9th and 11th; Retford 3rd and 14th; York 4th; Newcastle 10th; Arbroath 1st and 5th (Fed); Thurso 7th.

1897 Retford 9th, 5th; Arbroath 1st, 1st, 3rd; Essendine 1st, 4th, 5th, and 10th, 3rd, 1st, 2nd, 2nd, 1st; York 13th, 1st; Claypole 2nd; Newcastle 4th and 5th.

1898 Essendine 5th, 4th, 1st, 3rd, 3rd; Claypole 4th, 7th; Retford 3rd; York 2nd, 3rd, 3rd, 4th, 5th, 7th; Newcastle 1st, 2nd; Berwick 4th, 5th, and 7th; Arbroath 1st, 3rd; Perth 4th.

1899 Essendine 25th, 4th, 7th, 9th; Claypole 2nd, 3rd, 4th, 5th, and 6th, any age birds; Retford 2nd, 11th, 2nd, 7th, 1st, 4th, 1st, 2nd, 11th, and 6th; York 14th, 10th, 13th, and 4th, 5th, 2nd; Newcastle 3rd, 8th, and 3rd, 8th, and

Mr R Mattock.

1st and 3rd, 4th, 2nd, 1st, 5th, 6th, and 3rd; Perth 6th, 7th, 1st, 3rd, 3rd, 4th; Banff 4th, 5th, 6th, 3rd, 4th, 6th, 9th; Lerwick 3rd, 3rd, 1st, 6th; 2nd Newcastle with young birds, and also 2nd in record Perth race.

I do not know that I can add much to these details. Mr Mattock has only just begun, and if he continues to 'play the game' as he has been doing for the past few years he is bound to develop into one of the most successful long-distance racers we have in the South of England.

North Staffordshire Lofts
Mr P W M Bate

Although comparatively young in the fancy, P W M Bate, of the Stone and Staffordshire North Road Clubs, occupies a prominent place, not only on account of the success already achieved, but also on account of the rich promise of future champions his system is bound to evolve.

Perhaps it will clear the ground a bit if I state what his system is. It is simplicity itself. He just gets the best blood procurable irrespective of cost, spares no trouble or expense in looking after it, and the product he works easily up to about 150 miles for a couple of seasons before asking them to do the longer races. The result is that he has as grand a lot of promising birds as anyone needs wish to have.

Mr Bate himself has an interesting personality. A devotee of the summer game, he is usually to be found wielding the willow in club and ground matches on the country cricket ground. At football he is carrying a bit too much weight to be able to bustle 'em about at half-back as he used to do for the Stone Club, but he can still teach a forward how to part with the ball rather than try to pass him with it. With the gun and the bike he is quite at home; also photography and – pigeons.

He is also an enthusiastic Volunteer, and holds the Queen's commission as Captain of the 1st Shropshire and Staffordshire Volunteer Artillery.

He says his only hobby is pigeons, but I can't see where that comes in.

There is certainly plenty of evidence that the pigeons have swallowed up chickens, fowls, and ducks.

Two pairs of prisoners have two large houses joined together for their benefit, with a wired outlet big enough for about 50 hens. The birds can literally go fielding in it, so they should take no harm.

The main flying loft is an ingenious affair, being a railway carriage resting on pillars 5ft high.

It is 22ft long, and has a gangway 3ft 6in wide running its entire length. A special feature of this corridor is that it is continued about 2ft beyond length of loft, and has a window overlooking trap. This window is glazed with dark glass, so that the birds are unable to see through it from the outside, while a very snug look-out is provided for the watcher inside.

Trapping arrrangement is very good, and the whole has a very smart and

Mr P W M Bate.

workmanlike appearance.

There is also a young bird loft, and a main prisoners' loft to accommodate 20 pairs of breeders, with a good outlet.

Plenty of fresh air and cleanliness is the watchword of the lofts.

I can only touch on a few of the birds.

There are about 60 flying out, and almost all trained more or less.

Foremost among the 1899 bred ones is a chequer pied hen, almost a counterpart of her dam, a direct descendant of 'Old England' and 'Miss Midwell';

Mr P W M Bate's loft.

her sire of pure Barker strain.

This young hen was Mr Bate's best in 1899, and is likely to add still more honours to the old blood.

Amongst the old birds, pride of place is held by his Nantes mealy, 'General'. This bird flew Nantes as a yearling, after winning 1st Guernsey in 1898.

In 1899 he has again won 1st Weymouth, 1st Guernsey, and 1st Nantes.

The success of this bird is rather a staggerer to the easy working system, but Mr Bate now intends giving him a full year's rest, when he hopes perhaps to equal 'Old 86' work.

Nantes mealy combines the strains of Stanhope, Delmotte, Colville, and Gits, and was bred by the owner.

Another grand bird is 'Young Mealy' son of the Nantes mealy. This bird was put on the North Road in 1899, and won a prize in each race, viz, 2nd Bradford, 1st Northallerton, and 3rd Newcastle-on-Tyne. Certainly the distance flown is not very great, but considering the grand birds that went down in going through the ordeal, 'Young Mealy' may well be prized. He is a good-sized bird, but his wings seem to have been made for a bird twice as heavy – they are immense.

There are two other Nantes birds and a great number of other good birds that I cannot go through in this sketch, as I wish to describe a few of the pigeons.

One of the first amongst them to catch the eye is red cock, bred 1895 by Mons G Gits' 'Soffli's' best blood, and described by Mons Gits as a very valuable bird, always amongst the earliest prizes. His hen is a black chequer bred by Gits and specially recommended to pair with the red.

This pair should do Mr Bate some good, as they are a grand type for the long races.

Mr Bate has also a lot of N Barker blood – direct descendants of 'Irish Giant', 'Wild Irishman', 'Paddy', 'Biddy', and the 'Marica' hen.

Another hen purchased at the sale of the late Mr Gorsuch, and dam of two La Rochelle birds to his loft, is greatly valued. Her sire pure N Barker, her dam sister to the 'Marica' hen.

Another pair that have bred promising youngsters are of pure Wegge strain, bought through Mr Darbyshire in 1899, extremely well-feathered, likely birds.

There is also Delmotte's strain from Bancroft and a pair of stock birds from Ince.

A hen now paired to the Nantes mealy takes my eye, bred 1894, purchased from G Crosbie Dawson. She is sister to Mr Moss's first bird from Bordeaux in 1898.

Another hen, purchased from Crosbie Dawson and bred by G H Whitfield, combines the strains of Stanhope's Grand National hen with Wormald's 'Successful'. This hen is a grand-fathered one. Every time I come across the real good old blood I am more and more impressed with its excellence. It seems that the deeper you dig down in it the richer becomes the lode.

Mr Bate's ambition is to breed real long-distance birds.

He has had a fair share of success, winning outright in 1899 the 12-guinea old birds average cup, also the Continental cup and old and young birds average challenge vase, and many other specials. On the North Road he won two 1sts, two 2nds, and two 3rds in six races, so that he was generally somewhere about. On the South Road he took four 1sts and a 2nd in five races (old birds).

Young birds he cannot look after very well, as he is generally in camp with his corps at the time of the young bird races.

Genial and open-handed, he is a general favourite, and is one of the first and foremost in anything to protect and develp the racing pigeon, and that he takes a very active part in the pigeon racing world in Staffordshire will be seen from the fact that he is president of the North Staffordshire Federation of Homing Societies, the president of Stone and District Homing Society, and vice-president of the West Midland Centre of the National Homing Union.

Hints to Beginners on Training Young Racing Pigeons

by J KENYON (Latham nr Ormskirk)

The first thing the novice has to do is to realise that there is no royal road to success with racing pigeons, but that, as with most other sports, patient plodding is the real road to success.

It is true that occasionally a fancier has a passing success without having taken any pains to deserve it, but the exception only proves the rule. So that I do not hesitate to say that the most consistently successful fancier will be found to be the man who takes most pains. Therefore let the beginner be at great pains in laying the foundation of his loft, and as, sooner rather than later, every man who gets enamoured of the sport craves to possess birds to do the very longest distances, let him be sure to start only with squeakers bred from long distance birds.

Always bearing in mind there are thousands of birds that can do 300 miles for every ten that can do 500 miles and upwards, and it is only piling disappointment upon disappointment endeavouring to breed 500-milers from birds only capable of doing 300 miles. Having secured them in an aviary or buying from reliable sources, let the beginner see to it that he rears to maturity only such birds as are from the very commencement healthy subjects; that is to say, that no matter how much the parents may have cost, or how much the individual squeakers may have cost, or of whose never-to-be-lost blood they may be, do not let them cumber the loft space one single hour longer than they continue healthy and vigorous, for to keep any but the very healthiest and most vigorous is only to court disaster in the future. Mended and patched-up constitutions may and often do carry the bird over the first year or two if not put to too severe a test, but depend upon it the patch will give way eventually, and most probably at the very time the fancier most desires success, namely at a hard trying race on a bad day over a long course; so why court disaster and disappointment by keeping any but the very best and soundest?

Having got the right birds, there are three essentials which in my judgement make for success at the game over and above all others:–

(1.) A realisation that a pigeon sees, and does not feel his way home.

(2.) Exercise of plain common sense in training.

(3.) Being possessed of the magic of patience.

This is not the place to go into the question of sight v instinct, beyond say-ing this, that whatever it may be that brings a bird home, he cannot and will not come home, even from the shortest distances on a day when he cannot see to distinguish the surrounding country. So all I aim at is that the beginner should grasp this fact, for fact it is as can be proved, as it has been proved over and over again. Then plain common sense should tell the novice to *edu-cate, educate, educate* his young birds.

Let him begin with them after they have been flying around home for about six weeks. If he lives in the country let him take them two or three, or if he has plenty of time and leisure three or four tosses around home in different directions, so as to let the birds know what is expected of them, namely, that they are to look for and return to their home. This is surely common sense (if the loft is situated in a town all these preliminary tosses are not necessary).

Then let him take them a couple of miles in the direction he is going to train, then four miles in the same direction; let them do the four mile toss twice to make sure (common sense again) that the birds have got what my old friend J O Allen called 'a line,' that is to say, that they fly the four miles straight, and don't go round by way of some landmark on the other side of the loft and they learnt when being tossed around home; then let him send them eight miles, then twelve miles, then twenty miles, making sure at each toss that the birds have 'a line'.

Common sense should tell a fancier that one place may be much easier for a pigeon to home to than another; for instance, a bird used to flying around a town so conspicuous as Brussels is, in the midst of a comparatively level country, would need far fewer tosses for the first ten miles than a bird that had to home to a little cottage snugly nestling on the sunny side of one of the Derbyshire hills. In fact, on a clear day the ten miles to the Brussels bird would be as nothing, whereas the Derbyshire bird, if taken ten miles the first toss, might – indeed most probably would – spend many weary hours in finding home, at the same time learning nothing useful for future quick work, and possibly taking harm.

The common complaint is that many birds are lost at or under twenty miles. How often have I heard fanciers declare the twenty mile point of liberation to be a difficult place for birds to get away from, whereas the real difficulty is not in the locality, but is created by the want of 'common sense' in not making sure that the birds are carefully trained on 'common sense' lines up to that point. After twenty miles, send thirty, then forty, then fifty-five, then eighty, then 100, and finish with 125 or 130, when the novice will have done all that need be done towards educating his young birds, and if he will be advised he will stop all the best and choicest at 100 miles.

I said earlier on in these few hints, that the 'magic patience' would achieve much at pigeon-flying.

As in every other pursuit in life, most things come to the man who knows how to wait, so it is with pigeon-flying.

What a hurry to be sure some beginners are in. No sooner are the squeak-ers in the loft than they are put on the road, and 'hail or shine', as the Yankees say, are kept going, for the novice says 'What is the use of keeping birds that cannot face all weathers?' In my humble judgement no greater mistake can be made than either beginning too soon, that is before the youngsters have made a fair good start moulting the flights (it is not a bad rule to wait until third flight is moulted) or tossing the young birds in bad weather.

Let the beginner very carefully pick his weather for the first 30 miles, so that the birds may learn their business and the country near to home thoroughly, and even if doing this necessitates missing the first year's racing, I am convinced that the fancier who takes the most pains for the first 30 miles will beat his impetuous neighbour in the long run.

By carefully selecting the weather I do not mean waiting for a tail wind (indeed, for choice, I prefer a gentle breeze against the birds, for then the fancier knows his birds will have had to work their passage home), but wait for a settled clear weather with a clear atmosphere, so as to avoid, if possible, the youngsters being tossed on a day when they can't see half a dozen miles in front of them, and therefore should they take a wrong turn, as the best may do, have to fly all over the country drifting aimlessly in the dark, serving no useful purpose, for it is not around the country they are wanted to fly but as straight home as they can go.

This result of always homing straight and quick can only be assured by the exercise of common sense, by being patient enough to give the youngsters, so far as the fancier possibly can, a clear atmosphere when tossed, so that they can *see* to make the fullest and most economical use of their mental and physical powers.

Roupell Park Lofts.

Mr. T. PEED

Will, during the Breeding Season, be able to spare

SQUEAKERS

from the following choice strains :—

BARKER, BONAMI, COLVILE,

HANSENNE, LOGAN, JURION,

THIRIONET, DELREZ, OLIVER,

LUBBOCK, etc., etc.   .   .   .

Four of my birds flew BORDEAUX, 1898 National race.

SUCCESSES IN 1901.—1st Federation, 1st L.U.F.C Chard O.B. ; 1st Federation, 1st and Special L.U.F.C. Penzance O.B. ; 2nd and 3rd Federation, 2nd Club, 1st and 2nd Produce L.U.F.C. Chard Y.B. ; 8th Federation, 4th Club Tisbury ; 4th Open Race L.U.F.C. Semley ; 1st and 2nd Emsworth Y.B. C.P.F.C. ; 2nd Yarmouth C.P.F.C.

Prices, 7/6, 10/-, 15/-, 20/- each.

Lists of Full Pedigrees and Particulars on application.

T. PEED, Norwood Rd., West Norwood, Surrey.

The Preparation of Racing Pigeons for Great Races
by J JANSSENS

The preparation of a pigeon destined to participate in long-distance racing is an operation of long duration, or, in other words, a question of beginning in time.

Let us suppose we are dealing with a pigeon that we intend to compete with in a race of between 500 and 600 miles in 1902.

In the month of November of the year preceding that in which the race is to take place, it will be necessary to carefully watch our subject as to its moult. *All* the tail feathers must be moulted and renewed before January 1st. The same remark applies to the wing. The ten exterior or primary flights must have been replaced by new ones. [Note. – One knows that the ten feathers of the wing *nearest* the body (the secondaries) are only partially replaced each year, about one-half.] If the bird is not found to have had a full and complete moult as described, it is not normal, and to attempt to race it the following season would be imprudent if not ridiculous.

For young fanciers who cannot well distinguish the new flight feathers from the old ones in the wing at the side of the body, it would be well in the month of August to mark all those not moulted; they will then see which have been replaced.

During the whole of the moulting period I am of opinion that it is necessary to feed well, because at this time a pigeon has need of nourishment in order that the new feathers may be healthy and vigorous.

The moult being complete in the month of December or beginning of January, a wholesome food should be given, but in less quantities, so as to avoid needlessly heating the birds, the cocks being separated from the hens.

Towards February the 15th the food supply should be increased, and on March 1st proceed to mate up.

Opinions differ as to the advisability of rearing one or two youngsters in the first nest by those birds intended for the most trying races.

In Liege a great number of fanciers do not mate the birds at all intended for long-distance races; others let them rear a single young one, and then remove the hen from the loft, so that during the whole of the flying season the cocks are alone. I have tried this system, but must confess I did not find it successful.

This is how I proceed, and I follow the lines laid down by my late esteemed friend, M Charles Wegge, whom I considered one of the very best Belgium ever knew.

My pigeons destined for the 'Grand National' race rear their two young ones of the first nests; then, so as not to be contrary to nature, I let them bring up one in each successive nest.

If the moult is not as far advanced in the wing in the month of June, ie, if, at this period, the pigeon has not lost more than two feathers (a single one would be preferable), I so manage the pigeon that the hen lays her eggs towards the beginning of July, the Grand National taking place about July 15; thus, the pigeon to be engaged sits from eight to fifteen days before being basketed.

In order to have it in this state, I remove the youngster directly the cock commences to drive his mate to nest, or to induce it to take this course. To my mind, the pigeon is in the very best condition when it has been sitting from ten to fifteen days, the flesh then being firm and the plumage tight.

I know that the heat in the basket and vans will cause the bird to drop another feather in the course of the journey, but as it will still have seven large ones left, it will be quite able in this state to accomplish the journey.

If the moult is farther advanced, it will be necessary to so treat the pigeon that he sits several days, at the same time feeds a youngster at least twelve days old.

In my opinion a pigeon should never be sent to a long race when in the following condition:

    (a) A hen chased to nest.

    (b) A cock chasing.

    (c) When feeding as youngster from one to seven days old.

    (d) When casting the feathers which form the covering of the wing, or when too far advanced with the large flight feathers. [N.B. – It is often found that these two latter conditions are found together.]

    (e) When ill, or when in such a state that one thinks it is not 'normal.'

To judge this latter point it is necessary that the fancier has an observant eye.

If the eye of the pigeon is not bright, the plumage dull, the male bird not lively, these are so many indications that it is unfit for the task we contemplate giving him.

It is also necessary during the period of preparation that fortifying food be given, but not that of an exciting nature, such as 'hemp'.

I think the pigeon should be at the very least two years old before participating in the long races with any real chance of success.

Dax being about 550 miles from Brussels, I send my birds intended to compete to:

        First stage ....25 miles

        Second stage ....70 miles

Third stage ....200 miles
Fourth stage ....300 miles
Fifth stage ....400 miles

At the three last stages I participate in races with birds intended for the great race, and which are not more than four years old, at which age they are still quick enough to have a good chance of being in the prizes.

The last stage mentioned should be flown a month prior to the basketing for Dax, so that ample time is given for rest. Fifteen days is sufficient, but I allow a month, for fear the bird should encounter bad weather and be compelled to sleep out, then ten days would not be sufficient time to recoup itself.

The above remarks are my own opinions on the subject, and may not be those of others. I say this because I do not wish to enter into a controversy. I do not pretend to be infallible.

I will now offer a few remarks under the heads you ask me.

The first question you ask is: 'How far a youngster should be trained?'.

For youngsters born in March, April or May, I commence the work of training in July, and see that all that are fit go to every stage.

I am of opinion that to fully develop the instinct of orientation in the youngster, it must have a thorough training, particularly at the short stages. The first distance I send from the loft is three miles; second six; then I jump to 20, 30, 40, 55, 76, 100 and 140 miles. From this stage I jump up to about 200 or 210 miles.

Here I divide the birds under consideration. Those from my stock birds which have bred me good youngsters previously, I consider their first year's apprenticeship duly finished. The other section I continue to *train* up to between 250 and 350 miles, but *not* taking part in all the races unless in a perfect state of health, and being in such a condition of the moult that it will be safe to send them.

I mostly make a practice of resting a youngster intended for the 300-mile race for a month, then three or four days before being put into the basket he is sent for a short toss of from 30 to 40 miles. This system has always given me satisfactory results. I must add that for long distances it is well to have youngsters that are mated and sitting, and if possible rearing a youngster.

So as not to exhaust young hens, I do not let them lay. When they have been mated from ten to fifteen days, and *have made a nest* of twigs, straw, or anything, I put in it some artificial eggs. Almost without exception they sit on them, and 18 days later, I give them a youngster, which they will bring up.

"How far a yearling and third season pigeon should be sent?'

I recommend with these pigeons the same training that I give to youngsters. Those of one year old should not be sent farther than 300 or 400 miles. Those of the third season to the 'bitter end'. It has been with these two-year olds that my grandest successes have been achieved.

My Bordeaux hen when two years old won the first prize from this point in the face of a north wind, the bird second to her being timed 30 minutes later. Fifteen days later she won the 26th from St Jean de Luz, 597 miles.

My second prize from Dax was also won with a pigeon two years old.

Every year we have a race from Libourne, 460 miles from Brussels, reserved specially for two-year olds, *carrying special rings.* This race is termed our Derby. Again this year one of my birds was in the prizes at this race, and later took 11th St Vincent, about 600 miles. I think that at two years old birds which are of the pure Lierre or Antwerp strains are at their very best, and can then with confidence be expected to compete successfully in races up to 600 miles.

As to the question which is the better for long-distance racing – cocks or hens, there are more good cocks, but when one drops on a good hen, generally she is more reliable than the cock; in other words, she is more regular. I have had many good cocks, but never one as good as my best hens. The difficulty is that the cocks are more often fit, whereas the hen when being driven, when laying or after having laid from three to four days, or having just hatched, cannot be sent to a race with any great chance of success.

One can send a cock to a race that is not paired, but to send a hen under such conditions would not be possible. It is on account of all the above points that there are always more cocks then hens entered in he races.

On the other hand, many fanciers think that the hen is not suitable for racing purposes, and never enter them. It is a complete mistake. I found this assertion on the facts that my 'Bordeaux Hen', previously mentioned, has *never* been out of the prizes up to the day on which I lost her. My Red Hen won *seven prizes* in 1900, of which two were 1sts, three 2nds, and two 3rds, when she took part in the race from Dax, 557 miles, in 1900, and took 3rd place and King's prize; she had been sitting ten days and was then three years old, I not having sent her in 1899, considering her too good then to risk. At every race she either won 1st, 2nd, or 3rd.

In 1901 she won four prizes, amongst which were a 1st and a 2nd; six weeks before the Dax race her mate was taken ill. I was then compelled to mate her to another, so was unable to so arrange her condition I wished for the great race. She had a youngster of eight days old, 'and I lost her.'

My blue hen, La Cheminee, which has three years in succession taken prizes from Dax, was also two years old when she won her first prizes from the 550 miles race, and has always sat on eggs ten days when she has won prizes. She was at her best when, in 1901, she won 12th Burgos (Spain).

My mealy sat 12 days when he won 2nd Dax in 1899, the same in 1900, when he won 25th prize.

I could give a host of similar instances, but as I said in the early part, 'when the pigeon is sufficiently advanced in the moult it is necessary to arrange in such a manner that he has a young one from ten to fifteen days old, if not, it will be risking a complete moult in the basket, and the success will be nil.'

. . THE . .

STAINCLIFFE LOFT.

Mr. E. H. CROW,

School House, Staincliffe, Dewsbury,

WINNER OF

33 FIRSTS and over 20 SECONDS

from all stages up to and including Rennes, offers Squeakers during 1902 from his stud of birds, which comprise the very best blood from the Lofts of

JANSSENS, BONAMI, JURION,

WEGGE AND WIELEMANS . .

amongst them are **several pairs bred by M. Bonami and six bred by the late C. Wegge.** They comprise some of the finest birds in the fancy and have produced winners in all parts of the British Isles.

Amongst my many successes are **three 1sts Bournemouth, 1st Ventnor, 2nd Cherbourg, 2nd Jersey, two 1sts and two 2nds Rennes,** all races over 200 miles.

LISTS ON APPLICATION, 2d. EACH.

Sole agent in England for the Habicht Timer

The Formation of a Loft
by GUSTAVE BLAMPAIN

In answer to your request for me to to contribute an article to your work –
Squills' Diary – I do not think, in order to describe the formation of a loft,
that I can do better than give you description of my own, which may be cited
as a model of its own. It has not always been as at present; improvements
have been gradually made, whilst studying the habits and characteristics of
the birds I have introduced.

In order the better to attach a pigeon to his loft, it is necessary to render his
sojourn there as agreeable as possible, taking good care to suppress all obstacles which would disturb his re-entry, so that a pigeon shows no hesitation at
any time to take possession of his home.

I think the elevation of a loft should be as high as possible. My own is 50
feet above the level of the soil; the exit, instead of being on the roof, where
the pigeon finds means to amuse himself on his return, is placed in the centre of the gable facing South West. The opening measures 5 feet high, and 4
feet 6 inches in width. No board for rest is placed outside, but inside of the
opening to the loft there is a platform 3 feet 6 inches in depth. In this manner
the pigeon flies directly into the bay, and if the pigeon stays on the platform
in the interior, I at once let fall a metallic curtain placed at the level of the
external opening. My loft is 22 feet in length, and 15 feet 6 inches wide and
7 feet high. It is divided in two, that is to say, in the centre towards the opening or exit, a cage or passage is formed of 7 feet long, in which I catch the
pigeon upon arrival. This passage consists of trellis-work partitions, and is
about 3 feet 6 inches wide. I should add that at the end on a level with the
platform, through which the pigeon has to pass into the loft, are a set of bolting wires. Before entering the loft and when passing over this platform a
pigeon sets in contact an electric bell, which announces to me its arrival.

The nest-boxes are upon each side of my loft along the walls; they are
about 2 feet wide by 18 inches high and 18 inches deep, and a shelf is placed
in every one. I have tiers of four nests, one above the other, which can be
closed by means of a lathwork door and by means of these doors I can, if I
desire, shut any pair of birds up and imprison them in their cage.

During the moulting season, that is to say in August, I pull down the large
doors and take out the floors of the nest-boxes, so that the birds cannot use

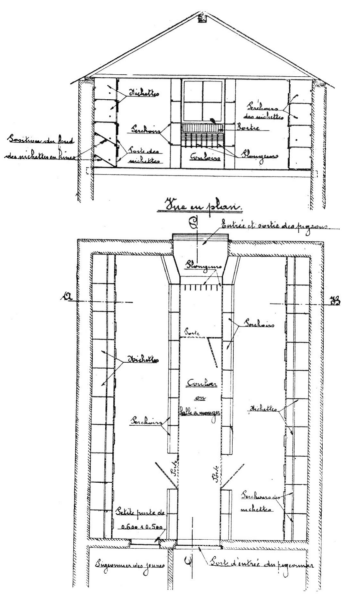

PLAN OF M GUSTAVE BLAMPAIN'S LOFT

them for breeding, and having the free use of the loft they terminate in this manner a good moult, and I consider it absolutely essential that pigeons should have a good moult in order to be able to represent their owners in proper condition the following season.

My loft is whitewashed with lime in March and in October each year.

During the time that the hens are sitting the cocks do not care to remain in their nests. It is therefore necessary that the loft should be well supplied with perches. I have fixed my own perches in four rows facing the nests. They are of wood, about eight inches long and are separated by wooden partitions, so that each bird has a perch of its own, and consequently little or no fighting takes place.

A pigeon loves light, therefore one cannot have too much light in the loft. In the roof of my own loft I have placed four windows of 2 feet 6 inches square, two on each side of the roof, and which in summer time can be opened during the day when the weather is fine, and closed in the evening.

It is necessary to take every precaution against humidity in a loft, because many pigeon illnesses arise from this cause, such as wing disease, but this does not mean to suggest that a loft should not be well aerated. The floor of the loft should always be made of wood.

When it is too hot in the summer it is as well to replace the door of entry by one of wire mesh, in order to establish good ventilation, for the fact must not be lost sight of that hygiene plays an important part in getting pigeons into condition to train them with success. It is necessary that the fancier should keep his loft clean, especially the nests, where the young deposit their excretions, which, if allowed to remain, give off a bad smell. During breeding the nest-boxes should be cleaned at least twice per week. After having cleaned the loft, some clean, dry sand should be thrown down.

The birds should always be fed in the entry hall, or passage, I have described. This should be kept perfectly clean, for the pigeon loves cleanliness, and eats with a better appetite under such conditions. The water in the drinking fountain must be renewed each day, when it has iron added it is much better. Take care to thoroughly clean the drinking fountain from time to time. It is also necessary at least once a week in the summer time that the pigeons should have a bath. I place my bath near the entrance door. It is of the width of this entry, and measures about a yard in length. A waste pipe allows the water to run away, and it is renewed by a small syphon. It is useful to place at the disposition of the pigeons some grit, such as old mortar and gravel, which is necessary to grind their food and help the better to nourish them. It is also very good to give them some broken egg shells dried and crushed into a powder, which helps the hens to form a good shell to their egg and to lay regularly; and I may say it is very necessary and indispensable for the hens to lay regularly in order to get the cocks in condition for competing in races with success.

With the object of keeping birds in the best of health, it is indispensable that pigeons should be placed upon a *regime,* that is to say, that the food must be distributed each day at the same hour. During the training and racing season, three meals per day are necessary; the first at 8 o'clock in the morning, the second about noon, and the last about 4 o'clock. The quantity to be given must be regulated. There must not be any left on the floor of the loft, for it is better that a pigeon should be on the thin side in preference to being too fat. In the winter time two meals are sufficient, not too copious, in order that the birds do not become too ardent. It is very necessary not to lose sight of the fact that until the moment they have terminated the moult, it is necessary to keep them well nourished, as the food plays an important part in the production and growth of the new feather. The food which I employ is wheat, tares, maize and beans. The grain must be sound and healthy, and in nowise mouldy. The two first meals are composed of a mixture of equal proportions of the three first-named grains, the last meal consists entirely of beans.

During the moult I give to my birds, before the first meal, a portion of linseed. I have always found, by following this prescription, that my pigeons have been in good health and well disposed. Towards February 15th, as the cocks commence to become very amorous, in order to avoid the natural coupling, I remove the hens from the loft and keep them shut up during a month in a well-aired and ventilated place for the purpose. This separation enables me to change easily my couplings. When I put the hens with the males I shut up a pair in each compartment, and in view of the ardour of the pair effected by the separation, in a couple of days each pair has taken possession of their nest-box; but you must avoid allowing too many to come out of their nest-boxes into the loft at the same moment. Five or six pairs at different points will be sufficient. As soon as you have seen them take to their place, shut them up and allow other couples to come out. It is necessary as little as possible to change the nest-box of the cocks. Let them have the same as in former years.

My young ones are weaned at five-and-twenty days, for it is necessary to avoid exhausting the old birds. They are placed in a special loft close to that of the old birds, where they remain all the year until November. At this date, in order to introduce them with the old ones, I open the door of communication between the two lofts and suppress all food in the young bird loft. By this means they are obliged to become habituated to the other loft to take their meals, and after six weeks time I close the entry to the young bird loft, and they enter with the old ones. A separate department is necessary for the young ones, in order that they may be tranquil and not bullied by the old birds, which makes them despair. Take good care to have the young birds' loft well supplied with separate perches, so that each bird has a place of its own.

I do not train or race my young birds before they have arrived at the age of one year, believing that their development should not be arrested by the fatigue of training. In order to watch the flight of my birds when on the wing, I have had constructed at the summit of my loft an observatory which measured 6 feet, and is 7 feet high; the four fronts are glass, and I can thus follow my birds with the eye in all directions.

The preparation on the pigeons for long-distance races is a special study in itself. It must be noted with care by the amateur the different dispositions his pigeons are in when they are engaged in the preliminary trials, for it is often that they will then show the form that will classify them with success. The best classical condition is to engage the cocks at the moment when they have young of from 19 to 20 days old, and are just commencing to drive the hen to nest, but not with too much ardour, because then the pigeon becomes feverish, does not eat, and is weakened. You must, therefore, be careful to mate your most trusty cocks with hens that you know will lay regularly, this is an important factor, that being so, you must carefully note the time of the previous hatching, and by doing this you will have the required formula to know exactly at what date your pigeons will be in the best condition. Some cocks home exceptionally well when the hen has just laid her first egg, there are also cocks which give their maximum of velocity when sitting from ten to eighteen days, but they are rare, I possess one which this year in this condition has won two 1st prizes from Charteauroux, 281 miles, and Limoges, 350 miles.

For the hens there are two conditions in which to engage them, so as to hasten their return home; the first is when they have eggs from fifteen to eighteen days – there are even fanciers who await the hatching of the first egg – but this condition, to my thinking, is dangerous, because at this moment the hen prepares the soft food which nourishes the young in its first days, and which she must necessarily have in her crop during the time she is confined in the baskets, which must make her sick. The state in which I prefer, and which I adopt, is to engage the hen when she has a young one of from ten to twelve days old, at the moment of sending to the race. This is my favourite condition that I prefer to all others. The long races always take place in July and August – a period in which, properly speaking, the moult begins. It is, therefore, necessary at that time that the pigeon should have the young ones, which keeps them well in feather and stops the moult, in order that they may be in the best possible condition to brave the elements, but, as I said previously, it is in carefully noting the performances of one's birds and their state during the preparatory stages, that one is able to know the best form in which to send them to the races in which he desires to compete.

But I must say that is not sufficient merely to prepare pigeons by such and such a method in order to succeed. It is necessary, and of primary importance, that, at the start, you must have pigeons of race and of class that will endure fatigue, and have the gift of orientation very much developed. In order to get the pigeons fit for a long race, providing that you have bird of good race, you must give them a training at progressive distances. At the age of one year my birds are trained 6, 15, 30, 90 and 120 miles maximum. The second year, which forms the best year of their training, they make the same journeys, then are sent 250, 375 and 500 miles. It is during this year, and these last stages, that one discovers the pigeons of capacity which must be reserved, and one may have confidence in for the future.

It is only in the third year that my own pigeons show their rare quality of endurance and maximum of velocity. It is not therefore necessary then to fatigue them by too much training. I give my own 15, 160, 300, and to finish, 500 or 560 miles. It is necessary to have fourteen of fifteen days' rest between the two short stages, and a month or six weeks' rest before the great last trial, which is for us the National race of Belgium.

During the absence of the males or the hens, one must take away the one which remains in the loft, and close its nest-box in order that the pigeon which is on the road would find its nest free when it returns as well as its mate, either cock or hen, which is at home and which you will take good acre to put back as soon as the voyageur returns. At the return of the pigeon from a race give him some bread crumbs, of which they are very fond, and during the racing season, after their last meal, give them a little millet seed, which adds fire to their zeal without exciting them, for the best condition it must be remembered is a natural and not overheated one. It is in following the principles that I have described that I have always taken part with success in our National races.

I hope my efforts to contribute this article for your readers will give them pleasure.

T. H. Burton,

THE ELMS, SEAFORTH.

I SHALL BE ABLE TO SPARE A

FEW PAIRS OF SQUEAKERS

DURING THE BREEDING SEASON.

My Stock Birds have been bred by **Messrs. W. G. Orchardson** (including a daughter of **28** and **27**), **T. W. Thorougood, Hansenne** (direct), **Pletinckx** (direct), **Mons. Reay** (direct), **Pure Wegge's** (from birds imported from Mons. Dellmans), **Councillor Slater, H. J. Longton, A. H. Osman, J. Makeague, Stanhope, E. R. Edwards, J. Meadows, Clarembaux, T. H. Harrison,** etc. Besides these, I have the valuable strains of **Logan, Gits, Clay, Toft, Taft, W. C. Moore, Colvile, Jurion, Swigger, Delrez, Gibson's White Tail,** etc.

——:o:——

My birds, in addition to many other prizes, won in the Breeze Hill H.S. **The Young Bird Cup,** for Best Average in all Young Bird races.

——:o:——

Up to Bournemouth I easily led for Old Birds Average, but meeting a bad day at Jersey, and only having yearlings to send, I failed.

——:o:——

My Birds won 28th from BORDEAUX, in M.F.C. Open Race 561 competitors, also Jersey, in the La Rochelle Club.

I can recommend my birds to do good and consistent work.

FULLER PARTICULARS UPON APPLICATION.

Management of a Loft of Old Birds for Racing during the Breeding Season
by W T RIMER

The above covers such a multitude of *pros* and *cons* that any remarks of mine must be treated as merely embodying the cardinal points and special features, which, in my opinion, are essential to success, fanciers generally having to fill in the necessary details of management best fitted to the peculiar conditions dominating their loft according to locality and surroundings. So far as my district is concerned, Northumberland and Durham, the lateness of Spring season, the frequent and prolonged E and NE winds make success only possible by an all-the-year round preparation of the birds, necessitating only the housing of the most robust and pluckiest of racers – a maximum applicable to all.

With these few mems I will dip into my subject and assuming the birds have had a good and unchecked moult, sexes separated, lofts disinfected and generally prepared for breeding purposes, and by a careful scrutiny of past records, the various matings decided upon, I, about a week before 14th March (pairing-up time), give the birds two good purgings by putting about half an ounce of Epsom salts to one quart of water, at intervals of four or five days. To be effectual the trough should be emptied on previous night and all corn removed so that the birds are given same fasting and thirsty. After sufficient time has elapsed for all to drink the trough should be emptied and fresh water supplied. This treatment is most beneficial to all, more especially to hens, particularly yearlings, securing regular and easy laying, and in the case of the very amorous retarding too early laying and consequently assuring a greater certainty of fertility of eggs.

Daily treatment – Fresh water twice daily, in very hot weather thrice; a little Condy's, sufficient to produce a pale claret colour, is beneficial; a continuous supply of grit, Jenkinson's and Rd. Woods, Mansfield (the latter I used with great success for conditioning when 'Show Homers' were my hobby) and hopper feeding. This I advocate strongly to those whose profession does not admit of their feeding regularly, and, furthermore, during breeding season some birds (close sitters) will not leave their nests and consequently endure unnecessary fasting, a thing to be deplored, resulting in their gorging when next food is available.

My staple food, all the year round, is Maples and Tares; the latter, I submit, contains all the muscle-forming properties of Beans, with the great advantage of being much more easily digested. Beans have never been a success with me, although, for the 300 miles and longer races, I give a few of the smallest last thing at night to such birds as will take them. Mid-day is the time for the tit-bit, Canary Seed, but only in moderate quantities. I have all corn put into calico bags to remove any dust, etc, and then hand picked before being put into the hoppers.

Every bird is turned out thrice daily, about 6am, mid-day, and an hour before dusk, but no great amount of exercise is imposed so long as they take a minimum of 10 to 15 minutes.

A bath twice a week is, I think, essential. I generally give same the day before basketing, and on Sundays as a pick-me-up after racing. Another little fad of mine is that every Sunday during the racing season, I give my birds a teaspoonful of Epsom salts in one quart of drinking water as a 'cleanser' after confinement in panniers, and the consequent risk of their having picked up some foul corn or become contaminated by some 'comrade in distress'.

Stock birds and yearlings are paired up first, those intended for the long races being held back as long as possible and invariably paired to a 'stay-at-home.' Nothing in my opinion takes 'the edge' off a good racer quicker that to home, maybe after a hard race in which the brave little warrior has so gallantly fought the elements, only to find its partner missing.

I like all to rear their first round up to about three weeks old, when I generally kill one. I advocate the rearing of first round because of its steadying and conditioning effect on the birds and its unquestionable benefit in endearing them to their home. After the first round *everything is sacrificed for the preparation and carrying out of the long distance races*–that goal to which we all so ardently aspire and which so few, alas, attain!

When the second round of eggs is laid fanciers find themselves in the throes of excitement for the time of training has arrived. Here let me counsel fanciers to start their preliminary tosses in plenty of time, say three weeks before first race, thereby giving themselves ample time to pick their weather, for birds must have such 'schooling' performed under the most favourable climatic conditions. A barometer, needless to say, is essential, and by close study is invaluable to fanciers effecting their training stages successfully.

Such birds as I allow to hatch second round have their young killed at 10 to 14 days (as soon as hen lays again) but the majority are given pot eggs so that training can be carried on uninterruptedly.

By the time the third round is laid we are sailing merrily along to the critical time when *the* races are at hand. Now is the tug of war, and the time when fanciers have to resort to every ingenuity to cheat nature, *ie,* preventing the moulting of primary flights, for unless by artificial means same can be retarded, but small will be our measure of success in the long and trying Continental races. Some are more amenable than others (I have had birds which have not dropped a single flight until well into July), but where, unfortunately, the renewal of primary flights exceeds two, if such birds have worked well, and are of a reliable strain, they are stopped, and where second series races are available then I race them in such up to 250 miles, but only when weather conditions are favourable.

In hot weather, fanciers must give regular – at least weekly – attention to keeping down vermin, and I have found a generous sprinkling' of 'Keatings' under and in nest pans, and an equally generous spraying of the lofts with turpentine or petroleum most efficacious.

I send all birds to following training stages, as I look upon these not merely for the purpose of training as for the far more important work of conditioning. A bird that homes showing signs of fatigue is given an extra day or two rest with the accompanying extra jump. Distances are 5, 10, 20, 45, 60 then to first race, 100 miles. The old stagers intended for Continental races I jump from 60 to 140 miles race point, thence to the Coast (300 miles); after that regular exercise at home during the 14 to 21 days rest which I think necessary before asking *the* question. I have never adopted, nor seen the necessity of, intermediate cross-channel races, contending that if a bird has done 300 miles without undue exertion it is then ready for any distance, and certainly my experience has confirmed such theory. All my racers from Jersey 400, St Lo 420, Rennes 475, La Roche 560 and Marennes 630 miles have been trained as mentioned *ie* jumped from the coast. Yearling cocks I stop at 220 to 250 miles, Yearling Hens at 300 miles and frequently they are sent along again up to these distances in the Second Series Races.

During training and up to 140-miles stage, I strongly advocate fanciers who can witness arrivals to carefully record same, and by close observation they will easily detect those birds which are shaping best, and homing consistently; furthermore, it is during this period I closely note whether the birds destined for Continental races are keenest when sitting or rearing Young Birds, and I *invariably basket them for such races under the breeding conditions most acceptable to them.*

I must admit my partiality for hens, my best work for years having been done by such with seven days squabs in nest, up to the period that close sitting is necessary to cover their young.

I never race cocks that are driving, hens within two or three days of laying, nor either sexes for five to seven days after hatching young. On the other hand I generally pool or nominate a cock which has a youngster in the nest, and *just looking for his hen,* or a hen with a seven days old squab in nest. True this is not a golden rule, but I have had invariably the largest measure of success under these two principles applied to the respective sexes.

This last season I found all my racers showed a tendency to moult primary flights prematurely – at least to suit my requirements – and so in the case of the racers destined for St Lo and La Roche I substituted seven days old squab for their eggs about four days prior to basketing, and in each instance the birds took kindly to them, and in their delight were all sitting close at time of dispatch. Having only been sitting eggs a week no question of soft food arose, and I assisted the feeding by a handful of small maize night and morning to each pair. Result:– Retarding of moult, and in spite of disastrous race from St Lo, had three home out of four, and one out of two from La Roche.

Quick trapping is essential, and, to ensure this, birds at home should be exercised early in the morning (fasting), called in, given their morning meal, and afterwards kept hungry pending arrival of racers. On their arrival, the familiar call, and a little canary seed, will generally prevent sitting out. I never allow birds on arrival any grain   other than canary seed and small maize (in limited quantities), and then the usual feeding in the evening, thereby allowing them ample time to recuperate, and minimising the risk of gorging with its attendant risks of foul crop etc.

In conclusion, to those fanciers who would be successful – I am addressing my advice to the ever increasing recruits to the Homing Fancy – I recommend them to acquire the virtue of patience, indomitable pluck ("the harder the nut the sweeter the kernel"), and a fixed determination to reach the goal of all true fanciers' success in the long distance races. Remember that as we expect sacrifices from our brave little warriors so also must we be always ready to make sacrifices on their behalf by constant solicitations for their welfare and comfort; rest assured the investment will be a good one, for sooner or later will we be rewarded with success incomparable to any labour on our part.

Mr. R. MATTOCK,

THE KENTWINS,

Enfield Road, Old Southgate, London, N.,

Will have for disposal during the season a limited number of high-class Squeakers.

This loft has raced with success from Banff, Thurso, and Lerwick, gaining prizes from every racepoint up to and including Lerwick. In 1907 winning 1st Thurso in the open race, and 1st Lerwick in the North Road Championship Club Race. The old strain for many years has gained successes in all parts of the country and at all distances, particularly in North-West Lancashire. Nothing but birds of the very choicest quality are kept or bred from, all are real long-distance, tried, reliable, workers. Youngsters bred from this strain have flown Aberdeen; 388 miles; and Banff, 427 miles; winning 2nd prize and all pools when only two birds homed before the close of the race.

Only Strong, Healthy, Reliable Squeakers sent out, at Moderate Prices.

Address all communications as above.

Long Distance Racing
by FELIX REY

I have much pleasure, in accordance with your request, in contributing an article to Squills' Diary.

My connection with the sport of long-distance racing dates as far back as 1873 – 31 years. During this time I have had considerable success, especially in the long-distance races.

I have always been adverse to consanguinity, feeling convinced that it conduces to degeneration in a strain. In order that my views may be the better understood, I will endeavour to deal with the subject briefly under short subheadings.

The Selection of the Birds.

The choice of the pigeons with which the fancier intends to found his loft of birds for long-distance racing ought to be made from amongst known and reputed birds of the most valuable and best long-distance strains, such as have the instinct of orientation and homing faculty highly developed, and the attachment and love of the pigeon-cote most deeply rooted, and being as a consequence superior to their fellow pigeons. For long-distance racing the fancier should give preference to pigeons hatched towards the end of April or beginning of May. These birds, having been hatched at the most favourable season of the year for breeding, have a greater prospect of making a perfect moult, and consequently, being of a most robust constitution, unite in themselves those conditions essentially necessary to develop into valuable long-distance birds.

Concerning Training.

The training of those birds that it is hoped will prove good long-distance pigeons, ought to be characterised by great moderation, so as not to wear out the subject. The distance flown should not be more than 150 miles for the first year; the training stages ought to be commenced towards the 15th of July for birds of that year, and a choice should be made of clear days favourable for flying, in order to facilitate the return of the pupils and to minimise the risk of their loss, for the easier the work they have to perform as young birds, the better their chances of developing.

At the age of one year the distance can be extended to trials of 220 miles, always, however, bearing in mind that it is better to act with prudence and great moderation, taking constant care to avoid putting *en route* a pigeon which, not possessing the quality for successfully undertaking the projected trial, makes a faulty return, thus spoiling, perhaps, a valuable bird, of which great hopes had been formed, and ruining its physical and sporting qualities.

This second year's training commences, as a rule, about the 1st May, and is continued until the end of June.

At two years of age the trials should be extended to 373 miles, under the same conditions as for one-year-old birds.

After each year's training there should be a drastic weeding out of all birds which, by reason of illness, bad moulting, or other causes, no longer fulfil the conditions requisite in birds to be relied upon for long-distance work. At the same time, it would not be prudent to too rapidly dispose of a bird for the simple reason of its not always returning in front of its competitors, as consistency and regularity in their return is often times more valuable than mere speed in long-distance birds.

At three years of age the birds have acquired all the necessary qualities to enable them to make a good show in the big competitions; their education being completed, it is no longer necessary to repeat the short-distance tosses and easy trials that they had to undergo during the preceding years, which would only have the effect of unduly exhausting them for the moment of the great trial of stamina to come later, and at which time the pigeon ought to have all its strength, vigour and vitality, in order to be able to make a good show in the trial in which it is going to take part. Thus for a trial of 560 miles five preliminary flights of 10, 30, 60, 125 and 250 miles I think quite sufficient preparation.

Coupling.

The coupling of pigeons intended for long-distance racing ought not to take place before the end of March, in order that the moult will not commence before the end of May, thus enabling the birds to fulfil their July engagements (when the long-distance competitions take place with us in Belgium) when the wing still carries the seventh long flight feather and enables the bird to compete with a perfectly full wing and thus fly with the greatest ease and least possible exertion.

State for Competitions.

In what state ought the pigeon to be in order to compete in a long-distance race?

It is with regard to this point above all others that it is necessary for the fancier to know his pigeons and to make use of his knowledge to the best advantage.

A certain pigeon will give excellent results if sent to a race when its eggs are just upon hatching, whilst another will not fly with speed consistency unless it has a young in its nest of from 12 to 15 days old.

Mr C Clark's 'Goldfinder II'
1st North Road Championship.

Mr Tomlinson's 'Marennes Queen'
1st Marennes Great Northern.

Mr Matten's National Winner.

Messrs Baron & Seed's 'Combine
Jewel', 1st Nantes Combine.

The Champions of 1905.

For cocks the most favourable moment for competing, in my opinion, is that condition when its young has attained about three weeks of age, and when it commences very gently to follow its hen to the nest, or it will fly equally well with a young one of four or five weeks and sitting on eggs about a week old.

For hens, I think it is preferable to enter them with eggs of 12 days, if the bird does not moult too rapidly or too freely. Many birds fly well in this condition, but for choice I prefer them with a young one of from about 12 to 13 days old in the nest.

In fact, I always prefer to enter my pigeons for the long-distance races when they have young ones in the nest, for consequent on the fact that they are feeding and breeding, the moult is somewhat retarded, and the bird will be in the best possible condition for flying, having all the necessary qualities most fully developed.

What to Avoid.

There are, however, two conditions when pigeons should not be entered for long contests:– 1st, at the moment when the cock is driving his hen to nest. At this time he is in a state of over-excitement to such an extent that he will not partake of his food, thus reducing his strength and putting him in a condition of undoubted inferiority. At this period the hen is in an equally unfit state to compete, owing to the closeness of her period of laying, for it is never safe to compete with hens shortly after or before laying. 2nd, when the pigeons have young of less than ten days of age, by reason of the secretion of the fluid which is produced in the crop, the non-discharge of which would lead inevitably to the most grave disorders, such as the putrification of the crop.

Putting into Baskets.

Some few days before the birds are to be basketed for a long-distance race, in order not to weaken those that are intended for the long-distance race by allowing them to feed their young too much, it is advisable to assist them by putting these under foster parents, and to put them back in the nests of the racing birds after having been fed by them. In this manner, the racing bird is called upon to perform less work in connection with rearing the young bird, at the same time is sufficiently attached to it to hasten its return to the loft on the race day.

It is also advisable to give the birds regular exercise as early as possible each morning, in order to accustom them to the morning and to being up and about at the earliest possible moment, a distinct advantage in connection with long-distance racing, especially when the race is not likely to close on the day of liberation.

The day before basketing for the race I think it is preferable that the food given to the pigeons should be composed of small seeds so that it may be the more easily digested when travelling in the panniers to the race point.

A Wrinkle Worth Knowing.

In order that a bird may not throw one of its long flights in the baskets it is a good plan to cut from off a piece about half an inch long, in order to allow the air to penetrate the tube or main quill of the feather which will thus be retarded and not thrown for some time longer than usual.

Return.

On its return from a race the pigeon ought to be the object of constant care and attention on the part of the owner; the food given should be composed of small, easily digested seeds, and during the three or four first days that follow baths should be regularly given, which are most useful in enabling them to build up their lost strength and to recover entirely from the exertion undergone in the big fly, wherein, it is to be hoped, it has won coveted honours for its owner.

THE NATIONAL HOMING UNION.

President: F. G. SKINNER, 11 and 13, Albury Street, Deptford, S.E.
Vice-Presidents: E. H. CROW, Staincliffe, Batley; GEO. YATES, Clayton Bridge; G. W. SINGLETON, New York Mill, Holcombe Brook, Bury.

The name of the Union shall be "THE NATIONAL HOMING UNION." It shall consist of Homing Societies within the United Kingdom, who shall apply for and be elected to Membership in accordance with Rule 7. The objects for which the Union is established are:—
1.—The improvement of the Homing Pigeon.
2.—The Protection and Advancement of the interests of its Members in all matters connected with the Sport of Homing Pigeon Racing.
3.—The Settlement of Disputes amongst its Members, and between its Members and others.
4.—The taking of such steps as may be found necessary to obtain redress for its Members in all cases of malicious shooting, unlawful detention, misrepresentation and fraud.
5.—The establishment of a uniform system of measurements for all flying routes, a uniform running allowance, and to secure a uniform system of timing.

NORTHERN CENTRE—
comprises the whole of Scotland, and the Counties of Northumberland and Durham, and such part of the County of York as lies North of a line commencing at Saltburn and running to the Western boundary of the County.
Secretary—John Handy, Nags Head, Houghton-le-Spring, R.S.O.

NORTH-WEST COUNTIES CENTRE—
comprises the Counties of Westmoreland and Cumberland, and such parts of Lancashire as are North of a line drawn from Crossens to Croston, thence in a North-easterly direction to the boundaries of the North-East Lancashire Centre, joining that Centre at Blackburn, and following the boundaries of the latter Centre to the Western boundary of the Yorkshire Centre.
Secretary—Geo. S. Wylie, Ladysmith Terrace, 362, Newhall Lane, Preston.

NORTH-EAST LANCASHIRE CENTRE
comprises such portions of Lancashire as are within a line drawn from Barnoldswick to Gisburn, thence to Clitheroe, Whalley, Blackburn, Darwen, and Ramsbottom, back to Todmorden, and along the borders of Yorkshire to Barnoldswick.
Secretary—Austin Holden, 11, Harold Street, Burnley.

WESTERN CENTRE—
comprises the whole of Ireland and such portions of Lancashire and Cheshire as are South of a line drawn from Crossens to Euxton, and west of a line drawn from Euxton to Halmerend Road.
Secretary—Charles Payant, Ainsdale, near Southport.

NORTH-WESTERN CENTRE—
comprises such portions of Lancashire as are South and East of a line drawn from Todmorden to Ramsbottom, thence through Halcome Brook and Darwen to Cherry Tree, and from that place in a South-westerly direction to Euxton, and from that place to Halmerend Road, and the whole of the county of Derby.
Secretary—George Yates, Clayton Bridge, Manchester.

EAST MIDLAND CENTRE—
comprises all places South of the Yorkshire and North-Western Centres as are East of a line drawn from Uttoxeter to Stafford, and from Stafford to Chipping Norton, and North of a line drawn from Chipping Norton to Harwich.
Secretary—C. J. Merrick, 76, Rothesay Avenue, Lenton Sands, Nottingham.

WEST MIDLAND CENTRE—
comprises all places South of the Western and North-Western Centres as are West of a line drawn from Stafford to Chipping Norton, and North of a line drawn from Chipping Norton to St. David's.
Secretary—Charles Harrison, Metchley, Prospect Hill, Redditch.

LONDON CENTRE—
comprises such portions of England as are South of a line drawn from Chipping Norton to Harwich, and East of a line drawn from Chipping Norton to Portland Bill.
Secretary—H. C. Bowden, 5, Campdale Road Tufnell Park, London, N.

SOUTH-WESTERN CENTRE—
comprises such portions of England and Wales as are South of a line drawn from Chipping Norton to St. Davids, and West of a line drawn from Chipping Norton to Portland Bill; also the Channel Islands.
Secretary—F. H. Beswick, 28, Jenkin Street, Aberdare, Glam.

WESTMORELAND AND CUMBERLAND CENTRE—
comprises all places in England North of Morecambe Bay, and West of a line drawn from Sandside through Windermere in a N.W. direction to Abbey Junction.
Secretary—Mr. J. C. Percy, 23, Sunderland Terrace, Ulverston.

YORKSHIRE CENTRE—
comprises such parts of the county as are situated South of a line drawn from Saltburn to the western boundary of the county.
Secretary—S. Green, 30, Green Lane, New Wortley, Leeds.

Secretary, CHARLES C. PLACKETT, Glenmore, Gledhow, Leeds.
Ring Marks, NU, Letters and Numbers.

STUD REGISTER.

In introducing this feature in the ANNUAL the idea has been, as far as possible, to bring the history of famous pigeons known in this country down to date. These famous birds have often been spoken of, but heretofore there has been no authentic record kept or work compiled containing these details. SQUILLS' ANNUAL each year will, we hope, in future contain particulars of the most successful birds of the year. The entry of good birds in the ANNUAL should save fanciers a great amount of work in the compilation of their stud lists, for it will only be necessary to send the details for entry in the ANNUAL, and in compiling a list a fancier need only briefly give details and refer to the stud number of his birds in SQUILLS' STUD REGISTER.

Subject to any modifications the Proprietors may deem necessary, the following are the proposed

RULES AND REGULATIONS.

1. The Register is intended to be a reliable record of proved racing pigeons and their performances, with particulars of their strain, and the names of the owner and breeder.

2. The following birds are eligible for entry, and can be entered upon payment of a fee of 1s. at any time by either the owner or breeder.

    A. Pigeons that prior to the formation of the Register have by their well-known performances established their reputation as long-distance fliers or producers of long-distance fliers.

    B. Pigeons that have won a prize or prizes in any race organised by any recognised society or federation in the United Kingdom of 200 miles and upwards, or that have flown in and their performance been verified before the close of any race of 400 miles and upwards

c. The sire or dam of any bird registered or entitled to
be registered in the Register.

D. The progeny of any bird registered, or entitled to
be registered in the Register, that has flown in
the year it was hatched in any race organised by
any recognised society or federation in the United
Kingdom of not less than 100 miles and upwards,
or as a yearling in any such race of 150 miles
and upwards.

3. A separate form for each entry will be supplied to owners
or breeders intending to register, upon payment of the necessary
fee of 1s. to *The Racing Pigeon* office, 19, Doughty Street, W.C.

4. Any additional performances by any registered bird
can be duly registered from time to time upon payment of a
fee of 1s., and special forms for the purpose are supplied. The
original registration number will be always retained in any
re-registration under this rule.

5. All original entries, registrations of additional performances,
and transfers of ownership will be published from time to time in
The Racing Pigeon, and every year will be incorporated in
Squills' Register.

6. The Proprietors of the Register reserve to themselves the
absolute right to refuse any entries or to remove any bird from
the Register on proof to them that registration has been obtained,
in their opinion, under false pretences or in any improper manner.

7. The Rules and Regulations will be subject to any alteration which from time to time it may be deemed advisable by
the Proprietors to make, and any alteration will be from time to
time published as occasion may require.

The Proprietors have under consideration a scheme for
offering prizes in connection with birds entered in the stud
register for their owners and breeders to compete for, but as the
object in compiling the Register has simply been to serve the best
interests of the sport, before offering any such prizes, if it should
be decided to offer them, the Proprietors intend to give the matter
mature consideration, and will make the announcement relative to
such prizes, if offered, in the columns of *The Racing Pigeon* at a
later date. The idea of offering the prizes would not be with the
object of inducing registration for the purpose of competition for
these prizes, but to secure the due recording and registration of
such well-known performers and producers as will be found in
the present issue, and be a source of interest to readers.

SQUILLS' STUD REGISTER.

Stud No.

1 **Rome I.**
Owner, J. W. Logan.

Black cheq cock, bred by Gaspar Heutz, of Aix la 'Chapelle. *Performance:* First bird to return in the great race from Rome in 1878, organised by the Society Union and Progress, of Brussels. He was stamped No. 2113, toss took place 5 a.m., 23rd June, 1878, and Rome I. arrived at 1.20 on July 2nd at Aix la Chapelle. Rome I. was not allowed to compete for the prizes, as Aix la Chapelle is not in Belgium, but Mr. Logan holds the certificate from M. Delmotte, the president of the Union and Progress, that the bird was the first to return, for which he was awarded a special diploma.

2 **B6, Montauban.**
1873.
Owner, J. W. Logan.

Blue cheq cock, bred by N. Barker. Bought with the whole of N. Barker's loft in 1878. *Performances:* Winner from Montauban, 14th in Belgian National from Langon, 9th in Belgian National from Mont de Marsan. *Strain of sire:* Hannot, of Brussels. *Dam:* Marica. Strain unknowm, as she came to Barker's loft as a squeaker.

3 **B16, sister to Montauban.**
Owner, J. W. Logan.

Blue cheq hen, bred by N. Barker. Kept for breeding. *Strain of sire:* Hannot, of Brussels. *Strain of dam :* Marica.

4 **D1.**
Owner, J. W. Logan.

Red cock, splashed black, bred by Debue, of Uccle, near Brussels. *Performances :* 31st in National from Auch in 1878. *Strain of sire:* Debue. *Strain of dam :* Debue.

5 **D2, 1879.**
Owner, J. W. Logan.

Red cheq cock, bred by Debue, of Uccle. A son of D1, bought by Mr. Logan as a young bird, and paired with B16 produced Mr. Logan's 105 and 163. *Strain of sire :* Debue. *Strain of dam :* Debue.

6 **664.**
Owner, J. W. Logan.
Strain of sire : Gits. *Strain of dam :* Gits.

B. cheq hen, bred by G. Gits. Mr. Logan obtained this hen through Mr. J. O. Allen in 1878. Her only known performance was to produce Old '86.

7 **Dardenne Pied.**
Owner, J. W. Logan.

B. cheq pied hen, bred by Dardenne, of Verviers. Bought from Dardenne in 1893. A daughter of his well-known pied cock, Le Bariolé, winner of following prizes in long races :—1887, 2nd Bordeaux ; 1888, 50th St. Sebastien (in Spain) ; and 51st Dax ; 1889, 299th Dax. *Strain of sire:* Dardenne. *Strain of dam:* Dardenne.

8 **Old '86.   1879.**
Owner, J. W. Logan.

B. cheq c., bred by owner. *Performances :* 1882, flew Arras ; 1883, 3rd Rennes, 309 miles ; 1884, 1st Rennes, 309 miles, 1st La Rochelle, 444 miles ; 1886, 1st Ventnor, 135 miles, 1st Rennes, 309 miles, 1st La Rochelle, 444 miles, arriving 4.48 p.m. day of toss against a North wind. *Strain of sire :* Barker and Hannot. *Strain of dam :* Gits.

9 **156.**
Owner, J. W. Logan.

B. cheq cock, bred by owner. Kept for breeding, and, paired with 163, produced 199, the dam of Mr. Logan's 538, that flew the first National from La Rochelle, 1894. *Sire :* Rome I *Dam :* Mr. Logan's 129

Stud No.

10   **22.**   Blue c.. Imported from Antwerp as a squeaker, Owner, J. W. Logan. and flew Ostend to Sheffield twice 1873.

11   **170-1881.**   Blue cheq cock, bred by owner. *Performances :* 1884, 3rd Granville : 1886, Cherbourg ; 1887, 3rd Owner, J. W. Logan. prize La Rochelle, 444 miles, arriving 6.40 p.m. day of toss against a North wind. 170, in the eight years he was in the loft, never spent a night out until he was lost at La Rochelle in 1889. *Sire :* Logan's 105. *Strain of dam :* Goossens, of Antwerp.

12   **105=1880.**   Red cheq cock, bred by owner. Kept for Owner, J. W. Logan. stock ; only performance was to breed 170. *Sire :* D2. *Dam :* B16.

13   **163.**   Blue cheq hen, bred by owner. *Performance :* 1884, 2nd La Rochelle, 444 miles. *Sire :* D2. Owner, J. W. Logan. *Dam :* B16.

14   **129.**   Red cheq hen, bred by owner. *Performances :* Flew 135 miles as young bird,and then kept for stock, and Owner, J. W. Logan. was the only hen Mr. Logan ever paired to Rome I. *Sire :* Logan's 53, a son of Logan's old 22, that flew Ostend to Sheffield twice in 1873.

15   **538=1189.**   Blue cheq cock, bred by owner. *Performances :* 1892, 6th Nantes ; 1894, flew La Rochelle in first Owner, J. W. Logan. English National race. *Sire :* Logan's 170. *Dam :* Logan's 509. 509 by Old '86, out of 163.

16   **9512=1895.**   Blue cheq pied cock, bred by owner. *Performances :* 1899, 1st Shetland C.C., 1st Shetland Owner, J. W. Logan. M.H.L,, 8th National, arriving 4.24 p.m., day of toss. *Sire :* Logan's 942, 942 by 538 out of 509. *Dam :* Dardenne pied. 509 by 86 out of 163.

17   **964=1896.**   Blue cheq cock, bred by owner. *Performances :* 1898, 8th Shetland; 1899, 3rd Shetland and cup. *Strain* Owner, J. W. Logan. *of sire :* Logan and Tempest, of Verviers. *Strain of dam :* Logan's Old '86 and Rome I.

18   **RP28-1900.**   B. cheq cock, bred by owner. *Performances :* 18th prize National race from Marennes in 1905, Owner, J. W. Logan. arriving 7.53 p.m. day of toss, 2,277 birds competing, distance 463 miles. *Strain of sire :* Logan's Old '86 and Rome I., and going back to Logan's 22 that flew Ostend to Sheffield twice in 1873, grandsire on father's side won 8th prize from Lerwick C.C.F.C. in 1898, and 3rd prize from Lerwick C.C.F.C. in 1899. *Strain of dam :* Logan's Old '86 and Rome I. and going back to Logan's 22 that flew Ostend to Sheffield twice in 1873. grandsire on dam's side won 1st prize Lerwick C.C.F.C., 1899 ; 1stprize Lerwick M.H.L., 1899 ; 8th prize Lerwick National 1899.

19   **9838=1898, no ring.**   B. cheq cock, bred by owner. *Performances :* 31st prize National race from Marennes in 1905, arriving at 8.23 p.m. day of toss, 2,277 birds competing, Owner, J. W. Logan. distance 463 miles. *Strain of sire :* Wegge's Vendome. *Strain of dam :* Logan's Old 86, and Rome I., and going back to Logan's 22.

20   **RP126-1901.**   Mealy cock, bred by owner. *Performances :* 101st prize National race from Marennes in 1905, arriving Owner, J. W. Logan. 6,4 a.m. second day, 2,277 birds competing ; also flew Marennes in National in 1904, taking 168th position, distance 463 miles. *Strain of sire :* Logan's Old '86. *Strain of dam :* Jurion and Logan's Rome I.. Grandam of 126 was bred by Mr. Biggs, of Chapmanslade.

The late J W Logan.

Stud No.

21 **RP103-1901.**
Owner, J. W. Logan.
B. cheq cock, bred by owner. *Performances:*
Flew Marennes in National race 1905, arriving 8.20 a.m.,
2nd day, 2,277 birds competing, distance 463 miles ; in
1904 won 7th prize from Rennes, 307 miles. *Strain of
sire :* Logan's Old '86, Rome I. and 22. *Strain of dam :* Delrez's, of Verviers,
1st Bayonne, winner of 11,000 francs in pools.

22 **RP33-1900.**
Owner, J. W. Logan.
B. cheq cock, bred by owner. *Performances :*
Flew Marennes in National race 1905, arriving 2 p.m.,
second day, 2,277 birds competing, distance 463 miles.
Strain of sire : Logan's Old '86 and Dardenne·
Strain of dam : Delrez's, of Verviers, 1st Bayonne, winner of 11,000 francs in
pools.

23 **Red Aberdeen,
1877.**
Owner, H. Stanhope.
Red cheq cock, bred by owner. *Performances:*
1877, Didcot five times, London ; 1878, Birmingham
3rd prize, Derby 3rd prize, Leeds 1st prize, Newcastle
2nd prize (equal) ; 1879, Birmingham, Derby, Leeds
1st prize ; Newcastle, Berwick, Aberdeen first home,
these three private tosses, not races. *Strain of sire :* Servais, of Ixelles. *Strain
of dam :* Servais, of Ixelles.

24 **Mealy Aberdeen,
1877.**
Owner, H. Stanhope.
Mealy cock, bred by owner.   *Performances :*
1877, Didcot three times, London ; 1878, Derby,
Leeds, Newcastle equal 2nd ; 1879, Sheffield, Newcastle, Berwick, and Aberdeen second home. *Strain
of sire :* Cloerts, of Bruges. *Strain of dam :* Servais,
of Ixelles, sister to the Red Aberdeen.

25 **Mealy Thurso,
1887.**
Owner, H. Stanhope.
Mealy cock, bred by owner. *Performances:* 1887,
Derby 7th prize, Sheffield ; 1888, Derby 1st series
prize, Ripon 2nd prize, Newcastle 1st series, Berwick
1st series ; 1889, Ripon 1st series, Newcastle, Berwick,
Arbroath 1st series, prize ; 1890, Newcastle, Berwick
twice, winning 4th prize North Middlesex club ; 1891, Newcastle, Arbroath,
Thurso second home, very disastrous race. *Strain of sire :* Stanhope's Red
Aberdeen, *Strain of dam :* O. Grooters, of Brussels.

26 **Old No. 1,
1884.**
Owner, H. Stanhope.
Red cheq cock (part white flights), bred by owner.
Performances: 1884, Derby ; 1885, Cherbourg,
Rennes ; 1886, Cherbourg, Granville, Rennes, Nantes;
1887, Cherbourg 2nd prize, Granville, La Rochelle,
1st prize open race 1888, Derby 2nd prize, Newcastle
1st prize, Berwick ; 1889, Newcastle 1st prize, Berwick, Arbroath ; 1890, Newcastle, Berwick, Banff. *Strain of sire :* Stanhope's Red Aberdeen. *Strain of
dam :* Mealy Aberdeen, N. Barker, and Trullemans.

27 **Red Banff,
1884.**
Owner, H. Stanhope.
Red cock, bred by owner. *Performances :* 1884,
Derby, Sheffield 2nd, and 2nd series prize, Leeds ;
1885, Cherbourg open 6th prize, Granville ; 1886,
Cherbourg, Rennes open 5th prize, Nantes 2nd prize,
La Rochelle ; 1887, Cherbourg, Rennes, La Rochelle ;
1888, Derby, Ripon, Newcastle ; 1889, Sheffield 1st series, Newcastle, Berwick,
Arbroath, 2nd and 1st series ; 1890, Newcastle 1st prize ; Berwick, Banff 1st
and silver cup. A very beautiful bird, much admired by the late J. P. Jones,
of Cardiff. *Strain of sire :* Red Aberdeen, Gilson, N. Barker, Mealy Aberdeen, Sluys. *Strain of dam :* Red Calf, Vekeman's of Antwerp.

28 **Marennes Ruby,
RP6943-1901.**
Owner, H. Stanhope.
Red hen, bred by owner. *Performances :* 1901,
Ashby, Banbury, Didcot ; 1902, Rugby, Didcot,
Bournemouth ; 1903, Guernsey 5th, Wakefield Fed. :
1904, Guernsey, Nantes 3rd prize in club, and 3rd in
Fed. ; 1905, Rugby, Didcot, Bournemouth, Guernsey,
Marennes National race, 110th prize and £5 R.P. special, in over 525 miles.
Strain of sire : Owner's old strains, inbred, Logan, Calf, *Strain of dam :*
Taylor, of Bristol.

H Stanhope, Esq.

29 **457-1888.**
Owner, **John Wright**

B. cheq pied c., bred by owner. *Performances*: 1888, Bridgenorth ; 1889, La Rochelle ; 1891, Nantes; 1893, Cherbourg. *Strain of sire*: 403 by 321, out of 320, James Colvile H4, Carrier, Logan hen. *Strain of dam* : 402 by 315, out of 324, J. P. Taylor, Jurion, J. O. Allen.

30 **486-1889.**
Owner, **John Wright.**

B. cheq pied h., bred by W. C. Moore. *Perform ances* 1890, Bournemouth, 3rd prize, Farnworth club, and 6th prize Manchester F.C. *Strain of sire* : Brother to Moore's Iron Duchess. *Strain of dam* : Logan's 379, by 168, out of 169.

NOTE.—The above 457 and 486 are Mr. John Wright's old stock pair. The following are a few of the winners from their offspring, with their performances in Mr. Wright's own words as far as possible : In 1896 a grandson flew La Rochelle as a yearling. In 1897 he flew Marennes, winning 2nd prize, beating all the birds in the Southern Section 4½ days. ↳ sold this bird to Mr. Cunliffe ; he bred Mr. Cunliffe's 1st bird home from Marennes, winning 1st New Mills F.C., 1st Stockport Federation, 2nd M.F.C., and 4th Great Northern Open, after 9½ minutes lost in trapping and running. I also sold him a son bred in 1898 from 457 and 486 (No. 275) and he bred from him 6th Worcester New Mills F.C., 2nd Swindon New Mills F.C., 3rd Stockport Fed., between 4,000 and 5,000 birds competing, 1st Nantes M.F.C. open race, 1902, as a yearling, beating 2nd bird and club winner by 148 yards per minute, only two birds homing the same day. No. 275 is also a grandsire of 1st and premier Worcester young birds race, 1903.

Mr. Cunliffe's white pied cc ‹k, his best bird in 1903, is a great grandson of 457 and 486, and won 5th Rennes, M.F.C., when ten months old ; another white pied cock, also a great grandson, flew Marennes 1904, and came home minus two flights in one wing the morning after toss. A granddaughter of 457 and 486 flew Rennes and Marennes 1904. She was bred in 1903, and is own sister to 1st Nantes, M.F.C. My first bird home in the Great Northern race, 1904, which won 36th prize and 14th in the M.F.C., hatched April 27th, 1903, and my second bird in the Great Northern race in 1903, hatched in 1902, are also granddaughters.

Mr, Swain flew a son of 457 and 486, winning for him in 1898, 1st Crewe, 1st Stafford, 1st Worcester, and 1st Bournemouth, and also the special oil painting ; in 1899 he won two 1st prizes. Mr. Swain had another son of 457 and 486 ; it bred him three hens that won numerous prizes up to Jersey. A brother flew Nantes at ten months old ; another brother won 2nd from Swindon for Mr. Spencer, of Burnley. Mr. Swain also has a grandson that flew Nantes three times. He also had a granddaughter which flew Jersey and bred his 1st prize Stafford, 2nd in Fed. and five-guinea cup for 1st nominated bird ; the year after won four 1st prizes and five-guinea cup.

Mr. Gladstone, of Neston, bought two cocks of Mr. Swain off 457 and 486 ; one bred for him 1st Bath and 1st in Fed., and 2nd prize Nantes ; the other bred for him 3rd prize winner from Nantes. These two cocks and grandchildren also bred winners.

Mr. Armistead, Morecambe, bought a granddaughter of 457 and 486 from Mr. Swain. She flew Jersey and bred for him his No. 799, who flew Jersey, and bred birds to fly from Jersey and Nantes from five different cocks. His No. 14, bred in 1894, crossed the Channel th ːteen times, Jersey six times, Nantes four times, Rennes twice, and Marennes once. In 1896 she won 1st Rennes, Morecambe club and 58th federation open ; in 1897, 2nd Bournemouth, 1st Rennes, velocity 1023 yards, only bird home in Morecambe. Cock off 799 flew Jersey four times, winning prizes, and Nantes three times.

Mr. Padkin had a granddaughter from Mr. Swain off 457 and 486 ; she bred for him four birds that flew Nantes in the disastrous toss in 1903 ; one of them won 1st Jersey and 1st in federation in 1903

I sold Mr. Wildgoose a cock off 457 and 486 ; he bred birds to fly from Nantes, etc., and won prizes. Mr. Kershaw bought a cock from me which bred prize winners for him. Dr. Barker and many others have won prizes with offsprings. Stowe Bros. A23 is a son, and is sire and grandsire of many winners. Many good performances of the progeny of 457 and 486 since 1904 **not** recorded here.

ʌu i No.

44

189. **Late bred, 1897.**
Owner,
H.M. The King.

Blue cheq cock, bred by His Majesty The King,
Performances : 1898, 3 miles, then jumped to Boston
across the Wash, 20 miles, Dewsbury, 115, Whitley,
177 ; 1899, 2 miles, then Boston, Lincoln, York,
Whitley, Banff, and Lerwick 1st G.N. race. *Strain*
of sire : A pure Jurion, bred by the late Mr. Geo. Chisem, of Dewsbury.
Strain of dam : Bred by H.R.H. The Prince of Wales from a pair of birds,
the gift of Monsieur F. Duchateau.

45

HM621=1902.
Owner,
H.M. The King.

Blue cock, bred by His Majesty The King.
Performances: 1902, untrained ; 1903, Newcastle-on-
Tyne ; 1904, Newcastle-on-Tyne ; 1905, Nantes.
Strain of sire : 189 above. *Strain of dam :* Pure
Jurion.

46

HM704=1903.
Owner,
H.M. The King.

B. c. cock, bred by His Majesty The King. *Per-*
formances: 1903, London ; 1904, London ; 1905,
Nantes. *Strain of sire :* Pure Jules Janssens.
Strain of dam : Duchateau, Jurion.

47

824=1893.
Owner,
H.M. The King.

B. c. cock, bred by Monsieur F. Duchateau.
Performance: 1895, Dax. *Strain of sire :* Pittevil,
Janssens and Wegge. *Strain of dam :* Pittevil and
Janssens.

48

D335=1899.
Owner, H.R.H.
The Prince of Wales.

Blue cock, bred by owner. *Performances :* All
stages to York, 1899 ; 1900, Newcastle-on Tyne ; 1902,
Newcastle-on-Tyne ; 1905, Nantes.

49

802=1894.
Owner, H.R.H.
The Prince of Wales.

B. c. cock, bred by Mons. F. Duchateau. *Per-*
formance : 1898, Dax. *Strain of sire :* Pittevil and
Janssens. *Strain of dam :* Delmotte and Menier.

50

832=1895.
Owner, H.R.H.
The Prince of Wales.

Black cheq cock, bred by Mons. F. Duchateau.
1897, G.N. Dax ; 1898, G.N. Dax ; 1899, G.N. Dax ;
1900, G.N. Dax. *Strain of sire:* Vanderhaeghen.
Strain of dam : Gits.

51

Le Criquet,
1877.
Owner, the late
Alexandre Hansenne.

Dark cheq bronze cock, bred by owner. *Per-*
formances : 1877, Paris 1st, Toury 3rd ; 1878, Issoudun 1st, Ville Perdu 4th ; 1879, Poitiers 10th, St.
Benoit 1st ; 1880, La Souterraine 4th, Bazas 24th ; 1881,
Morcenx 10th, Lectoure 35th ; 1882, Toloso 5th, Nar-
bonne 14th ; 1883, Bilbao 20th, Libourne 21st, Toloso
3rd. *Strains :* Dedoyard, Hansenne, Vekemans.

52

Le St. Vincent,
1877.
Owner, the late
Alexandre Hansenne.

Dark cheq cock, bred by owner. *Performances :*
1877, 2nd Port de Piles, 8th Angouleme ; 1878,
Issoudun 5th, Bordeaux 35th ; 1879, St. Benoit 1st,
Mirande 8th ; 1880, St. Vincent 2nd, Poitiers 4th ;
1881, Auch 37th, Argenton 7th ; 1882, Poitiers 17th,
Murat 6th ; 1883, Agen 1st, Bayonne 28th ; 1884, St.
Jean-de-Luz 46tn, Libourne 21st. *Strains :* Dedoyard and Vekemans.

53

Calvi,
Owner, the late
Alexandre Hansenne.

Dark cheq bronze cock, bred by owner. *Per-*
formances : 1st Calvi, 1888 ; 1889, 19th Niort. The same
year as flying Calvi also flew Dax in Grand National.

NOTE. Le Bruxelles, a grandson of Le Criquet, won 2nd Grand National
in 1894, 5450 birds. Le Bon Bleu bred 1898, Le Bonneval, 1903 ; Le Vieux
Fonce, and the renowned Le Nouan, which Hansenne sold for £44 to M.
Pirlot, of Liege, were five of Hansenne's most noted winners, and were all
descendants of the famous St. Vincent and Criquet. M, Alexandre Hansenne
died in 1903, and his loft was sold by auction the 6th December, 1903. The
birds were mostly of the Verviers type, with thick round heads, abundantly
feathered.

Stud No.

54
Voliere.
Owner,
Mons. N. Delmotte.
B. c. cock, bred by owner. *Performances*: Won 1st Creil, 1st Chantilly, 1st Orleans, etc., and was sire of 1st Chantilly, 1st Arras, 4th Orleans, 2nd Orleans, 2nd Le Mans, 8th Tours, 5th Dover, 2nd St. Jean de Luz, National F.C., and many others. This famous pigeon sired most of Delmotte's best birds, and his blood still runs through the lofts. *Strain of sire*: Vanschingen. *Strain of dam*: Vanschingen.

55
La Pale.
Owner,
Mons. N. Delmotte.
Mealy hen, bred by owner. *Performances*: Won nearly £1,000 in prizes, including 2nd Claremont, 1st Chantilly, 3rd Etampes, 5th Paris, 5th Chartres, 24th Salbris, 1st Le Mans, 5th Le Mans, 1st Denis, 1st Angers, etc., etc. She was dam of L'Hirondelle, the Grand National winner of 1893, and grandmother of the Grand National winner of 1900 (I think it was 1900 when Delmotte won his Grand National a second time). *Strain of sire*: Vanschingen. *Strain of dam*: Vanschingen.

56
Le Male Pale.
Owner,
Mons. N. Delmotte.
Mealy cock, bred by owner. Cannot give performances, as he has been dead about 14 years. Was the best sire Mons. N. Delmotte ever possessed. Was lent to Mons. Jurion, for whom he sired many famous birds. *Strain of sire*: Vanschingen. *Strain of dam*: Vanschingen.

57
Le Premier Revenu,
1891.
Owner, A. Jurion.
Red cheq cock, bred by owner. *Performances*: 1893, when only two years old, 7th prize Dax G.N., 2,917 competitors; 1894, 72nd St. Jean-de-Luz G.N., 5,548 competitors; 1895, 30th St. Jean-de-Luz, G.N., all pools, 4,157 competitors; 1896, the 14th prize after 8 minutes on loft, St. Jean-de-Luz, G.N., 3,794 birds, again winning all pools. He only competed in the G.N. races, and M. A. Jurion describes him as his best bird, really a champion. *Sire*: Le Revenu, who flew five times and won in the G.N. twice, winning pools. He is by Vieux Roux, Delmotte, and pure Diable hen. The Vieux Roux, Delmotte, was a present from M. Delmotte, a brother of his 1st Vierzon. Diable was from Premier Prix de Liege, a great winner in the long races. Premier Prix de Liege, was from Vieux Castel, one of the best birds M. Jurion ever had; he died in 1892, when 24 years old. *Dam*: Le Crombe Carcasse, bred in 1890, and trained every year 200 or 300 miles, a pure Diable.

58
Le 1st prix de Bayonne, 1894.
Owner, A. Jurion.
Dark b. cheq cock, bred by owner. *Performances*: Won easily in 1896 the 1st prize in the race of Gand from Bayonne, 540 miles, 494 birds competing, also 1st Malines in double. *Sire*: Le Jeux Blancs, a Petit Diable, a great winner, bred from Diable and La Vielle Bayonne. La Vielle Bayonne was a marvellous racer, for ten successive years she won a prize in the G. National. She was a daughter of Diable. *Dam*: Le Poreau, an excellent stock bird. She was from a Delmotte cock and La Vielle Bayonne.

59
Le Vierzon.
Owner, J. Pletinckx.
Cock, bred by owner. *Performances*: Flown Vierzon, 8th prize of honour Limoges, Orleans, Poitiers 9th prize, 1,193 competitors, La Mothe National race, 500 miles, Toulouse prize, Vendome prize; 1885, Bilbao (Spain); 1886, Bilbao 3rd; 1888, Dax prize, National race.

60
Barbiche.
Owner, J. Pletinckx.
Blue pied cock, with white, beard bred by owner. *Performances*: Winner of 100 prizes, including 7th Bayonne, National race, 560 miles, over 1,800 competitors; 8th Morcenx, 550 miles, National, 1,935 competitors; Dax 142nd prize, National, 560 miles, 2,923. Recognised as J. Pletinckx's best bird, and the founder of his strain.

61
Hippolyte, 1884.
Owner, J. Pletinckx.
Blue hen, bred by owner. *Performances*: 1890, Dax, Colombe Fidèle, 25th prize, 3,748 competitors; 1891, St. Jean dè Luz, Roue d'Or, 79th prize, 2,376 competitors; 1892, St. Jean de Luz, Pigeon d'Argent, 23rd prize, 2,638 competitors; 1893, Dax, Colombe Nord, 163rd prize, 2,917 competitors; 1894, St. Jean de Luz, Union Progrès, 145th prize, 5,543 competitors; 1895, St. Jean de Luz, Aleona, 244th prize, 4,157 competitors; 1896, St. Jean de Luz, Grand Colombier, 70th prize, 3,794 competitors; 1897, St. Sebastien, Trompette Gand, 54th prize, 340 birds.

The late MONS. A. HANSENNE.

MONS. N. DELMOTTE.

MONS. A. JURION.

MONS. J. PLETINCKX.

A Quartette of Famous Belgians.

Stud No.

68 **Champion Hen.**
        180-84.
        Owner, J. O. Allen.

Blue mottle nen, bred by owner. *Performances*: Cheltenham, 1884; Cherbourg, 1885; 1886, flew Nantes, 459 miles, in 9 hours 45 minutes, velocity 1380, was 5th in race. *Strain of sire*: Flown Northampton, a present from N. Barker, *Strain of dam*: Blue mottle, pure Gilson, flown Willesden Junction in 1880.

69 **Julienne.**
        Owner,
        The late J. P. Jones.

B. cheq hen, bred by Jurion. *Performances* to the loft of M. Thirionet, who raced her: 1886, 4th Tressy; 1887, 1st Vendome, 1st Morcenx, disastrous Grand National, 1st Morcenx, N.F.C.; 1888, untrained; 1889, 6th Poitiers; 1890, 3rd Angouleme and prize in Grand National from St. Jean-de-Luz. She was the dam of following winners for M. Thirionet: 9th Bretigny, 10th Deurdin, 9th Orleans, 2nd Labans, 2nd Argentil, 4th Chateau Rigault, 15th Poitiers, 1st Vendome. *Strains*: Julienne was a granddaughter of Jurion's famous Diable and of Grooter's noted Vieux Bœuf.

Julienne was imported for the sum of £32 by the late J. P. Jones. She met with an unfortunate end, as, we believe, when on the way to a purchaser, she escaped from an office in London, and was not heard of again.

70 **9-1890.**
        Owner,
        The late J. P. Jones.

B. cheq hen, bred by the late J. P. Jones. *Performances*: 1890, 45 miles; 1891, Sheffield, Ripon and Newcastle; 1892, Berwick and Banff, 440 miles, tossed at 4 a.m. and wired in 8.19 p.m. same day with No. 3. *Strain of sire*: Pure Debue, bred from Logan's Debue I. out of Debue VI. *Strain of dam*: A pure Boon, of Uccle.

71 **3-1888.**
        Owner,
        The late J. P. Jones.

B. cheq hen, bred by J. W. Logan. *Performances*: 1888, 20 miles; 1889, 108 miles; 1890, 20 miles; 1891, 40 miles; 1892, Stafford, Ripon, Berwick and Banff, 440 miles, liberated 4 a.m., wired in 8.19 p.m. *Sire*, Logan's '86. *Dam*, 163.

The late J. P. Jones died early in 1893, and his birds were sold at the Cottonwaste Exchange, Manchester. He tried for many years to fly birds from Banff to Cardiff, but Nos. 3 and 9 were the only two that he got to perform the feat. He imported the famous Julienne from Thirionet.

72 **Old Boley, 1879.**
        Owner, Wm. Kaye.

Red cheq cock, bred by Cruttenden. *Performances*: 1881, Derby, Leicester, Bedford, London, Dover, and Folkestone, Bristol, Exeter; 1882, Plymouth; 1883, Cherbourg; 1884, Cherbourg; 1885, Cherbourg, 1st and silver cup Rennes, and 1st best average all races to Rennes, his son 5th Rennes same race; 1886, all stages to Nantes, 450 miles, in 9 hours 42 minutes, one of his sons being 2nd same race; 1887, all stages to La Rochelle, 525 miles, afterwards kept for stock. *Strain of sire*: Oliver, of Bexhill. *Strain of dam*: Oliver, of Bexhill.

73 **Clay.**
        1881-H10.
        Owner,
        John Armstrong.

Red cheq cock, clay colour, bred by L. Hasaer. *Performances*: 1881, Paris, Etampes, Vendome; 1882, Chateaudun, Mont Louis; 1883, Paris, 2nd prize Chateaudun, 20th Ville Perdu, prize of honour St. Benoit. *Strain of sire and dam*: Leopold Hasaers, of Antwerp.

74 **G528-1882.**
        Owner,
        John Armstrong.

Brownish silver hen, dun colour, bred by G. Gits' Stock. *Strain of sire and dam*: A granddaughter of the Donkeren and Queen, own sister to Gits' C.

NOTE.—The above pair of stock birds were bought by John Armstrong from Singleton Green, who imported them to this country. Mated together they bred the celebrated family known as the Clays, so called because of their peculiar clayish red colour.

75 **Successful,**
        1890.
        Owner, T. H. Hall.

Red cheq cock, bred by owner. *Performances*: 1890, 5th Didcot, 11th Newbury; 1891, 1st St. Malo, 1st La Rochelle; 1893, 2nd La Rochelle. *Strain of sire and dam*: Successful was own brother to Dublin and Old England, and was exported to Australia where he did so well.

Stud No.

76
Dublin,
1890.
Owner, T. H. Hall.

Dark red cheq cock, splashed black, bred by owner. *Performances :* 11th prize Didcot, 3rd prize Newbury ; 1891, Lymington, 17th prize Cherbourg, 195 miles, then jumped to La Rochelle, 440 miles, a jump of 245 miles ; 1892, equal 1st prize Cherbourg, 6th prize St. Malo, 264 miles, 5th prize Nantes ; 1893, 5th prize Cherbourg, 11th prize Rennes, 303 miles, 1st prize La Rochelle, 1426 velocity. *Strain of sire :* Stanhope's Red Aberdeen. *Strain of dam :* Red cheq hen, imported from N. Barker. Flew Arras, Chantilly, Etampes, and Orleans. Bred from a son and sister to Montauban and a Mausta-Debue.

77
Old England,
1890.
Owner, T. H. Hall.

Dark red cheq cock, splashed, bred by owner. *Performances :* 1891, 1st Cherbourg, 12th St. Malo, 7th Rennes ; 1893, 5th La Rochelle. *Strain of sire and dam :* Old England is own brother to Dublin.

78
Perfection,
1890.
Owner, T. H. Hall.

Red cheq hen, bred by owner. *Performances ;* 1890, Ventnor ; 1891, 2nd Plymouth, 4th Scilly Isles ; 1893, Cherbourg. Own sister to Dublin, Successful, and Old England.

79
Wellington,
1889.
Owner, T. H. Hall.

Light b. cheq cock, bred by owner. *Perform-ances :* 1889, not trained ; 1890, 11th Lymington ; 16th Cherbourg ; 1891, 11th St. Malo and Rennes. *Strain of sire :* N. Barker. *Strain of dam :* N. Barker.

80
Midland Leader,
1889.
Owner, T. H. Hall.

Dark cheq cock (part white flights), bred by owner. *Performances ;* 1889, 2nd prize, Reading ; 1890, 10th Cherbourg, St. Malo, and 6th prize La Rochelle ; 1891, Cherbourg, 3rd prize, St. Malo, 11th Rennes, and 5th La Rochelle ; 1892, Cherbourg extra prize St. Malo and 7th Nantes ; 1893, 5th Cherbourg and 3rd prize Rennes, Midland Leader crossed the Channel twelve times. His sire, Indian Prince ; his dam, Lovely. *Strain of sire :* Indian Prince, Logan b'ood. *Strain of dam :* Lovely.

81
Victor,
NMC1148.

Red cheq cock, bred by the late E. Reynolds. *Performances :* 1898, Chippenham, 9th Bridgend, 6th Llanelly, 1st (only bird home) Haverfordwest ; 1899, first home in four out of five races, 4th Tavistock, Penzance, Scilly Isles ; 1900, 3rd Alderney, 6th St. Malo, timed in day from Bordeaux, N.F.C. ; 1901, St. Malo, 13th Bordeaux, N.F.C., special for first bird into London ; 1902, 10th St. Malo, 2nd Bordeaux, N.F.C., 1st J. W. Logan's special ; 1903, 4th La Roche, G.N., winning £56 in money. *Strain of sire :* Rob Roy, bred by J. J. Barrett, the famous Rob Roy of the late E. Reynolds. Father of his loft. Bred from Eclipse (Rome and Excel) and hen from Competitor and Excel, the whole of the blood is Stanhope's and Julienne. *Strain of dam :* Jeannie Deans, bred from a son of Evangelisti's Sibley cock, 1st Retford, 6th Arbroath, Thurso, etc., and Hansenne, Calvi and Criquet.

NOTE.—The late E. Reynolds died in 1902. The whole of his birds were advertised in, and sold by, *The Racing Pigeon,* the list appearing November 4th, 1903. Victor is the bird that will ever be remembered in association with his name.

82
Excelsior,
909=5.
Owner,
Ernest E. Jackson.

Blue cheq cock, bred by owner. ·*Performances :* Worcester, Gloucester, Bournemouth, Guernsey, Rennes, and Marennes, 550 miles, when only 23 weeks old, winning 8th prize Marennes, Manchester F.C. Excelsior sired many birds to fly the Channel, and is grandsire to no end of Channel birds. An own brother to Excelsior flew Marennes as a yearling, and Bordeaux as a two-year-old, and four or five other brothers flew Nantes, and a half-brother flew Marennes at eight months old. *Strain of sire :* Offermans. *Strain of dam :* Offermans and Delmotte, bred by Mr. J. Cock, of Worcester.

Stud No.

83
Old Brighton,
1874-1888.
Owner, G. P. Pointer.

Red cheq cock, bred by owner. *Performances* : 1888, Holme, Essendine and Grantham Y.B. races, N.M.F.C. ; 1899, Newark 5th prize, Doncaster and Newcastle ; 1890, Newark, Doncaster, York, North-allerton and Berwick ; 1891, Retford, York, Durham, Berwick and Arbroath, winning 9th prize, vel. 981 ; 1892, Retford, York, New-castle, Berwick, and 4th prize Aberdeen. This bird flew in every old bird race of the N.M.F.C. from 1888 to 1892. He is a large, strong, handsome bird, and won the following amongst other prizes in the show pen for birds flown over 100 miles : 1st Barnet, 1st N.M.F.C., and 2nd Wood Green, 1890 ; 1st Wood Green, 2nd N.M.F.C., 1891. *Sire:* red cheq bred by F. Romer, Esq., and flew in the 19th Century races. *Dam* : a large blue hen which flew 100 miles in the 19th Century club races, and then kept for breeding.

84
Motor.
947-1893.

Dark red cheq cock, bred by G. B. Pointer. *Performances :* 1893, flown Essendine, Newark, and Doncaster ; 1894, Retford 10th, velocity 1364, York 21st, Newcastle, velocity 1252, and Arbroath 17th ; 1895, Retford, York, Newcastle 2nd, Berwick and Banff, taking 1st, 2nd, and 3rd prizes, being the only bird home before the close of the race in the A.P.H.S. ; 1896, Retford 7th, velocity 1260, York, velocity 991, New-castle, velocity 833, also 1st average cup for the above three races, and Thurso 1st, velocity 1454, A.P.H.S, and 1st federation, beating all English records for over 500 miles ; 1897-98, kept for stock ; flown Retford, *Sire:* Red cheq, bred 1889, known as the Young Brighton. 1889 flew Essendine ; 1890, Newark, Don-caster, Northallerton and Newcastle ; 1891, Retford, York, Durham, and Ber-wick ; 1893, Retford and York. Dam of Young Brighton is a red cheq, bred by Mercer Bros., and flew as youngster Essendine, Grantham, and Newark, being well up in all her races. Her sire flew Holme, Essendine, and Newark three times, Doncaster three times; Retford and York twice, winning 2nd and 3rd Holme, Essendine, 2nd and 9th, and Doncaster 1st. Her dam, a blue, flew Grantham 4th, Doncaster twice, 3rd and 6th. *Dam* of The Motor is a pure Stanhope, black cheq pied, bred 1892, trained 80 miles, then kept for stock ; bred by Mr. Stanhope, and bought by me when a squeaker from his No. 9 and 10, 1892 list. Her sire, No. 9, was a red cheq pied, and flew 1889 Derby ; 1890, Sheffield, Doncaster, York, Northallerton, Newcastle, and Berwick, 1st in the N.M.C. race ; 1891, Sheffield, Retford, York, Durham, Berwick, and Arbroath. Sire, a son of 2nd Banff, No, 10, blue cheq pied, flew 1886 Plymouth and Penzance, taking 2nd series prize ; 1897, Cherbourg, Rennes, and 2nd La Rochelle ; 1888, not flown ; 1889, Cherbourg and Granville ; 1890-91, not trained.

NOTE.—The Motor, Old Brighton, and Young Brighton, were well-known birds in London on the North Road from 1888, until the time this loft was dis-banded.

85
745-1895.
Owners,
Orchardson Bros.

Red cheq cock, bred by owners. *Performances* : 1895, Swindon ; 1896, 10th Wellington, 25th Jersey, 11th Rennes ; 1897, 3rd Bournemouth, Jersey, 55th Marennes ; 1898, 8th Jersey L.H.S., 12th Jersey W.L.S. Fed., 2nd Jersey W.L.S. pool, 1st Marennes W.L.S. Fed., 1st Marennes La Rochelle Club, 1st Marennes L.H.S., 745 in the Marennes race, beat the next bird by over 3½ hours (in this race he won £22 in cash and the 15-guinea cup, presented by W. H. Bell, Esq., to the Rochelle Club) ; 1899, he won 1st prize Swindon L.H.S., this being the first race in which he competed. He was lost in the Guernsey smash, and thus ended the life of one of England's champions. *Sire* : 28, blue cheq cock, 1894, pure Bonami imported ; sire flew 562 miles, dam flew 269 miles. Performances of 28 : In 1894 and 1895, not flown ; 1896, owner's 1st bird home from Worces-ter and Swindon, winning 6th and 7th respectively, also flew Bournemouth and Jersey, winning 27th prize. *Dam* : 27, red cheq hen, 1894, pure Bonami ; sire, a very fast bird up to 375 miles, having won 1,000 francs at this distance ; dam flew 269 miles as a young bird. Performances of 27 : 1894 and 1895, not flown ; 1896, Bournemouth ; since kept for stock.

The name of Orchardson and Bonami will always be associated with the above pair of stock birds, 28 and 27, and they, in addition to the 23 hen, were the foundation of the lofts which were sold through *The Racing Pigeon,* December 6th, 1902.

Stud No.

86
23=1893.
Owners,
Orchardson Bros.
..red by .i. Bonami, Antwerp. *Performances*:
1893, not flown; 1894, Ventnor, disastrous race:
1895, Bournemouth; 1896, Jersey and La Rochelle.

87
Nabob.
Owner, W. H. Cottell.
Dark cheq cock, bred by owner.[a] *Performances*:
The first pigeon that flew Bordeaux to England, 500
miles. Bred in 1882, flew S.W. route; 1883, Ventnor; 1884, Cherbourg, Granville, Napoleon Vendee
3rd, 350 miles (his brother being 1st); the same year he won the cups, and 1st
and only Bordeaux, L.C.S., afterwards kept for stock and lived to a great age.

88
150.
Owner, Bert Baron.
Blue cheq cock, bred by Barnes and Farrington.
Stock bird, and only trained short distances. *Strain*
of sire: He is by 56 out of 527. 56 was a blue cock,
bred by Mr. E. L. B. Bower, from 1227 and 337. 1227,
brother to Mr. Bowers' Shot Cock, from 593 and 4593 from 8 and 10, 8, pure
Colvile. 10, bred by Mons. Van den Broeck. 4, granddaughter of Gits' C
cock, the mother of Mr. Bowers' loft, and dam of Southport Pilot, the first bird
to fly Nantes to Southport in the day, winning 5th prize Liverpool H.S., 1892.
She is dam and grandam to dozens of prize winners in Continental races. 337,
blue hen, bred by Mr. A. Hilton, Wigan, from his 1 and 2. 1, red cheq cock,
bred 1887 by Mr. J. Armstrong, Wigan, from his H10 and G528, the celebrated
Clay pair. 2, blue hen, white flights, bred by Mr. Hilton from Mr. Gainer's (of
Gloucester) Nos. 1 and 2 on his 1888-9 list, which were good winners in the
Continental races. 527, the dam of 150, was also bred by Mr. Bower from 1158
and 168. 1158 from 666 and 13. 666, bred by T. W. Wilson, from 354 and 974.
354 from his 7 and 8. 7, pure Debue. 8 from a son of Mr. Logan's Old Smal
and a granddaughter of Rome I. 974 from his 3 and 2. 3 from Mr. Logan's
326 and 258B. 326, son of the Mausta Mealy cock. 258 from a son of Mr.
Logan's Rome I. cock and 187 hen. 2 from an N. Barker cock and Oliver and
Cruttenden hen. The Oliver and Cruttenden hen, own sister to grandam of
Iron Duchess. 665, not trained. 354, own nest bird to the first bird to fly
Nantes to Liverpool in the day, winning 2nd prize L.H.S. 1892 for Mr. T. W.
Wilson. 13, pure Carpenter. 168, bred by Mr. Bower from 710 and 1009.
710 from 131 and 148. 131, son of Mr. Thorougood's 42 and 47. 42, pure
Wegge blood. 47, pure Gits' C blood. 1009, red cheq hen, bred 1894 by Mr.
P. Verdon. Sire, red cock, bred 1892, Wegge and Grooters strain. Flown
Paris and Vendome. Dam, red cheq hen, bred 1893, from Vekeman, Wegge
and Debue blood. *Strain of dam*: 527, bred Messrs. Barnes and Farrington's
1st prize Nantes, Birkdale H.S., 1st prize Southport United Ring Club, 1st
Southport Dis. Fed., 5th prize West Lancashire Sat. Fed. open Nantes race,
1903; also 5th prize Jersey, 1903, and 2nd prize Swindon, 10th Fed.; also 5th
prize, Bournemouth. She is grandam to the 2nd prize Jersey and 1st pool,
Mount Street F.C.; also 5th prize Birkdale H.S., 1905.

89
147.
Owner, Bert Baron.
Blue cheq hen, bred by Barnes and Farrington. Stock hen. Her sire, 861, Old Danby;
her dam, 1027. 861 was a red cheq cock
called Old Danby, bred by Mr. E. L. Bower,
Pletinckx and Carpentier blood; flown Jersey three times. Dam, 1027,
blue hen, bred by Mr. Bower, kept for stock, from 57 and 148. 57
from Hilton's 11 and 12, Clay blood. 148 from Thorougood's 42 and
47. 42 pure Wegge blood. 47, pure Gits, C blood.
NOTE.—150 and 147 are the parents of Combine Jewel, the remarkable hen
that won the Nantes Combine disastrous race. They have also bred numerous
other winners.

90
Combine Jewel,
18=01BNE.
Owner, Bert Baron.
B. cheq hen, bred by Baron and Seed. In 1901
she flew all stages to Worcester; 1902, all stages to
Bournemouth (hard race) and timed in: 1903, all
stages to Jersey (found in loft, very fast race); 1904,
Crewe, Worcester, Bournemouth, and Jersey; 1905,
Crewe, Stafford, Worcester, Bournemouth, Nantes, 1st Lancashire, Nantes
Combine disastrous race, pools, 1st East Lancs. Fed., etc. *Strain of sire*:
Owner's 150. *Strain of dam*: Owner's 147.

91
RPS1904=7670.
Owner, L. Coates.
Dark cheq hen, bred by owner. *Performances*: 1905
1st and special Jersey, Kidderminster Central H.S.,
only bird home day of toss, a very hard race. *Strain*
of sire: Bell (of Badsey), Moore, Bennett's three times Bordeaux. *Strain of*
dam: Clay, Logan, Barker, Gits.

Stud No.

92

**Iron Duchess,
bred 1886.
Owner, W. C. Moore.**

Dark blue cheq hen, bred by owner. *Perform-ances :* 1886, flew Cheltenham, 104 miles ; 1887, flew Cherbourg, 260 miles, and Rennes, 367 miles, getting 14th position from Rennes, Manchester club, 118 birds competing ; in 1888 flew Cherbourg, Rennes, and La Rochelle, 504 miles, winning the following prizes : Cherbourg 1st prize, Farnworth club, 156 birds competing ; and 7th prize, Manchester club, 422 birds competing ; Rennes 1st prize, Farnworth club, 17 birds competing ; and 1st prize, Manchester club, 49 birds competing. (On this day it rained in torrents from 6 a.m. to 5 p.m., and it was 8.53 p.m. when she arrived at her loft, being the only bird hnme on the day of toss). La Rochelle 1st prize, Manchester club, beating every other bird by a clear day, 20 birds competing ; birds in eight days owing to bad weather. In 1899 again flew Cherbourg, Rennes, and La Rochelle, winning as follows : Cherbourg : 5th prize, Liverpool club, 125 birds competing ; 8th prize, Farnworth club, 167 birds competing ; 38th prize, Manchester club, 695 birds competing. Rennes : 1st prize, Farnworth club, 28 birds competing ; 3rd prize, Manchester club, 148 birds competing ; and 1st Federation pool prize from Rennes, 17 competitors. La Rochelle : Manchester club, arrived too late to win a prize. In 1890 flew Cherbourg at two tosses, making a velocity of 1,100 yards. Not tossed afterwards. *Strain of sire :* John Wright's Old Paris strain, Gits, Gilson, Le Cleer, and N. Barker. *Strain of dam :* N. Barker, Oliver, and Cruttenden, Brighton.

93

**Red Prince,
bred 1890.
Owner, W. C. Moore.**

Red cheq cock, bred by owr,er. *Performances.* Flew Cherbourg, Sottevast, and Avranches. Winner of 1st prize Cherbourg, Manchester club, 1893, on a wretched day, wind North-East, velocity 837 yards, 536 birds competing ; and 1st prize and gold medal Cherbourg, Farnworth club, 1893. *Sire :* Blue Rochelle, bred by me. Flew Cherbourg twice, Sottevast, Nantes, and La Rochelle, winning 8th prize Nantes, 1891, and 9th prize La Rochelle 1890, Manchester club. He (Blue Rochelle) sired my Nantes three times hen, which flew Nantes to my loft three years in succession ; also sire of Young Blue Rochelle. Red Indian, the sire of my 1898 Rennes winner. The dam of Sirdar, etc., etc. The parents of Blue Rochelle I purchased from the late Mr. William Heap, Manchester, many years ago, and were the grandparents of his Queen Victoria. They were as follows : Blue cheq cock, bred by Mr. Dejaiffe, of Tamines. St. Quentin a prize, Orleans a prize, Langon 31st, and 8th prize in the Grand National from Auch. Blue hen, bred by Mr. Perillieux, of Brussels. Delmotte strain. As a young one she did Valenciennes, Tergnier, Corbeil, and Salbris. *Dam :* Red Princess, bred by me. Flew Cherbourg and Rennes in 1899, winning 6th prize from Cherbourg, Liverpool club, 13th prize from Rennes, Manchester club, and 2nd Fed. pool from Rennes when only a year old ; Cherbourg and La Rochelle 1890, Sottevast (in two tosses), and Granville 1891, Cherbourg, Avranches, and Nantes 1892. She was an inbred Iron Duchess.

94

**Skylark,
18XX96=17.
Owner, W. C. Moore.**

Dark blue cheq hen, bred by owner. *Performanct :* Flown 1897, Jersey ; 1898, Jersey and Rennes, my 2nd home from Rennes ; 1899, flew well to Weymouth, then jumped to Nantes, my 2nd home, winning 2nd prize Warrington club, 11th prize La Rochelle club, 300 birds competing, and 15th prize West Lancashire Federation open race, 755 birds competing, velocity 1,110 ; 1900, Guernsey, Rennes, and Nantes, my 3rd home from Nantes ; 1901, flew Weymouth, St. Malo, and Bordeaux, Manchester club, being my 2nd home from Weymouth, winning 12th prize, 791 birds competing ; my 2nd home from St. Malo, winning 20th prize, 855 birds competing, and my 1st home from Bordeaux, winning 3rd prize, 191 birds competing, and 12th prize open race, 574 birds competing ; 1902, Weymouth, then jumped to Bordeaux with the National F.C., winning 50th prize, 1,601 birds competing. *Strain of sire :* Iron Duchess, Heap's Queen Victoria, and Pletinckx. *Strain of dam:* Moore's Ragman and Silver Streak. Ragman flew Avranches three times and Nantes. Silver Streak flew Nantes, 1892, winning 4th prize and homing before my 2nd prize winner ; and Nantes again, 1893, winning 2nd prize Manchester club, doing the distance, 430 miles, in 8 hours and 20 minutes, timed in 8 hours and 25 minutes ; velocity, 1,500 yards.

95
Good Chequer,
bred 1894.
Owner, W. C. Moore.

Dark blue cheq hen, bred by owner. *Performances:* Flown 1895, Mortain, winning 6th prize Manchester. club, velocity 1273 yards, 139 birds competing, and Angers, winning 13th prize Manchester club, 106 birds competing ; 1896, Weymouth, then jumped to Rennes, winning 1st prize and gold medal Manchester club, 166 birds competing, and 1st prize open race, 205 birds competing, flying the distance, 367 miles, in 7 hours and 16 minutes, velocity 1483 yards ; 1897, Weymouth and Rennes, being my third bird home from Rennes ; also winner of three 1st prizes in show pen. *Strain of sire :* Iron Duchess and Red Prince. *Strain of dam :* Logan's 326, a son of Mausta Mealy, 280 (Iron Duke) and 173.

96
Young Gem,
NUooMA100.
Owner, W. C. Moore.

Blue cheq hen, bred by owner. *Performances :* Flew as a youngster Worcester, Gloucester and Swindon ; 1901, Bournemouth and Weymouth twice ; 1902, Bath, Weymouth, St. Malo (the disastrous race) and Nantes, being my 2nd home from Bath, my 1st from **Weymouth**, winning 4th prize Manchester club, 1,157 birds competing ; my 4th from St. Malo and my 2nd from Nantes ; 1903, trained privately to Weymouth, then jumped to Marennes, arriving at 6.45 p.m. day of toss, winning 6th prize Manchester club, 10th prize La Rochelle club, and 39th prize Great Northern open race, 973 birds competing. *Strain of sire :* Bred by Mr. Gould, Lynm, from his 13th prize winner from Marennes, Manchester club Southern Section, 1899. Contains the blood of Iron Duchess, Red Prince, Young Ocean Queen, and Heap's Queen Victoria. His dam bred by me from a Delmotte bred cock, out of a daughter of Red Prince and Iron Duchess. *Strain of dam :* Gem, bred by me, one of the most consistent workers I ever had, never spent a night out of her loft after liberation. She flew Cherbourg twice, Avranches twice, and Nantes, winning amongst other prizes, 3rd prize Avranches 1893, Manchester club. Her sire was my first bird home from La Rochelle 1890, winning 7th prize Manchester club. Contained the blood of Mons. Bekeman's Rome I., Gits' Van Loon, and Michiels. Her dam flew Cherbourg. Bred from a pure J. O. Allen cock and a granddaughter of Logan's '86.

97
Prince I.,
bred 1891.
Owner, W. C. Moore.

Blue cheq cock, bred by owner. *Performances :* Flew Swindon as a youngster ; 1892, Cherbourg ; 1893, Cherbourg, Avranches and Nantes, winning 1st prize Cherbourg, Liverpool club, 2nd prize Avranches and 7th prize Nantes, Manchester club, Southern Section, flying the distance, 430 miles, in 8 hours and 55 minutes, timed in 9 hours 17 minutes, velocity 1373 yards ; 1894, Cherbourg, Avranches and Nantes, winning 3rd prize Nantes, Manchester club ; 1895, Cherbourg and Angers. *Strain of sire :* Red Prince. *Strain of dam :* Ocean Queen, bred by me in 1900, and flew Swindon same year ; 1891, Bournemouth, winning 1st prize Liverpool club, 443 birds competing, velocity 1173 yards ; 1892, Bournemouth, Cherbourg, Avranches and Nantes, winning 2nd prize Bournemouth, 510 birds competing, velocity 1258 yards ; 7th prize Avranches, Liverpool club, and 1st prize Nantes, Manchester club, 127 birds competing, velocity 1049 yards. Her brother, Ocean King, won 2nd prize Nantes 1892, Manchester club ; whilst a sister flew Nantes very well the same day to my loft. Strain : Logan's Old Smal, Iron Duchess, Red Prince, etc.

98
756-1902.
Owner, T. H. Burton.

B. cheq cock, bred by owner. *Performances :* 1902, Swindon ; 1903, Bournemouth ; 1904, Jersey and Nantes, W.H.S. ; 1905, Jersey and Marennes with the National F.C. *Sire,* 195 ; sire, 397, Thorougood, Wegge, Orchardson's 333; dam, 1871, Bonami, pure Debue and Clay. *Dam,* 364 ; sire, 397 ; dam bred by Mr. W. G. Orchardson from 602 and 445. 26A, 42-47, Wegge, Gits and Bonami.

99
863-1901.
Owner, T. H. Burton.

Blue cheq hen, bred by R. Slack. *Performances :* 1901, untrained ; 1902, Swindon ; 1903, Bournemouth ; 1904, 2nd Jersey, W.H.S. ; 1905, Jersey, 1,714 birds, Liverpool Dis. Fed. ; 1905, Jersey, timed in, and 48th Marennes Great Northern open race, 1,140 birds competing. *Sire,* 903. *Dam,* 265. 903, from 842, Rome I., Hansenne, Logan's 105, and Old '86 blood ; and 3070, from Thorougood's 500, and 26A blood. 265, from 3057, Bonami, and 776, sister to 314 and 1223, Gilson and Gits.

Stud No.

100   **24-1901.**
Owner, T. H. Burton.

Dark cheq cock, bred by owner. *Performances.* 1901, Swindon ; 1902, Bournemouth ; 1903, Rennes, 11th Marennes, La Rochelle Club, 47th Marennes, Great Northern open ; 1904, Jersey, 14th Marennes, National F.C., 1st £1 pool, 1st 10s. pool, 1st 5s. pool, Northern Section, 525 miles, and £10 match against Mr. J. L. Baker's No. 1063 : 1905 Weymouth smash, then jumped to Marennes, 100th National race, 2,277 birds. *Sire :* Blue cheq cock, 1059-1900 1900, Swindon ; 1901, Bournemouth ; 1902, Jersey ; 1903, Rennes, 20th Marennes, La Rochelle club ; 1904, Jersey, lost at Marennes. His sire combined the blood of Gainer, Offermans, and T. W. Thorougood. His dam, No. 2 (14 pair), sister to Mr. Orchardson's Grit, which crossed the Channel several times, including Bordeaux twice. *Dam :* Dark cheq hen, 343-1898. Flown Bournemouth in 1899, since kept for stock, being the dam and grandam of many good birds, including my No. 24 (No. 1 pair), the 47th Marennes in the Great Northern, 1903, and 14th National race and pools, 1904. Her sire, 397 ; her dam, 2.

101   **1990-1898..**
Owner, T. H. Burton.

Blue cheq hen, bred by owner. *Performances :* Flown 1898, Swindon ; 1899, Guernsey, 10th Jersey, Liverpool H.S. ; 1900, Jersey, 7th Nantes, L.H.S. ; 1901, 6th Weymouth, L.H.S., 28th Bordeaux M.F.C. ; 1902, Jersey ; since kept for stock. *Sire :* 573, Van Walcheroun and W. C. Moore. *Dam :* 314, J. W. Toft, T. W. Wilson, Logan. Brothers and sisters to 1990 are winners of 1st Guernsey, 7th Nantes, 7th Jersey, 1st Gloucester, 1st Monmouth, and 3rd Swindon, West Lancs. Saturday Fed., dam of owner's most consistent birds.

102   **243-1902.**
Owner, T. H. Burton.

Blue cock, bred by Wilson. *Performances :* 1902, Swindon twice ; 1903, Bournemouth, 2nd Rennes, W.H.S., 4th Rennes, L.H.S. ; 1904, 4th Rennes, W.H.S., 8th Rennes, Liverpool Dis. Fed., 47th Marennes, National F.C., and £5 R.P. special, over 525 miles ; 1905, flew Weymouth smash, then jumped to Marennes, taking 182nd National race, 2,277 birds competing, and Marennes again with the Great Northern, making three times Marennes. Sire, 271 ; dam, 871. 271, a pure Wegge, sire of 1st prize Nantes in the Liverpool Dis. open race. 871, sire, a pure Delmotte, from birds imported direct ; dam combining Logan, Gits and Posnaer blood.

103   **13. Bred 1896.**
Owner, J. W. Toft.

Red cheq cock, bred by owner. *Performances :* 2nd Rennes, 1897, at 10 months old, W. L. Fed. ; 6th Rennes, 1898 ; 4th Jersey, 1899 ; 1st bird in Liverpool Dis. Bordeaux, 1900, timed in 6.54 'day of toss ; Bordeaux, 1901 ; Bordeaux, 1902. Crossed the Channel 14 times, did not spend night out till 1901 Bordeaux race. *Strain of sire :* Posnaer, Marica. *Strain of dam :* Gits, Logan's '86.

104   **182-1893.**
Owner, J. W. Toft.

Blue cheq hen, bred by Wilson. Not flown. Dam of 707, 1st Bordeaux ; 708, Bordeaux ; 728, Wilson's first bird in five races ; 2626, 1st. Nantes ; 1099, 2nd Marennes, 4th Nantes. Grandam of many winners, and one of the founders of my loft. *Strain of sire :* Janssens and Logan. *Strain of dam :* Daughter of Logan's '86.

105   **707-1896.**
Owner, J. W. Toft.

Red cheq hen, bred by owner. *Performance :* 1st and cup Bordeaux, National F.C., 1899. *Strain of sire :* Logan. *Strain of dam :* 182.

106   **347-1895.**
Owner, J. W. Toft.

Strain of dam : Logan.

Red cheq hen, bred by owner. *Performances :* 1897, Marennes, my first bird home but did not trap ; 1898, 5th Weymouth, 6th Rennes ; 1899, Nantes ; 1900, Bordeaux ; 1901, Bordeaux. *Strain of sire :* Logan.

107   **229-1897.**
Owner, J. W. Toft.

Blue cheq hen, bred by owner. *Performances :* Nantes 3rd, 1899 ; 2nd Nantes, 1900 ; Nantes, 1901 and 1902. *Strain of sire :* Logan. *Strain of dam :* Hansenne.

Stud No.

108 **2626.**
Owner, J. W. Toft.

Blue cheq cock, bred by owner. *Performances:* Nantes 1st, Marennes, 1905. *Strain of sire:* Gits and Logan. *Strain of dam:* 182.

109 **351-1895.**
Owner, J. W. Toft.

Red cheq cock, bred by owner. *Performances:* 1st Marennes, Liverpool Wednesday H.S., 1898; 4th Marennes, 1897, L.H.S.; Nantes, 1899. *Strain of sire:* Moore. *Strain of dam:* Logan.

110 **Lytham Emperor, W96L688.**
Owner, D. Hedges & Son.

Red cheq cock, bred by S. Gibson, Huyton. *Performances:* 1897, flown all stages to Jersey; 1898, all stages to Marennes, 530 miles, winning 5th Liverpool H.S., 12th La Rochelle club, 17th West Lancs. Saturday Fed. open race, 298 birds competing; 1899, all stages to Guernsey; 1901, flown Jersey; 1902, flown Bordeaux, 600 miles, jumped from Bournemouth, 409 miles of a jump in the National F.C., 1,601 competing. *Strain of sire:* Bred by W. H. Bell, off 1587-1581, J. W. Logan. etc., blood. *Strain of dam:* Flown Marennes, Collignon, Logan's Iron Duke and 326, 301 blood, Mons. Thys, Janssens and Trulleman's famous White Pigeon blood.

111 **Her Majesty, 18A98-52.**
Owner, D. Hedges & Son.

Blue cheq hen, bred by G. W. Henson, Aberdeen. " Longest distance champion of Great Britain." *Performances:* 1898, flown all stages to Girvan, 170 miles; 1899, Stanley, Stirling, Kilmarnock, and Portpatrick, 200 miles; 1900, the same stages, and Navan, 300 miles; 1901, Stanley, Kilmarnock, and Portpatrick, Navan, and Skibbereen, 483 miles, winning 1st and special and Lord Provost silver cup, beating next bird by two clear days; she also won 18th diploma in Scottish National, Skibbereen; 1902, trained to Portpatrick, then sent to Bordeaux, 854 miles 97 yards in the National Flying club, a jump of 654 miles. *Strain of sire:* Pure Sirjacob; flown Thionvil, Arton, Luxembourg, Criel, and Bretizing. *Strain of dam:* Bred by E. H. Crow, off his 434-87. 434 contains the blood of Logan and Heap's Little Red Boardman and Duerinck. 87. Sire, bred by Mr. Thewls; dam, pure Van Muylem.

112 **Lytham King, 18B98-170.**
Owner, D. Hedges & Son.

Blue cock, flown by H. P. Calvert, Walton-le-Dale. *Performances:* 1899, flown Bournemouth; 1900, all stages to Jersey, then jumped to Marennes, 551 miles, winning 4th prize Manchester F.C. open race, velocity 1232, 2nd Chorley and 1st Preston Dis. H.S. for Mr. Jackson's £5 special best average three Continental races. *Strain of sire:* Pure W. C. Moore, contains the blood of Young Blue Rochelle, Iron Duchess, Nantes three times, and Thirionet. *Strain of dam:* Logan, Stanhope, Gits, Hetinckx, and Heap.

113 **Lytham Wonder, NU99B711.**
Owner, D. Hedges & Son.

Dark blue cheq hen, bred by owners. *Performances:* 1900, flown all stages to Bournemouth, then jumped to Bordeaux, 622 miles, a jump of 409 miles, and holds the record as the longest distance yearling of England. Granddaughter of Lytham Star, the first bird that ever flew 534 miles to England, which he did when only nine months old from La Rochelle in 1888. *Strain of sire:* Lytham Star, Black Gilson, and N. Barker. *Strain of dam:* Pure E. E. Jackson, from 333, won 6th Marennes; B78, flown Nantes.

114 **359C1894.**
Owner, J. Wormald.

Blue cheq hen, bred by owner. *Performances.* 1894, Banbury; 1895, Cherbourg; 1896, 10th prize Ashby, Bath, Weymouth, Jersey, and Rennes; 1897, Winchester, 3rd prize Jersey, and 4th prize Rennes; 1898, Bournemouth; 1899, Durham and Newcastle, 7th prize Didcot, 3rd prize Bournemouth, 7th equal Jersey, and 10th prize Rennes; 1900, Winchester, and 55th from Bordeaux, 630 miles, in National F.C., 1103 yards velocity; 1901, Bournemouth, 18th prize Bordeaux in National F.C., winner of J. W. Logan's £5 special, and winner of H. Stanhope's £5 special for first bird into Yorkshire; 1902, Bournemouth. *Strain of sire:* Van Muylem and Delmotte. *Strain of dam:* Mausta Mealy, Jurion, and Mills' Madrid strain.

Stud No.

181
40.
Wilts 1896-285.
Owner, J. Compton.
Dark cheq nen, bred by owner. *Performances*:
1896, Haverfordwest; 1897, Guernsey, 1st Crystal
Palace F.C., St. Malo, Nantes; 1898, Guernsey, St.
Malo; 1899, Guernsey, St. Malo, 9th Southern
Counties F.C., Nantes, Bordeaux, 39th Nat. F.C.;
1900, Alderney, 7th club and 5th open C.P.F.C., Nantes, and Bordeaux, 9th
S.C.F.C.; 1901, Alderney, 4th club and 3rd open C.P.F.C., Nantes; 1902,
Guernsey, St. Malo, Bordeaux. *Strain of sire*: Heap. *Strain of dam*:
Hall.

182
70.
Wilts 1897-657.
Owner, J. Compton.
Mealy hen, bred by owner. *Performances*:
1897, 10 miles; 1898, Haverfordwest; 1899, all stages
to Tralee; 1900, Guernsey, 5th Southern Counties
F.C., St. Malo, Nantes, and Bordeaux, 4th Nat. F.C.,
and 1st Southern Counties F.C., velocity 1289. *Strain
of sire*: Delmotte. *Strain of dam*: Delmotte.

183
8.
Wilts 1893-108.
Owner, J. Compton.
Mealy[4] hen, bred by Owner. *Performances*:
1893, not trained; 1894, Maidstone, Deal, Ostend,
Brussels 2nd; 1895, Maidstone 2nd, Deal, Lichter-
velde, Brussels 6th. Always home on day of liberation. *Strain of sire*: Heap. *Strain of dam*: Heap.

184
A39.
NUooJL26.
Owner, J. Compton.
Mealy cock, bred by owner. *Performances*:
1900, all stages to Normanton; 1901, all stages to
Berwick; 1902, Tamworth 7th, Ripon 1st, Wilts F.C.,
and 4th South-Western Fed., Newcastle 5th, Perth
9th club and 11th Fed., and Banff 2nd club and 3rd
Fed.; 1903, Perth; 1904, Chesterfield; 1905, Berwick. *Strain of sire*:
Heap. *Strain of dam*: Heap, No. 8 above.

185
A6.
NUooBS914.
Owner, J. Compton.
Dark cheq hen, bred by owner. *Performances*:
1900, Tamworth 10th, Derby 7th, Chesterfield; 1901,
not trained; 1902, Tamworth, Chesterfield, Ripon,
Newcastle 5th; 1903, Newcastle and Banff 5th; 1904,
Newcastle 4th, Berwick 4th, S.C.F.C., Banff 3rd,
Wilts F.C.; 1905, Ripon 2nd and Berwick. *Strain of sire*: Heap and Kaye.
Strain of dam: Heap and Hall, No. 40 above.

186
C4.
1902RPG7909.
Owner, J. Compton.
Red cheq hen, bred by owner. *Performances*:
1902, all stages to Normanton, Wilts smash; 1903,
Chesterfield; 1904, Newcastle; 1905, Tamworth,
Ripon, Berwick, Banff 1st, Wilts F.C.; and has never
had a night out. *Strain of sire*: Mackenzie. *Strain
of dam*: Sawtell.

187
A30.
CP1900P166.
Owner, J. Compton.
Blue hen, bred by owner. *Performances*: 1900,
Alderney; 1901, Guernsey, St. Malo, 9th Southern
Counties F.C., Nantes; 1902, Guernsey, St. Malo,
Nantes, and Bordeaux National F.C.; 1903, Guernsey
and St. Malo; 1904, Alderney, St. Malo, and Maren-
nes, 32nd prize and 2nd 2s. 6d., 1st 5s., 1st 10s., 1st £1 pools, Section C,
National F.C., and 1st Crystal Palace F.C.; 1905, Weymouth, Alderney, and
Marennes. *Strain of sire*: J. J. Barrett. *Strain of dam*: Heap and Kaye.

188
1898-261.
Owner, M. Howells.
Mealy cock, bred by owner. *Performances*: 1899,
1st Dover, 208 miles, 145 birds competing, Porth and
Rhondda Valley H.S. *Strain of sire*: Late J. P.
Jones, of Cardiff. *Strain of dam*: R. Williams, Ely.

189
The Grooter Cock,
bred 1891.
Owner, John Wones.
Blue cheq cock, ticked white on head, bred by
Flavill. *Performances*: 1891, flew Exeter, competing
in three races, was my first bird home in every race
but would not trap until another bird arrived; 1892,
flew Southampton; 1893, flew Ventnor 2nd prize,
Cherbourg, Rennes, La Rochelle 5th prize; 1894, flew all races to La Rochelle,
winning 5th prize in the Grand National organised by the Manchester F.C., 581
birds competing, also 1st and 1st special Central Counties, and 1st Worcester
Dis., also taking 1st, 3rd and 4th Brierley Hill Dis club, only one other bird
getting home before close of race. Not tossed afterwards. *Strain of sire*:
Grooters. *Strain of dam*: Grooters.

Stud No.

Three Times Bordeaux Hen,
140 **1385, bred 1897.**
Owner, John Wones.
Blue cheq hen, bred by owner. *Performances*, Flew 1897, Bournemouth ; 1898, Bordeaux ; 1899, Salisbury, 1st prize Central Counties, Bordeaux 1st prize and Mr. Logan's silver cup ; 1900, Bordeaux, jumped from Weymouth. This bird took many other good positions in races, but no record of them has been kept. Not tossed after 1900. *Strain of sire :* Colvile's F14 and Old Blue Pied Hen. *Strain of dam :* Gits and La Rochelle, Grooters.

Bred 1898. Unrung.
141 **Owner, John Wones.**
Blue cock, bred by Flavill. *Performances:* 1898, flew Weymouth ; 1899, flew Rennes, winning 1st prize and Mr. Showell's silver cup, also flew Bordeaux ; 1900, flew Bordeaux, winning 13th prize Grand National and a member's cup, also 5th prize Central Counties club ; afterwards sold to Mr. Wood, Derby. *Strain of sire :* Grooters. *Strain of dam :* Jurion.

The Jurion cock,
142 **bred 1891.**
Owner, John Wones.
Mealy cock, heavily splashed black, bred by owner. *Performances:* 1891, flew Exeter ; 1892, Southampton ; 1893, 3rd prize Cherbourg, Central Counties, also flew La Rochelle in this year ; 1894, all stages to La Rochelle, winning 3rd Central Counties, 8th prize Grand National, organised by Manchester Flying Club, 581 birds competing ; also 2nd prize Brierley Hill and District club, being one of the only two birds home before close of race. Not tossed afterwards. *Strain of sire :* Jurion. *Strain of dam :* Jurion.

A1901E117.
Owner,
143 **Dr. G. W. Dowling.**
Blue cheq cock, bred by T. A. Bradley, Esq. *Performances :* 1st club and 1st Manchester Dis. Fed. Swindon, 1902 ; 1st club and 2nd Manchester Dis. Fed., Guernsey, 1903 ; 1st club, 1st nomination, and 5th Manchester Dis. Fed., Guernsey, 1904 ; 2nd club, 9th Manchester Dis. Fed., Nantes, 1904 ; 9th Manchester F.C. and 12th open Rennes, 1905. Also flew Marennes, 1905. *Strain of sire:* W. H. Wilson. Toft. *Strain of dam :* Toft.

RPH1902-3752.
Owner,
144 **Dr. G. W. Dowling.**
Red cheq cock, bred by owner. *Performances :* Flown La Roche as a yearling, all stages to Rennes in 1904 ; all stages to Marennes in 1905, winning 11th Manchester F.C., 67th Great Northern 1905. *Strain of sire :* J. Millward's Ragman and Logan. *Strain of dam :* Toft.

A1901E178.
Owner,
145 **Dr. G. W. Dowling.**
Blue cheq cock, bred by owner. *Performances :* Equal 1st Swindon Alderley Edge H.S., equal 1st Swindon Manchester Dis. Fed., 1903 ; 7th Bourne-mouth A.E.H.S., 1903; 2nd Worcester Manchester F.C., and all stages to Rennes, 1904 ; lost at Nantes, 1905. *Strain of sire :* J. Millward's Ragman ; *Strain of dam :* Logan's Old 86.

RPT1903-76.
Owner,
146 **Dr. G. W. Dowling.**
Blue cheq hen, bred by owner. *Performances :* 2nd Bournemouth, A.E.H.S., 1904 ; 2nd Manchester Dis. Fed., Southern Section, 3rd Nantes, A.E.H.S., 1904 ; 11th Manchester Dis. Fed. Nantes, 1904 ; all stages to Rennes, 1905. *Strain of sire :* W. H. Wilson, Toft. *Strain of dam :* Toft. Pletinckx, Gits and Kaye.

RPA1904-6360.
Owner,
147 **Dr. W. G. Dowling.**
Red cheq hen, bred by F. Boughey. *Perform-ances ·* 3rd Worcester, Manchester F.C., Southern Section, 1904 ; 2nd Guernsey, A.E.H.S., 4th Man-chester Dis. Fed., Southern Section, Guernsey, 1905. *Strain of sire :* De Groote. *Strain of dam :* Toft and Logan.

448. NUo1HK.
148 **Owner, W. I. Oakey.**
Blue cheq cock, bred by owner. *Performances :* 1901, trained 30 miles ; 1902, 4th Weymouth, 1st Guernsey (premier velocity on day in two Worcester clubs), Rennes ; 1903, 7th Bath. 6th Weymouth, and Guernsey, 6th La Roche G.N., Mr. Logan's and R.P. specials, value £15 ; 1904, Templecombe ; 1905, Granville. *Strain of sire :* T. W. Thorougood. *Strain of dam :* T. W. Thorougood.

Stud No.

149

360MC1893.
H. J. Longton's
No. 23.

Red pied hen, bred by owner. *Performances:* 1893, Worcester, 30th Swindon ; 1894, Swindon 21st, Cherbourg ; 1895, Jersey, my 1st bird home, La Rochelle, 1st prize open race. *Strain of sire:* John Buckley, N. Barker, Wright, Heap, and S. Gibson. *Strain of dam:* N. Barker, owner's No. 53, red pied hen.

150

5, 290SHS1894.
Owner,
H. J. Longton.

Blue cheq hen, bred by owner. *Performances:* 1894, Gloucester, Swindon ; 1895, Worcester, Swindon ; 1896, Bournemouth, Jersey, Rennes 10th, La Rochelle 1st prize open race. *Strain of sire:* N. Barker, owner's No. 64. *Strain of dam:* N. Barker, owner's No. 29.

151

267, 266SHS1894.
Owner,
H. J. Longton.

Dark cheq hen, bred by owner. *Performances:* 1894, 6th Gloucester, 10th Swindon ; 1895, Bournemouth, Jersey, 1st prize and silver cup Rennes ; 1897, 13th Jersey, 41st Marennes open race and won £10 match ; 1898, 2nd home Jersey, 8th Bordeaux, N.F.C. open race, and won Northern stakes, £6 6s. ; 1900, Guernsey, 14th Bordeaux N.F.C. open race, and the record into North of England, velocity 1248¾. *Strain of sire:* John Armstrong's Clays. *Strain of dam:* sister to owner's No. 5.

152

65, 346NU01JA.
Owner,
H. J. Longton.

Black cheq pied hen, bred by owner. *Performances:* 1903, Guernsey, La Roche, my 1st bird home ; 1904, 2nd Guernsey, Marennes, timed my 1st bird home ; 1905, 48th Marennes N.F.C. open race, 2,277 birds, 1st in three pools, winning £36. *Strain of sire:* Owner's No. 520, flown Bordeaux. *Strain of dam:* Owner's No. 501, flown Marennes three times.

153

1900NUJL427.
Owner, J. P. Taylor.

Red pied cock, bred by owner. *Performances:* 1900, not trained ; 1901, Weymouth ; 1902, Jersey ; 1903, Jersey ; 1904, Jersey ; 1905, 1st Weymouth ; Newcastle club, also Jersey. *Strain of sire:* owner's 570. *Strain of dam:* Logan, De Groote.

154

1901NUP4811.
Owner, J. P. Taylor.

Red cheq hen, bred by owner. *Performances:* 1901, Bristol ; 1902, Bristol, 1st N.E.C. Fed. yearling produce, £13, 1,200 birds competing, only 10 timed ; 1903, 3rd Tyneside open race, Oxford, 140 competing, only seven timed, 1st Newcastle club Ventnor, 3rd two-year-old produce, Ventnor, N.E.C. Fed, ; 1904, 3rd Tyneside open Selby, 700 birds competing. *Strain of sire:* Owner's P29, Pletinckx. *Strain of dam:* Owner's 64, Dardenne.

155

1903RPS8180.
Owner, J. P. Taylor.

Blue cheq hen, bred by owner. *Performances:* 1903, Bristol ; 1904, Weymouth ; 1905, 1st Seaham Harbour club, N.E.C. Fed., Jersey, also R.P. £5 in Up North Combine. *Strain of sire:* Bovyn. *Strain of dam:* Owner's 64, Dardenne.

156

The Old Mealy.
1901.
Owner, H. M. Flint.

Barless mealy cock, bred by owner. *Performances:* Many good positions English races. 1904, Rennes and 41st National and also won match ; 1905, Guernsey and Marennes, 2nd Buxton, 12th Manchester F.C., 62nd Great Northern, and also winner of £5 match against Messrs. Stow Bros.' Brigham. *Strain of sire:* T. Whittle's 1st La Rochelle hen. *Strain of dam:* Bovyn.

157

King of the Peak.
1902.
Owner, M. Flint.

Red cock, bred by owner. *Performances:* 1904, 1st Nantes Derbyshire Federation, 1st Nantes Buxton Dis. H.S. *Strain of sire:* W. Pickford. *Strain of dam:* W. Pickford.

158

Queen of the Peak.
1902.
Owner, M. Flint.

Blue cheq pied hen, bred by W. C. Moore. *Performances:* 1904, Guernsey, Rennes, and Nantes ; 1905, Guernsey, 18th Derbyshire Federation, and 1st Nantes Derbyshire Federation, and 1st Nantes Buxton Dis. H.S. *Strain of sire:* N. Barker. *Strain of dam:* H. J. Longton's two La Rochelle winners and Wielemans.

Mr. J. WRIGHT.

Mr. N. BARKER.

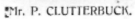

Mr. P. CLUTTERBUCK.

Mr. F. ROMER.

Popular Fanciers representing North, East, South
and West.

Stud No.

159
Jack,
BFC1899-145.
Owner,
F. T. Halewood.

Mealy cock, bred by W. Elsden. *Performances*: 1899, Essendine; 1900, York; 1901, 1st Newcastle by three hours, winning Mr. Doll's silver cup; 1902, 1st St. Malo, 1st Nantes, 107th Grand National Bordeaux; 1903, Alderney, St. Malo, 1st Grand National La Roche, winner of £77 10s. and King's cup; 1904, Alderney, 2nd St. Malo, Marennes in National. *Strain of sire:* Barker's Ghost. *Strain of dam:* Pure Stanhope.

160
Velocity,
RPH05-9142.
Owner, C. Middleton.

Dark cheq cock, bred by owner. *Performances:* Grateley, Templecombe, Tisbury, Chard, winning R.P, cup and £20 cheque for highest velocity, 1st London Fed., 1st Lewisham and New Cross Dis., velocity 1847 yards. *Strain of sire:* Gladstone and Logan, *Strain of dam:* Redoubte, Old Rome, and Iron Duchess.

161
Young General,
RP05D9329.
Owner, E. Coates.

Blue cheq cock, bred by owner. *Performances:* 1905, 1st club, 1st produce, and equal 3rd North Lancs. and West. Fed. Worcester, 936 birds, velocity 752. *Strain of sire:* Logan's 242, Smal, Gits, Moore's Young Duke, and Old '86. *Strain of dam:* Gits, Posnaer, and Marica.

162
Old General.
1809-1900.
Owner, E. Coates.

Red cheq cock (splashed), bred by T. Wadsworth Wilson. *Performances*: 1901, Bath; 1902, 2nd N. Lanc. and West. Fed. Bournemouth, 1,170 birds, 8th Jersey; 1903, Jersey; 1904, 6th Jersey and Nantes; 1905, 1st club and 1st pool, 1st N. Lanc. and West. Fed. Bournemouth, 1,778 birds, vel. 964 yards. *Strain of sire:* Logan's 242, Smal, Gits, Donkeren and Queen. *Strain of dam:* Moore's Young Duke, 242, Smal, and Old '86.

163
Monarch,
NU03JC413.
Owner, E. Coates.

Dark blue cheq cock, bred by owner. *Performances:* 1904, 1st North Lanc, and West Federation Bath; velocity 1075, 1,294 birds, and Bournemouth; 1905, 1st N. Lanc. and West Federation Jersey, velocity 981, 1,010 birds, and Nantes Combine smash. *Strain of sire:* Rome I, and Logan's 242, Smal. *Strain of dam:* Gits, Posnaer and Marica.

164
Coronation.
RP1902-8362.
Owner,
Fredk. Hambleton.

Dark cheq hen, bred by owner. *Performances*: 1902, 1st and 1st Coronation sweepstake Worcester, L.C., 1st and special Gloucester, after flew Weymouth; 1903, 1st and special Bath, 6th and pool Weymouth, after flew La Roche in N.F.C.; 1904, 1st, special and pool Worcester, L.C., equal 2nd N.S. Fed., 1st, and special and pool Guernsey, L.C.; 3rd prize winner in open race N.S. Fed., 2nd Nantes, L.C., then sent to Marennes, N.F.C., winning 37th prize and one of the R.P. £5 prizes; 1905, 3rd Weymouth, afterwards flew Marennes in the N.F.C. L.C. denotes Leek Club; N.S.F., North Staff. Fed. ; N.F.C., National Flying Club. *Strain of sire:* Colvile, F14, Iron Duchess, and Heap's Queen Victoria. *Strain of dam:* Colvile and Delmotte, Mills' D22 and W34, sire, Colvile.

165
Dreadnought,
RP03T3337.
Owner, A. Williams.

Light cheq frill cock, bred by owner. *Performances:* 1905, 1st and Dewar cup Perth, Pontypridd F.C., flew disastrous Banff race, Welsh C. club, Perth hard race, only two home on the day in club. *Strain of sire:* Wegge, Pletinckx, Gits, inbred. *Strain of dam:* Wegge, Pletinckx, Gits, inbred.

166
Vanessa, 1905.
RP2339.
Owner, A.E. Breach.

Dark blue cheq hen, bred by owner. *Performances:* 1905, Weymouth 2nd South Coast Fed., also *Racing Pigeon* cup, and 1st Eastbourne C.H.S., hard race. *Strain of sire:* H. P. Vanderhaeghen, Carcasse. *Strain of dam:* Logan, Barker, Jurion, Orme.

167
Bonny Blue,
1902Z232.
Owner, J. Eaton.

Blue cock, bred by owner. *Performances:* 1st Jersey West Lancs. Sat. Fed., hard race, 1st open, 1st Lymm F.C.

Stud No.

219
Lady Dainte,
NU03WS20.
Owner, J. E. Field.

Red cheq hen, bred by owner. *Performances*: 1903, Oakley, Tisbury 3rd prize, Chard 2nd pri. e, vel. 1,124 yards per minute, joint winner of best average young bird races ; flown in the North Middlesex F.C. 1904, Essendine, York, Berwick, and Perth, 371 miles ; flown in the Woolwich Dis. H.S. 1905, Tisbury, and Exmouth, taking 1st, 2nd and 3rd prizes and pool, only bird timed in on the day, and joint winner best average O.B. races. *Strain of sire* : Pure Stanhope. *Strain of dam* : Wegge cock, Key hen.

220
Comet,
19EX00-139,
bred 1900.
Owner,
W. H. Mattinson.

Blue pied hen, bred by V. Crompton. *Performances:* 1900, 3rd Worcester, 2nd Cheltenham, and 2nd Swindon, 11th Manchester Federation ; 1901, 1st and gold centre medal, velocity 1562, Eccles H.S., 2nd Patricroft H.S., Bournemouth ; 1902, stock. 1903, 5th Worcester, 5th Bournemouth, Guernsey, dropped at her old loft 25 mins. before my 1st prize winner, and would not leave, with Eccles H.S., and Marennes with Great Northern O.R. ; 1904, 1st Worcester, 4th Bournemouth, 9th Guernsey, 8th Nantes, Eccles H.S. ; 1905, Worcester, Swindon, Bournemouth, 4th Guernsey, and Nantes with Combine, the 4th bird home in club, Eccles H.S. *Strain of sire :* Barker, Offermans, Vekeman, and Longton's 468 and 23. *Strain of dam :* Turner and Brewer.

221
Eclipse,
RP11904-1453,
bred 1904.
Owner,
W. H. Mattinson.

Blue hen, bred by owner. *Performances :* 1904, Worcester, Cheltenham, and Swindon ; 1905, Stafford, 2nd home, 1st Swindon, Eccles H.S., Bournemouth and Nantes with Combine, 3rd home in Eccles H.S. *Strain of sire :* W. H. Wilson. *Strain of dam :* 513 and 4, W. H. Wilson, Logan, Gits, Thorougood and Barker.

222
Ocean Queen,
1902RP.
Owner, G. Stubbs.

Red cheq hen, bred by owner. *Performances :* 1902, Bournemouth ; 1903, 7th Alderney, C.P.F.C. ; 1904, Alderney 1st and 1st 6d. pool, Hampshire Dis. H.S. ; Southern Section, St. Malo, Rennes and Marennes, N.F.C., 110th prize and winner of R.P. £5 special, first bird home to Portsmouth by five hours ; 1905, Alderney, St. Malo, being held over five days, then Marennes N.F.C. *Strain of sire :* Colvile, Barker. *Strain of dam :* Owner's Old Duchess.

223
Reliance, NU01DP.
Owner, G. Stubbs.

Red cheq cock, bred by H. J. Horton. *Performances :* 1901, untrained ; 1902, Banbury ; 1903, all stages Newcastle, 300 miles, vel. 1018, being one of the first birds to fly Newcastle into Portsmouth on the day of liberation : 1904, 2nd Derby, vel. 1196, Eastney Columbarian, 2nd Newcastle, Hampshire Dis. H.S., vel. 798 ; 1905, Derby. *Strain of sire :* Holtom, Heap's Victoria, Moss. *Strain of dam :* Victoria, Stanhope, Logan, Wellington and Perfection.

224
1903RPC.
Owner, G. Stubbs.

Red cheq hen, bred by owner. *Performances :* 1903, Alderney 4th, E.C.S. ; 1904, Alderney, St. Malo, 1st H. and D., 7th 1s. pool, Southern Counties Fed., 1st Rennes and gold medal, H. and D. ; 1905, Alderney, St. Malo, Marennes, N.F.C., 60th prize, 7th Southern Counties Fed., 344 miles. *Strain of sire :* Barker. *Strain of dam :* Barker.

225
1903RPC6016.
Owner, C. Trapnell.

Blue cheq hen, bred by owner. *Performances :* 1904, trained 40 miles ; 1905, 5th Yeovil, 6th Salisbury, 5th Winchester, 3rd Horsham, hard race, 1st Tunbridge Wells, 2nd and medal Folkestone. *Strain of sire :* Debue, Gilson's 26A and Clay and Logan's '86. *Strain of dam :* De Groote and Jurion.

226
Jewel,
RP1902.
Owners,
Johnson Bros.

Red pied cock, bred by owners. *Performances :* 1902, Swindon ; 1903, Bournemouth ; 1904, 3rd Bournemouth, 18th Northwich Federation, 1st Sandbach Dis. H.S., and 1st Northwich Dis. Federation, Jersey race ; 1905, Rennes. *Strain of sire :* Hill, New Mills ; *Strain of dam :* Jurion, Barker.

Stud No.

227

La ly Molly,
RP f1903-32.
Owner, Dr. Moore.

Blue cheq hen, bred by owner. *Performances:* 1903, trained 60 miles ; 1904, 1st St. Malo, 201 miles, Reading Dis. S.R.F.C., 100 birds, and 16th Southern Counties Fed., 529 birds, also flew Rennes, 233 miles ; 1905, jumped to Rennes from Weymouth, then sent to La Roche, 304 miles, winning 1st Surrey Fed. open race, 1st 1s. and 2s. 6d. pools. *Strain of sire*: Barker. *Strain of dam* : Barker and Debaillie.

228

1898-130.
Owner,
J. A. Thorburn.

B. c. hen, bred by Jas. Moss, Congleton. Stock. Dam of many winners, including two 1st prize winners from Nantes, 432 miles, for owner in Liverpool H.S. and Liverpool Wednesday H.S. She is the foundation of owner's loft. *Strain of sire*: Moss' old strain.

Strain of dam: Moss' old strain.

229

Champion of the
North,
Co + 2-632.
Owner, H. Whitlow.

Blue pied cock, bred by owner. *Performances*: 1902, Coleford ; 1903, 5th Guernsey and 1st La Roche, West Cheshire Federation, Cheshire Dis. H.S., beating all birds flying into the North of England with the National F.C. ; 1904, Bath ; 1905, Guernsey. Never spent a night out. *Strain of sire*.

Owner's Regular. *Strain of dam*: Croft, Buckley.

230

Co + 2-564.
Owner, H. Whitlow.

Dark blue cheq hen, bred by W. F. Youde. *Performances*: 1902, Bath ; 1903, 14th Bath and Guernsey ; 1904, 15th Bath, 27th Guernsey,. 2nd Nantes, Chester Dis. H.S., only two birds home same day, 2nd West Cheshire Federation and pool club ; 1905, Guernsey and Marennes, with the Great Northern. *Strain of sire*: Armstrong's Clay.

Strain of dam : Whitlow and Moore.

231

Marconi,
1899.
Owner,
M. Evangelisti.

Blue cheq cock, bred by owner. *Performances*: In 1899 he flew up to York, N.M.F.C. ; 1900, Banff 1st prize in C.M.F.C. and 2nd in A.P.F.C. ; 1901, Lerwick, Shetland Isles, 600 miles ; 1902, not flown ; 1903, not flown ; 1904, Lerwick 4th prize, N.M.F.C. ; 1905, Thurso 19th prize, and all pools in the North Road Championship club. Marconi is the sire of the 8th prize winner Thurso, N.R. Championship club. Still in loft. *Strain of sire*: Mons. E. Clarembaux's Le Wet, and my own La Guida. *Strain of dam* : Dardenne, bred by Mons. Leon, Sirjacobs.

232

Cup Winner.
Produce ring, 9E,
1899.
Owner,
M. Evangelisti.

Black cheq cock, bred by owner. *Performances*: In 1899 flew all races in London North Road Fed, up to York, and up to Retford in North Middlesex F.C., and was awarded 1st prize and Lord Mayor's cup at the Dairy Show by Mr. John W. Logan ; 1900, all stages to Perth, 360 miles ; 1901, Lerwick, Shetland Isles, 600 miles, winning 7th prize in the North Road Championship club, and 6th prize N.M.F.C. Since reserved for stock. He is the sire of my noted Banff Mealy and of Holbein, winner of 11th prize in last year's National race from Marennes, 2,277 birds competing. A brother of Holbein also flew Marennes in Last year's National. Still in loft. *Strain of sire*: Mons. E. Clarembaux's Le Wet, and own La Guida. *Strain of dam* : Jurion and own La Guida.

233

Holbein.
RPB1903-5583.
Owner,
M. Evangelisti.

Black cheq cock, bred by owner. *Performances*: 1903, untrained ; 1904, Yor c ; 1905, Guernsey, St. Malo and Marennes, winning 11th prize, a R.P. £5, and £3 members' sectional prize in the National race, 2,277 birds competing sent by 853 owners. A brother of Holbein also flew Marennes in the National. Still in owner's loft. *Strain of sire*: Cup Winner, flown Shetland ; Mons. E. Clarembaux's Le Wet and own La Guida. *Strain of dam* : Schreiber hen, flown Shetlan l, winner of R.P. cup, 1902 ; a pure Jurion bred by the late Mr. W. F, D, Schreiber.

Htµd No.

819
RP05-9887.
Owner, A. Wheeler.
Blue cheq cock, bred by owner. *Performances*: 1905, Axminster, my 2nd bird home, Teignmouth, 1st prize Reading Dis. S.R.F.C., and 2nd prize Reading Federation, 474 birds. *Strain of sire*: Evangelisti. *Strain of dam*: Owner's Ivy.

820
RP05-2287.
Owner, A. Wheeler.
Blue cheq hen, bred by owner. *Performances*: 1905, Yeovil, 1st prize Reading Dis. S.R.F.C., and 2nd prize Reading Dis. Federation, 703 birds competing, Axminster and Teignmouth. *Strain of sire*: Maunder's Gits and Taft, Price's Ajax. *Strain of dam*: Evangelisti.

821
Black Prince,
1898G1086.
Owner, A. C. Ilett.
Black cock, bred by H. Marshall. *Performances*: Flown 1899, Selby, North route; 1900, Jersey; 1901, Bournemouth, 7th club and 17th Fed., Jersey 3rd prize and 92nd Fed., Rennes, 361 miles, 3rd prize and 24th Fed. ; 1902, Rugby 3rd prize, Banbury 2nd prize, nominated special and pool, and 3rd Fed., Guernsey 2nd prize and 10th Fed., Jersey, 289 miles, 2nd prize, special, and 9th Fed. ; 1904, Guernsey, Marennes, 520 miles, 3rd prize and 5th Fed. *Strain of sire*: Bred from a son of Whiteley's Marennes pair, Logan, Lambeane, Mills' Madrid and Van Muylem, our of my old 88, Gits and Logan, flown Rennes. *Strain of dam*: Winner of 1st Rugby and Rennes, 3rd Weymouth, 6th Ventnor, 11th Cherbourg, etc., J. O. Allen's champions, and Thorougood's Clay blood, sister to Nigger.

322
Peter,
NU01-310.
Owner, T. Whittaker.
Dark blue cheq cock, bred by owner. *Performances*: 1901, flew Weymouth twice, 110 miles ; 1902, Guernsey ; 1903, Guernsey and Granville ; 1904, Guernsey and Marennes ; 1905, Guernsey, and kept for stock. Show pen winnings : 1st and special 350-mile class at Gloucester Show, 1904, J. L. Burgess, Esq., judge ; 2nd prize Malvern Show, 1904, L. R. Halstead, judge. *Strain of sire*: Pure Jurion. *Strain of dam*: Pure Jurion. Both bred by P. Clutterbuck, Esq.

823
Young Roger Hill,
NUJW03-19.
Owner, T. Whittaker.
Dark blue cheq cock, bred by owner. *Performances*: 1903, Weymouth twice, 110 miles ; 1904, Guernsey, 190 miles, 132 birds, 8th, and Rennes, 284 miles, 61 birds, 15th ; 1905, Guernsey, 190 miles, 118 birds, 11th, and Rennes, 284 miles, 81 birds, equal 15th, also Marennes, 433 miles, 16 birds, 1st and special Worcester United F.C., and only bird wired to take a prize, 218th National F.C., beating over 2,000 birds. Show pen : 3rd prize, Class 1, Worcester United show, 350 miles, under A. H. Osman, Esq. *Strain of sire*: RP99-2017, winner of R.P. £10 cup and 44th Bordeaux, 1902 ; Johnstone cock and Gainer. *Strain of dam*: Delmotte and Queen Victoria.

324
Favourite Edward.
RPZ1903-5200.
Owner, W. Heighway.
B. cheq cock, bred by owner. *Performances*: 1st and gold medal, Radcliffe Fed., 1st and special, Heaton Park Dis., from Guernsey, June 17th, 1905, vel. 1134, 283 miles. *Strains*: Jones, Didsbury, and Ollier, Knutsford.

325
NU03-389SW.
Owner, A. Walton.
Blue cheq hen, bred by owner. *Performances*: 1903, all stages to Bath ; 1904, 4th Worcester, 6th Swindon, 1st Jersey, hard race ; 1905, flew Worcester and Swindon, then won Guernsey 1st, Flixton (Urmston) Dis. H.S., would have been 5th Stretford Dis. H.S., but verified to disadvantage after taking 10 minutes to trap. *Strain of sire*: Logan and Hansenne. *Strain of dam*: Hansenne.

326
1903RPF9230.
Owner. H. Parkins.
Blue cheq cock, bred by owner. *Performances*: 1903, Cheltenham ; 1904, Guernsey ; 1905, 1st Weymouth, Slaithwaite H.S., 3rd West Yorkshire Federation, Guernsey, and then Rennes in Yorkshire Combine, winning 6th Rennes Combine, and one of H.P. fivers. *Strain of sire*: Midland Leader, Queen Victoria, Iron Duchess, and Thorougood's 6 and 19. *Strain of dam*: Gits and Van Muylem.

Stud No.

327
Dean Swift,
SPS54-1903.
Owner, A. G. Aspland.

Red cheq cock, bred by owner. *Performances*: 1903, untrained ; 1904, all stages to Newcastle, 250 miles ; 1905, all stages to Perth, 365 miles, winning 3rd S.P.H.S. and 7th L.N.R. Fed. *Strain of sire*: Wegge, Jurion. *Strain of dam* : Jurion, Stanhope.

328
Grace Darling,
OO1902-289.
Owner, J. W. Miles.

L. b. cheq hen, bred by owner. *Performances* : 1902, Guernsey ; 1903, 2nd Ventnor, 3rd St. Lo, first bird into Yorkshire by three hours in the disastrous Rennes race ; 1904, Marennes ; 1905, 5th Banbury, 1st club, Rennes, 13th federation, and 2nd 2s. 6d. pool. *Strain of sire* : Logan. *Strain of dam* : Noi, of Antwerp.

329
Wonder, 581-1897.
Owner, W. T. Rimer.

Blue and red cock, bred by owner. *Perform-ances* : As a stock bird he is responsible for the following : 1898, 1st Sheffield Y.B., Tyneside F.C., 5th N.E.C. Fed., 547 birds competing ; 1899, 2nd Doncaster Y.B., Tyneside F.C., 31st N.E.C. Fed., 781 birds competing, 1st Notts Y.B., Tyneside F.C., 1st N.E.C. Fed., 618 birds competing. No. 214, True Blue II., sire of 1st Bristol Y.B., Tyneside F.C. sweepstakes 1901. No. 245, 3rd Bournemouth, 291 miles, N. of England H.S., 36th N.E.C. Fed., 554 birds competing, Jersey, 398 miles ; 1900, 5th Bristol, 243 miles, Tyneside F.C., 35th N.E.C. Fed., 2,039 birds competing, 1901 ; then Bournemouth, and lost at Rennes. 245, dam of 2nd Doncaster Y.B., Tyneside F.C., 17th N.E.C. Fed., 1,608 birds competing, 1900. 640, flown La Roche. 9738, 13th T.F.C., Notts Y.B., 1903. *Strain of sire* : G. G. Richardson. *Strain of dam* : Jurion, Barker.

330
Reliance, 178-1897.
Owner, W. T. Rimer.

Blue cheq hen, bred by owner. *Performances* : 1897, Leeds, Sheffield, Buxton ; 1898, Normanton, Burton, Exeter, 4th T.F.C., 9th N.E.C. Fed, ; 1899, Doncaster, Notts, Oxford, Bournemouth, Jersey and T.F.C., 2nd N.E.C. Fed. ; 1900, Doncaster, Notts, Oxford, Bournemouth, 3rd T.F.C., 5th N.E.C. Fed., 554 birds competing; Marennes 29th, Manchester F.C., 205 birds competing ; 1901, 80 miles ; 1902, Doncaster, Notts, Ventnor, lost at Bordeaux, disastrous G.N. race. *Strain of sire* : From G. G. Richardson, pedigree unknown. *Strain of dam* : Debaillie.

331
Production,
221-1899.
Owner, W. T. Rimer.

Blue cheq hen, bred by owner. *Performances* : 1899, Doncaster, Notts, 2nd T.F.C., 2nd N.E.C. Fed., 618 birds, 1st produce stake homing with his 1st T.F.C. and 1st N.E.C. Fed. ; 1900, Doncaster, Notts, Oxford, Bournemouth, 4th T.F.C., and 7th N.E.C. Fed., homing with his 3rd T.F.C. and 5th N.E.C. Fed., 554 birds ; 1901, Doncaster 1st arrival, Notts, Bristol, 1st T.F.C. and 5th N.E.C. Fed., 2,038 birds, Bournemouth ; 1902, Notts, 16th T.F.C., 270 birds, Bournemouth ; 1903, Notts, 2nd N.W.F.C., Oxford 15th, 2nd pool, N.W.F.C., Ventnor, 10th T.F.C., St. Lo, 1st T.F.C., 1st and gold medal N.F.C., 1st N.W.F.C., 1st N.E.C. Fed., members and open 1st in 1s. and 5s. pools, 1st Fed. any-age produce ; 1904, Doncaster, Notts ; 1905, Selby, Notts, Bournemouth (private toss), Marennes G.N., 636 miles. *Strain of sire* : Grandson of his 468, and closely related to his noted birds, 23, 5, 114E, and 141. *Strain of dam* : N. Barker, Captain, Paddy, Marica, Rusty, Grooters, John Wright's Old Paris blood, Cruttenden, Heap, and S. Gibson's White Tail.

332
NUo1AL449.
Owner,
H. Longworth.

Dark blue cheq hen, bred by owner. *Perform-ances* : 1901, flew Worcester, Gloucester and Swindon, 137 miles ; 1902, flew Worcester, Swindon, Bournemouth and Guernsey, 281 miles ; 1903, flew Worcester, Swindon, Bournemouth and St. Lo, 308 miles ; 1904, flew Worcester, Swindon, Bournemouth, Guernsey, and 3rd prize Nantes in S. and D.H.S., 437 miles, vel. 891 yards per min. ; 1905, flew Worcester, 1st prize Bournemouth, vel. 999 yards, and 1st prize Nantes, vel. 686 yards, 21st Lancashire Combine, 11th 1s. pool (£5), and 6th 2s. 6d. pool (£5), 80,209 birds competing, only bird home in S. and D. H.S. before close of race, and 4th Lancashire Fed., 1,166 birds. *Strain of sire* : John Wright, Manchester. *Strain of dam* : Not known.

Stud No.

362
N326-1894.
Owner,
S. P. Griffiths,

Blue he̟ e· bred by J. Bell. *Performances*: Won 6th Swindon's and 7th Cherbourg, 1895, since kept for stock. A remarkable hen. Dam of 1st prize winners at all stages from Worcester to Rennes, and of many prize winners from Nantes, Marennes and Bordeaux. *Strain of sire*: T. H. Hall's Dublin and Miss Midwell. *Strain of dam* : N. Barker.

363
N749-1896.
Owner,
S. P. Griffiths.

Blue cheq cock, bred by C. Preston. *Performances:* Flew 180 miles as a yearling ; since kept for stock. Sire of winners at all distances. *Strain of sire :* Delmotte (Bancroft's 23 and 24), J. W. Barker's Young Gladstone and Goose hen. *Strain of dam :* Bancroft's 23 and 24.

364
GB112-1901.
Owner,
S. P. Griffiths.

Blue cheq cock (white flights), bred by W. Wright. *Performances :* Winner of several prizes in English races. A remarkable stock bird. Sires winners every season. *Strain of sire :* Delmotte, 23 and 24. *Strain of dam :* John Wright's old Paris strain.

365
N369-1895.
Owner,
S. P. Griffiths.

Black cheq hen, bred by E. Hewitt. · *Performances*: Winner of 8th Swindon, 3rd Salisbury, 1st Jersey, etc. Dam of winners at all distances. *Strain of sire*: Armstrong's Clay blood. *Strain of dam :* Armstrong's Clay blood.

NOTE.—The above five birds are the foundation of Mr. S. P. Griffiths' lofts, and descendants from them have won, during the last five seasons, over 100 money prizes in races alone.

366
FEB402.
Owner, F. E. Blagg.

Red cheq hen, bred by owner. *Performances* : 1903, Sheffield ; 1905, flying with Portsmouth Dis. F.C., 2nd Didcot, 1st and special Rugby, 1st and special Derby, 2nd Northallerton. *Strain of sire :* Gits and Lieutenant Barrett. *Strain of dam* : F. A. Key's Thurso birds (Thorougood).

367
FEB431RP.
Owner, F. E. Blagg.

Blue cock, bred by owner. *Performances* : 1903, Alderney ; 1904, 1st Alderney, flew St. Malo and Rennes ; 1905, St. Malo in the disastrous race of the S.C. Fed., then lifted to Marennes, winning 87th Grand National and R.P. prize. *Strain of sire* : Mackenzie's Bluebell and Gits. *Strain of dam* : Mackenzie.

368
FEB1900-126.
Owner, F. E. Blagg.

Mealy cock, bred by owner. *Performances :* 1900, Alderney, with Crystal Palace F.C. ; 1901, 5th Guernsey, Southern Counties F.C., St. Malo and 1st Nantes with Southern Counties F.C. on a very bad day, doing over 1000, about 200 in front of any other, then 1st Bordeaux with S.C.F.C., and 20th Grand National. *Strain of sire :* Lieutenant Barrett. *Strain of dam* : Piercy.

369
North Star,
1173LCS1897.
Owner, E. H. Lulham.

Blue cheq hen, bred by owner. *Performances :* 1897, 50 miles ; 1898, 100 miles ; 1899, Perth ; 1900, Perth, winning a prize ; 1901, a prize Newcastle, Perth twice, winning a prize, and 3rd prize in the North Middlesex from Lerwick, 605 miles, being the 4th bird home in all England ; 1902, 1st prize Essendine, Peckham Dis., York, Newcastle ; 1903, Berwick ; 1904, Berwick. *Strain of sire* : Jurion's Premier Revenu blood. *Strain of dam* : Jurion's Vielle Bayonne blood.

370
Tar Tub,
5967RP1903.·
Owner,
E. H. Lulham.

Red cheq cock, bred by owner. *Performances* : 1903, Essendine ; 1904, 60 miles ; 1905, 1st Peckham Dis.; 1st County of Middlesex, 1st London North Road Federation, 1,990 birds, Newcastle 4th, and pools Berwick in the North Middlesex. *Strain of sire* : Jurion's Premier Revenu blood. *Strain of dam* : Jurion's Veille Bayonne blood. ·

Stud No.

371
5965,
Red Dawn.
Owner,
E. H. Lulham.

Red cheq hen, bred by owner. *Performances·* 1903, Essendine ; 1904, Newcastle ; 1905, 1st Peckham Dis., 1st North Middlesex, 2nd A.P.F.C., 6th London, Banff Combine, and R.P. cheque from Banff. *Strain of sire* : Jurion and Delmotte's 1st Dax blood. *Strain of dam* : Delmotte's 1st Dax blood.

372
Storm King,
1900.
Owner,
E. H. Lulham.

Black cheq cock, bred by owner. *Performances :* 1900, Claypole smash ; 1901, York ; 1902, Newcastle smash ; 1903, Perth smash and Lerwick smash, winning 1st County of Middlesex, 2nd N.M. 1st London North Road Federation, and the Championship club's South E. cup. *Strain of sire:* Ruhl, of Verviers. *Strain of dam* : Ruhl, of Verviers.

373
Showman,
NU99C7540.
Owner,
W. J. Shinner.

Blue cheq cock, bred by owner. *Performances :* As a youngster flew all stages to Weymouth, 121 miles ; 1900, flew all club races to Guernsey ; 1901, flew club races up to Rennes, and Bordeaux in the Grand National, taking 14th prize ; also flew Bordeaux in the Grand National, 1902. *Strain of sire :* T. Hall's Indefatigable. *Strain of dam :* Clarembaux.

374
No Fool
(hatched April 1st),
NU99C7543.
Owner,
W. J. Shinner.

Mealy cock, bred by owner. *Performances :* As a youngster flew Weymouth, 121 miles ; 1900, Weymouth, Guernsey, and Rennes in club races, and in the Grand National from Bordeaux, 522 miles, winning 1st prize with a velocity of 1297 yards per minute. Since kept at the stud. *Strain of sire* : Bred by Cools. *Strain of dam :* Bred by Cools.

375
Wanderer,
WU1898-344.
Owner,
W. J. Shinner.

Red cheq cock, bred by owner, *Performances:* As a youngster flew Weymouth, 121 miles, and Bournemouth, 112 miles ; 1899, Templecombe, Guernsey, and Rennes ; 1900, Weymouth, Guernsey, Rennes, taking 1st prize from Guernsey, and in the Grand National Bordeaux race took 5th prize and members' silver cup for nominated birds with a velocity of 1285 yards for the 522 miles ; 1901, all stages up to Rennes in club races, and from Bordeaux in the Grand National. *Strain of sire :* Bred by Hellemans. *Strain of dam* : Bred by De Bruyn.

376
454NU02J.
Owners, Salt Bros.

Dark cheq cock, bred by owner. *Performances :* 1902, all stages Weymouth, 166 miles ; 1903, Guernsey, 266 miles ; 1904, 2nd Bath and 2nd Nantes, 400 miles ; 1905, 1st prize Nantes Hanley Working Men's Society, beating next bird two hours. *Strain of sire:* Dr. Cohen, Rome I., Mausta. *Strain of dam* : Our own strain.

377
Haydock Queen,
NU00AM88.
Owner, J. Ward.

Red cheq hen, bred by owner. *Performances :* 1900, trained 20 miles ; 1901, Bournemouth ; 1902, Swindon twice ; 1903, Nantes, 21st and special Ashton H.S., 3rd prize Wigan Fed. pool club, 5th prize Fed. 6th pool ; 1904, Nantes ; 1905, Nantes, 1st and special Ashton H.S., 4th Fed. 6d. pools, 8th Wigan Fed., 25th open, 87th Great Combine. *Strain of sire* : Armstrong, Clay, and Barker. *Strain of dam* : Barker, Bonami, Thorougood.

378
Lily, 19NP02-108.
Owner,
James Hindmarsh.

Mealy pied hen, bred by owner. *Performances :* 1902, 2nd Rugby, 4th Northumberland Fed. ; 1904, 1st Oxford, 1st Ventnor ; 1905, 1st Selby. *Strain of sire:* E. E. Jackson, etc. *Strain of dam :* E. E. Jackson and Red Admiral and Pebble.

379
416AHS1901.
Owner, A. Pickston.

Blue cheq hen, bred by owner. *Performances :* 1902, 1st Rennes, Altrincham H.S. ; 1903, 2nd Bournemouth, 6th Marennes ; 1904, 4th Marennes, A.H.S., 71st Great Northern, 7th 2s. 6d. pool, 18th Stockport Fed. ; 1905, 12th Bournemouth. *Strain of sire :* W. C. Moore, Carpentier, and Thirionet. *Strain of dam :* Jurion, Barrett, and Iron Duke.

MR. F. G. MARSH,
LANGTON LOFTS, SURBITON.
Fast Youngsters for Sale.

In 1908, Flying in four Clubs and with the Surrey Federation, my birds won
the following positions, in strong competition, viz. : 11 FIRST, 13 SECOND,
8 THIRD, &c.   Strain: Bonami, Janssens, Hansenne, Gits, Barker, Jurion,
Grooter, Bancroft, Moore.

MR. F. G. MARSH'S LOFT AT SURBITON.

"Dear Sir,—The blue cock has been a good one indeed; he bred me in 1908 birds to win,
as youngsters, in Selby race, distance, 79 miles, 3rd and 4th club.   In the Nottingham
race, distance 145 miles, 1st club, and Fed., beaten by one yard, 952 birds competing.
This year, with yearlings bred from him, in Rugby race, distance 175 miles, I won 1st and
3rd club, and 3rd and 20th Fed., 1,684 birds competing. With a youngster bred from son
and daughter from the blue cock, I won the Up North Combine from Oxford, distance
219 miles; this bird was only 4 months old when it won.—E, Dawson, Sunderland."

"Dear Mr. Marsh,—288 has flown well for me, and was first bird into Blackpool on
Wednesday, from Gloucester, in the West Coast Wednesday H.S.   It was like pea soup,
and an easterly wind; four hours and 21 minutes, so you can guess the day.   She goes
to Bath next Wednesday.   That is, if I can have the first pair bred same way next
season.   If not, she is staying at home.—Tom H. Beverley, Blackpool."

If you want a good youngster for racing or crossing, write early.   Every youngster sent out
will be strong and healthy, as it is my wish for them to do well wherever they go.

Rung R.P. and H.P.   All one price, **10/6** each, or **20/-** a pair, specially selected to breed
winners.

In 1905—1906, I won the longest Surrey Federation Young Bird Race, the
latter from Alderney (smash) with a youngster only 13 weeks old.

F. G. MARSH, "Langton," 340, Ewell Road, SURBITON.

Early Days of the Sport

Some Famous Long Distance Races of the Past

Many fanciers of the present day have no knowledge of the early days of the sport and of the many successful races that were flown twenty odd years ago *(ie in the early 1880s – Ed)*.

Some of the first races that really founded the sport of long distance racing in this country were those organised under the guiding hand of Mr J W Logan and his brother, Mr Geo Logan, in connection with the old United Counties FC.

The races used to be on the SE route, from Dover, Amiens, Paris, and Rheims.

The London Columbarian Society was the first to undertake the Ventnor and Cherbourg route. The performance of Mr Lubbock's birds and others created so much surprise that Midland and Northern fanciers hardly deemed the work genuine. Still this was soon put at rest, and in 1883 all sections of the United Counties went South to Cherbourg and Rennes, followed by races from Granville, Rennes, and La Rochelle in 1884.

But it was not until 1886 that efforts to fly 400 miles in the day were crowned with success.

In this year, the famous race won by 'Old 86' took place. The race was from La Rochelle, the birds being liberated July 17th, 1886, at 4.36 am, wind light NW, with the following result:

Name	Distance	Velocity	Address
J W Logan	444	1012	East Langton
G H Logan	444½	929	Church Langton
J W Logan	444	925	East Langton
J W Logan	444	909	East Langton
Colvile	457¾	896	Lullington
Colvile	457¾	805	Lullington
Colvile	457¾	805	Lullington
G H Logan	444½	733	Church Langton
J W Logan	444	660	East Langton
Colvile	457¾	641	Lullington
G H Logan	444½	629	Church Langton

| J W Logan | 444 | 547 | East Langton |
| J W Logan | 444 | 543 | East Langton |

Hardly had fanciers time to recover from the delight that this race had occasioned, when, on July 20th, a further record was made from Nantes by Northern men. It was the Nantes race that established the fame of J O Allen's champion cock and hen. The birds were liberated at 5am, beautifully fine and clear, wind NW by W, the detailed result being as follows:

Name	Distance	Velocity	Address
J O Allen	447	1385	Lytham
W Kaye	440	1361	Stacksteads
J R Schofield	438	1348	Whitworth
W Kaye (2 birds)	440	1345	Stacksteads
J O Allen	447	1344	Lytham
W Kaye	440	1320	Stacksteads
W Kaye	440	1293	Stacksteads
W Kaye	440	1278	Stacksteads
J O Allen	447	1242	Lytham
W Kaye	440	1226	Stacksteads
W Kaye	440	1108	Stacksteads

		July 21st	
T Berry	450	5.50 am	Barrowford
T R Schofield	438	5.42 pm	Whitworth

So much for the first important one-day records from La Rochelle to the Midlands and Nantes to the North. I doubt if any better performances have ever been accomplished since those days.

Bordeaux was first attempted by members of the London CS in 1884. The rough passage of the birds, which were convoyed by Bordeaux by boat under the tender mercies of good-natured sailors, no doubt militated against anything like a decent performance, Mr W H Cottell owning the winning bird – his Nabob, Squills' Register 87 (1906).

The first attempt to fly Bordeaux to the North was in the year 1889. The birds were liberated August 6th, at 5am, and the following arrivals timed:– J Fair, Rochdale, 613 miles, August 15th; W Pilling, Rochdale, 613 miles, August 16th; A Taylor, Burnley, 625 miles, August 20th; J Pilling, Rochdale, 613 miles, August 26th.

So much, then, for the early records on the South route.

It was not until 1897 that a one-day 500-mile performance on the South route was recorded.

The following are brief particulars of the 1897 Marennes race. I think this is almost as good a race as we have ever had for the distance, and proves that as far back as 1897, given anything like decent conditions of weather and a favourable wind, birds could get over the ground.

The race was flown on June 29th, and carried out by the West Lanc
Saturday Fed, distance, approximately, 530 miles. Mr John Buckley liberated 244 birds at 4am, clear, light, the first day arrivals being as follows:–

Medcalf, New Mills	1213	Thorougood, Sefton	1085
Wilson, Cressington	1187	Swann, Earlestown	1084
Wilson, Cressington	1175	Skeen, Earlestown	1059
Evans, Earlestown	1165	Beverley, New Mills	1059
Longton, Earlestown	1148	Lowe, St Helens	1058
Moore, Lymm	1148	Bradshaw, Crosby	1055
Wednall, Warrington	1145	Toft, Garston	1044
Pickford, New Mills	1142	Royden, Sefton	1035
Wilson, Cressington	1140	Burrows, Warrington	1034
Wightman, Earlestown	1138	Eccles Bros, Ainsdale	1017
Nelson, Leyland	1137	Matthews, Liverpool	1015
Pickford, New Mills	1131	Toft, Garston	1014
Toft, Garston (2)	1128	Eccles Bros, Ainsdale	1011
Howarth, Stacksteads	1128	Wilson, Garston	997
Toft, Garston	1119	Mealor, Ness	996
Wilson, Cressington	1109	Aspey, Wigan	995
Pollard, Gt Crosby	1101	Jolley, St Helens	994
Longton, Earlestown	1101	Darling, Liverpool	983
Harrison, Rockferry	1099 1101	Moore, Lymm	973

This brings us down to the date of the formation of the National FC,
formed to carry out an annual long distance National race open to fanciers in
the United Kingdom.

The race was to be a continuation of the race that was flown in 1894, which
was the first National, and was flown from Marennes. Mr Geo Yates, of the
Manchester FC, acting as secretary, and A H Osman, the present secretary of
the National FC, secretary for the Southern portion.

Although not a record race, it will perhaps add interest to these details if I
include a report of the 1894 event.

Grand National race from La Rochelle, 1894. 384 competitors, 610 birds
competing. Liberated by Messrs Hartley and Buckley, at 4.30am, July 23rd.
Atmosphere clear; wind SW. Result:–

1.	£15 13s 8½d, and Mr Logan's timepiece, Bell Northwich	497 miles, 12.24 pm, July 25th
2.	£12 10s 11d, Mawdsley, Wigan	518 miles, 1.35 pm, July 25th
3.	£7 6s 4½d, Bishop, Skipton	548 miles, 4.9 pm, July 25th
4.	£5 4s 6½ d, Kinman, Stratford-on-Avon	424 miles, 11 am, July 25th

5.	£4 3s 7d, Wones, Wood-setton	447 miles, 12.43 pm, July 25th
6.	£3 2s 8d, Walker, Red-dish	509 miles, 4.35 pm, July 25th
7.	£2, Sayer Bros, Ashtead	360 miles, 11.30 am, July 25th
8.	£2, Wones, Woodsetton	447 miles, 5.10 pm, July 25th
9.	£2, Kaye, Stacksteads	527 miles, 5.53 pm, July 26th
10.	£2, Ingham, Mytholmroyd	527 miles, 7.17 am, July 26th
11.	£2, Pickford, New Mills	522 miles, 6.36 am, July 26th
12.	£2, Gibson, Huyton	510 miles, 8.31 am, July 26th
13.	£2, Wilson, Hayfield	504 miles, 1.35 pm, July 26th
14.	£2, Strachan jr, Horsham	341 miles, 10.5 am, July 26th
15.	£2, Sayer Bros, Ashtead	360 miles, 5.4 pm, July 26th
16.	£2, Ince, Ashton-in-M'field	514 miles, 10.20 am, July 27th
17.	£1, Waine, Haydock	512 miles, 12.46 am, July 27th
18.	£1, Ashcroft, Mossley	513 miles, 1.33 am, July 27th
19.	£1, Harris, Shrewsbury	463 miles, 11.41 am, July 27th
20.	£1, Yates, Clayton Bridge	512 miles, 5.23 pm, July 27th
21.	£1, Colvile, Lullington	459 miles, 3.24 pm, July 27th
22.	£1, Whittingham, Embsay	550 miles, 8.47 am, July 28th
23.	£1, Houghton, Sankey Bs	507 miles, 7.34 am, July 28th
24.	£1, Stanhope, Pool	539 miles, 3.2 am, July 28th
25.	£1, Gregory, Slough	376 miles, 4.40 pm, July 28th

The first race from Bordeaux under the auspices of the newly-formed club took place June 29th, 1898, N Barker acting as conductor, 166 birds competing. The result was a record race for distance in the time.

1.	£20, J Ward, Oldham	744
2.	£10, F Romer, Wargrave	681
3.	£5, J Whitmarsh, Kidderminster	601
4.	£1, C Harris, Shrewsbury	575
5.	£1, W G White, Woolhampton	563
6.	£1, G Hoare, Brighouse	503
7.	£1, P Clutterbuck, Stanmore	459
8.	£1, H J Longton	1.7 am, July 3rd
9.	£1, T W Thorougood	8.1 am, July 4th
10.	£1, G Groves	12.23 pm July 4th
11.	£1, Webster and Pegg	6.53 am, July 5th
12.	£1, P Potts	4.54 pm, July 5th
13.	£1, Cottell	1.43 pm, July 5th
14.	£1, C Harris	5.15 pm, July 5th
15.	£1, Thorpe	7.24 pm, July 5th
16.	£1, G Hoare	7.4 am, July 6th
17.	£1, W Gainer	9.10 am, July 6th
18.	£1, G Wilkes	6.32 pm, July 6th

19.	£1, J Ward	1.29 pm, July 7th
20.	£1, W Gainer	6.13 pm, July 7th
21.	£1, Larner	6.46 pm, July 7th

There were one or two features of interest in connection with this race. Barker's hotel caught fire, and he had to escape in his nightshirt. The winning bird was a yearling hen; and last, but not least, Champion 555. Mr F Romer's second prize winner was lifted off the West route into Marennes.

The second race from Bordeaux, organised by the same club, took place July 11th, 1899, when 240 birds competed. This proved another record, the winning velocity being 908 yards per minute, against 744 in 1898. The following being the official result of the race:–

J W Toft, Garston	529 miles 880 yards, 908 vel
J Lissaman, Coventry	523 miles 856 yards, 814 vel
F Wheatley, Streatham	453 miles 1300 yards, 740 vel
J L Baker, Sandyfield	534 miles 1595 yards, 681 vel
J W Toft	529 miles 880 yards, 606 vel
J P Hill, New Mills	591 miles 638 yards, 602 vel
T Turner, Gloucester	490 miles 1100 yards, 594 vel
J W Toft	529 miles 880 yards, 582.9 vel
P Clutterbuck, Sarratt	472 miles 1300 yards, 582.4 vel
J Triner, Biddulph Park	579 miles 1330 yards, 570 vel
T Turner	490 miles 1100 yards, 556 vel
P Clutterbuck	472 miles 1300 yards, 499 vel
H Page, Swindon	474 miles 1320 yards, 488 vel
W and C White, Woolh'mpt'n	457 miles 1320 yards, 454 vel
E E Edward, Dunster	464 miles, 423 vel
R T Harris, Exeter	412 miles 1020 yards, 422 vel
W C Moore, Lymm, 2.37 pm July 13th	589 miles 1100 yards
Lythgoe, Warrington, 4.7 pm July 13th	590 miles 1150 yards

This brings us down to the year 1900, when on July 10th Bordeaux for the first time in the annals of pigeon racing was accomplished in the day, Mr Shinner, of Bromsgrove, making a velocity of 1297, the following being a list of the cash prize winners in this exciting race:–

1.	Shinner, Bromsgrove	522 miles 848 yards 1297 vel
2.	Birch, Walsall	538 miles 1649 yards, 1295 vel
3.	Baker, Sedgley	536 miles 425 yards, 1291 vel
4.	Compton, Bradford-on-Avon	457 miles 234 yards, 1289 vel
5.	Bennett, Kidderminster	527 miles 126 yards, 1286 vel
6.	Shinner, Bromsgrove	522 miles 848 yards, 1285 vel
7.	Harrell, Evesham	505 miles 410 yards, 1279 vel
8.	Birch, Walsall	538 miles 1649 yards, 1278 vel
9.	Marston, Walsall	539 miles 36 yards, 1273 vel

10.	Small, Gloucester	491 miles 172 yards, 1271 vel
11.	Fisher, Coseley	536 miles 1217 yards, 1268 vel
12.	Showell, Birmingham	529 miles 927 yards, 1262 vel
13.	Wones, Woodsetton	536 miles 366 yards, 1253 vel
14.	Longton, Earlestown	602 miles 899 yards, 1248 vel
15.	Thorpe, Altrincham	595 miles 978 yards, 1245 vel
16.	Smith, Pinxton	570 miles 1492 yards, 1242 vel
—	Baker, Sedgley	1241 vel
—	Baker, Sedgley	1241 vel
—	Beard, Bloxwich	1241 vel
—	Beard, Bloxwich	1240 vel

One of the most noteworthy performances in connection with this race was that Mr J W Toft, of Allerton, the previous year's winner, wired in four birds flying 598½ miles on the day of the race, and several birds flying over 600 miles were wired in on the day, establishing up to this day not only a one-day record from Bordeaux but a 600-mile record as well.

I said at the time that I thought this record would stand for many years, and I venture to think that the 1900 race from Bordeaux still remains as one of the best one-day races on record in this country.

My task, so far as placing on record a brief history of some of the most successful races that have counted as records on the South route is at an end.

I shall now give a few brief details of the races that call for mention from the North.

Successful long distance racing on the North Road dates back even earlier than on the South.

It was in 1875 that Mr Jesse Oliver flew Newcastle to Bexhill, over 300 miles.

Many good Channel racers owe their descent to Oliver's old North Road blood, Kaye's old Boley being of this strain (see 72 'Squills' Register, 1906).

The members of the Southern Counties club, of which Mr H Stanhope when living at Stroud was practically the founder, later on turned their attention to North Road racing, and in 1884, Thurso was first flown by Mr Gainer, the distance being 476 miles and his blue hen (see 457 'Squills' Register, 1906) was the only bird home at close of race.

It was also in connection with the Southern Counties club that the late J P Jones made the Banff record to Cardiff with two birds – his 9 and 3. (See Squills' Register, 70 and 71, 1906.) The distance was 440 miles, the birds were liberated at 4 am, and timed in at 8.19.

The first 500-mile race accomplished on the North Road was in 1894, by the North Middlesex club. The difficulties that had to be contended with are best described in the following report of the race:–

"North Middlesex FC flew the fifth and final old bird race from Thurso, 510 miles, on July 16th, the birds being held over ten days; 13 members and two non-members sent 36 birds, under the care of Mr C Wathern, who liberated at 4.45 am; clear, slight W, very changeable, and accompanied with thunder and lightning and heavy showers. The following days and up to close of race the weather was almost as bad as could be, the wind blowing from the SW, almost a hurricane the whole time. Result: A H Osman timed 11.6 am, July 18th; W Kilby, 8.5 pm, 18th; Pointer, 7.26, 19th; Evangelisti, 1.25 pm, 19th; Mercer Bros, 6.40 pm, 19th.

It was on the 29th day of June, 1896, that the first 500-mile race in the United Kingdom was accomplished in the day. It was in many ways a remarkable race, as birds lifted from 80 miles homed in the day, showing that on a fast day, with the wind directly behind the birds, they will fly blindly almost any distance, proving the influence of the wind and the manner in which it is certain to drag the birds in any race.

The report of this race is as follows:–

"The London North Road Federation flew their final old bird race from Thurso on June 29th, 1896. Four clubs competed with 35 birds. Mr Wathern liberated at 4.21 am, NNW. Telegrams were received from Arbroath, and Mr Taylor, of Newcastle, that the weather was perfect. The same weather prevailed London end, a 500-mile record being the result. — R Page, Hon Sec."

G P Pointer, AP	501 miles, 1454 vel
J Rix, CM	505 miles, 1433 vel
W Wakelam, NM	506 miles, 1429 vel
F Hasluck, NM	497 miles, 1428 vel
P Clutterbuck, AP	497 miles 1421 yards, 1425 vel

This 500-mile record had not long been established before Lancashire men the very next year put up a similar 500-mile one-day record from Marennes, which has already been referred to.

The 500-mile one-day record from Thurso was followed in 1898 by a one-day record from Lerwick, both by Midland fanciers and Mr Clutterbuck, of Stanmore. This sensational race created greater interest in the North Road, so much so that a North Road National was organised in 1899, but first let me give the result of the 1898 Lerwick race.

LONDON NORTH ROAD FEDERATION

Lerwick (Shetland Islands) race, flown June 27th. Six clubs sent 30 birds; liberated by Mr Bayley, convoyer to Central Counties FC, at 3.30 am; weather good; wind North. Six diplomas. Result:–

1st day –	
Clutterbuck, A P	591 miles 1,020 yards, velocity 1098.
2nd day –	
Clutterbuck, A P	591 miles 1,020 yards, velocity 848.
Clutterbuck, A P	591 miles 1,020 yards, velocity 720.
Clutterbuck, S C	591 miles 1,020 yards, velocity 700.

Pointer, A P	593 miles 291 yards, velocity 504.
3rd day –	
Hickling, A P	598 miles 920 yards, velocity 338.

This performance created a record up to this time for one day distance on any road.

The same year that Mr Clutterbuck made the one day record from Lerwick, the North Middlesex, jointly with the Midland Homer League, organised a race from Ollaberry, Shetland Islands, when seven competitors sent nine birds, which were liberated at 4 am on July 13th, 1898, by Mr H C Howden. Bright, clear, wind NNW, the result being: Delderfield, 622 miles 530 yards, velocity 903, timed in at 7.11 am the day after toss; Key, 622 miles 410 yards, velocity 845 – a record velocity for any birds flying over 620 miles in this country.

In connection with this race, it is worthy of note that the Midland Homer League timed in six birds on the day out of 12 sent, flying 523 miles, Wass, of Nottingham, heading the list with a velocity of 1405.

The race from Lerwick on June 27th, 1899, when HRH the Prince of Wales (our present King) won 1st prize, can hardly be classified in any way as one of the North Road records, except for the number of birds competing up to that date, there being 132 birds entered. The winning velocity was 1307 yards; distance to the Royal lofts, 510 miles 1,075 yards.

It was in 1902 that the most sensational race ever flown on the North Road took place; on July 2nd, from Lerwick, 144 competitors sent 246 birds. These were liberated at 4.7 am in bright, clear weather, wind NNW. The prize winners being:–

Pulley, London	598 miles 1330 yards, 1459 vel
Clutterbuck, Sarratt	587 miles 168 yards, 1448 vel
Pulley, London	598 miles 1330 yards, 1424 vel
Lewis and Prosser, Kettering	536 miles 431 yards, 1408 vel
Terry, Ilkeston	496 miles 872 yards, 1406 vel
Page, London	593 miles 1497 yards, 1404 vel
Clark, Marlpool	494 miles 640 yards, 1404 vel
Richardson, Derby	499 miles 872 yards, 1402 vel
Clutterbuck, Sarratt	587 miles 168 yards, 1353 vel
Pickering, Burton-on-Trent	509 miles 1716 yards, 1351 vel
Clark, Rushden	543 miles 88 yards, 1325 vel

The greatest distances flown on the day in this race were accomplished by Williams, Cardiff, 605 miles 665 yards, velocity 1306; Burns, Cardiff, 605 miles 633 yards, velocity 1242; John, Cardiff, 605 miles 498 yards, 1242; and Lulham, Sydenham, 605 miles 920 yards, velocity 1153.

The famous Old Albert's velocity in this race was greater than the Motor's from Thurso in 1896, notwithstanding that he had 98 miles more to fly. Truly a marvellous velocity, having regard to the distance, and one that will stand as a record, I feel sure, for many years.

The longest distance flown by a pigeon in the United Kingdom was that of Her Majesty, from Bordeaux to Mr Henson's loft, Aberdeen, in 1902, 854 miles 97 yards, the birds being liberated on the 8th July, and Mr Henson timing his bird in at 8.10am on August 2nd.

But I look upon the performance of Mr Moody's bird in flying from Bordeaux to Seaton Delaval as the finest performance for a bird flying over 700 miles. The birds were liberated at Bordeaux on July 18th, by Mr Warren, at 4.40, NW wind. Mr Moody timed in at 12.46 pm July 21st, winning 5th prize in the race, birds only reaching the Midlands the night before. Fanciers will appreciate the pluck that must be necessary for a bird to keep pegging away unceasingly three days in succession to get home in like manner to Mr Moody's hen.

A different matter to a performance extending over weeks, when a lift might be obtained on the way, or a call made at a loft or two.

The performance of the Moody hen, which was subsequently bought by Mr Logan, will give some idea of what we may expect from Spain in 1907, providing the races are flown in anything like suitable conditions.

I have not attempted to include all the best long distance performances, more particularly of local interest, such as Mr Lowe's performance from Marennes to East Bolden, 632 miles 774 yards, in 1905, velocity 771, the same bird flying the distance again in 1906, nor to Deignan, of Jarrow's performance in 1906, flying from Marennes, 634 miles 1681 yards, liberated at 4.30 am, and timed in at 10.41, velocity 815.

These and other performances all support the view I have so often expressed, relative to the quality of our English birds and their capability of well holding their own with anything bred in Belgium.

There may be many races of interest and importance that I have missed, or even performance far more meritorious than some of those I have chronicled. Still I have done my best to make brief reference to those that have struck me as likely to bear on the influence of the growth and popularity of the sport.

On the eve of undertaking bigger things, which it is hoped to accomplish from Spain, I thought it might prove of interest to put in an easy form of reference, some past long distance records.

Pigeon Diseases
by SQUILLS

During the long period that I have been associated with pigeons and pigeon breeding as a fancier I have received many and varied applications for advice.

Upon the aptitude of a fancier to know when his birds are healthy or when they are diseased depends his success as a pigeon breeder.

First let me describe the appearance of a healthy pigeon.

The most important consideration is the plumage.

The feathers should be close-fitting and the ground colour clear and clean.

In self-coloured birds the bars should stand out clear and well, and to find any dullness of one colour against the other is an indication that all is not right.

The feet are a good indication as to the health.

Hot feet are unnatural. Pigeon's feet should be moderately cool, and the colour of the legs and feet clear and bright. When the colour of the feet is pale and dull, with dull plumage as well, it is a sure sign all is not well.

The eye is another important factor from which health or condition can be judged.

The experienced fancier can learn much from the eyes of his pigeons. They appear to speak to him and tell him just how the subject feels.

Immediately a bird becomes sick the eye indicates the trouble sooner than anything.

A clear, bright eye is the surest indication of health I know, and as soon as a fancier can understand reading the health of his pigeons through the optic organs he has learned much.

Clear white wattles, and hard, clean cere are also good indications of health.

Birds that keep clean in the wattle when feeding their young are generally all right inside, but birds that go greasy on the beak and dirty in front when feeding must be watched, particularly if, in addition, the plumage loses colour and sheen.

So much for outward appearance of health, now let us examine the mouth and throat.

In young and old the most common disease amongst pigeons is throat trouble.

Open the beak, examine the throat well, and look for small yellow spots. If these or yellow growths are to be found anywhere in the mouth it is a certain indication that the blood is diseased.

In young birds, of from four to five weeks old, canker is very common either in the pharynx or mouth. When the surface can be got at it can be treated and cured, but if the growth is in the pharynx treatment is very difficult, and it is best to kill the subject.

Examine the beak. It should be dry and free from any stain.

Should there be any indications of stain on the beak under the wattle, carefully press the wattles and see if there is any mucous discharge. Catarrh and one-eyed cold show themselves first in this form.

Birds in good health are lively, vivacious, their feathers tight and close-fitting, appetites good, droppings firm and clean, nicely tipped with urine, which is the white against the coloured ground of the droppings.

Too much importance cannot be attached to the examination of the excretions in order to judge as to the health or diseases of subjects.

Green, watery, foetid droppings are a sure sign that the health of the bird is below par.

Small, clear pebbly excretions from youngsters in the nest and old birds is a sign that the food is right, and that they are thriving. Pigeons with the least semblance of diarrheoa are not fit to race, and must be guardedly watched.

I have penned this brief introduction in order that the beginner can examine a pigeon he may be about to purchase, and generally form an opinion as to its health and soundness.

Once learn exactly how the subject should be in its normal state, and you will readily appreciate when disease is present. A pigeon is diseased when any of its functions are not carried on in a normal manner, when there are unusual growths, injuries or parasites affecting any of the organs.

One of the most important habits to acquire is to look at a bird, not as an individual object, but as an individual made of many parts, each of which has its special function to perform.

Thus the beak, the tongue, the oesophagus, the crop, the ventricle succulent, the liver, the gizzard, the gall, the duodenum, pancreas, the small intestine, the caecum, the large intestine, the ureters, the oviduct, and the rectum – all performing their respective parts to give nutrition to the subject.

The health, the condition, and the very life of a pigeon depend upon the organs of nutrition doing their work well and effectively.

In order that fanciers may the better understand these organs the accompanying sketch will, I think, be of service.

After being eaten, the food passes into the crop.

The crop is simply a store, to enable a bird to carry food from its feeding ground to its nest.

From the crop the food passes into the stomach (ventricle succulent).

Viewed from the front this has the appearance of a small subterranean pas-

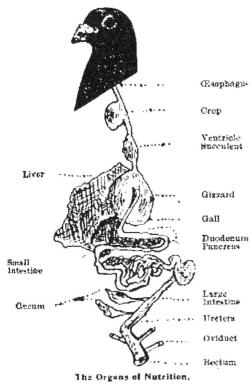

Œsophagus

Crop

Ventricle
Succulent

Liver

Gizzard

Gall

Duodenum
Pancreas

Small
Intestine

Cæcum

Large
Intestine

Ureters

Oviduct

Rectum

The Organs of Nutrition.

sage leading from the crop to the gizzard.

In the crop the grain becomes softened and swollen by the water mixed with it. After this, and during its passage through the stomach to the gizzard, it is impregnated with gastric juice. It then passes into the gizzard, which is really the second stomach or last stage of corn in its complete form, for although swollen in the crop and mixed with gastric juice in the stomach, it remains whole corn, and affords no nourishment to the birds until it reaches the gizzard. The gizzard is the true jaw of the pigeon, and here it is that digestion has its seat of action. In the gizzard, by the help of grit and its contractions, corn is reduced to a pulp, and afterwards discharged into the intestines. The intestines are composed of two parts, the small intestine and the large intestine. The large intestine is the last stage of the food. The digestive canal terminates at the rectum.

I look upon the organs of nutrition as amongst the most important in pigeons. Let these be deranged in even so slight a manner, loss of power immediately follows, for the blood becomes disordered, leading to inanition.

A free, easy digestion is the surest sign that birds are in good health.

A full crop at night and an empty crop in the morning, with sound, healthy excretions in the nest boxes or on the floor of the loft is a happy augury for the fancier, for his birds are fit.

The very first symptoms that the organs are deranged is the remaining of food in the crop beyond the normal time.

Take, as an example, the hen that has just laid her egg.

I have constantly warned fanciers against sending hens to training stages or tosses for two or three days after laying.

Generally it will be found that for forty-eight hours after laying the digestion is slow and the crop will contain food in the morning partaken of the previous night. As soon as the digestion is again normal your hen is safe for the next stage, but not before.

Besides the organs of nutrition we find in a bird's body, the nostrils, larynx, trachea, lungs and air sacs, which together constitute the respiratory apparatus.

The principal functions of this apparatus, are to supply oxygen to the blood and receive in return carbonic acid gas and watery vapour.

Then we have the circulatory apparatus composed of the heart, arteries, capillaries, veins and lymphatics, which carries the nutriment and the oxygen to every part of the body, and brings away the waste and cast-off material.

There is also the urinary apparatus, made up of the kidneys and the ureters, which separates and removes from the blood the greater part of the waste of the body. In addition there is the genital apparatus, consisting of testicles in the male and ovaries and oviducts in the female, their purpose being the reproduction of the species.

Having briefly explained the different organs that make up the whole pigeon, the reader will be better able to appreciate how minute these organs are, and with what mechanical exactitude all must do their work to keep the subject in health, and if any one organ becomes deranged, the damaging effect it must have on the whole.

In order to be able to cure a disease we must know something of its cause, and in order to appreciate to the full the cause we must know the agents generally liable to injuriously affect the diseased organ.

In my long experience with pigeons the greatest difficulty has been to correctly diagnose the disease from which the patient was suffering.

Once you can do this with certainty, the remedy of the trouble is less difficult.

The veterinary surgeon who attends to the suffering animal or bird has a much more difficult task than the healer of human sufferings. We poor mortals know where we feel a pain or ache, and can tell the healer our feelings and sufferings, but not so with the animal or bird. Examination and symptoms are all that the diagnosis can be based upon.

Here I should like to state that in many cases I have found pigeons respond readily to treatment, but in others drugs would have little or no effect.

The very formation of the organs of nutrition makes treatment all the more difficult, as pigeons have a nasty habit of vomiting anything objectionable before it passes from the crop into the stomach. For this reason I advise generally, as far as practicable, the administration of drugs in capsule form.

I commend to all my readers De Lacy's recent contributions to *The Racing Pigeon* on the subject of 'The Economy of Food'.

I entirely agree with him that improper food and improper feeding are answerable for more disease in pigeons than all the other causes put together,

In fact, when I am consulted on the subject of birds suffering from ailments, I at once enquire into the nature of the food used, its quality, and method of distribution.

The racing pigeon, being of such an active nature, the bird digests more food in proportion to its weight than any other bird or animal I know; it breathes more rapidly, its blood circulates quicker, its temperature is higher, and it makes a proportionate increase in weight in a shorter time.

A racing pigeon may, therefore, well be compared to a very perfect and delicate machine moving at the highest possible pressure.

Well managed, well cared for, with naught to cause the slightest derangement, such machine gives marvellous results; but let there be the least neglect, let there be irregularity in the proper oiling of the parts or the banking of the fires, one of the wheels goes wrong, and the machine collapses.

It is easier to prevent that collapse than it is to mend the break, and my object in penning these notes has been rather to warn the beginner and fanciers against the pitfalls likely to cause each of the diseases I shall briefly refer to, than to cure them.

The chief causes of disease in pigeons are – food, water, atmosphere, contagion, hereditary.

If pigeons have the hopper containing food always before them I do not think they will take food to excess, but if they are fed at intervals, especially if these intervals are too great, then I think obstruction of the crop and stomach will take place, and irregular digestion cause many diseases of the organs of nutrition. In fact, most pigeon troubles are traceable to this cause.

I have from time to time warned fanciers against the use of maize, because of its danger and because I have traced so much pigeon disease to its constant use. The danger is not to be feared from *good* maize, but from soiled maize, that may have been contaminated by sea water during shipment or exposed to moisture, causing fungoid growths.

The like warning is necessary with regard to tares.

Good, sound tares make the finest nutrition for pigeons, but they are very expensive and so subject to fungoid growths that fanciers cannot be too careful in the use of this grain to ensure its being of good quality.

Bearing in mind that grain in the least manner contaminated by fungus will cause incalculable harm to pigeons, fanciers cannot be too careful to obtain dry, sound corn, and to store it in such a manner that it always remains so.

Barley stored in a low temperature will become rampant with weevil; stored in a high and dry temperature, the weevil cannot hatch or live.

I have kept samples of the same corn, and stored it in high and low temperatures, and the effect on pigeons fed on it has been most marked.

Therefore, to ensure the food retaining its good qualities, it should always be stored in the dryest possible part of the house.

On the appearance of any disease in the loft at once look to the food and water, and see if the cause is traceable to this source.

Many fanciers use covered fountains to store the water. These will keep it cool, but bear in mind that closed fountains, unless gritty matter is used to clean and thouroughly scour them out, are apt to accumulate slime, and therefore it is absolutely essential that they should be thoroughly cleaned. For this reason I like fountains in two parts, in order that you can thoroughly clean the part holding the water regularly.

I have running water in my own lofts, but some object to this, believing that it is too cool for the birds, and that it is better for them to drink water that is of the same temperature as the atmosphere, which they would do if drinking from rivers or lakes in the ordinary way. Still, I have never traced any trouble in my own loft to this cause, so I do think the objection is a very serious one.

Either water or food fouled by excretions is the worst thing you can possibly give to pigeons.

Fanciers in the habit of throwing small seeds on the floor of the lofts promiscuously to their pigeons should bear this in mind.

The seeds fall on the top of recent deposits of excretions, and eaten by the birds, do serious injury.

Invariably clean the floor or that part of the loft you use as a feeding ground, before distributing rice or small seeds.

Dampness in the loft is also fatal to success, and will cause wing disease, rheumatism, catarrh and various diseases of the respiratory organs, in addition to feather rot.

A dry, well-ventilated loft is absolutely essential to success.

Another important matter is the necessity for proper air space for the number of birds kept.

Sun, light and air must permeate the loft, but even this in plenty is useless if the air within is vitiated with too great an abundance of pigeons not kept properly clean.

Twice the number of birds can be kept in the same air space if they are cleaned out regularly.

The secret of the success of some of those fanciers who appear to keep far too many birds in lofts big enough apparently to accommodate but half the number, is that they spend half their days with their pigeons, and droppings are never allowed to lie on the floor for a minute. If others dared to keep the same number in the same space, giving less attention, the birds would be rife with disease and useless for anything in a very short space of time.

The number of birds to be kept in lofts of various sizes depends generally upon the fancier's aptitude to attend to them, therefore I cannot define with exactitude the size of lofts and the number to be kept therein. Still, it is always safest to err on the side of amplitude of room and never overcrowd.

The Moult
by SQUILLS

So important do I consider the moult, that I propose to deal with the subject at some length.

Many run away with the idea that the moult is a disease. It is nothing of the kind; it is simply a natural function.

Nature ordains that almost all animals and birds shall, partially or totally, annually renew the fur or feathers that serve as a coat or covering.

If this renewal did not take place, after a time the feathered tribe would be denude of feathers, and unable to fly.

Birds suffer from many ailments during the moult, as it is a very trying period, as the drain on the system to supply the necessary sap or blood to the growing feathers is very great. In consequence of this, when the time is due, ailments are contracted, but this in no wise proves that the moult is in itself a disease.

In the case of the racing pigeon, I look upon a good first moult in the case of a young bird as an augury that the pigeon is sound, and will probably make a name as a racer if the blood is good. On the other hand, any check, tardiness, or irregularity in the moult on the part of the birds, the physique of which is so important, must condemn them as doubtful subjects.

Climatic influences very considerably affect the moult. An enormous percentage of parrots imported from abroad die during the first year of their importation into this country. This is due to the fact that the first moult they make in this country falls at a difficult season of the year. When once, however, they have passed through this first moult, little or no further difficulty is experienced.

For this reason, I find that the best pigeons to export to the colonies are late bred subjects that are more likely to come into line with the colonial moult, which, in the case of birds bred in the spring, starts at about the period ours is ending.

Birds do not actually die of the moult, but the drain on the system is so great that they will contract diseases during the moult which at once check the moult, and this check, with the illness combined, will result in death.

As I have stated, a good moult is the best sign of perfect health.

Now let me deal with the moult as it naturally proceeds in the bird.

The first feather to be moulted in a natural way is the shortest of the ten primary or flight feathers.

The wing is composed of 22 feathers, of which the ten outside, or long ones, are called primary flights, and the twelve small ones secondary flights. I have known birds with eleven primaries and eleven secondaries, but these are not common.

The commencement of the moult is governed by the time of the year a bird was bred, and the period at which it is mated.

Generally speaking, the moult will commence about five and a half weeks after the hatching of the first nest of young of the year that is to say when the hen is on eggs for the second nest, and the first youngster has been taken away.

An examination of the wing just before this period will show that the flight feather is lifting. It will have the appearance of being slightly longer than its natural length, and eventually there is a gap in the wing from which the feather is thrown.

In the case of a strong, healthy bird, the new feather steadily takes its place, first in the shape of a small bulb protrusion from the follicle, which gradually, as it lengthens, unfurls the webbing, and a new, tight, strong feather takes the place of the old one. Meanwhile, the next feather to it has been thrown, when the first was at half its length, the period dividing the fall of the two feathers being about a month. The subsequent flight feathers or primaries fall in succession, according to condition of breeding and health of the subject.

After the fifth or sixth flight has been thrown, the moult extends to the humerus and the shoulder feathers. About the time when the sixth or seventh flight has fallen, the moult becomes general.

Pigeons moult more freely when they are sitting than when they are feeding young.

The food the young require, and the drain of feeding, will retard the moult. In fact, whilst feeding young, they will often hold the next feather to moult until the young are old enough to be taken away, but directly the youngster is removed, the next feather is thrown.

Sometimes, through a check in the early part of the year, the moult is retarded, and I have known cases of two or three flights thrown at once, as well as the whole of the tail. Still, this need give no cause for alarm, so long as the birds are keeping strong and well, as they will renew the growing feathers without difficulty.

Sometimes, when a new feather is grown, it will be shrivelled in comparison to its fellows. This may arise from various causes. An injury to the follicle, want of sap at the time of its growth, or general debility. It is a mistake to pull out such a feather while the moult is still in progress, for the reason that a worse one will generally grow in its stead. I have, however, removed such partially deformed feathers when the moult has been thoroughly completed, and a perfect feather has grown in its stead, but I do not recommend their removal, owing to the risk entailed.

If, upon the renewal of a flight or tail feather, it is found to be split in the

centre, this is a deformed feather arising through an injury to the feather follicle, and I have seldom known a good feather to grow in its stead. Generally
speaking, it may be taken for granted that when once a deformed feather
grows it will always grow.

The natural moult of the tail is rather curious. There are 12 tail feathers in
most varieties of pigeons, but some have more, this being a special feature of
the variety.

The tail feathers are called rectices, and consist of six on each side.

I have possessed birds with 14 and 16 tail feathers. It is these 'sports' that
have led to the production of different varieties, as Darwin points out.

The moult of the tail usually begins when the 6th or 7th primary has fall-
en, and takes place in the like manner as the flights.

The first two tail feathers to fall are those immediately next to the medians
or two feathers, in the middle of the tail. When these have reached three-
fourths of their length, the medians, or two middle feathers, are thrown. After
these the third and fourth feathers fall on each side, counting from the centre.
The two outside tail feathers fall after the fourth and the fifth, or next feather to the outside, falls last.

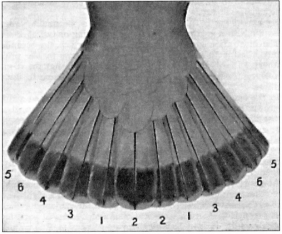

Outstretched Tail

**The numbers show
the natural order of
a normal tail moult.
The feathers fall in
pairs.**

In order to show clearly how the tail moults, I have numbered the feathers
in the illustration. The feathers fall in pairs, just the same as in the wing. For
instance, in a natural moult, the tail feathers numbered 1 should fall at the
same time, likewise No 2, but the moult of the tail is much more erratic than
the wing. Half the tail will sometimes be thrown at the same time, particularly the middle feathers, 1, 2, and 3. Many race young birds almost denud-
ed of their tail, but I think this is a mistake, and many a good bird and future
champion may be lost through doing so. Not only is the tail the rudder of the
bird, but it adds to the buoyancy, and I look upon good tail rectices as a useful acquisition to a good racer.

A gradual, steady moult and regrowth of the feathers is the best possible sign of health, and does not drain a bird's strength so much as irregular, broken moults, with bare patches of feathers missing at a time.

As breeding will retard the moult, so will its cessation when once it is started, hasten it. Successful exhibitors at the early shows know this so well that they separate their exhibition birds early in the year, stop all breeding from them, and thus force the moult, in order that they may be in the best possible plumage during the show season, and moreover, the moult taking place in the warm, summer weather, is much more easily accomplished.

The Moult in the Young Birds

This commences shortly after they leave the nest, and are able to take care of themselves. The moult proceeds in the same manner as in the old birds, the first feather to fall being the first short primary flight, or counting from the outside of the wing, the tenth. It is a good sign for this to be thrown without difficulty or delay.

If any difficulty is experienced in throwing this feather, it often goes bad with the subject

The least difficulty takes place in the moult of youngsters hatched in April, May and June. January and February youngsters sometimes are very tardy in the moult, and youngsters bred the latter part of June, and after the end of that month, seldom moult completely the same year. They will moult the small body feathers but not the flights or tail.

The whole of the secondary feathers are not moulted each year. There are some authors who contend that you can judge the age of your subjects by the number of feathers moulted. This method of judging the age might have been more useful before the introduction of the marking ring, but the latter is a better means of denoting the age that the uncertain indication of the secondaries. The new secondaries are broader and more rounded at the ends that the original nest feathers. There is not as much wear on the secondaries as on the flights and tail, and hence, I suppose, nature has not ordained the necessity for their regular moult.

The moult is comparable with the fall and growth of the leaves of the tree. The new feather is the leaf, the blood that feeds it is the sap, and when the plant is healthy, so it grows a healthy leaf. A stoppage in the moult of late bred birds indicates that there is no sap left to feed the growing leaf.

A good moult depends upon the health of the subject. An irregular moult indicates want of health.

If a pigeon is not making a good moult, nothing will denote this more plainly than the colour of the plumage. If there is a lack of lustre, you can depend upon it the subject is wrong. If it is found on examination there is stoppage in the moult, or it is late, or slow, means must be taken to assist nature. I have found that close confinement in a warm pen in the house, with some moist hay to stand on, will often be the means of assisting a sluggish moult. The bird should also be fed liberally on hemp seed and small seeds such as rape, canary, and a little linseed, but all of good quality; added to this in some cases I have found a capsule of linseed oil 5 minim, very effective.

The moult may be retarded by mating late in the spring, and it should be borne in mind that June youngsters do not moult as a rule early in after life as those bred earlier in the year. Moreover, late bred birds, when in their third season, as they do not moult completely until the end of the second retain their flights much longer than early bred ones. This information is useful in the case of racing pigeons, in connection with which a full wing means so much if they are to accomplish long journeys.

During the general moult, I find a mixed diet as good as anything but there must be no stint of good food. Linseed, buckwheat, and hemp seed assist the moult. A little new wheat is also good. In a state of nature, birds get an abundance of food in the fields during the moult, assisting them to renew their plumage without difficulty in the autumn, and yet food is new and fresh, and thus acts as a purge, and keeps the blood cool, but rich. A piece of rock salt in the loft at this period is a useful adjunct, and a regular supply of green food, such as lettuce and watercress, which if sprinkled with a little salt, the birds will eat with avidity, will all help to bring about the renewal of the feathers.

The most dangerous and trying period in the life of a highly-bred bird is the first moult. After this has been well and successfully accomplished, it is seldom that further trouble takes place, except where hens have been allowed to lay too great an abundance of eggs or cocks to feed too many young. The racer is also affected, and may have a bad moult, through a great strain in racing.

After a good moult, a pigeon is at its best. The colouring should be light and clean, the sheen on the neck sparkle like diamonds in the sun.

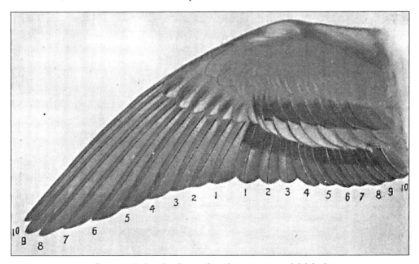

Outstretched wing of a three-year-old bird
The numbers show the order in which the primaries and secondaries moult.

Any pigeon that suffers from a bad or faulty moult will seldom breed sound young the following year, and in the case of the racer, if given any difficult trials, failure and loss will result. The only course open in the case of a valuable bird that has a faulty moult, is to lay it aside until the next moult has been accomplished, taking care meanwhile that nothing can drain its resources or sap its strength, to interfere with a good and thorough moult on this occasion. If this is successfully accomplished, the bird will show its old form again, both as a breeder and a racer.

As I have already stated, the first feather to fall, showing that the moult has commenced, is the shortest flight feather, being the tenth from the outside of the wing. The illustration given shows the natural order of the moult of the primaries in the case of a normal moult.

The danger after the second feather has been thrown, is that sometimes the third and fourth may be thrown in conjunction, especially if means have been taken early in the year to retard the moult. It is very unwise to send a bird to a race with a wide gap in the wing from this cause, and the only course open to a fancier who finds that any of his birds throw two feathers at a time in this manner, is to reserve then until they have re-grown the feathers, and give the bird a short toss of 20 or 30 miles subsequently, and then loft it into a race of some distance away. Strange to say, the habit of throwing two feathers at a time (by this, of course, I mean four in all, as in a natural moult they throw two, one in each wing) is a regular habit in some birds, and occurs annually at about the same period. They must be treated accordingly.

A good, free moult is a good augury, and a sign of health, although causing inconvenience to the racing man. In the case of birds I have rested whilst growing the feathers as moulted, they have generally come up well afterwards, and as a rule hold the next feather somewhat longer than usual.

As I have pointed out, the moult varies in individuals. The usual course is for a bird to start to moult after it has sat the first nest, and hatched and fed the first round of young, and laid again, the first feather being thrown three or four days after sitting starts, even though a bird may be feeding young, but nevertheless, May or June bred birds, or those that come under the category of late bred birds, will frequently hold their first flight feathers much later than this period, and moult more slowly than the early bred birds will do. But here, again, no guarantee can be given that this will be so. A fancier must endeavour to learn to know his birds in this respect.

For the reason that late bred birds moult more steadily in after life than the early bred ones, we frequently hear of birds that in their first year did little or no training, and in their second year have only been trained about 100 miles with the young birds, making champion racers. In fact, this class of birds in Belgium is very much prized, owing to the good condition the wing can be kept in for the long races.

It is because such birds keep in much better feather towards the middle of July than those bred at an earlier period of the year, that those that survive their second season do so well, and when fanciers know the real value of birds of this class, and know how to treat them properly, they will take grater

care of them. I am aware that some fanciers advocate as a method of retarding the moult, letting their pigeons sit the first round of eggs without hatching, so that when the second round is laid they do not moult in the ordinary
course, as they would have done had they hatched and laid again. In fact, by
this means the first feather, it is contended, is retained in the wing until they
have reared the young from the second nest, and they start sitting on the third
round of eggs.

I have carefully tried this method, in fact, it was more than 20 years ago
when I first did so, and found that if I adopted the practice, my birds went
somewhat stale in the early part of the year, and it seemed to disorganise their
systems.

In some cases, I must confess they held their feathers a little longer, but
where fanciers adopt this practice, I should advise them, in the case of old
birds, not to put them in training until they have actually hatched the second
nest, and reared the young one therefrom until it is seven or eight days old.
Both Wegge, who was one of the most famous racers in Belgium, and
Janssene, strongly advised that it was best to allow birds to hatch their young,
and to partially rear them, in order that they might use the soft food, and any
means adopted such as I have referred to, to retard the moult in an unnatural
way, can only be done at the expense of the health of the bird. I cannot understand why those who advocate this system do not follow the practice of the
Belgians, and keep their old birds separate until well into April, which would
have the same effect, and in my opinion would not be so likely to cause any
ill effects.

It is natural for a bird to moult at a certain period after being mated, and it
seems to me if it does not do so that the blood is likely to be deranged. In fact,
when I find any of my birds are not moulting in a normal manner, or holding
on to their feathers beyond the normal time, I am uneasy with regard to them.

I am merely recounting my personal experience on this subject for what it
is worth, as I have not found the practice of retarding the moult in this manner pays me.

It is a mistaken idea to think that a pigeon cannot accomplish a good journey even growing its fourth, or even its fifth flight. In fact, I have seen birds
that have been prizewinners in hard long-distance races, that had a re-grown
the fourth feather, and were flying with the fifth flight out. Still, for my part,
I like to have a bird on its second or third feather, because there is more danger, as I have stated, of two feathers being thrown at a time when they get
beyond this stage.

It is not so much the actual feather that a bird has thrown that is important
in the case of long and hard races, but that the wing should be fairly full, and
that you should be able to send a bird to a race growing a feather in the basket; that is perhaps a third up, with no possibility of the next feather being
thrown during detention.

I do not know if others have noticed the same thing, but I think I can safely say that a very warm, badly ventilated loft inclines to a fast moult, and
moreover, in such conditions, the newly-grown feathers will not be as hard

and healthy as they might be, whereas a cool, dry loft in itself, quite apart from other considerations, helps to keep back the moult.

Strange to say, I have also found that in competing from the north, there was much less danger of birds throwing feathers in the baskets when sent to the long-distance races, than is the case when they are sent to the south, for the reason that the temperature, being lower in the north, has a tendency with birds sent from the south, to retard the moult, whilst the birds are waiting for liberation in the baskets, whereas the temperature in the south, being higher, has a tendency to make the birds inclined to throw their feathers more freely

I have, in describing the feather, endeavoured not to use any technicalities that the beginner could not understand.

The secondary feathers do not moult each year. Many Belgian authors, as I have already mentioned, contend that you can tell the age of the birds by means of these. I do not agree that such is any conclusive evidence.

My experience is that with the secondaries the moult varies in individuals to a greater extent than in the wing or tail.

The illustration given shows the order in which the secondaries moult, that is to say, in the opposite direction to the primary feathers. It will also be seen from the example, that the re-grown secondary feathers are of somewhat different shape to the nest feathers,

In the wing illustrated, three secondaries have moulted and been re-grown, and according to the Belgian theory referred to, the age of the bird, the wing of which is illustrated, would be three years, but as before mentioned, this theory cannot be accepted as an established fact.

If fanciers wish to experiment with respect to it, they can mark the secondaries in such a manner as to distinguish them, and thus be able to see how many are moulted each year.

At any rate, let me here say that I think it unwise to send birds to races that are minus a secondary feather, as well as a flight feather. I do not know what other fanciers views on the subject may be, but I like to find birds with well-developed secondaries of good length and fair width, in fact I think these are of considerable value, and although I do not, perhaps, attach as much importance to breadth of flight as some fanciers do, breadth and length of secondaries I do attach importance to, and find many of the best birds I have owned have possessed this quality.

There is one matter in connection with the retarding of the moult that I ought to refer to in passing. That is the cutting off of the small end of a flight feather to delay its being thrown. This practice is very general in Belgium. Just the tip of the flight is cut about half an inch down the quill. Owing to the air thus being able to enter the barb it is contended that this prevents the natural flow of sap to the root, and delays the feather being thrown, the feather remaining in the wing sufficiently long for the bird to accomplish a journey it would be called upon to perform. I cannot say that this practice gives positive results, but mention it for what it is worth.

I have not dealt with the question of repairing broken flights, as that is dealt with in my book of diseases. In the case of bent feathers immersion of the feather in boiling hot water will at once straighten these, but where birds have a bath, and can lay in the sun's rays, a crooked feather will straighten of its own accord, this being nature's remedy.

There is one matter of importance in connection with the moult that I consider most essential, and that is that the small body feathers, as well as the feathers of the humerus, should be clean when regrown, without any frayed edges, or the semblance of fret marks. The cause of fret-marked feathers is no doubt due to the fact that a strain tells on the system of a bird. The result is that the corpuscles of the blood required to build up the system are not sufficient to feed the growth of the feather at the same time and there is a check in the feather growth, and subsequently when the growing feather starts to re-grow after the check, a stain more or less marked, appears across it.

I have endeavoured to give an illustration of a fret-marked feather in order that the beginner may see the nature of such a deformity at a glance.

A practice common amongst many Belgian fanciers is to train their young birds a very short distance during the year in which they are bred, as they contend that by following this practice the birds moult better, and it allows them to develop good feather physique, and constitution, with the result that later on in life they can the better accomplish the long journeys, and do the work they are called upon to perform. In fact, M Vassart and many of the most successful Fleurus long-distance racers attach very great importance to the first moult being allowed to take its natural course through no drain on the birds.

Frequently birds are found with blood quills both in the wing and in the tail.

This, in my opinion, is due to lack of vitality, and the treatment that I have found best in the case of blood quills is to hold the affected feathers under the tap, so as to give the bird a gentle shower bath. In fact, during the growth of the feathers, I know of nothing so important as a regular bath, and to improve the feather and enrich the blood, particularly when in the moult, I believe a free life in the fields, with plenty of green meat, most essential. In the case of town lofts, the green-meat should be served regularly in the loft, and when lettuces are in season, I know of nothing better.

If fanciers will bear in mind the importance of the moult, they will have no difficulty in keeping their birds in the best of health.

A fret-marked feather
The stain or faint line in quill and across feather shows where its check in moult occurred.

During the moulting season not only do birds renew their feathers, but it is the period at which they throw off any effete matter from the system.

Bearing in mind the importance of good feathers to a bird whose success depends upon its power of wing, I cannot too strongly urge upon fanciers the necessity of good, sound food during this period, and when the moult is at its full, let there be no drain in the system by late breeding. The career of many a good bird is ruined through the anxiety of its owner, when it has proved a good one during the season, allowing it to reproduce in too great an abundance.

It is often found, on the other hand, that a bird that has done badly, been discarded early in the year, and allowed to rest without mate or breeding, the following season proves a phenomenal success, soley due to the fact that it has a reserve of strength through the good moult it has thus been enabled to make.

SQUILLS' DIARY AND ANNUAL

INTRODUCTION by SQUILLS

It is with much pleasure I present to readers the eleventh edition of my Diary.

A feature that I have added to the Diary this year is the continuation of the register of famous pigeons. The 1906 edition of the Annual contained particulars of practically all the famous racers known and talked about in England from the time the sport started. Later editions have brought these particulars up-to-date, and this year will be found included in the register practically all the 1st prizewinners in what may be termed the classical events of the year.

During the past year, generally speaking fanciers have had no cause for complaint. Serious disasters have been few, whilst the most important long-distance races turned out fairly successful.

During the past season the National FC organised its second race from Mirande. A few years back such a distance would not have been dreamed of, let alone attempted. It was no doubt the organisation by that enterprising club, the Manchester FC, of a race from San Sebastian that led others to emulate their example.

After the Manchester FC San Sebastian race came the Harborough and District FC race in 1907. The management of this venture had the advantage of knowing the pitfalls, Vost, the MFC conductor, had had to contend with, with the result that these birds had an easier journey to the racepoint, and although the returns were few, some of the winning birds did remarkably fine performances, showing how well the distance could be done. But what to me was the most interesting feature in the race was the remarkable condition in which the winning birds arrived home. Although they had spent a night out, they did not show nearly the distress the birds liberated early in the morning that arrive home the next day.

Long-distance racers have, no doubt, to thank the organisers of the Manchester San Sebastian race for their lead, for I cannot help thinking it was the fact that pigeons could successfully home these distances that led to the organisation by the National FC of the first race from Mirande in 1908, with a late liberation.

The first National race was organised in 1894 by the Manchester FC from La Rochelle. We have seen many changes in the sport since those days, but nevertheless there is one matter of passing interest that I cannot help referring to, and that is in connection with the rules of the old 1894 National.

The committee elected to decide upon these rules by the whole body of fanciers of the United Kingdom consisted of Messrs John W Logan, John Wright, George Yates and A H Osman, and many of the rules propounded the day when these gentlemen met at East Langton have stood until this day, and are used in connection with some of our most important Open races.

With a north wind, the birds competing in the first National race from Mirande were liberated at 10 am. Messrs Highnam Bros of Bradenstoke, timed in the first arrival in the race at 10.41, and won 2nd prize, whilst Mr John Wones, who had won the San Sebastian race the year before, timed in at 12.50 pm and won the premier prize with a velocity of 940 yards per minute. Mr Wones' distance was 634 miles 334 yards, Messrs Highnam's distance 563 miles 1,475 yards.

I ventured to prophesy at the time that I thought it would be a long time before the fine performances of the winning birds and the returns in this race would be beaten, and see no reason to change my opinion.

There were many who contended that the 1908 race would have given better returns and a different result if the liberation under the conditions that prevailed had been at a somewhat later hour.

I contended, and shall always contend, otherwise. The fast birds that it was alleged reached the middle of the Channel the night of the first day, in my opinion, did not fail in the Channel, they failed just because they were fast birds. My own experience of these fast birds is that directly and immediately they have no pals to help them they are hopelessly lost. They have the speed and will keep going as long as there is a flock in which they can pack, but break up the flock, compelling birds to fly a given journey on their own, and you can generally bid adieu to these fast pigeons as was the case in 1908.

I claim that the 1909 race has helped to prove this theory correct. Many of those who claimed that their fast birds were lost in the Channel in the 1908 race were again amongst those who did not time in in 1909, although the conditions were almost identically the same, but the liberation was effected two hours earlier.

Just these two hours, in my opinion, made all the difference, for had the National race birds been tossed at 10 am, with a north wind instead of 12 noon, we should have had a much finer percentage of returns the second day.

I cannot help thinking that the birds in this year's National race did not reach beyond Marennes the first night, and think the north country birds had over 500 miles to fly the second day, making the performance of Doctor Barker's pigeon all the more meritorious.

At the same time this late liberation meant a good deal for the majority of the birds, just a little more help from the flock would have landed them at Nantes or it may be Rennes the first night, and some of the less-experienced birds could have picked up their line from there.

However, the 1909 race was just the race to find out the real good ones, it has silenced those who complained in 1908 about the hour of liberation, and the birds that accomplished Mirande in race time in 1909 have undoubtedly proved their intelligence and grit. It only remains for me to congratulate Messrs Saunders, Field and Barker, the winners of the first three positions in the race on their splendid performances, a credit to the sport and something English fanciers need to be proud of.

There is just one matter in connection with the National race that I think is of more than passing interest. That is the sight and intelligence theory.

I well remember reading Mr Logan's little book *The Homing Pigeon Fanciers' Guide* in 1885, and how forcibly I was impressed with what he wrote on the sight theory. The longer I live the more convinced I am that we shall always get better results if we allow for the intelligence of our birds and cultivate this intelligence than if we place blind and implicit confidence in the instinct theory.

Those who believe in the instinct theory will often point out to you the remarkable performances of some inexperienced yearlings in a smash and the downfall of what might be termed the old stagers.

I will try to explain my idea as to these smashes. My opinion is that when a good pigeon has been over the ground once it takes the same course a second time, it knows the road it ought to come, sticks to the line gamely and probably cannot get through.

Up to this point the uneducated pigeon is with the flock, it doesn't know the road and beats about until by chance a lucky break is found and it gets through.

Some birds in racing from 400 or 500 miles get an entirely different line. It may be a good line or a bad one. When the race is one in which the winning birds take a bad line, you will seldom hear of the winners again. By a bad line I mean a line off the usual route that the major part of the flock should know and ought to travel. If, on the other hand, the race is one in which the birds have travelled the usual and normal route, the pigeons that win will win again and can be backed with confidence.

From the above it will be seen that in smashes I do not think the best pigeons always win. Luck is a much more important factor than when just normal conditions of flying prevail and all birds stand a reasonably fair and equal chance.

I have had young birds take four hours to fly 20 miles on a fine day because they had joined a flock and flown a roundabout course home. These self same youngsters have taken almost the same time to fly subsequent journeys. For the reason they have kept to the roundabout way home.

The same thing applies to winners in races that win under exceptional circumstances when the usual course that experienced birds would take is a bad one. They learn a bad line, attempt to come that line another year, and stand a hopeless chance against the educated pigeons that knows its way over the proper and regular course.

That the National has proved and will prove the finest possible race for finding out the intelligent pigeon, I feel firmly convinced.

A yearling like Mr J Wones' yearling hen may do the distance, but the odds are very, very great against birds of this class getting home. The more education the pigeon has had, providing it is not a broken down wreck, the more intelligent it is, the more certain are its chances in a race of this class. Every pigeon sent to Mirande, in my opinion, ought to have had at least some previous experience from Marennes in previous years.

When Mr J W Toft, and subsequently Mr Willie Orchardson won the National race from Bordeaux, I remember a long chat I had with them on the subject, and both of them made use of the same expression, "I have won because I sent my best pigeons".

A man who wants to win 500-mile races today has got to do the same thing – send his best. For only the best can or will win.

Not fast ones. To the man who has fast ones that can win in 300- or 400-mile races, I say keep on winning with them as long as you can. The more often they go over the course the more experience they will gain for what you want of them later on, but unless those same fast birds are backed by a long-distance ascendency, don't be too cock-sure they will win for you when the extra distance is tacked on.

The above subject is one in which I am deeply interested, and therefore I make no excuse for placing these observations before the readers of my Diary. Then Nantes Combine race of 1909 was only a qualified success. Of the 8,000 odd birds that were sent to compete, I do not think the wind and weather alone accounted for such a large number of absentees.

It is with satisfaction and pleasure I note that in 1910 arrangements will be made to retain the vans in which the birds travel to Nantes at the racepoint. By proper arrangements this is easily possible, it may be that a small cost is incurred, but of what moment is this compared to the comfort, welfare and safety of birds of so much value to their owners.

The greatest confusion was caused throughout Lancashire by the telegram of non-liberation having miscarried. This should give food for thought, and I think will prove the necessity of never relying upon one wire alone, whether of the liberation of the birds or their non-liberation.

Then again conductors should never fail to promptly answer telegrams sent them by those in charge of the race, no matter how many previous wires they may have dispatched. Send a plain, simple and immediate answer to any question that is asked from the home end. The cost is small, and it may be the means of allaying the tension of those awaiting news that has gone astray.

In my anxiety to obtain news from conductors who have delayed answering messages I have suffered many anxious moments.

Thanks to the information I obtained from the Chief of Police at Nantes, I was able to inform Lancashire fanciers that the birds were not liberated on the Wednesday and to obtain the publication of the news in some of the Manchester papers.

At the same time, it is a very great pity that the incident occurred, and I am sure its recurrence can be prevented by the sending of telegrams in duplicate to the secretary at two addresses that will find him. There is safety in the system, and in later years I have always adopted this course in the case of telegrams of liberation and non-liberation.

One cannot help saying that the performance of the winners in this great race was something to be proud of. There is no county except Lancashire that could muster 8,000 birds for a race with every competitor flying over 400 miles. The winners Messrs R H Cawley, Prestwick, 926; W Waddington Trawden, Colne, 909; and J A Lord, Bacup, 915. Mr J A Lord's performance was a particularly remarkable one, as he has been 3rd for two years in succession. Will he win next year? Who knows!

I think it would have been a very great pity to have sectioned the race, as any system of sectioning would have reduced interest in the competition, and in races of this length, where the whole of the birds are homing to a given area, as is the case in the Lancashire Nantes Combine. I don't think sectioning is of great moment.

The withdrawal of the Stockport Federation from the joint management of the Marennes race had a marked effect on the entry, with the result that its plucky promoters, the Manchester FC, lost nearly £100 over the venture. Nevertheless the race proved a splendid 500-mile one, again showing what good birds can do when well conditioned and liberated in favourable weather. When news reached me on the morning of liberation, I prophesied that the race would be a splendid success thanks to the weather conditions in France and over the greater part of the course. The winners were Messrs Maddocks & Collins, Macclesfield, 1139; Ollerenshaw, Denton, 1133; F Bibby, Biddulph, 1131.

It is a pleasure to note that there is a tendency on the part of the Stockport Federation to join hands once more in the conduct of the race, and I hope that it will receive increasing support. We have too few races of this class, but the difficulty that ever presents itself is the expense of carrying them out. There is also a tendency for some of the best fanciers in Yorkshire to promote a race from Marennes. It would, I feel sure, be cheaper and better for the three bodies to combine and again have a united Great Northern. The original promoters of the race were the three bodies named.

It was with feelings of pleasure that I saw a better understanding arrived at between Yorkshire fanciers during the past season, and a United Yorkshire Combine Race. The result must have been very gratifying to its promoters, although some disappointment has been felt and expressed at the slow velocity the official result made before publication. Pickard of Wakefield, 1158; Sutton, 1141; and Pickard, 1139, were the winners.

The 13th and 14th July last were memorable days for racing from France, and it will be interesting to bear in mind the conditions of weather that prevailed to lead to such magnificent results.

The delay in publication of the result was due to the old trouble, measurements.

The time has arrived when delay from this cause ought no longer to be possible.

After pegging away at the subject year after year, and proving beyond a shadow of doubt how erratic the measurements were through the use of different racepoint locations the National Homing Union has at last put its house in order, and now issues a set of racepoint locations agreed to be used by all calculators.

This removes many obstacles that previously existed in the interchange of measurements and their use in Open races such as the Yorkshire Combine Rennes race.

There is, however, one reform that is needed, and that is the official checking of the loft locations of every fancier in the country.

Many of these loft locations have been made haphazard, and I believe are incorrect. In some Federations I know a proper survey has been made and the locations correctly marked, but in most cases no proper system exists. The time will come, I feel sure, when the Union will keep a proper register of every fancier's loft location against his name and address and issue certificates of the location.

At present we must be thankful for small mercies, and grateful to think that racepoint locations have been properly adjusted.

Another very fine long-distance race was that promoted by the Up-North Combine from Rennes. 3,706 birds competed in this event, and bearing in mind that over 8,000 birds were sent by fanciers in these parts to Bournemouth, 300 miles, it will give some idea of the popularity of the sport. The Rennes race was well organised and well carried out; the winners were Crossley, 1057; Wilson, 1033; Leack, 1032.

Although such a splendid team of birds went to the race, one disappointing feature was the small entry in the pools. Is this lack of entry due to the fact that it is believed that certain sections have an advantage in position over others, or is it because our friends in these parts have had so many disasters they are obliged to rely on new faces to represent them, and hence, very wisely, won't back birds that have not been over the course before.

In reviewing the races of the past season my task would not be complete did I not refer to the races on the North Road.

It is needless for me to say that I am one of the oldest North Road racers in this country; consequently I know the pleasures and disappointments the North Road offers as well as most fanciers.

The want of combination on the North Road, the small number of birds proportionately that reach the final stages, and the laxity of support the pools receive, all tend to put a damper on the spirits of the most ardent North Road racers.

No conductors ever left for Lerwick with greater hopes than Merrs Pulley and Clarke in 1909. I do not think a finer lot of birds, with a greater percentage of Lerwick birds amongst them, ever competed in a Lerwick race, and yet so far as the long-distance men are concerned at any rate, once more Lerwick laid many a champion low, and many a loft desolate. The two best performances in the race were those of Messrs Fred Houghton of Wellingboro', and Lulham of Sydenham.

The treachery of Lerwick as a racepoint has once more been proved. Lerwick remains the unsolved mystery it proved in 1900. Here we have a pigeon like Mr Matlock's champion of the 1907 race, that flew Lerwick again in 1908, sent in the finest possible condition in 1909, and it is heard of no more. The same thing applies to other real good pigeons, lost after doing the journey well on previous occasions.

This is one of the greatest drawbacks to the North Road. With a north wind over the whole course and anti-cyclonic conditions, pigeons with a pair of winds of any sort will get home. The 1902 Lerwick race proved this, and the 1908 race was little better.

In these easy blow home races on the North Road, the good, bad, and indifferent score alike. Pigeons with no experience jumped from 150 to 600 miles, doing it with ease. I remember in the 1896 race from Thurso, Mr Rix lifted a bird from Essendine to Thurso, and timed it in on the day, winning 2nd prize.

But directly trouble is met on the North Road, especially from Lerwick, we get the most mysterious results.

In the past season's race men flying to Yorkshire timed in all, or nearly all, the birds they sent, whilst the percentage home in London before the close of the race was meagre in the extreme.

My explanation of this is that Yorkshire fanciers can train their birds single up to Banff or Thurso the year of the race, without distressing them too much for the long journey afterwards. Having been over the ground the same year they know the road better, I am convinced that the trouble we get on the North Road is at the further end of the course. No men could have been more slavishly devoted to the birds than Messrs Pulley and Clarke last year. Previously conductors have been blamed, and want of telegraphic communications along the line of flight has been blamed. All these things have been tried, and we are no forrader. The mystery of Lerwick remains unsolved.

Banff gives fairly good results, but here again we suffer from dearth of numbers. The Wilts, Exeter and West of England men at one time competed with Londoners, but want of special prizes and lack of enterprise resulted in the organisation of local competitions. The result is, we have the Welsh Championship Banff race, the Northamptonshire Federation Open Banff race, the Gloucester and Bristol Federation Banff race, and the London Banff race. I made a special journey, and asked the Gloucester men why they did not support Lerwick. The answer was, "It's too risky". "The prize money too small, having regard to the entry fee, and we can race from Banff for better prizes at a cost of a little more than 1s per bird."

Interest in the North Road might be revived if a good championship prize was offered for competition; the whole of the Banff races to be flown on the same day.

An effort is being made by the North Road Championship Club to organise a popular Thurso race. Had the Lerwick race been abandoned for a year, and Thurso substituted, this effort, I believe, would have been fruitful of good results, but I do not think there are enough fanciers or enough birds flying the North Road to support Banff, Lerwick and Thurso races.

One of the worst set-backs to North Road racing, I cannot help thinking, is in the dearth of generally good and experienced conductors. On the South route considerable advancement has been made in this direction, but none on the North Road. Amateurs are requisitioned who know little or nothing of the game.

The great difficulty is to find competent steady conductors, who can be thoroughly depended on to carry out their work in a trustworthy manner.

I still maintain that you may have the most competent weather experts you like to control the races, but unless you have a conductor whose first and every care is his birds, your long-distance races will be certain failures.

I obtained the weather charts from the Meteorological Office many years ago – used them for what they were worth, as well as the forecasts, but am still of opinion that it is impossible to give positive instructions to conductors from the home end, especially where the course is a long one.

Take, for example, the telegram I sent to Lukeman in connection with the Reading and Southern Counties Nantes race. My telegram merely gave him in plain English the official forecast, he acted on it, used his own judgment and liberated the birds in his charge, practically every bird in the race sent fit getting home, whilst other conductors at Nantes the same day held over, saying the weather was too awful to contemplate starting their birds.

I must not omit reference to the Midland Combine Marennes race, and congratulate Messrs Flint and Guy on their success in winning 1st and 2nd prizes in the race. That two yearlings should get through on the day, and leave the old stagers out for the night, once more proves that good yearlings will often find a way when the old birds that know the road pull up for the passage in the right line to get clear. The Midland Combine Marennes race is gaining in popularity. It is about the same distance as Nantes to Lancashire, and one noticeable feature of the race is the amount for which the birds are pooled, having regard to the number competing. I think the pool average has been as good, if not better, than at any race flown.

The Marennes race flown by London Federation was a poor testimonial of the quality of southern birds. I fear, however, the difficulty of training stands in the way of the successful flying of the Channel route by men in the south. They are not far enough from the coast. The birds are trained from the SW, and then when sent across the Channel, have to make an acute angle.

Having a good deal of faith in what I call "educated direction of flying", any method of zig-zag training is bad, except in the case of old birds that know a wide range of country.

Take as an example the young bird race of the London Columbarian Society last year. It was a complete disaster, not because the day or the weather was bad, but because Granville was off the direct line of flight the birds had been trained; when they struck off at Granville they flew in the direction they had been educated to fly, and only a lucky few, probably the slowest, managed to strike the right line later and get home.

I will guarantee that if you liberate young birds off their line of flight, they will first strike off in the direction they have been taught to take. I do not suggest that they will all be positively lost, but the risks are greater and the difficulties greater than if kept to one given line and hence the necessity for training as straight a course as possible when educating inexperienced pigeons.

There has been a falling off in the number of shows held during the past season, likewise a considerable falling off in the entries.

The fact of the matter is that racing men generally are beginning to find it useless to exhibit unless they specialise. In the majority of cases when the show specialists have swooped down on the show with their entries, the poor racing man has got little or nothing to show in return for his entry.

Dr Barker was one of the most successful exhibitors in modern times, and he had the courage to confess he kept an entirely different stud for racing. Some exhibitors do this, but will not own it.

Of this I am convinced that breeding to type for show ruins the chances of the bird for racing.

It is over a quarter-of-a-century ago when George Cotton and others started the exhibiting businesses. I was a regular attendant at the shows in those days and have been ever since. I bought some of Cotton's, Palmer's and J W Barker's best show birds, and found them useless as racers.

When *The Racing Pigeon* was started it was given a distinctive name to disassociate it from the ordinary homer, but had it continued to foster the showing business by offering fivers for showing and a £25 Championship, the name of racing pigeon would have been a misnomer.

The bulk of the entries at the shows came from the racing man, the bulk of the prizes carried off by the show specialist.

The strongest arguments I have heard in favour of the show specialist is that he has just as much right to specialise as the young bird racer, the yearling racer, or the long-distance racer.

This cannot be denied, and as long as he exhibits his birds and competes amongst his fellows with this known object, no one will complain. The racing man will in time gain wisdom and refrain from finding the money and naturally the entries become less and less. Members' shows with entries limited to members will then become popular. It will be possible to prove that the birds have actually qualified by the club race sheets.

To increase the entries, there has been a tendency to accept birds in the old bird classes that have flown the qualifying distances to any loft and at any period of their lives. This, again, is a retrograde movement.

The evil effect of showing in the past has made itself felt. Carriers, Antwerps, Dragons were at one time purely sporting pigeons.

Judging to a type, breeding to type can and will only lead to loss of quality in the true racer, and the wider it is known the better that the racing pigeon is a purely sporting pigeon the better it will be for the sport.

The beginner will be wise if he leaves birds possessing a pre-disposition to a show type severely alone if he wishes to gain success in the sport.

I feel particularly fortunate this year in being able to include in the DIARY articles from Messrs W Sheldon, R H Cawley, W Saunders, S P Griffiths, and Doctors Barker and Tresidder.

Every one of these fanciers is a gentleman of standing in the sport who has not gained his success by mere flukes.

Both the beginner and the old hand may learn much from the articles they contribute to arouse interest and instruct.

The man who is a fancier at heart can always find something to interest him in reading the views of those who have gained success and who he knows to be sincere.

But here let me once more warn the beginner against making continual radical changes in his system of management.

Many a loft owes its downfall to changed conditions of management. Once a system is giving good results, continue on the same lines, introduce any change with caution.

I am a strong believer in a well-ventilated, cool, dry loft. The more fresh air the better.

I don't know if others read Leiut. Shakleton's remarks relative to his experiences in the cold regions. He said those that coddled up and kept indoors were sick, but those that kept on the move out in the open were free from colds and hardy.

Those who turn horses out to grass seldom find they catch cold when turned out in the open, but directly they are brought back into the stables, trouble often begins and they want tonics.

Keep the loft as near as possible the same temperature as the air outside and your birds will be all the better for it.

The question as to the number of birds a fancier can keep in a given space is always a moot one. A man like farmer Langton, who spends every spare hour of the day with his birds can house twice the number that another man can, who is only able to clean them out once or twice per week.

Roup and kindred complaints are greatly the outcome of dirty lofts and overcrowding. The least symptom of disease should be a warning to the fancier to look to his ventilation, the number of inmates of this loft and his corn.

With regard to the corn used I cannot lay too much stress on the importance of sound grain of the best quality.

It must not only be sound when it is purchased, but must be kept sound.

I like the Belgian principle best of spreading it out on a clean dry floor, and turning it over regularly with a wooden rake or hoe.

It must, of course, be kept free from attacks of vermin. The excretions of rats or mice in food will cause fungoid, and this is most dangerous to the health of pigeons.

Last summer was a particularly damp one in my own locality, and I attribute my own birds doing badly entirely to my grain getting moist, although it showed no outward effect either in smell or colour.

Another season I shall keep my store room artificially heated if we get a very wet summer had have fitted a stove up for the purpose. Some districts are, of course, better situated in this respect than others.

It is a mistake to store grain in metal bins.

Some years ago I designed wooden drawers with net-work bottoms so that a current of air could pass through them. I gave a rough sketch at the time in the 'Fanciers Gazette', and drawers somewhat similar have since been designed and put on the market that serve the purpose admirably.

I have been struck with the success of many bakers in our sport and attribute it to the fact that they are able to store their own corn under such favourable conditions.

Unless a man has a good place to store his corn and keep it thoroughly sweet and dry he had better buy in small quantities from the merchant when it is bound to be well stored, for he knows if it is not it soon becomes unsaleable.

A good plan is to store it in half sacks and turn it over occasionally. If bins are used wooden ones are the best.

During the past few years there has been a tendency for many good and successful long distance fanciers to give up racing young birds. The losses in racing youngsters become more and more serious each year.

So convinced were men like Verssart and many champion long-distance men in Belgium of the folly of racing well-bred long-distance youngsters that they seldom train young bred from their best long distance racers any distance.

The fact of the matter is that there are so many pigeons on the road and so many taking part in our huge Federation tosses that the youngsters get bewildered, and may be flying miles off their line before knowing where they are.

My own experience is that youngsters that have flown from 60 to 100 miles as youngsters are quite as good as those that go through the whole of the programme. In fact, you will often find your best old birds prove to be those that have only had some easy training up to 50 or 60 miles as youngsters.

Later in life when they have grown, work for the workers is my policy.

When you find you have a good one, treat it accordingly, but the way to find out the good, bad, and indifferent is the training pannier.

For breeding purposes I prefer birds of one, two or three years old, with a good ascendency, and that have themselves done some work. The youngsters bred from old pigeons want more time to grow than those bred from the younger generation. The young bred from the yearling children of your best racers should not be despised. Many a champion owes its parentage to yearlings of this class.

The amount of training on the road birds should be given entirely depends upon a fancier's method of home training, the food he uses and its distribution.

Hopper-fed birds want more exercise at home and more tosses on the road than those sparsely fed.

More pigeons are lost through being given too much food than too little.

A fat pigeon cannot race against a lean, hard one. At the same time when pigeons are sent to a long race of 500 miles they must carry sufficient flesh to pull them through and allow for waste.

It is only from experience that a fancier will know if his pigeons are or are not fit. Once this detail is mastered, and thoroughly mastered, the road to success is easy.

Many birds fail to reach home in race time simply because they are physically unfit to do the journey at the time they are called upon to perform it.

Home exercise twice a day or half-an-hour I think enough to keep any birds fit. Some give more, others less, but if birds fly half-an-hour freely there is not much amiss with them.

A full-winged pigeon is better for a race than one with a flight out, but I have won many a good pool with pigeons minus a flight, even the fourth, out.

A tardy moult in the case of a bird that has usually moulted freely is a bad sign.

The most important eye test I know is colour. That is to say, the ground colour against the chequering, or in self-colours the bar against the under colour.

When these are dull, seemingly running into one another there is something amiss, but if they stand out clearly one against the other, there is plenty of sheen on the neck and the wattle is spotlessly white; with keen, bright eye, there is not much amiss with your candidates.

The merry joyful appearance of your bird, combined with the above signs, are more to me than any amount of handling. Ninety-nine out of every hundred pigeons handle differently, and some are very deceptive in this respect.

I am not so much averse to sending old cocks slightly on the drove, or calling to nest as some folks. Some hurrying home in this condition when they know the way. But hens just about to lay or getting eggy should never be sent to a race. Never take hens off the nest until a clear three days after laying second egg.

Some are much opposed to sending birds feeding small youngsters to the long races. I am not, as they will often eat well and thrive in the baskets, whilst others fret and won't eat.

The question of the condition in which birds work best is entirely a question of the individual. Each pigeon has its peculiar traits that want finding out; when once you have found them out make the best use you can of the knowledge.

Don't breed too early, don't breed too often.

Breed for intelligence and staying power. These are the most valuable attributes of the true racer.

In conclusion, I desire to thank those who have written from time to time suggesting additions or improvements to this Diary. I am only too pleased to consider these suggestions some of which have proved useful.

I look upon the sport of pigeon flying as a delightful hobby, for we ever find new phases cropping up, new records being made, and new theories enunciated. There is always a degree of uncertainty to be contended with, and it is just this uncertainty that lends charm to our sport. The uncertainty of who may be champion of to-morrow. Readers, if you but persevere your turn will most assuredly arrive.

HOW MANY NORTH ROADERS ARE TRAINED

by E H LULHAM

Having in a weak moment promised to write a short account of how I train, race and generally manage my birds, I must, I suppose, set about my task, but when I think of the many splendid articles written by *Squill*'s, Dr Barker and others, I do so with much hesitation, and wish it to be clearly understood that I address myself, only to the young fancier and beginner.

I will start with the first day of the year, assuming that the birds have made a perfect moult (without which, no bird should be put into serious training, or sooner or later it will mean an empty perch from this cause alone), and also that there are birds of all ages, from the yearling to the old warrior of ten years' standing, or flying, perhaps I should say.

With the New Year the birds intended for racing will have fully grown their last primary, and all are put on short commons until the end of the month, and turned out, cocks at 11 o'clock and hens at 2 o'clock, and allowed to take what exercise they like, usually from one to one and a half hours, each sex being fed on entering the loft by hand, and to the accompaniment of the call employed during the racing months.

By February the birds will have become hard, buoyant and on the light side, and it will be found necessary to daily cast a very careful eye over them; and those of three years, or who have flown 300 miles, should receive more food than their younger relations. Special care is given to birds of six or seven years and upwards, who after years of severe racing never incline to corpulence (would that we humans could say the same!).

During February birds are turned out twice a day, weather, of course, permitting, cocks at 8 and 12.30, hens at 10 and 2.30, and all are fed twice a day, but although they receive more than in January (when flying only once a day) are still kept to a more or less short allowance, according to the weather. With the 1st of March, hoppers containing peas are placed in the loft and birds are fed up, prior to being mated about the middle of the month.

The mating of the birds is the result of many weeks of careful thought and study of the physical characteristics of the pigeons, that I may reproduce as nearly as possible the lovely type of birds to be found in the loft of my friend Jurion – birds of wonderful intelligence, pluck and constitution, together with a beauty of outline and strength of body that is only possible to the real racer.

In nearly all cases where a bird is intended for the 500 or 600 mile race, it is mated to one which will not be trained until the old bird races are over, so that a good pigeon is not upset by the possible loss of its mate.

When paired, great care is taken to prevent fighting, and birds are trained to their nesting boxes, old birds being allowed to retain those occupied by them in previous years.

The loft is now always open, and until they have been sitting a fortnight the birds fly when and as much as they like, and have free access to both. When all have comfortably settled down to domestic life, they are again turned out to the loft at regular intervals, but on no account are birds made to leave their nests. In this way they are gradually made fit and hard for training, which begins as soon as the second round of eggs is laid, with a short toss of 3 miles. The youngsters are then ready to be removed to the young bird lofts, training begins in earnest, and birds are turned out three times a day, and with two or three days' interval receive tosses of 10, 18, 25, 37 and 60 miles, with the exception of some of the more experienced, who perhaps miss a stage or two, according to their condition and the races for which they are intended. "Old Frill', for instance, going to the 10, 25, 50 mile training stages, and then to the second race, Retford 135, Darlington 220 miles, and a 25 mile spin a few days before going to Lerwick 606 miles, feeding the first youngster he has been allowed to hatch, about 9 days old, and always carrying $9^3/_4$ to 10 *full flights*. Yearlings go to all training stages and races up to 250 miles when possible, and occasionally only is one picked out and sent to Banff 445 miles. Third season birds go to Berwick, Perth and Banff.

When birds are feeding their first round of youngsters they have peas and as much green food as they will eat, and a handful of mixture composed of tares, wheat, rice, dari and groats is placed in nesting boxes each night. After youngsters are removed, the old birds have two nippers, one of peas and one beans, two years old. The bath is allowed after each race during May, but is then forbidden until the long races are finished.

For many years I have had no sand in my loft, but each morning a spade full of burnt ballast is broken up to the size of buckwheat, which birds eat up very readily.

As soon as races are over birds are allowed to go to nest and rear one squeaker, those which have pleased me most, having one of their youngsters put under less successful birds, or yearlings which have had perhaps less chance to prove their merit. This is the time when birds want change of diet and very generous feeding, and as soon as this last squeaker is taken from them every effort is made to induce a good and steady body moult; a little small seed early in the morning, plenty of green food and a bath when they like. With October the sexes are separated and the hens are shut up, having little or no exercise until they are well over their body moult, while the cocks have an open loft, but are not encouraged to fly. The amount of forced exercise (if forced at all) birds should take during the training and racing depends much upon the situation of the loft, for if kept on the top of a house or stables, they will, if fit and well, put in enough flying on their own account; but if the loft is on the ground and shut in by trees or buildings, they must be turned out and made to fly for a given time.

Never shall I forget, whilst spending one of my many pleasant visits to my friend Jurion, sitting out in his garden and reading during a broiling summer's morning, no wind and intense heat, but his loft is high up on the top of the house, and to my great surprise the birds were continually flying off in batches of 15 or 20, taking wide sweeps round his house and settling, only to be off again in a few minutes. Had his birds been kept in a loft on the ground, I do not think they would have passed the morning in pleasurable exercise on such a day.

My experience with the Jurion blood may be interesting, because through the personal friendship and kindness of Messrs Jurion I have had many birds bred from their most noted birds, together with two wonderful stock hens of the same blood from my friend, Mr J W Jones of Sandringham, and have been able to try them in this country, much of my success being due to crosses from this blood. As young birds they do not in many cases seem as large and as well developed as many other strains, but my experience of the strain is that is a slow-growing strain, developing year by year, and after this development they show continual improvement; in fact, one finds frequently that the best birds of this blood are sometimes slow but reliable as young birds, as yearlings they show considerable improvement, and in their third and fourth year they are very courageous, developing great flying and staying powers, with more feather and bone, and seem to come on wonderfully with greater age; in fact, this, in my experience, is one of the traits of the Jurions, that they improve very considerably the longer you keep them, and moreover will race to a great age, maintaining their speed and staying powers. Bearing this in mind, fanciers who have the Jurion blood should not be disappointed if, in its first and second years, it does not fly quite as well as some of the more quickly developing strains; in fact, I know of no strain in my loft that improves so much year after year as the Jurions do.

During the months of November and December my birds are fed on a generous diet to encourage in every way the completion of the moult, which brings them to their most unpleasant period of short commons with the New Year. Above all things, it is important for the beginner to read some of the valuable articles before mentioned, think out carefully a line of treatment suitable to the environments of his birds, and apply it year in and year out. Success must follow.

THE SECRETS OF SUCCESS

by S P GRIFFITHS

In giving my short article under the above heading, I feel that I have stated something that is not quite an actual fact. Because really and truly there is nothing in it, at least nothing in the nature of secrecy. There is nothing in it the average fancier does not already know, and there is no earthly reason why one and all should not attain more or less equal success to myself.

All that is required is an average amount of brains and intelligence, a fair share of patience, birds of good strain and constitution, the happy knack of selection in the matings, strict attention to small things, and a natural love for pigeons and the sport generally.

The above combination, to put it briefly, is the sole and simple reason of my success, and I feel in my own mind that I have written and told all there is to tell. But, I suppose, if only for appearance' sake, I must go on and extend this simple epistle and inflict something already known.

Still, I will do my best to set down here exactly what I do myself and how I do it.

To begin with, it is absolutely essential to possess birds of undoubted physical capabilities. By that I mean sound constitution, good feather, and their general outward appearance the correct shape and build. It is not my intention to give an extended description of the word "pedigree", although I am quite aware to be successful one must have birds of good blood or strain. That, providing one uses judgment in buying, is not a difficult matter to obtain. But, to me, the word as generally applied, is a very much overrated asset. Too much is made of the past and not enough of the present. It is all very well for advertisers to give a glowing account of what the progenitors of the particular birds they advertise did years ago in Belgium or somewhere else. But the question is – at least this is what I should want to know if I was an intending purchaser – what is the strain doing today? At least, what is it doing for the particular individual who advertises it? If well, then that is a certain fact that the strain has not deteriorated. If, on the other hand, the seller is relying simply on the reputation of the founder or the successes of someone else, and has himself been a continual failure as a racer, then the best advice I can offer is to give it a "miss", because it is a certain fact there is something wrong somewhere and it does not require a great deal of time to ruin a strain of birds in the hands of a non-fancier.

I am sure fanciers would do much better from their matings generally if they paid more attention to the outward structure and general appearance of the particular birds they intend to couple together than simply relying on "pedigree". Do not run away with the idea that just because so-and-so mated a couple of pure Jimmy O'Goblins and produced a little "Gold Mine", all you have to do to attain success is to procure a pair of birds similarly bred, mate them together, never minding their outward appearance, and there you are. Oh! dear no, you will most likely find you are a long way off, and the best result will be a very poor counterfeit.

I think there is no harm in copying so-and-so, and good results might follow, providing the birds are structurally suited to one another. I trust I have made my meaning clear, because, personally, I attach the greatest possible importance to mating birds together that I consider structurally suited and to this more than anything else I attribute my success as a breeder. Another thing, I never mate two light coloured birds together.

It is the close and strict attention to small things that make you beat your neighbour more often in a season than he beats you. By small things I mean clean water three times on a summer day, instead of twice; feeding at regular intervals with the requisite quantity, cleanliness, observation, no delaying till tomorrow what really ought to be done today. For instance, in having a look at one of your best birds, you find a badly twisted flight quill. Go at once and dip it in boiling water, that will put it straight. Do not leave it until next day, for the chances are that by then it will be beyond repair.

Any keen fancier is observant enough to notice the least detail out of place in his own loft, to see at a glance any bird that may not be looking quite itself.

In my own lofts, at least where the racing birds are kept, no sand or anything else is used on the floors – just the plain boards. The reason I do not have anything put on the floors is because I think there is an entire absence of dust, and in consequence the birds carry more bloom. These floors are scraped over twice each day by "Blackie", a man who at the present time is quite an expert at the game, at least, he has almost scraped the entire floors away. No less a personage than "Squills" himself, when over at my loft, noticed the dilapidated condition of the floor, and in addition he observed that "Blackie" had muscles on his arms quite out of proportion to the other parts of his anatomy. "Ah!" thought he, "now I have discovered the secret of Griffiths' success. It's this continual scraping out". Well, that may not be everything but it certainly is one of the parts of the whole.

The *regime* my birds are subjected to during the year is this: They are separated from December to the end of March, and during this period they are fed mostly on barley, a few peas, beans, or cinq corn, just according to the weather temperature.

When mating once takes place, I afterwards let nature more or less take its course. The birds rear their first nest completely up and during this period they have practically the open hole. Of course they are back and fro to the fields a good deal, but I don't altogether dislike this, because for one thing it is an agreeable change for them, it tends to lower their condition, and although they are not quite fit in the early races, I think they receive the benefit later, and are at their very best when the long and hard work has to be done. When the hens lay their second eggs, I allow them to hatch out. They rear both youngsters for five or six days, then one is destroyed and they feed the other until they lay again. During the racing periods my birds have a free hole each day up to noon. They are then shut in and have a fly in the evening of from half-an-hour at the commencement extending to an hour or more as the long races come on. After the races are over each pair are allowed to rear one nest of youngsters.

My young birds that are intended for racing the year they are hatched are never allowed out with their parents, for the reason that if allowed to do so they develop the habit of going off to the fields. They will not do this if kept separate.

In choosing birds for particular races or more especially for pooling, the very best way to judge condition, I am firmly convinced of this, is to watch the bird or birds outside the loft. It is a fatal mistake to think you can tell when a bird is in tip-top form just by holding it in your hand. The oft-quoted phrase, "as hard as nails" is a misnomer, it might mean anything or nothing, and is more a guide to feather quality (tight and silky) than actual condition. Of course a bird's flesh must be solid. The times I watch my birds, when I am home, are these: The hens, when they come off their eggs in the morning and the cocks in the evening. A few minutes will soon convince you if your birds are all right and whether the birds you fancy most are worth having a "bit on".

Although I am away a good portion of the week, I try to put in all the time possible with the birds during the racing season, and on many occasions I am afraid I do this at the expense of my business. There is hardly a night that my man and I leave the birds before eight or half-past. I always feed them myself, in fact, on most occasions I do this when I am at home. When I have to be away they are fed according to instructions by a man, I am thankful to say, who will carry them out to the letter.

There is never a bird goes to any race during the season that I do not pick out and basket myself, not that I couldn't trust my man to do it, but it is one of the pleasures I look forward to. In fact, if I had nothing else to do, I should soon become equally proficient to "Blackie" at the scraping business.

In conclusion, let me say that I never handle or mess my birds about more than is absolutely necessary. I am very often in the loft, and in consequence the birds are fairly tame, but I never go peeping at hens that are sitting, or feeling if the eggs are alright, lifting youngsters out of the nest, or doing any little thing likely to annoy them. Birds do not like these unnecessary attentions, and if you have a loft full of "squabs" and never hear a sound, you can take it for granted they are coming on alright.

SUCCESSFUL MANAGEMENT, A FEW OF MY OWN OBSERVATIONS

by H G EDMONDSON

To start with, let it be clearly understood that it is not my desire or intention to pose as an authority on pigeons or their management, but receiving a most cordial invite to write a few lines for *Squills*, it is with pleasure I do so in recognition of the many tips I have picked up from "Food for Novices" that have been applicable to my own management.

For many years I have found it a wise policy to carefully read and re-read the grand and useful articles in *Squills* contributed by various fanciers, not to scan it over, discard it, and forget all about it, but to deduct therefrom any hint that appeals to my own management, and to apply it persistently, not try it for five minutes, and then toss it over.

My first advice, therefore is – read all the articles, learn them, and apply to your loft with persistent determination anything that appeals to your own way of management.

I don't pretend to know a lot yet, and hope to be able to keep on learning useful little wrinkles as long as I am in the fancy. It is just as a result of this willingness to learn that I attribute my many successes. The last seven years my birds have won hundreds of prizes, in fact it is an exception for them to be outside of them. In 1912 over 80 prizes were won – this in the very strongest of competition, and they include 18 firsts in races, 24 firsts in pools, 5 specials for averages etc. Therefore, heaping together the total firsts won, as I have noticed many do, the total first prizes reach about 45 – these including 1sts in Federation and Championship. At the "fall" of one season, with myself begins the guarding for the welfare of the next. At this time of year fanciers find themselves with a loft containing valuable birds that have valiantly come through all the races, and usually in the tough rubbery state of muscle brought about by racing condition. Races all finished, therefore, a general reparation to a more natural condition is sought after. A free hole is given them at this period, quietness reigns throughout the lofts, and they are let sit peaceably on a final round of eggs. A few may be allowed to rear, others sit till tired, according to disposition of the bird. Whilst this period of quietness and peace proceeds, there is also a change in diet started. Having been worked up to concert pitch on solid food, such as peas, beans etc, and exercise, they are now got on to a diet inclined to ease them down, a few handfuls of wheat, barley and maize is introduced with a little linseed in the mornings. Greenmeat is indispensable at this period. The birds now begin to tire of their bad eggs, and the other few have finished rearing, and a nice moult is beginning to make itself evident. The desire to nest ceases as the nest pans are removed at once after they leave eggs etc, in another fortnight or so they are separated entirely.

They are moulted on a good mixture containing a fair amount of barley and wheat. Every morning they get a few handfuls of linseed. They get a free hole in turns, the cocks one day, hens the next. No fads, plenty of fielding, bath and greenmeat, and they are fed just sparely if I find they have returned from fields with fair good crops.

Gradually, the peas etc, are lessened, and more barley substituted, till at last I have them on their staple winter food, barley, with a mere handful of peas.

I believe there is no evil so prevalent and common amongst fanciers as over-feeding. Neither at moulting time (which I am aware is contrary to lots of sound opinions), or any other time, do I overfeed. This is a choice which rests with every fancier himself. I am confident six times as many birds are ruined through over-feeding as under-feeding. After all the grain is cleared from the fields this winter, 1 ounce per bird per day of good sound corn is sufficient. In the racing season they get more, but are always made to clear up all and look round for a bit more.

I have many a time given them a complete day's rest from food, and it is surprising how it brightens them up, and helps to bring them in good colour. On such occasions they are just fed very sparely the following day, the third day are back on ordinary rations, looking deeper in colour, brighter in eye and cere.

I have particularly noticed that where birds are overfed, and are raced to eggs there is a great tendency to go "stale" for the rest of season. A well-nourished vigorous lively bird you must have, but over-feeding and over-taxing of all the inner organs will never produce it. By the time spring is here my birds are handling very like the ordinary wild birds, not like heavy pie meat, but hardy light and buoyant. A little on the lean side certainly, but clean and healthy inside.

This is where, in my opinion, the man who shows is at a discount in racing season. With the inner organs of my birds clean and healthy, it is amazing how beautiful they look and how they put on healthy flesh to be worked up into muscle, as soon as the diet is altered and the barley dropped out, increasing the peas. These form the staple food from March to August, and in the racing season a little seed and rice is given.

I pair up about middle of March, and prefer setting them away at varying intervals from March 15th to April 1st, thus set away for the season it is much easier having something always "hot" instead of a whole loft falling on favourite form at the same time.

I "play" with them according to their own disposition, sitting, feeding, or whatever it may be, and I pool on two things only, viz, favourite form and colour depth

The latter I consider perhaps the most important of all factors as to when a bird is at top form. Treated on Nature's lines right through the winter, every bird will be found to come on into magnificent natural top form in summer. Some come on sooner than others, and as each comes to its zenith, and its colour stands out beautifully clear and distinct, you will find eyes bright, cere clean, and your bird in top form. Given its favourite condition, sitting or whatever it may be, whilst this form lasts you can pool up with full confidence. For type of bird, I have no fads. So long as it is an easy moving bit of machinery, handling perfectly sound, it will do. I prefer a few different types in my loft, so that I have generally something to fit in with the prevailing conditions on the race day.

I never pool or place confidence in birds, because they are consistent racers or old favourites. They are all on a footing, and as each comes to favourite condition and form, it is expected to do its bit; if it fails a time or two I try it in other conditions before condemning it as a laggard. In nine consecutive races in the past season I won heaps of pools, and this with strong clubs sending 400 to 490 birds per race, also twice 1st in Combine Pools, 6d, 1s, 2s6d, etc, with very strong competition.

Breeding season is the only time my birds ever see the hopper, This is put in overnight, filled with good maples, and removed at 8.30 am. After that they get hand-fed and a free hole. I am not an advocate of forced flying. If mine put in 15 to 20 minutes' good free flying twice per day, viz, 6.30 am, and at evening I am quite content, but if they had not a free hole then I would give them more.

No bird, if it is a game hard tryer, should be flogged along every week. That's the road to make "plodders" of them, not winners.

The strength, build, and constitution of your bird, and the races it has already come through that season, should be the deciding factor as to the amount of work you expect it to do in winning style.

Some are naturally built for more work than others, and it's only a fool's game to try and get a bird to race hard and fast by sickening it. There are so many "favourite conditions" to get birds to win in that it is no use detailing them, but one of the writer's favourite winners is a bit out of the ordinary, and has never yet failed me when paired to a fresh hen four to five days. It's a condition so easily got; just remove his hen and leave a few spare ones about, and it also enables you to get two crosses from him by different hens in the one season. It does not follow that he has to rear them himself.

Trapping is nowadays a winning factor. Let me impress that strongly on the beginner. It's enough to give a chap the "horrors" to see the many preposterous arrangements for trapping nowadays, and quite enough in themselves to spoil a bird once and forever. Peeping round corners, catching birds in traps, and unusual rattling of corn tins won't do.

Out of 60 birds in my lofts I can pick any one of them up, just start right away, feed them gently at the loft door, pick one gently up now and again, and when race day comes just walk up bold and feed them in the same way. I have no trap; they are down and in the loft like winking, and I can pick them up, clock them, and have a good trapper left again for next time. A tit-bit in the shape of ordinary rice, which is generally Bassein, and a little seed is always given on return from races or training tosses.

In YB races I have scored also very heavily, and the management of these is simplicity itself. Once on the wing they love exercise, get them away scouring round the country; and after half-an-hour's good flying, when they pitch, immediately feed them in. Regularity in this detail will ensure you youngsters with which prompt trapping will become habit, and immediately they home you have them in. They never get a free hole with me. I feed youngsters on Maples' with just a little mixture now and again. Training stages, old birds old birds of all ages Durham, 10 miles; Bradbury, 20; Stockton, 30; Thirsk, 50 (twice); then the first races, 100 miles. YB stages also late breds, which I always train with them, Plawsworth, 7 miles; Durham, 10; Bradbury, 20 (twice); Northallerton, 40; Thirsk, 50; and from here some of them go to Beningboro, 66, and the others to Selby (first race) 82. After the 50 mile stage I always divide them, lifting half to each of above places.

I never chance all my lot away at one time. I like to make a pleasure of training, and not kill myself, carrying the lot at one time. Just a nice handy basket – packed nice and smart with a few chips – makes it a pleasure to both birds and owner.

If there is one fad that I have above others it is for fresh air.

The entire front of my range of lofts are absolutely open from top to bottom, no woodwork, but entirely wire bar work exactly the same as common breeding canary cage. They are situated in one of the most exposed situations possible to find in the whole of England. They get the full blasts of winter, and the entire sunshine of summer. Being in such an exposed position wood lofts I found could not stand it, and quite recently I have been obliged to have them rebuilt. The back is an 18 inch stone wall, the roof solid cement, with boarding beneath it; the two ends are brick walls, and the floor 1 inch flooring deals. The front is absolutely open, and always has been so. My nest boxes run right along the back of loft, and have no fronts fixed on them, and never have had. The birds are quite used to it, and it is not inclined to make them so nervous as the lofts that have fronts on. With open boxes the birds get used to you, sit still when you approach them, and you can have a word with them when you go in and out of the loft every time. From experience I vastly prefer them. I know some readers will say it's enough to kill anything with about 1 oz or 1¼ oz of food per day in winter, and an entirely open loft with the cruellest of hill top winter blasts and snow driving in. However, there it is, and it's plain facts. The rain, hail and snow does get in, nothing to stop it, but so does plenty of fresh, ever-circulating air, and the loft floor is generally much dryer than the horrible stuffy damp humid affair got up on scientific ideas. The birds can always get a good dry airy perch or box, and the splendid recompense I get in the summer time, seeing the whole row of nest boxes with birds on eggs, youngsters in nest, etc, basking and enjoying the sunshine, whilst in their nests, and the abundance of fresh air they get, amply repays me for the loft getting a nice wash out now and again in the winter. In closing this humble article, and wishing all readers greater enjoyment from their hobby in 1913, and the future I cannot conclude without saying I have never been a follower of any "flash-in-the-pan" class of fancier, preferring to always follow the winning fancier year after year, and also no matter what the strain of birds stick to that which suits your own management, and make it the base of the loft.

HOW THE GREAT ROME RACE WAS WON

by C H HUDSON, the Winner

'KING OF ROME'

When I saw the announcement in *The Racing Pigeon* inviting fanciers to compete from Rome, I thought I should like to send, as I think English pigeons, like English racehorses are as good as it is possible to get. I knew my bird was a very experienced old stager and I knew he would do the task, or die in the attempt. Let me add, I do not think it was the distance that caused so many to fail. If Rome had been in the line of flight with Bordeaux and Dax I am sure we should have had many more birds home, and had we had good weather, I feel sure we should have done the distance in much less time.

'King of Rome' is on the small side, and was bred from a yearling cock, and a two year old hen. He was from the first nest. I like first nest birds, and I like to breed my racers from birds that are not too old. If I have a good old bird I prefer their grandchildren for racing as I have not found old birds breed good racers for me. But they breed champion stock birds.

Last season I paired 'King of Rome' on March 21st. I took the first pair of eggs away from him after he had sat 10 days. In the second nest I let him rear one youngster. He had been sitting 17 days in the third nest when I sent him to Rome. He was exercised around home for 30 minutes night and morning until after he had flown Guernsey. After this distance I gave him one hour night and morning until five days before he went to Rome, and during the last five days he was turned out of the loft three times a day, and flew as long as he wished. The tosses he had before he went to Rome were 9 miles, 20 miles,

30 miles, 59 miles, Templecombe 139 miles, Guernsey 249 miles, winning 1st and special and all pools, then Rome 1,001 miles. He was fed on the best maple peas until Templecombe was reached, then half beans and peas for the Guernsey race, and all beans after this until he went to Rome. After he homed from Rome, he was fed on canary seed for a few days, then he had rice and canary. After he had been home seven days I let him have a few peas, and then all peas as I saw he was all right. I believe this careful feeding has had a great deal to do with him having such a real good moult, as he cast his last flight feather on November 3rd, and has only one fret-marked feather. When he went to Rome he had one new flight, one half grown, and eight old flights. When he homed he had a full wing, eight old flight and two new, not having cast a flight while he was away.

Prior to 1913 'King of Rome' flew as follows: 1907, 50 miles; 1908, Weymouth, 165 miles; 1909, Guernsey, 249 miles; 1910, Rennes, 333 miles; 1911, Guernsey, and Bordeaux, 561 miles; 1912, Guernsey, winning 1st and special, and Dax, 645 miles, winning 15th Midland Section.

This gives a fairly full record of the preparation and career of 'King of Rome' during 1913. As to my own career as a fancier I may say my father was a short distance racer when I went to school at Nottingham where he won a great many prizes. I was always very fond of the pigeons and many a time I begged a half day off from school to take the birds for a toss. In 1887 we left Nottingham, and came to Derby, bringing our pigeons with us. We settled some of them, and began to train them. In this short distance racing we were great winners right up till the time I gave up this kind of sport.

My father died in 1901, and then all the birds were brought to my home. In 1903 I used to chat with a Mr G Ray of Derby, on long distance racing, and I began to think I should like it better than short distance as we had won from all the two mile spins, and it was getting a bit stale. So I put the matter to my brother, telling him I was going to try long distance and inviting him to stand me in. But says he "No long distance for me". I might say he was just about pairing up himself, and had not much time for pigeons, anyhow he left the matter entirely to me. The result was all the short distance birds had to go, and I had some old favourites I did not like to part with, I can tell you.

In 1904 I began with long distance. I joined the Derby Homing Society, and I won 1st prize Mangotsfield, in my first season, with a mealy youngster, bred from a pair of birds I had from a fancier in London. Each year since I have had my share of prizes and specials.

Experience has taught me valuable lessons and in buying my stock birds I try to buy the best my pocket will allow. I always buy one or two each season to cross with my own. As a rule I do not separate my birds till the New Year. But the young hens I keep in the young bird loft as they are more trouble than older birds if kept together. I keep one or two quite young cocks with them, as I think I am not so likely to have a flyaway as if separated. I take all nest boxes out of the loft after breeding is over, leaving the birds only the box perches. I clean out each day, give clean water twice a day and in warm weather three times a day. I always use a little sand under the perches for the droppings. I pair my stock birds together early in February and my racers in March.

My works always brings me home at daybreak, and upon arrival go into the lofts, and when the birds are rearing youngsters I give them some rice, followed by peas, as I like an early morning feed for my youngsters. I am very keen on maples peas for rearing youngsters. As we get to the Channel I add a few beans and when we get across I give them beans only. When training I like to give my birds fair room in the baskets with some cut straw at the bottom for litter, and when I get to the station to send them away for a toss I give them a little rice and canary seed to pick at as I think it keeps them from fighting. When they home from a toss or race, they have a light feed of rice and canary seed and you should see them trap. I can strongly advise fanciers that have bad trappers to try this. They will have no bad trappers. I think bad trappers are a fancier's own fault. I have never had one and I hope I never shall. I do not give my birds seed at any other time.

I like careful training up to the first race with plenty of tosses on nice days if possible for yearling and youngsters. I give as many single up tosses as I can possibly find time for, using my bike for the first 10 miles. I give them a bath twice a week. Watercress or lettuce on a Sunday when I have been racing on Saturday. Health grit each day in the breeding season, and once a week in the off-season. I do not give them any the day they go to a race. I like a little barley for winter feeding after the birds have done their moult till about 14 days before I pair up when they have all peas. When very cold they have peas and barley.

The 'King of Rome' has always been on the same shelf when breeding, and he has the same box perch when he is not sitting. If any other bird is on his perch he fights until he gets it. No other will do, but that one and I have never seen him beaten for his shelf or his perch.

When sending my birds away to a long race I shut the shelves up so no other birds can take possession, each shelf is then ready for its owner when it homes without having a fight for it. I think many a good racer is spoilt through having to fight for his nest box after a long journey.

I would like just to warn young fanciers always to see the birds have plenty of clean water as it is useless buying the best of corn, and letting the water be dirty. Also in the winter after a sharp frost see that your water is not ice. If you see it is going to be very sharp the best plan after the birds have been fed and watered is to empty the water pot. It is better than having to thaw the ice in the morning.

Just a final word, I was very pleased to win Mr Doll's cup. That there was £20 left after all expenses of the race had been paid, also came as a pleasant surprise. Mr Doll's evidently is a master hand at organising. I wish him every success with the race from Miranda and good health to see it through. I have always been, and I hope I always shall be a pigeon fancier, and if this article should assist any fancier striving in the initial stages of the sport, I shall be more than pleased.

HOW I MANAGE TO CONDITION MY PIGEONS

by W SAVAGE

One of the most necessary items to ensure successful racing is a clean loft, plenty of light and fresh air. My loft is a wooden structure, 18 feet long, 8 feet wide, 7½ feet high at back, 6½ feet at front, built on brick pillars 18 inches off the ground. The whole front is wire netting from roof to floor and end to end; it also faces north east. To get all the sun in the loft possible, I have two large skylights fixed in the roof, viz 8 feet, and are 3 feet 6 inches wide thus allowing the sun to flood the loft the greater part of the day.

It makes no difference what the weather is, I never close my loft in any way. It is very cold in winter, but the birds take no harm. I very very rarely have a sick pigeon.

The birds are cleaned out twice a week in winter, and a good layer of sharp sand is put down each time. My loft is built with partitions, to fold back; I can close them and make three lofts. The trap is divided in centre, in order that I can separate birds and liberate, alternately, the cocks and hens. I am convinced you cannot have too much light, fresh air and sun, if birds are to keep fit and fly well.

The closer you can keep to the birds' natural state the better health they enjoy. They can stand any amount of cold; no coddling at any price. So much for the loft.

GENERAL MANAGEMENT

As to the management of the birds.

My birds are separated at the end of September, and have now finished a good moult. My plan is to let the cocks out one day and hens the next, allowing them their freedom as long as they like to be out. The bath is kept full before them from 1st January to December 31st, the water being changed three or four times a week (in the summer oftener); the birds will bath in the coldest weather in winter if water is kept clean.

After my birds are separated they are fed once a day only. Always just before dusk and allowed all they can eat, the food consisting of beans one day and peas next.

The birds keep on the poor side by this treatment, but my hens put in a lot of flying continually dashing off the house and flying freely while at liberty.

My cocks never fly so well as the hens when separated. This method of feeding is continued up to the first week in March, no matter what the weather is like. About the end of the first week in March they are put on two meals a day, about 9am and as late in the evening as they can see to pick it up. I always give a regular supply of grit (Record).

I pair my birds the middle of March. When paired I let them have a free entrance to loft (thus preventing many clear eggs, not disturbing them to fly until the youngsters are ready to remove to the young bird loft, but as soon as the first nest of youngsters are removed I then commence in earnest to get the old birds ready for racing. The loft is open in the morning, birds are turned out and left to themselves until midday. The loft is then closed, and the birds remain inside until evening, late enough to give them the fly and feed them before dark.

I turn every bird out of the loft; hens and cocks start their flying, and I begin by making sure they do not drop before they have done 20 minutes; they soon increase this to 35 or 40 minutes. This I continue right through the racing season, always timing them.

When the time is up I shake the corn in a box and open the door of the loft; it is surprising how some of my old birds know when they have done full time, and drop immediately they see me with the corn. I have birds in my loft that have had this treatment for 7, 8 and 9 years. I always fly them before feeding them.

Training Birds

I have found it useless to try and get birds to fly freely if you have other birds sitting about outside. All must fly or none.

As the distance of the races increase and evenings get lighter, I turn them out later. I think my flying the birds late in the evening has something to do with getting my birds home up to a late hour on hard race days. Having got them fit, the training commences. As a matter of fact I have little trouble to get them into racing trim after the winter treatment. I give them single-up to 5 miles, then entrain them to 15, 25, 40 and 60 miles. I do not allow the birds too long rest before getting them racing.

I like to send all yearlings to the first race; older birds, after doing the 60 miles, I give them Retford (120), Newcastle (240), then any race on the day, but yearlings are more frequently sent, and do more racing along the line.

I am convinced that holdovers will kill the best pigeons in the world, and I have had my worst knock-outs after birds have been held up some days in the baskets, and my best wins have been on so-called bad days, when the velocity has been low and birds liberated at time stated. The better condition a bird is in the more severely is he punished if made a prisoner too long.

The Age I Like Best for Good Work

It does not matter at what age up to 7 or 8 years old, I have found hens are capable of putting up the performance of their lives up to this age.

Cocks I do not think keep their speed as late in life. My favourite age for cocks is 3 years (the age of my Perth winner this year); after 4 years I think they slow up in my own experience, in fact, I cannot remember winning at any time with an older cock than 4 years.

I am rather partial to a yearling cock up to 350 miles. I have won some hard races with them at this distance, but you want to know your bird. Yearling cocks I think when first paired become excited and agitated, and with some this excited state lasts longer than with others, especially when parted from their hens, but time and experience will cure this.

I have found it a good plan to get yearlings paired no later than March 12th or 14th, and my plan is to let yearlings hatch and leave the training of them as late as you reasonably can, in order that it may be in a more settled state of mind, when they will work all right, but if paired late and you start training them before they are settled down and you lose them at distances that make you wonder what is amiss.

I have had a number of yearlings fly Banff, 430 miles, treated in this way, in fact, I sent three this year and had all three home, clocking two of them in even time. I think cocks need more careful treating than hens, but I treat both alike.

I never hatch more than one pair of youngsters from any birds I am flying, but I like all to hatch and rear both the youngsters if both hatch, and I look upon eggs as a bad omen for the racing season.

The number of birds I keep are 24 pairs, they are fed by hand all the year round.

My favourite number of birds to send to a race is seven, this being the number on three occasions I sent to Perth, and each time winning LNR Fed. Twice I sent seven to Banff 1st Combine, also 2nd Combine, so you will agree with me it is my lucky number.

The strain of bird I fly – the foundation of my loft was Jurion (direct), N Barker (direct), Bonami-Jones (Sandringham).

I am fortunate as a rule and do not lose many old birds, although I keep them going right up to Lerwick. I do not encourage birds in my loft unless I know all about them there is to know.

FAVOURITE CONDITIONS FOR RACING

It has always brought me the most success to race my birds while sitting 8 to 14 days, this condition I prefer both for cocks as well as hens, and always endeavour to arrange this.

The condition of the wing is not such an important matter with me as with many fanciers as long as I satisfy myself that only one flight in each wing is being moulted at the same time and new flights are growing healthy, that is, as far as I trouble. I do not believe in trying to check the flights falling in any way, you are only preventing Nature taking its course.

Small seeds should be used very sparingly during racing. I prefer good rice, and think it best for the birds, but do not use even this too freely.

I hope I have made myself clear as regards treatment of yearling cocks. With careful handling I have found they do their best work after being mated 12 or 14 weeks, and will fly very gamely when comfortably setting about 10 days.

When my birds have returned from a hard race I do not trouble them to fly the next day, but allow them their liberty, so they can bathe, but in a couple of days they are doing their evening fly regularly till they go again.

I do not value a bird so much for a 1,400 or 1,500 yards velocity win. I prefer to breed from birds that have flown at a velocity of about 800 yards for 12 hours and won.

I attribute my repeated success to mating my gamest birds together, and not over-rating a win with a fast and easy velocity.

At the end of the racing season the old birds are allowed to rest, and not given any forced flying.

I have endeavoured to put on record the treatment of my birds for the last 18 years and trust they may be of some benefit to younger members of the Fancy.

THE IMPORTANCE OF CONDITION

by J WORMALD

Mr Squills has again invited me to write an article for his Diary. Some years ago I had no compunction about doing this, but now the sport has become so scientific that I have grave doubts as to the wisdom of accepting his invitation! Not only because my methods of racing lack the scientific finish of the really up-to-date fancier, but also because of this lack I may be accused of not practising that which I preach. Let me then say at once, that to practice what I preach can be summed up in one word – condition, which can only be carried to a really successful issue by the most careful and continual personal attention to your birds. This, I am afraid I do not always give; or by providing a professional loftman to look after them – which I also do not do.

I feel that I dare not attempt to explain how birds ought to be conditioned. This has so often been done, and opinions differ so much as to the best methods, that I shall confine myself to emphasizing the great importance of condition.

We have only to think of certain fanciers, who shall be nameless, who have spent endless sums on buying the very best material to work with, and who failed year after year. Then they procure a really good loftman to look after and condition their birds, and they at once begin to win. The same birds, the same old strains – with perhaps a few more added – and they now win. Why do they win? Condition. The fact is that pigeon racing is becoming so very scientific that those fanciers who only go in for it as a hobby have no earthly chance against the serious fancier, who misses no point, or the professional fancier, who spares no expense. I use the word professional advisedly, because they are becoming more plentiful, especially in those parts of the country that happen to be on what I will term the 'line of flight', and consequently the places where the money goes to.

Now I am not writing this with any wish to belittle what I have termed the professional fancier; possibly many will say I have no right to use the term, and it is obviously an absurdity to say that those who are rich enough should not have a loftman of their own. What I want to emphasise is this: that these men win, and they win on condition, and if you are a competitor, unless you are prepared to give the same amount of trouble and scientific training to your birds you will be beaten. Don't think they have a better quality of birds than you have; in nine cases out of ten in these days that is all rot. The birds are all right; it is the handling that is wrong. Possibly not the handling of the few weeks during the racing season – this may be scientific and good. It is the handling for the full twelve months of the year. What an awful bore! What a grind! I can hear many of the readers of this article say. Winning in these days can only be done by hard work. To the true fancier it is neither a bore nor a grind; to me, I have reluctantly to admit it is both. Oh, for the days of twenty years ago, when one could win without all this trouble.

However, here we are, scientific pigeon racing up-to-date, and likely to get scientifically worse before it gets better. I think I shall start an Amateurs' Club, but what the definition of an amateur is to be, goodness only knows! Perhaps someone will oblige through the medium of the Press.

Another consideration which has just struck me, or was I struck by it some years ago? If any fancier, after reading this article, decides to become what I have most improperly termed a 'professional', let him pause before going to the expense of engaging a professional loftman, and ask himself these questions: "Am I flying my birds on a route that is suitable for the part of the country in which I reside?" and, if so, "Am I on the line of flight?" If he can answer both questions in the affirmative, don't hesitate to either put in the extra work yourself, if capable; if not, pay for the best man you can possibly get. If the questions must be answered in the negative, go to no extra expense until you have got your birds racing on a route that will suit them.

If you are off the line of flight you may not do so much winning, but condition will tell. Don't be put off by people who say the old route is all right, and that the birds are at fault. I am quite positively certain that a route which suits one side of this country does not always suit the other, and when you have really tumbled to it after many losses and much disappointment, do not go on the lines of the voter who voted "Yaller because his father voted Yaller and his grandfather afore 'im", and so on. Get a move on, and try something else.

There are a great many other points about this scientific modern pigoen racing which bother me no end, besides the conditioning. The individualities of each particular bird is really too sickening to an amateur. No two birds seem to produce their best form under the same conditions; one races best to youngsters, another on eggs, and so on, and so on. Then even the experts differ, it having been lately asserted that our pigeons cannot be described as having intelligence. I have been talking this over with my American friend fancier, and he was of opinion that if his pigeons lacked intelligence some of them were 'mighty cute'. With this I agree. Some birds I have owned have stood out head and shoulders above the others. The 'cutest' bird I ever had was my old red chequer cock, 4025, who, in 1912, when returning from Bordeaux, had the misfortune to tumble down a chimney, but he selected that belonging to a seminary for young ladies. Talk about no intelligence!

Then there are the temperamental peculiarities of different birds, which require no end of study and thought to master, and which, if not properly mastered, may cause failure in getting the best out of a bird, even though you may have got him physically fit, bad trapping being one of the most common forms, and very often the fancier himself being responsible for a bird, in his initial efforts at racing, getting into this bad habit, although I am told there are strains that have habitually proved themselves unreliable. Be this as it may, I am of opinion that it is a temperamental nervousness, which, even if a strain possesses it, need not necessarily follow that every member of that strain should turn out to be a bad trapper; but I do think it requires management when the bird is young to prevent it. Some birds are so phlegmatic that nothing seems to upset them, and I am confident that breeders have not paid sufficient attention to temperament when mating up their birds. I do not say that the nervous, flighty pigeon should not be bred from, as some of the finest racers are very often the most highly strung, but to my mind the faults in temperament of a bird must receive just as much consideration, when mating, as the faults in physical proportion; and a strain that has running through it a calm and placid nature must score heavily over one that is excitable and jumpy, the latter being, of course, much more difficult to train and taking longer to recover from a race, but at the same time may put up a brilliant performance.

There is also the case of birds not bred in your loft, that you have had given or bought. I have never been able to quite make up my mind about the reason, but in my experience the bulk of these birds, however fit you may get them, do not race as well as one's own strain; in fact, I go further, and say that I lose them – not because they are bad birds, but because they require extreme care until they are in their third season. When I moved to my present abode last year I had three spare cocks in the spring, and owing to the war, I let the 1914 birds pair as they liked, and amongst them were three cocks which I had bought. These were not late hatched, all strong birds, and yet these three cocks were the ones left unpaired, all my own strain having mated themselves up. This may, of course, have been merely a coincidence, and I give it for what it is worth, but I myself do not quite think it was coincidence only. I think there is some disturbing influence which assails a bird of a different strain that is imported into your loft, and I am getting to be more strongly of opinion that a bird not bred in the loft should be bred from before asking it any very great question as regards training. Of course, there are plenty of instances to prove that birds bred out of your loft do race, and race well, but if the truth was known it is probable far more are lost.

In conclusion, let me assure my readers that when next we race, whether you race as amateur or professional, unless your birds are really fit you will not win. Condition will become more and more the dominant factor in winning races, and it really looks as if in years to come that a 'pigeon trainer' will be a recognised profession.

M. EVANGELISTI,

808, SEVEN SISTERS ROAD, TUTTENHAM, LONDON.

THE birds in my loft have flown successfully every season during the past 21 years, never failing each season to win one or more first prizes in both young and old bird races at all distances and in all kinds of weather, up to Lerwick, Shetland Isles (1st prize 1897), 600 miles, north ; and Bordeaux, 460 miles, and Marennes, 400 miles, south ; winning 11th prize from the latter place in the National race, 1905, 2277 birds competing.

I AM PREPARED TO BOOK A . . . .
LIMITED NUMBER OF YOUNG BIRDS

Price of squeakers from Marconi, flown Shetland ; from the Lord Mayor's Cup winner, flown Shetland ; from the Banff Mealy ; and from Holbein, winner of 11th National, **£5** each. From other pairs of the same strains, **£1** each. The principal strains in the loft are Pletinckx, Jurion and Vekemans.

A deposit of 5s. per squeaker will be required with each order.

AN OLD FANCIER'S CHAT WITH THE YOUNGER GENERATION OF FANCIERS

by JOHN W LOGAN

"At a time like the the present, it would do much to keep interest in our sport alive if some of our old fanciers would from time to time give the younger generation a hint or two."

Inspired by the above sentence in the "Racing Pigeon" of November 25th last, I will endeavour to respond, prefacing my remarks by saying that I shall write only as dictated by experience gained in the breeding, training and racing of the racing pigeon for the past fifty years.

That is to say, that the advice or hints which I shall give to the younger generation as to what I think they should do or should not do in connection with our sport, will be founded, not upon hearsay or theory, but upon what has been grounded into me by that cruel, though in the long run beneficent task master, bitter experience.

There is one subject in connection with our sport about which the majority of us find it difficult to give a satisfactory answer to the enquiring mind of the younger generation.

It is this: What is the power that enables our birds to return home from such long distances and in such quick time?

I believe I am correct in saying that it is the pretty general opinion of people who have never tried to train homing pigeons, that those birds are possessed of an unerring instinct whereby they are enabled when liberated at almost any distance from home, to immediately start off in the right direction for that home.

I have not the slightest hesitation in saying to the younger generation of pigeon fanciers that they will indeed be leaning upon a broken reed if they placed their faith in the belief that pigeons are possessed of any such unerring instinct as will enable them to find their way home with certainty, unless they have been most carefully trained, educated and conditioned.

The fondness for home which all animals have is, I believe, accentuated in the racing pigeon; but this fondness for home, for its own particular nest-box or perch, varies greatly amongst the various birds forming a stud of racing pigeons.

There are some birds that will not bear to be shifted from one nesting box to another, to say nothing of being moved from one adjoining loft to another, whereas there are others who do not seem to mind, if they are made comfortable in their new quarters.

But this great fondness for home is not what is meant and generally understood by "instinct", for that is a power we are told that does not vary in individuals of the same species.

Then what is this so-called "Homing Instinct"?

Let me answer this question in the words of the late Mr Tegetmeier, FZS, a well-known Zoologist, who made a life-long study of the homing or racing pigeon.

"The ability of the homing pigeon to find its home is usually attributed to some mysterious power or instinct that these birds possess, *that is not possessed by other varieties of pigeons*, but I believe that 'instinct' has nothing whatsoever to do with the homing power, but that the birds find their way solely by observation, and I ground my belief on the following facts:–

"(1) Any peculiar instinct, such as that of nest building, power of migration, etc, bestowed on any species, is equally bestowed on all individuals of that species, and not on a few only, thus all swallows migrate, but all pigeons do not return from 100 miles distance.

"(2) Instinct is the same in all cases. All swallows fly south in the autumn, but the homing pigeon can return from north, south, east or west, a variation in action that is incompatible with the notion of unreasoning instinct.

"(3) Pigeons must be regularly trained by stages or the best birds will be lost.

"(4) The best birds will refuse to fly in a fog".

And he concludes by saying: "It is much easier to cut short the question of the homing faculty of pigeons and call it instinct, than to investigate the facts of the phenomena.

I fancy I can hear the novice saying to me, "If this question of instinct is so perfectly clear as it seems to be to Mr Tegetmeier, namely, that it is education and observation and condition that alone gives the homing pigeon the power to do these long distances, why is it that there are many fanciers who still hold that it is mainly instinct that guides our birds in doing the extraordinary distances they constantly do?"

Let us therefore examine a few of the arguments generally used by those who support the "instinct theory".

Supporters of the instinct theory declare they are unable to believe that sight alone can enable a bird to find his way home if liberated 100 miles from the last point he was tossed at, and they scoff at the bare idea that a bird that has been carefully trained, say 310 miles, would be able to negotiate 470 miles, that is 160 miles further in the same direction, unless aided by unerring instinct.

A very able and informed correspondent who appears to lean to the instinct theory, in the "Racing Pigeon" of November 18th, 1916, quoted the following definition of instinct:–

"Instinct is an impulse to a particular kind of action which the being needs to perform as an individual, but which it could not possibly learn to perform before it needs to act; as a general term it includes all original impulses and that apparent knowledge and skill which all animals have without experience."

He admits this definition sounds sufficiently lucid till it is tried to make it fit our particular subject, and in that I certainly agree with.

Then he goes on to say: "My own idea is that homing pigeons have a sense of direction, how it has been acquired need not be entered into, as we know they have the faculty of finding their way".

Surely it would be more helpful to enter into the question of how the homing pigeon has acquired this sense of direction, and this is exactly what I propose to make a humble effort to do, though I admit it is much easier to cut short the question of the homing faculty and call it instinct, than to investigate the facts.

I propose, therefore, to ask the younger generation of fanciers to follow me while I try to show the scheme of education by training a homing pigeon gets at the hands of a man to equip him for the work he has to do.

The practice as regards the training of our birds varies slightly in method, but I shall not be far wrong if I say that 99 fanciers out of 100 start by training the young birds of the season by very easy stages.

Let me illustrate what I mean by "easy stages", by giving the practice of several well-known and very successful fanciers selected by me for this purpose of illustration because they have in their articles in Squills' Diary enabled me to do so by giving the particulars as to distances and methods that we are seeking.

Dr Barker, of Clitheroe, says in the 1905 Diary, that the tosses he gives his young birds up to 100 miles are 1, 2, 4, and 7 miles in a direct line for the race points; then after 7 miles his birds are sent 12, 20, 35, 55, 75 and 100 miles.

Mr Toft, of Liverpool, in the 1901 Diary, says: "I train youngsters the year they are hatched 130 miles, the second year up to 200 miles, the third year up to 400 miles, the fourth year any distance within reason".

Mr J L Baker, of Sedgley, in the 1902 Diary, says: "I personally hold the opinion that youngsters the year they are hatched cannot have too many tosses at the shorter stages (singly up preferred) as this gives them a good knowledge of the country over which they finish their races, and also teaches them to rely on their own intelligence".

Mr John Wones, in the 1909 Diary, gives his methods of training young birds the year they are hatched as follows: "For the first 10 miles I toss them myself, giving them about six liberations up to this distance, and then about four liberations more from the 10-mile stage, as many of them single-up tosses as I have patience to wait for; after this I take them four miles further on, afterwards giving them two tosses at 30 and 40 miles; after which they go with the club birds".

We may safely assume, therefore, that young birds the year they are hatched have had a most careful course of training for the first 75 to 100 miles from their loft in the direction for the race points from which they will have to compete.

Now I propose to follow the young bird in his future career, and as I have raced my birds from my present home for the past 41 seasons, I will use by way of illustration, the race points flown by our local clubs, down to the Channel edge, and after that, with slight modification, the race points used by the vast army of fanciers in the Midlands, Staffordshire, Cheshire and Lancashire.

We shall not be far wide of the mark if we say that the majority of youngsters are put into training at three months old, take part in the club young bird races from 80 to 100 and 135 miles, and in addition many of them have one or two additional tosses from the edge of the Channel, Weymouth or Bournemouth, if the club racing is over.

So we gather that with the youngsters infinite pains is taken as soon as they are three months old to teach believed-to-be future long distance champions the way over the first 100 to 135 miles, for which purpose they very frequently get as many as 15 liberations.

Notwithstanding this most careful education we must bear in mind that very large numbers of young birds are lost, and then a final selection of those left is made, for the successful fancier ruthlessly drafts the slackers and keeps only the best workers to strengthen his loft.

Next year, as yearlings, they go over the ground again down to the 135-mile stage, that is to the edge of the English Channel, and most probably get their first try at crossing the water by being sent to a race from Alderney (200 miles) or from Guernsey (about 230 miles) in company, be it remembered, of old birds that know the Channel, as well as they know the surroundings of their own loft.

I am not forgetful of the fact that some fanciers send yearlings on to Rennes (310 miles) or beyond, but the majority of fanciers are satisfied with 200 to 225 miles for second season birds.

Now, it will be noted that when sent to Alderney or Guernsey the yearlings get their first big lift of 65 miles in the case of Alderney and 95 miles if Guernsey, but it must be also remembered that they go to this race in company with a large number of old birds that have crossed the Channel many times, and I can speak from experience when I say that the presence of these older birds, with their experience when crossing the Channel is of very great assistance indeed in teaching the yearlings the way across.

So we see that the yearlings, or second season bird, by the time he is 15 or 16 months old has been twice over the ground by easy stages between his loft and the Channel edge, and has in addition, if he has stood his work, crossed the Channel once.

The third season, or two year old bird goes over the ground again down to the Channel edge, goes to a race from Alderney or Guernsey, and to a race from Rennes (310 miles), so that the two year old bird that stands his work will, by the time he is say, two and a half years old, have been three times over the ground between his loft and the English Channel, and will have

crossed the Channel three times.

The three year old, or fourth season bird, goes over the same ground again down to the English Channel, not of course as many tosses in this first 50 miles, then he is put into a race from Rennes (310 miles), and afterwards into a race from Marennes (470 miles), when he gets a jump of 155 miles.

Now, to enable us to understand what the task is that is asked of the birds that go to Marennes for the first time with 155-mile jump, we must know the ages of the various birds that usually make up a contingent for a race from Marennes, we must know what proportion of the birds usually going to a long distance race are old birds that have been over the ground many times, and what proportion of the birds sent are facing the 155 miles jump for the first time.

When we have this information before us we shall be better able to judge the task set the younger birds of the convoy that are facing the jump of 155 miles for the first time.

Remember I am out to show there is really no occasion to ascribe to any occult faculty the racing pigeon's power of negotiating in good time the jump of 155 miles, for one simple reason, if for no other, that generally speaking, indeed, almost invariably, when a two year old bird is asked to do the jump of 155 miles from Marennes to Rennes, he is asked to do it in company of other birds, very many of whom know the road thoroughly well.

So wonderful is the pigeon's memory and power of observation, after careful and systematic training, that it is found in practice that, after a pigeon has negotiated a long distance even once, but certainly if twice, he can generally be depended upon to do the journey again and again, always supposing that his owner gets him fit, that is, physically fit to stand the strain of the journey – a most important point which is, I fear, sometimes forgotten.

Let me digress for a moment and give an instance, a wonderful one it seemed to me at the time, of the marvellous manner in which a pigeon can remember a country.

Very many long years ago my old friend J O Allen, bought a yearling bird in Antwerp that had done very good work as a young bird in Antwerp over the usual Belgium training ground into France. This bird Allen brought to his loft at Smedley, near Manchester. It was bred from one season, then accustomed or used to Allen's flying loft; he then put it on the road with his young birds and it did very well indeed. So next season Allen put him on the road with the old birds and trained him down to London and Mid-Channel between Dover and Calais, and the bird made excellent time back to Manchester for two seasons running.

Next season (we were then training the SE route via Dover, Calais and Arras) Allen was tempted to send this bird on to Arras, which was very close to the line of country that the bird had travelled when flying from Paris to his home at Antwerp as a young bird. Immediately the bird (surely not guided by instinct, but by his wonderful memory) found himself over country that he knew, instead of going back to Manchester as he had done twice already from Mid-Channel, he slipped off back to his old home at Antwerp.

Surely it was memory and knowledge of country, not instinct, that took the bird back to Antwerp.

I remember one of the extra long distance fanciers of Belgium, when discussing the merits of a race from Spain, saying that once a bird of his had done Barcelona (I think it was), which means finding his way round, for I don't suppose they go over the Pyrenees, that he found he was justified in expecting the bird to do the distance again, the difficulty being to get the pigeon to do the big jump the first time.

Now let me get back to the ages of birds when sent engaged in long races. I am glad to be able to give some reliable figures in reference to this, for we are greatly indebted to "Nor'-West" in the "Racing Pigeon" of June 5th, 1915, for giving some most interesting and instructive tables analysing the entries of birds of various ages, and the number and the value of the prizes won by birds of various ages in five Great Northern Marennes Races for the years 1910 to 1914 inclusive.

These tables will well repay most careful study from the point of view as to which is the most reliable as a prize winner, the young or the old pigeon, and I earnestly advise the younger generation, for whom this article is penned, to make a careful study of them.

"Nor'-West' sums up his views on this particular point (which, however, is not the one that we are for the moment concerned with) that is, the value of old birds versus young ones as prize winners, in the following pithy sentence: "To me the tables prove very conclusively that it is not only a case of 'the old dog for the hard road', but rather the old pigeon for any sort of a day, good, bad or indifferent. They are the birds to rely on for 500-miles races".

Now let us turn to the tables and we shall then learn that in the five races analysed, 6,484 pigeons competed, 2,270 being two year old and under, and 4,213 three year old and over; so that supposing these figures are fairly representative, as I think they are, of the proportion of young to old birds going to a Marennes race, it follows that as very few birds are sent to Marennes before they are two years old, when they get their first big jump of 155 miles, they go for this trial of their powers in most excellent company, and, having regard to the fact that the natural instinct or impulse of the pigeon is to fly in company, they are also in the midst of very best possible environment to enable them to negotiate the 155 miles of fresh country that separates Marennes from Rennes.

This seems to me a quite fair deduction to make, because every 22 of these two year old and younger birds in the convoy, are accompanied by 42 older birds that know the road and so would not waste any time in starting off for home if the day is clear.

What takes place is, I believe, this: A convoy of pigeons when liberated at the race point, be it 100 or 500 or 5,000 in number, being gregarious in habit, obey their instinct of flying together in flocks; all keep together for a time and then break up in smaller flocks or groups, the chances being, in my judgment, that each flock or group contains old and young birds more or less in proportion to the composition of the entire body. So that if the average approximate number of old birds in the convoy be as 42 old birds to 22 two year old and younger birds, that is, birds out for the first time at a jump of 155miles, the chances are that each group as it breaks off from the larger body will continue both two year old birds and old stagers, and as the old stagers predominate in number as 42 to 22, and further, as the two year old birds do not know the way, the odds are that each group of birds is led by some of the four and five year old birds that are in their prime, and so they leave Marennes and get over Rennes where the two year old birds are over ground they have been over before.

A further careful study of the figures in the tables given by "Nor'-West" will, I contend, afford ample evidence that we have surmised with approximate correctness what took place when these various convoys were liberated at Marennes for England.

We know that each lot of birds, starting together, would keep together for a longer or shorter period, or possible breaking into groups almost immediately, at any rate doing so eventually, each group containing birds of the various ages of which the total convoy was composed, and so reached England in still smaller groups containing birds possibly of most of the various ages engaged.

The proof we get of this is that when we consult the tables we discover that the total number of prizes or early positions in time of arrival were distributed amongst the birds of various ages, and we learn that the total of 4,214 birds of three years old and upwards won 217 prizes, and that by the 2,270 two year old birds, that is birds doing the jump of 155 miles for the first time, no less than 102 prizes were won.

In one small convoy of race birds sent to Marennes, in 1913, I know for an actual fact that the two year old birds that were doing the jump of 155 miles for the first time and the older birds did keep together from the very start to the finish of a 465 miles course, although it took them 15$^{1}/_{2}$ hours to do the journey. In the Midland Homer League Race from Marennes, on July 16th, 1913, when the birds were tossed at 5.15 am, wind north, it is on record that seven birds dropped together on my loft at 8.46 pm the same day as liberated, four of these birds being mine and three of them belonging to my friend Dr Buckley, of Nottingham, where Dr Buckley's three birds landed early next morning, viz, 5.43 am, not being able owing to darkness to go on many miles after they were disturbed off my loft the night previously by my putting the light on to enable me to read the race marks on my birds.

My four birds were composed of one bird that had done Marennes before and three two year old birds which had been jumped from Rennes. Dr Buckley's three birds were his very light coloured RC Hen 3617 that had flown Marennes before and two others that did not know the way.

In the case of Dr Buckley's three birds and my four birds, we have evidence almost amounting to positive proof that the seven birds kept together the whole of the journey from Marennes to Langton, a distance of 465 miles, which journey they accomplished, we believe, without rest, in 15 hours 25 minutes, or at the rate of only a little over 30 miles an hour, the slow rate of progress being owing to the fact that there was a fairly strong north wind blowing against them the whole of the way.

We know these seven birds started together with the rest of the birds liberated for the race; two or three people saw them land together here at Langton therefore it does seem to me quite unnecessary to suppose that the two year old birds facing the jump of 155 miles for the first time were helped by any "instinct or propensity prior to experience or independent of instruction", which is Paley's definition of instinct.

In considering this question we must not forget (a) that homing pigeons are educated to fly together in large consignments from the first young bird races onwards: (b) that they are gregarious, that is, "having the habit of living in a flock, not habitually solitary or living alone".

All the nine two year old birds that I sent, also two of Dr Buckley's birds, had the jump of 155 miles. Had they been sent by themselves, I am convinced that so far from seeing three of mine home the evening of the liberation, on a very hard, slow flying day, and four more in good time the next day, commencing from 4.22 am, time of the first two year old bird's arrival, I should not have seen a bird the same day, and only a certain proportion of them the next. Therefore, I contend that it is unreasonable to suppose that it was not the guidance and help that those two year old birds got from the older birds at the start, and until they got on to ground known to them that enabled them to do so well.

I am sorry I cannot now obtain, after this lapse of time, the proportion of old and young birds in the club's convoy, but the lot I sent consisted of six old birds, that is three year old and upwards, that knew the road, and nine two year old birds that had the jump of 155 miles.

To complete my story of the doings of my own birds, for which I hope I may be pardoned, and to enable the younger generation to use their own judgment after a full presentation of the facts, let me say my two year old birds had been most carefully trained first as youngsters and then as yearlings. As yearlings they crossed the Channel by themselves from Alderney. The jump from Bournemouth to Alderney is about 65 miles; on some occasions, which occur oftener than are pleasant, this jump by themselves spells disaster, whereas on other days, when the atmosphere is clear and the wind is not strongly adverse, so that they can without distressing themselves keep flying long enough to hit their bearings off, after probably going to the French coast first, the majority of them manage the jump all right.

As two year olds they were trained down to the coast line, then flew Guernsey and Rennes, before being sent to Marennes, so that before going to Marennes they had crossed the Channel three times.

Again, I contend that instinct, which Webster defines as "an agent that performs blindly and ignorantly a work of intelligence and knowledge", had very little, if anything whatever to do with bringing these two year old birds from Marennes in decent time.

Another reason why I have never been able to believe in instinct is because I learnt very early in my experience as a fancier of these birds that the nature of the ground, that is the configuration of it, has a most important bearing on the bird's ability to find his way home. Surely if it is instinct that guides the bird, it would enable him to find his way home as unerringly over difficult ground, that is country without prominent landmarks, as he does over the country beyond Brussels on the way to Paris, which is flat so that Brussels, with its huge magnificent Palais of Justice, situated as it is on an eminence in the town, can be seen for an enormous distance.

In 1882 I started a loft of racing pigeons close to Carmarthen, almost in sight of the Bristol Channel. I found I should have to train the young birds in an easterly direction, and that I should be compelled to train them down the Great Western line which runs alongside the Bristol Channel up to Cardiff, a distance along the coast line of about 75 miles.

As soon as ever I commenced to train the young birds it was borne in upon me that beyond all doubt training 75 miles along the coast line of the Bristol Channel from Carmarthen to Cardiff was a soft job as compared with training on the south road from East Langton down to Bournemouth and the Continental route.

After Cardiff, I sent the birds to Bristol. To get there they had a dozen or 15 miles of the mouth of the Bristol Channel to cross, and so on to Reading.

I scarcely lost a single bird out of the big lot I put on the road, and it was a splendid guiding sign, or land mark, that helped them to find their way home without any difficulty.

Then again in 1893, when I first flew birds on the North Road from East Langton, I soon discovered that to get young birds to fly 230 miles from North Berwick to Langton was child's play as compared with getting the same birds to do Guernsey, almost exactly the same distance. In fact it is much easier to do the 230 miles from here on the North Road than to get the same breed of birds to do 130 miles on the South Road, the reason being the magnificent guiding land mark that the East Coast of England affords practically from Scotland downwards.

As showing how an easy configuration for flying may and does help the birds, I should like to point out to the younger generation that a bird in flying from Thurso to Yarmouth can do so without crossing anything worth calling a hill, and be guided by the coast line practically the whole of the distance of 460 miles.

I contend that if Lerwick had still been, as it undoubtedly was once, on a continuation of the mainland of Scotland, instead of being on an island, separated from Thurso by 150 miles, and also within 230 miles of another mainland namely, Norway and Sweden, the North Road route to Lerwick would have been the easiest 600-mile course obtainable by birds trained in the neighbourhood of London and the South of England.

Up to Thurso I maintain it is not only an easy, but a fair road to the birds, but owing to the big jump from Thurso to Lerwick, 150 miles being over ground that I believe the birds find it very difficult indeed to negotiate, unless the atmosphere approaches the marvellous clearness which obtains in the South of France when the Pyrenees Mountains can be seen with the naked

J W Logan, Esq.

human eye an immense distance away.

Then again, one must not forget that Norway is within 230 miles of Lerwick, and it was to Norway, in my judgment, that the big convoy of birds went in 1899, when almost every bird was lost.

Please do not misunderstand me. It is not the length alone of the big jump from Thurso to Lerwick that so constantly puzzled the birds at times, but it is the 150-mile jump on that particular configuration of ground that does the mischief, and to instance what I say, let me say I have found it quite easy for birds by themselves that had never been beyond North Berwick (230 miles from here) to negotiate the jump of 210 miles into Thurso, the reason that they are able to do so being that North Berwick is situated in the middle of the coast line between Thurso and King's Lynn, a distance of 420 miles, as can be seen on the accompanying map.

But had Lerwick been on a continuation of the mainland of Scotland we can see a reference to the map what a magnificent land mark would have been offered the birds, a land mark that would have stood them better than the "instinct" that is supposed to guide them on the journey over unknown land, but which frequently fails as completely as it did at Lerwick in 1899, when a mistake at starting cost the English North Road fanciers as grand a lot of birds as ever were sent to a race.

It may be said that it is only coincidence that the failure of the instinct theory invariably occurs when there is an absence of land marks, but so it is, it does then completely fail as we shall show a little further on in the comparison of the records of performances in the four classic races from Rome and St Sebastien.

So long as we believe the pigeon has to rely on a most retentive memory for places once seen, coupled with very keen sight and great powers of observation, we can at once understand that if liberated on quite strange ground a bird would have very great difficulty in homing quickly if his direct way, and, therefore, shortest way, home was blocked by hills frequently covered with mist; but if he were guided by an unerring instinct this obstacle would present but little difficulty.

As showing the difficulties thrown in the way of a bird's speedy return home by the difficult configuration of the ground, let me instance the case of the Lancashire and Cheshire fanciers, who, I am told, when they tried to train their birds across country to the East Coast of England en route for the North Road racing Mecca, the Shetland Isles, found that Yorkshire hills to be an insuperable difficulty, and so wisely gave up the job of trying to train across them.

Then again, ask the members of the Armadale and District HS, of which Dr Anderson is the President, what proportion of their young birds they lost when they first started training over what I believe to be the very hardest bit of flying country as yet negotiated in these Isles.

But, judge for yourself, get a contour map and cast your eye upon the line of country from Armadale to Dumfries and take stock of the hills between the two places.

Anyone who has tried, or will try, to fly birds over the bit of country in Lanarkshire and Dumfrieshire, that the Armadale birds have to negotiate, will, I am certain, support my contention that the natural configuration of the ground flown over does necessarily greatly influence the bird's powers of homing quickly, and that only the very best and gamest can fly that country season after season.

We know that birds belonging to the Armadale Club do get trained over the 70 miles between Armadale and Dumfries, but it presented very great difficulties at first, whereas had "instinct" blindly guided the birds they would have found their way across these 70 miles with just the same ease that my birds did across the 70 miles of training from Carmarthen to Cardiff.

Four classic pigeon races that have taken place within my recollection – two from Rome, organised in Belgium, and two from St Sebastien, in Spain, organised in this country – one by the Manchester Flying Club and the other by the Harboro and District SR Flying Club. These four races afford much food for reflection on the subject that we have been discussing, and I shall ask my readers to study the published results of these races with me, to enable them to come to a decision on this question of Sight versus Instinct.

The distance from Rome to Brussels … … … … … is about 735 miles
,,   ,,   ,,   Belgian frontier … …   625 ,,
,,   ,,   St Sebastien to Manchester …   715 ,,
,,   ,,   ,,   Wolverhampton   650 ,,

In the 1878 race from Rome, the birds were liberated on the 23rd June, and the first bird was reported at Aix le Chapelle on the 2nd July, thus arriving on the 10th day.

In the 1913 race from Rome, 1,557 birds were sent from Belgium, and 106 from England, and they were liberated on the 26th of June, and the first arrivals in Belgium were one bird on July 6th (that is the 10th day) and two on July 9th, and eight more up to July 13th. The first arrival into England was to Mr Hudson's loft, at Derby, on July 29th (that is the 33rd day), distance flown being 1,001 miles. The second arrival to Messrs Vester and Scurr's loft, some time in August, prior to the 16th, the distance flown being 1,100 miles.

The first race from St Sebastien, organised by the Manchester FC, was flown on July 10th, 1907, 150 birds competing, and they were liberated at 10 am, wind SE, and the first four prizes were won as follows:–

				Time occupied
1. Burton, Liverpool,	about 720 miles,	at 6.22 am, July 12th	…30hrs
2. Bentley, Knutsford,	,, 715	,, ,, 7.15 ,,	,, 12th …31hrs
3. McClelland, Old Trafford,	,, 715	,, ,, 1.30 pm,	,, 13th
4. Baxter Bros, Lostock Gralam,	,, 715	,, ,, 9.16 am,	,, 20th

The second race from St Sebastien, organised by the Harborough and District South Road FC, was flown on July 22nd, 1907, 283 birds competing, and they were liberated at 11.55 am, wind NW, and the first four prizes were won as follows:–

					Time occupied
1. Wones, Wombourne,	636 miles, at	4.54 pm, July	23rd	…21 hrs
2. Downing, Dudley,	634 ,,	,, 5.07 ,,	,,	23rd	…21 hrs
3. Compton, Bradford-on-Avon,	555 ,,	,, 2.57 ,,	,,	23rd
4. Brearley, S Molton,	538 ,,	,, 4.24 ,,	,,	23rd

After carefully analysing the result of these four races, I was at once seized with the fact that there must have been some vast difference in the difficulties of the task imposed upon the birds sent for the Rome races as against the difficulties of the task imposed upon the birds sent to the St Sebastien races, or there could not possibly have been such a vast difference in the time occupied by the winning birds in doing the journey in the one set of races as against that required in the other two races.

Why should the first arrivals in both of the Rome races have required nine days to do the journey, whereas in both races from St Sebastien into England, the time occupied in doing the journey was a question of hours, notwithstanding the fact that the distance from St Sebastien to Manchester (715 miles) is only 20 miles less than the distance from Rome to Brussels.

In the Manchester Club race the first two birds did the journey in 30 hours and 31 hours respectively, or at the rate of 24 miles per hour, after allowing 8 hours for darkness during the 24 hours.

And the first two birds in the Harborough Club race accomplished the task in 21 hours, or at the rate of 30 miles an hour; two other birds running the first two very close the same afternoon.

In the Manchester Flying Club race, the performance of Mr Burton's bird, the winner, and also that of Mr Bentley's, was an exceptional one, for though the weather was fine at the start, it was most uncongenial pigeon flying weather, and the birds must soon have encountered the worst possible weather for pigeons to fly in, as described in "Racing Pigeon" of July 17th, 1907: "Adverse conditions all along the line of route: intermittent rain on the first day; impossible conditions for liberation at Nantes on line of route".

Under these conditions the performance of the first two arrivals was a most magnificent piece of work and stamped the birds as being of the best.

The percentage of losses in the Manchester Club race was, I fear, very heavy, but I put that down entirely to the wretched weather conditions that prevailed all along the route.

The weather conditions for the race of the Harborough Club were very much more favourable, so that seeing that the birds were not liberated until noon, the performance of the four birds that arrived home the next afternoon, against a north wind, will take a bit of beating, and stamp the birds A1.

Then, as there were 62 arrivals before the close of the race, I shall expect in happier days yet to come to see great things from St Sebastien.

In comparing the Rome race with the race from St Sebastien, do not let the

"novice" imagine that there was any difference in the flying abilities or powers of the Belgium or English birds, for the fanciers of both countries would
send to all four races only the very best birds which in their judgment were the
best qualified to compete in such a trying race.

If any advantage rested with birds of either country, it was with the
Belgians, for they have so many more birds to select the best from than we
have in England, as evidenced by the Belgian fanciers in 1913 sending to
Rome 1,557 birds, whereas English fanciers could only muster 433 birds to
send to the two St Sebastien races.

It is evident, therefore, that there are some inherent difficulties in the route
from Rome to Belgium, as compared to the route from St Sebastien to
England, possibly some vast difference in the configuration of the country that
the birds have to fly over.

Let us study the map that accompanies this article and see what we can
make out of the problem.

A sight of the map at once shows us that pigeons liberated at Rome, to return
to Belgium, would, if they flew in a direct straight line, have to travel a distance of about 625 miles to reach the Belgian frontier, and the distance to Aix
le Chapelle, the home of the first bird that returned from Rome in the 1878
race, is 730 miles.

We shall also learn that a straight line from Rome to the Belgian frontier
means that Switzerland, and the Alps Mountains, stand between Rome and
Belgium – a barrier which precludes the birds from being able to take a
straight line for home.

In trying to solve a difficulty it is always wise to start by eliminating the
impossible, so let me at once say that having travelled through the Alps to
Paris, I have no hesitation in saying I do not believe it is possible for birds to
get to Brussels by travelling round the eastern side of the Alps, for the Alps
extend on the eastern side to opposite the right hand corner of the Adriatic Sea,
to say nothing of the fact that as they would most likely strike up the coast line
of Italy first, they would have the Appenine Mountains as well to cross.

Now, having eliminated the impossible routes, the only possible one is for
the birds to make West, go up the coast line from Rome in the direction of the
Alps, just skirting the Appenines above Florence, then following the coast line
round the Gulf of Genoa until they come to Toulon and Marseilles and so on
to Nimes, for they must follow the coast line, or possibly make a cut across
the gulf, until they can get completely round the mountains, for the Alps practically form the coast line round the Meditteranean at that point. Rome to
Marseilles is 378 miles, measured in a straight line; Marseilles to Nimes is a
little over 60 miles; Nimes to Brussels, through Lyon, is 480 miles; making a
total of 918 miles from Rome to Brussels, that the birds would have to fly.

I have a map before me expressly made for the use of the Belgian flying
clubs, which shows the course the Belgian clubs usually train, and gives the
names of close on 100 training and racing points and exact distances from
Brussels, and this map shows Nimes and the best line from Nimes towards
Brussels seems to be up the Rhine valley to Lyons and then up the Loire to

Orleans.

Nimes is 150 miles to the East of Montauban, Lectour, Auch, Mirande, and Toulouse; all five well-known racing points; all towns the mature long distance birds in Belgium know very well.

We now see the enormous difficulties which birds starting from Rome have to face before they can get over land with which they are familiar, having close on 500 miles of strange country to find their way over before getting on to ground they know.

All birds that could get to Nimes would have a very fair chance of getting home, for 150 miles in that country and climate does not present the same difficulties as that distance would in England, or up in the North Sea.

I remember the late Mr Alfred Lubbock writing from the South of France in 1882 describing the atmosphere in the South of France in these words: "It is quite extraordinary in this country how close the mountains sometimes seems, which in reality are many miles distant. Most days in the year the Pyrenees are visible (that is to human eye) at a distance of 70 miles, and I am told that from Bordeaux it is often possible to see them at 140 miles, and they even tell me that Mount Canejau, one of the highest in the Pyrenees, is often visible from Marseilles, a distance of 180 miles.

So the novice will at once see that if these Rome birds once get anywhere near ground with which they are familiar they would immediately get their bearings for home, as we saw in the case of Mr Allen's birds related a little earlier in this article.

Of course it is impossible to say the exact course that the birds take, but supposing they do take the course I have mapped out for them, which certainly seems the easiest, for it brings them on to their old training ground, it seems difficult for me to believe the birds could have been guided to find their way to these old training grounds by instinct, for if they have been, surely they would have done the journey in less than nine days, and, most certainly, instead of a few scores of birds straggling home after many weary weeks of hard work, we should have had quite a large proportion of the birds back in decent time.

Instead, both the two races from Rome have been long drawn out tests of the pigeon's intelligence and powers of endurance, in which only a very few birds, and those the very gamest, could hope to regain their homes, and I am bold enough to assert that in addition to pluck, staying power and intelligence, the element of luck must always be an important factor in these long trials.

Take one great source of danger to the birds in their gallant attempt to get home; they can get their living in the corn fields we know, but what are the odds they will escape the gun of the small holder who has more love for his crops than he has for stray pigeons.

Then again, birds of prey are a source of constant danger to a tired bird, and the question of finding fresh roosting quarters night after night is very likely to lead to trouble.

Be all this as it may, I still contend that the birds that struggle through from Rome, either to Belgium or England, none the worse in health, are treasures almost beyond price, for they have proved themselves possessed in the very highest degree of all the qualities that go to make a good long distance racing pigeon.

The champion stayers who have the grit to fight on over hundreds and hundreds of miles of unknown countries until they get on the line for home, are what we want to maintain and to improve the breed of racing pigeons.

My earnest wish, therefore, is that we may have enthusiasts amongst our fanciers in Great Britain to organise and support an occasional race for birds in this country, calculated to test the staying powers of British birds, as Rome does that of the Belgian birds.

Now let us once again examine the map to assist us in understanding the work the birds are faced with when liberated at St Sebastien to return to England.

Let us see whether the surroundings of St Sebastien and configuration of the ground is likely to have been helpful or obstructive to the desire of the gallant little travellers to reach their homes.

The most important consideration of all is to find out how far the birds tossed at St Sebastien were, when liberated, from ground with which they were already familiarised from having flown over it before.

Were the St Sebastien birds faced with any special difficulty beyond the one difficulty, enough in all conscience, of flying 700 to Manchester, having the English Channel, with its very varying and uncertain conditions of winds and atmosphere to face, after having flown 420 miles towards home?

It is very important to remember that the English birds flying the French route have been trained for many years now down the western coast line of France, via the Channel Islands, Rennes, Nantes, Marennes, and Bordeaux, so that the birds sent to St Sebastien by the two English clubs would know the south west coast line of France.

It follows then that any of these birds that had been tossed at Bordeaux and Marennes, and very many of them had been at one or both places, were, when tossed at St Sebastien, within a comparatively short distance of ground they had been over before, and ground of exactly similar character, with a well defined land mark along the whole coast line, so different in character to the 500 miles of absolutely new ground that faced the birds liberated at Rome.

The birds when starting from St Sebastien would find a low range of hills directly behind them from 700ft to 800ft high. On their east or right hand the high Pyrenees Mountains up to 9,000ft or 10,000ft high. On their west or left hand the very mountainous coast line of Spain running due west from St Sebastien, and within 60 miles of St Sebastien and close to the sea I notice there is one mountain 4,400ft high.

Now I contend that an intelligent pigeon is just as likely as a human being to eliminate the impossible when faced with a difficult problem, and would, therefore, after having a good look round, soon make for the only course that was open to him, which, in this case, happens to be the very course over which he has been trained, namely, the western coast line of France, which runs due north from the town of Biarritz, in France, which town is only 30 miles distant along the coast line from St Sebastien (refer to your map). When at Biarritz, a bird would be within 100 miles of Bordeaux and 160 miles from Marennes, so one can understand that it would not be very difficult to believe that some of the birds would soon be on ground with which they were familiar.

Now let me test the correctness of the surmise as to what it was quite made possible for an intelligent bird to do.

The Harborough Club birds were tossed at 12 o'clock on July 22nd and the wind was NW. The birds would have at the outside seven and a half hours in which they could fly, for I think I am right in saying it would be getting dusk in France at that time at the end of July, so that if we assume that the birds could do 280 miles the day they were tossed, it seems to be about as much as they would do, for they would certainly spend some little time before making up their minds which road to go.

Now 280 miles the first day would leave Mr Wones' bird 360 miles to do the next day after spending the night somewhere in the neighbourhood of Nantes, so as the bird arrived in Wombourne just before 5 o'clock in the evening of the day after liberation, it means that, supposing he started at 6 o'clock in the morning of the second day, he flew the distance from the spot in France where he rested the night, to Wombourne, at the rate of very nearly 33 miles an hour – that is very nearly 1,000 yards a minute.

In conclusion I contend:–

(1) That an impartial consideration of the accompanying map, with the explanations I have given, will help to make clear the natural difficulties owing to the configuration of the land that beset birds in a race from Rome to Belgium and very many times greater than in a fly from St Sebastien to England.

(2) That the comparatively easy task (that is easy as compared to Rome) that birds liberated at St Sebastien have to get in to or over familiar ground in the South of France is amply sufficient to account for the fact that 20 hours sufficed for birds to fly from St Sebastien to Wombourne, and 30 hours from St Sebastien to Liverpool (this latter race being held under very uncongenial conditions of weather for pigeon flying), whereas it took nine days in both the two races to fly Rome to Brussels.

(3) That no facts or figures in this analysis of these four races are at all calculated to make it easier to believe it was unerring and unreasoning instinct that enabled the birds to reach their homes.

Whately said: "Instinct is a blind tendency to some mode of action, independent of any consideration on the part of the agent of the end to which its action leads".

So personally I can find no tittle of evidence to justify us in denying to the homing pigeon the possession of such extraordinary powers of sight, memory, observation and endurance, coupled with intelligence more than sufficient to enable him to find his way unaided by any occult power to his dearly loved home from any distance in reason.

A FEW REMARKS ABOUT PIGEONS IN GENERAL AND PIGEON FANCIERS

by E E JACKSON

Dear Squills, – You have been good enough to ask me to write something for your Annual. Before the war I could, as a rule, sit down and write something about pigeons, but now I feel I know nothing at all about them.

At one time I used to think I knew a bit about birds, but now I have come to the conclusion I know nothing. In peace time, when I was training my birds, I always tried to select good days to send them away on; I don't think I shall ever take the trouble to do so again.

I have no doubt that you get more birds to do the 'far end', as it is often called, by selecting good weather, but after all, you get a lot of birds to go right through which would not face hard weather. The result is you breed from those birds, which are only soft if put to a very hard test, and all you can expect is soft young ones. Since the war started I have had, perhaps, a better opportunity than most people of seeing what pigeons can do, as I have had the training of birds in three different countries, all of which are quite different.

First of all I was in France. The country there is, in my opinion, very easy for pigeons to fly over, and from what I could see, a bird had every chance of regaining its loft should it make a mistake, as no one took any notice of a pigeon if it should drop in a field or on a house, as in most parts of France there are thousands of farm-yard pigeons, which live all the year round on what they can find in the fields.

In France you see very high, round towers, built out in the country for pigeons. I got to look in one of them, and it contained 2,500 nest boxes, and from the look of it, had never been cleaned out since it was built, some hundred years ago. I was told by the owner that it was cleaned out once a year, but I feel sure they must have missed at least one year, probably on account of the war, as I never saw a place in such a state in my life. The birds kept in these places are never fed; they just go all over the country as they like, and once a month the owner goes in and takes the young ones from the nests and kills them for market.

What I noticed most was the great amount of canker to be seen amongst the young birds. I should say at least 50 per cent of them had the canker. The only thing was, it was a different kind of canker to what I have seen in England.

As a rule you would see small lumps on the wattle, or on the eye cere; not in the mouth, as you often see it at home.

I was billeted in a farmhouse in a small village in France, where about three hundred pigeons were kept, and I took a great interest in these birds. I noticed that they always went to drink at a pond in the middle of the yard, which contained, without doubt, the filthiest water anyone could imagine. Although a stream ran past the farm, of clean water. Not only did the cows and horses go to this pond, and wade up to their haunches in it, but all the drainage from both the house and stables ran into it as well. On a hot day the stench from it was enough to knock you down.

I am quite certain that at least 80 per cent of the young birds in the nests in this loft had canker. From what I could see, they don't seem to develop the canker until they are about three weeks old, and it is not often the French people let them get much older before they kill them and take them to market, where they get about tenpence each for them.

I came across very few what we should call in England good fanciers. Nearly everybody keeps pigeons, especially on the borders of Belgium, but very few of them look after them like we do in England. You are shown birds which have flown this and that, and won prizes from here, there and everywhere, but when you take into account the flat country they have to fly over, and no water to cross, and the amount of prizes in each race, it is nothing wonderful, and I am of opinion that if old Squills, and some of our best racers in England, lived next door to the best of them, they would knock spots off them.

I know we owe, to a great measure, the success we have attained in England to the Belgians, but that is only because they have been at it longer than we have. Take N Barker. He was an Englishman, but went to live in Belgium. He soon began to show the Belgians that he could fly pigeons as well as any of them. We often see pure Barker's advertised. I wonder what is a pure Barker. The greatest boon to the fancy that ever happened was when Mr Logan bought N Barker's birds, and after breeding what he required from them, gave youngsters to Barker, but not before he had crossed other good strains into them, which are at the present day what we know as pure Barker's. I have a lot of them, and for the reason that I don't know what other strains Mr Logan put into them, I call them pure Barker's, and I don't want anything better.

During my stay in France and Belgium I only came across one pigeon fancier who kept a proper pedigree book, and he lived just in Belgium. Now this man, a postman, was a good fancier. He did not keep many birds, if I remember right – sixteen pairs, all told – every bird moulded alike, and they had won some races, and no mistake. He told me that he started with four pairs of birds about 30 years ago, and had never introduced a cross since. The only thing I can say is he must be an artist at the game, as finer and better shaped birds I never saw. I asked him how he managed to keep them to such a standard, and he said, "By flying them, every one, to the far end, and breeding from only those which did the work". It made me think that I had been going on the wrong track for my short time of pigeon flying, which extends to 33 years or so. I have always found that birds bred from sisters or brothers to the good racers are the best for racing. I have never yet found a hen bird which has been a hard racer breed a good young racer. They will breed them to fly the distance, but I never yet got one to show much speed.

As so many of the English and Irish pigeon fanciers have seen for themselves the good work the birds have done out at the front, it is useless for me to go into any details, even if I were allowed to do so, which I am not at the present.

It is quite evident to me, after what I have seen, that a good pigeon will stand up against almost any kind of weather, providing it is trained to fly in bad weather I should never have dreamed of asking a bird to fly in. The only thing is, if you try this system of flying pigeons, you must be prepared to lose a very high percentage for the first few years, until you weed all the ones out which will not face the music.

Pigeon racing nowadays has become a very fine art, and it is no use thinking, because you buy some expensive birds, which have done well for some other fanciers, that they will do well for you. They may or they may not, just as you manage them and how your loft is situated as regards line of flight from the race point. You must be situated just right; or if you are only a few miles wide it is seldom you will win, unless the wind happens to be blowing strong in the right direction. High hills play a great part in deciding which birds will win. I have noticed this more since I came to Ireland than I did in England. For some unaccountable reason, in some cases the birds will fly well enough over the hills from South to North, but not the other way about. Then again you find that birds do well time after time from places it looks almost impossible for them to find. This may be over the hills, which, even on a fine day are, as a rule, covered to some extent in clouds. I have one place in particular in my mind's eye which I had to fix a pigeon depot at. When I went to arrange matters and saw the country – nothing but high ranges of mountains for miles in every direction – I never thought we should get a bird to do the journey. To my great surprise and pleasure we have had wonderfully good results from that particular place time after time; and from other places, which look as though it ought to be quite easy for them to home, we have had nothing but disasters, time after time.

That is one of the reasons I have come to the conclusion I know far less about pigeon flying than I used to think I did. I am quite sure of one thing – that unless you let the birds imported to Ireland from any other country, get used to the climate before training them, you will lose a great many, because the climate in Ireland is very different to that of England. You can never depend on it keeping fine for a whole day, no matter how well it may look in the morning. Not only that; it may keep fine all day at the home end, and also at the point where the birds are liberated, but if they have any distance at all to travel you may be sure they will pass through two or three different kinds of weather on their journey. When it rains in Ireland it has quite a different way to that of England – it comes down as though the sky had burst, and whilst it lasts I defy any pigeon that was ever hatched to fly through it. After what I have seen this last six months of Ireland, I have come to the conclusion that birds which have homed from various parts of France to the north of Ireland, must be equal to any birds in the world, and I feel sure it is to a great extent because they are the survival of the fittest after years of hard work.

As to different strains of birds, for my part I don't think it matters very much whether they are Barker's, Logan's, Osman's, Van Cutsem's, or what they are, so long as they have been bred for generations from nothing but sound stocks of the very best racing blood, such as the above-mentioned fanciers have done for years, and hundreds of other fanciers also, or we should not have such a fine lot of really good birds in the country.

I have almost always found that yearling hens breed the fastest birds. A two-year-old cock paired to a yearling hen which has not been raced much and had no right hard races to fly, but own sister to right good racers, mind you, have given me the best results. In nearly every well-kept loft you will nearly always find a pair of birds which breed more winners than any other pair in the loft. You can pair a brother to the cock and a sister to the hen together, but seldom get the same good results. When you find you have a pair which stand out as breeders above the others, take my tip, keep them as stock birds, and most likely, sooner or later, they will breed you a champion, which will repay you for their keep. Then, again, you find some birds breed good young ones no matter what you pair them to; these are well worth keeping.

There is a very big difference between a good pigeon and a champion, and if you only have the luck to breed a champion and still have its parents, you may congratulate yourself, not only as a money-making concern, but on being able, if properly managed, to found a loft of birds which will keep you well to the front.

I have almost always found that hens a little on the small side bred better birds than big ones. I like a hen with broad shoulders, not too deep, good wide back, and short legs, something after the bulldog look about it, paired to a cock a little on the narrow side if anything, with a bit of length about him. I find this gives better results than having two either narrow racy-looking birds or two very stiff ones together.

When I was asked by Squills to write something for his Annual, I of course said "Yes, I will", feeling very much honoured at being asked, as I am no writer in any shape or form, but should any of the few tips I have found worked well for myself be of any use to new starters in the fancy, I shall feel well repaid for what little I have done.

WORKING MAN'S ESSAY

by CALDWELL BROS, Runcorn FC

We here endeavour to give an article which will be instructive and beneficial to all working men fanciers, as to how we founded our successful loft of racing pigeons. We have kept pigeons practically as long as we can remember, and commenced racing in the Runcorn Club, about 1903. We then had, probably, about thirty birds of an inferior class, and our management was also faulty. Of course, we had no success our first three years' club racing. We then set about to take our hobby more seriously, and work on a better basis, which, for the benefit of readers of this article, we will try to explain in three parts – viz: 'Loft Management and Breeder Pairs for Stock', 'How We Train and Race Old Birds', 'How We Train and Race Young Birds'.

Our lofts, of which we had three in two back yards, were built on the yard floor (but raised about one foot off the ground is much better). Our young bird loft we built on the top of the old bird racing loft. We would have the following number of birds in each loft: Four pairs of breeder pairs, about a dozen pairs of racers, and about twenty young birds. Our lofts we made with open fronts, these being wired through with quarter-inch wire rods to admit plenty of fresh air. Each loft we whitewashed out twice a year – before breeding season and after. We clean out twice a day all the year round; clean water twice a day; and the bath twice a week if the birds desired it. We have our bath arranged with a rubber tube on to the tap, which we could leave running while our birds were bathing. We always use sand for our loft floors and nest boxes; our birds delight in this, and eat it. We get it with a pick out of the soft red sandstone beds not very far from our lofts. We are great believers in cleanliness; we consider it essential for founding a successful loft of racing pigeons.

We commenced our present lofts with three pairs of squeakers. These we purchased from successful and reliable breeders and racers. The three pairs we kept for stock to produce our racers. Our advice to working men taking up the hobby is to purchase squeakers from successful fanciers, and keep them as a base to breed from. We think all working men fanciers should have two or three reliable stock pairs. Three or four nests of young from these each year would be ample. We mate up our stock birds the beginning of February. When rearing youngsters, we always see the nest pans are in good condition and clean. We use pitch-pine sawdust as a bed for the squabs, also clean straw, if the old birds care to carry it to build which they generally do. We never handle the young while in the nest, only for ringing purposes. We always put in the nest box a second nest pan. When the cock starts taking notice of the hen again, we remove the first nest pan for cleansing purposes, when the youngsters are moved away from the parents and weaned. We generally move them from three weeks to one month old; we think they come on much better than if left with the parents. If we see any youngster doesn't happen to pick up, we hand-feed it for a time, until it gets eating. We see that our stock birds, at all times, get a fair amount of food. We also teach the youngsters to eat in the nest boxes before moving them from parents. We also use good clean grit during the breeding season. Our birds delight in grit. We never use any kind of saltoats; just a little table salt (occasionally) in the drinking water during breeding season we find sufficient.

How We Train and Race Our Old Birds – Before putting our old birds in training, we exercise them round home half an hour morning and evening, for about a fortnight, before we commence to train them to get them into condition for the road, always feeding them into the loft after exercise. This teaches them to trap quickly when coming from training or race stages. We would probably put about twenty old birds in training, give them three miles single up first two tosses, then about eight miles twice, fifteen miles twice, twenty-five miles twice or three times, then forty miles same, always making a point of being at the lofts to trap our birds when coming from training tosses if possible, or arrange for someone to trap them and feed them in. We would then be ready for our first race. Of course, we see our panniers are in good condition before commencing training. We also give our racers a twenty or thirty miles' spin about middle of the week. We also mate our racers up about February 14th for the races up to 250 miles, about March 14th for 400 and 500 miles. The first year or two we stopped our yearlings at 200 miles stage. When we found we could hold our own, and win up to 200 miles, we commenced to try our birds at 300 and 400 miles, assuming that by holding our own and winning up to 200 miles, we had the cue our birds must be in good condition, after which we always sent our yearlings to our first Channel race about 280 miles. These would then be ready as two-year-olds for 300 to 400 miles. Finding we could hold our own and win at 300 to 400 miles, we then went on to 500 miles. We always like to see our old birds flying freely round home before going to the distance races about 40 minutes at 5am, same about 8pm. Most of our best performances have been done on eggs. We consider eight to ten days on eggs best condition for our hens. Our cocks also fly very well sitting. Another condition we like for cocks is feeding a big youngster and just taking notice of the hen. We have always done our best performances when our birds were growing up second or third flight. It is necessary to know your birds and study their moult, also mate them at the right time if you want to catch them on the second or third flight for the 400 or 500 mile races. Some strains we find moult much quicker than others. Our own birds moult very slow; we attribute this to having our lofts open at the front. Just to give an idea how our birds moult, our 'Pool Hen', when she won 2nd prize Nantes in the Great Lancashire Combine race, 1911, July 8th (an awfully hard race, velocity 882 yards per minute, very few birds home day of toss out of 8,404 birds), she was mated up February 7th, reared two youngsters, and was sitting ten days; she went to the race growing her second flight, and homed with practically a full wing. Our good Marennes cock (winner of 1st Section B National Flying Club Marennes race, 1914), mated up March 14th, reared two youngsters, was sitting seven days when sent to the race, July 14th, and growing up his third flight, and homed with 9¾ flights each wing. We always make a study of the moult of our racers, and make a note of each when sending to the 400 or 500 mile races; we also make a note of how our yearlings moult. When selecting our pool birds, we like to take our most consistent pigeon, or one that has done good work previously, also keep our eye on the bird we can see improving as the distance increases. Of course, we find there are certain birds which experience and, shall we say, instinct, tells us not to miss; and you cannot get away from them. We always note a big improvement in our birds as soon as they throw their first flight. We always like to see our cock birds get fairly on the move as we get to the distance races, and also with a good clear eye for condition. We also handle our racers as little as possible during racing season. We do not believe in under-feeding or over-feeding our birds (we like to catch the medium). We hand-feed twice a day, and our birds clean all up. Our principal feed is maple peas. As distance increases, we add food – beans, with a little rice and canary seed (occasionally) as a tit-bit. We believe in good corn, and any corn we get is looked over before given to our birds, and the unsound corn picked out. It takes very little unsound corn to upset your birds for the racing season. We always try to race through the season on the same sack of corn by this method. There is little risk of upsetting your birds system. Happy is the working man fancier who can lay in stock a sack of good corn in the winter for the racing season. We consider this a very important item.

How We Train and Race Our Young Birds – We would generally commence with about twenty young birds, and give them about the same training as our old birds, with additional tosses at about 50 or 60 miles. We believe in our young birds being well trained. We also like to see our youngsters fly freely round home, they help the old birds if put out with them, if you want to give them a final spurt. We have never taken seriously to young bird racing, although we have had a fair amount of success in club and Fed racing. We always feed our youngsters twice a day, except on a Friday previous to racing on the Saturday, then they get one feed. We generally give them just sufficient to keep them sharp, as we always like to see them on the alert. We rattle them into the loft after exercise with corn in a can, and we soon get them used to trapping. We never let them hang about, only on Sunday, when we give all our birds open loft, weather permitting. We always have someone to trap our youngsters from each training stage and feed them in on arrival. We generally race our youngsters up to Bath, 134 miles, occasionally, selecting several of the best out at 100 miles. It has always been our ambition to

improve our methods. With this end in view, we have visited other successful lofts, and also introduced fresh stock, to try and improve our own birds.
We believe in introducing a pair of good reliable squeakers occasionally from
successful breeders.

In conclusion, we shall be well repaid if any of our working men fanciers
derive any benefit from this article. We feel certain there are lots of good
fanciers who look well after their birds, and give them every possible attention, but fail to get the best out of them at the right time through want of forethought. We always believe in trying to look ahead, and we also realise this
is most essential if you want to found a successful loft of racing pigeons.

'' LYNDHURST '' LOFTS.

MR. G. L. TAYLOR, "Lyndhurst,"
Prospect Road, Summer Hill,
Sydney, New South ¦ Wales,

can offer to Australian Fanciers, Squeakers from his Famous Stud of Racers at prices

FROM £2 2s. TO £3 3s. PER PAIR.

The Stock Birds from which these Squeakers are descended were selected by Mr. Taylor during his visit to England and Belgium in 1908, from the Lofts of Mr. J. Wones, of Wolverhampton, closely related to his National winner ; Mr. T. Dobson, of Milnthorpe, a son of Gallant and daughter of Briton, Champion Albert, the famous Lerwick racer, an own son ; a son of Forlorn Hope, from Mr, A. H. Osman, a daughter of Mr. S. P. Griffiths' famous A. pair ; A. P. Taft ; Oliver Dix ; N. Barker —direct from his own loft ; Collignon ; John Wright ; J. Davidson, of Glasgow—his Osman Delmotte, and Stanhope cross ; John Wright, from 702, 708, 738, 728 Stock Birds, and Mr. J. Barcroft's best ; likewise Gits and Jurion of their most famous blood.

No Loft in Australia contains better blood, that has been more carefully selected, than that contained in the "Lyndhurst" Lofts,

The Squeakers can be thoroughly recommended as the blood has been tried at all distances.　ADDRESS AS ABOVE.

Gambling Through Pooling

by J BRUCE

The war has been responsible for many changes in almost every conceivable direction in life, and, if anything, the period after the war has been worse in its general effects than when hostilities were at their height. So far as the sport of racing pigeons is concerned, I must confess to a feeling of 'fright' at the tremendous increase in 'pooling' on *insignificant* races – or, to hit it straight in the mouth, and call it by its proper name, barefaced, naked 'gambling'. A single-up short race was flown from a 60-mile stage early last May, and one man won 1st and 2nd prizes and pools totalling over £86. In the Manchester Flying Club £1,000 was paid into the pools. In sundry open races, where some competitors flew less than 100 miles, and in a few cases not much over 200 miles, thousands of pounds were paid into the pools. Pooling or gambling has undoubtedly become rampant throughout a very large section of the Fancy with money to burn. I am a sufficiently old-fashioned pigeon flyer (who has seen the sport in a good many of its phases) to confess to a feeling of shame that birds have been used as tools to satisfy this terrific gambling appetite. Hands are held up in pious horror if a Sunday race is suggested, but apparently no effort is ever made to limit or restrict week-day gambling on the birds and their chances. Is not the sport of racing pigeons a sufficiently healthy hobby for a man in any walk of life, a splendid and exciting sport in itself, that some must needs satisfy a lust for financial gain by heavily backing their favourites in short sprints? I always understood that the average fancier looked upon these short hops as stepping-stones to greater events, the longer distance tests, where, after long and careful preparation and the weighing-up of form, a man in the old days of the sport was, to an extent, justified in standing on his toes with his tail well up and satisfying his craving and backing his confidence by having a 'flutter'. Much care, forethought, skill, and a very good pigeon, were necessary for those longer events, vastly different to that required for the short flights.

Looking back to my early days in the sport, and reading the records of its progress for years before that, pooling was unknown; prizes were flown for – sometimes small, sometimes very fair amounts in the longer races. The sporting spirit was keen to win, just as it is today, but the pooling element was missing. I do not say that the same spirit or desire to win with one's birds is any different now from what it was in earlier days, but I do believe that the heavy pooling or gambling on the birds creates a more unhealthy atmosphere. Bring an over-plus of money into a sport, and an increasing unhealthy desire to win is created in a good many breasts that would not be bred there if heavy financial baits were not dangling. Fortunately the sport is conducted in a wonderfully clean way, and the chances of fraudulent practice have been reduced to an absolute minimum. They will still further be reduced before many years have gone by.

Granted that the sport is very cleanly conducted, that is no reason why advantage should be taken of it to increase the pooling or gambling spirit, which has made in recent years such heavy progress. Within my own experience I have known of numbers of cases where men who could not afford it have been tempted to pool beyond the strength of their pockets. The tone of a good many men at club headquarters: "Aren't you going to pool; go on, put them in the pool," has been more than a good many members could stand up against. Greed to snatch some of the big plums in big open races has tempted many immature fanciers barely out of the novice stage (often in that stage) to send birds that haven't a ghostly chance of touching. You can see the class of pigeon at the race markings.

In the shorter races the play of the wind, loft positions, sections and all the rest of it, makes pigeon racing a sufficient 'gamble' in itself. Certain fanciers in certain races will always be in a bed of roses through their loft positions – that has always been so from the first race ever flown, and it always will be so.

Am I too pessimistic – too much of a wet blanket? I wonder. What I have been praying for is a return to the old Continental programme and long-distance racing; the cutting of these many war and post-war short races, the using of them as stepping stones only, to the far end; the re-arrangement of the 'pooling' (if pooling fanciers must have), cutting out heavy pools in the short races; that one day some gleam of wisdom will enter the big army of fanciers who chase the 'will o' th' wisp' and finance big open events in the scramble for large and succulent 'plums'. A man in his shirt would have as good chance of catching his shadow by waltzing round. Perhaps this new system of flying for big money in short races is a phase of the war period and the aftermath. Whatever it is, it has dug well in. Is it good for the sport? I sometimes wonder whether it is the outcome of the long months of inactivity from racing and a frantic haste to make all the hay possible whilst the sun is shining during the short period of the racing season.

Do new fanciers enter the sport now quite imbued with the same idea as the earlier novices used to tackle it, with the incentive and the glow before them of some particularly fine long-distance performances? Or are they attracted by reading or hearing of the big sums fanciers have won in races of shorter lengths? Is it a rush to a new goldfield? If that is the spirit there will be more than the usual stampede to back out when the hard bare rocks are hit. I think the golden path to the top of the prize list has been made too glittering for the making of good, sound fanciers, and by that I mean those who are able to stand the hard knocks they will undoubtedly get when a full long-distance programme is tackled. And there must be no mistake about it; the ultimate aim and ambition of every racing pigeon fancier must be sooner or later the supreme long-distance tests. That is what the sport has been built upon, that is what it will always depend upon; if not, it will degenerate into a vast army of more or less nondescripts. The flying of longer-distance races calls for heavy sacrifice – losses are heavy, and always will be; without losses through these tests – the stopping at shorter distances – will mean a big falling away. Cruel these long flies may be, but there is no other way to keep up the same standard or efficiency that has ruled throughout the sport in past decades. You can pool and gamble as much as you like in the shorter races, but the only gain will be a financial one. Neither you, nor your birds, nor your loft, will stand high in the estimation of racing pigeon men. You will have achieved nothing lasting. Reference to names of good birds or owners of them is not necessary here, but one of the pictures I always keep in mind is of one good game pigeon years ago that won a big open race, flying 494 miles, over 16 hours on the wing, against a head wind, flying 50 miles further than the majority of the competitors in the race, being timed in on the edge of dark, when a lamp had to be lit to get the bird to enter the traps. Whether the owner drew a penny or many pounds for a win of that description does not matter. That sort of performance is the essence of pigeon racing. The lucky owners of birds of that calibre, whether they win much or little, can indeed stand on their toes with their tails well up, for it is the culminating point in the career of a loft – achieved by the bird with the assistance of the man, for the two must work together. It is through the combination of the two that many such performances have been done in the past, and will be again in the future, but the guiding spirit on the one side is never upheld solely by what he is going to benefit financially by such a win. The fancier who won't play the game of pigeon racing unless big prizes are dangled before him is very far removed from the class of fancier who has built up the sport from its infancy. The training and racing of the birds themselves gives sufficient sport – and often quite exciting sport – to a true pigeon fancier, without unduly spoiling it with "How much can I win?" or "How much am I going to win?" The man who happens to strike a good quick youngster and then sets about entering it in every possible big event, in order that because of its speed he can gamble on it to win heavily, is not a good fancier, and does not deserve to have a real good one to satisfy his craving for big money wins.

I am not against any fancier winning prizes in a well-flown race, but the man who persistently uses his birds for gambling through heavy pooling, is not my idea of a good racing pigeon fancier.

I had thought of touching upon the present prices of pigeons and the amazing boom in the sport, together with the fascination the big-priced 'pures' have against the cheaper common or garden well-tried pigeon whose lineage is old-fashioned, but I'm afraid my poor old head won't stand more than the thousand and one bricks that will be shied at it as a result of my gamble on 'pooling'.

Milestones
by C E L BRYANT

Some few weeks ago I saw for the first time the well-known play whose title I have borrowed for the heading of this article. As SQUILLS' DIARY penatrates to regions too remote to be visited by theatrical companies, it may be as well to explain that the play covers a period of fifty years, the first act dealing with the sixties, the second with the eighties and the third with the first decade of the twentieth century.

By a merciful provision of Providence we grow old gradually, but the play is not so considerate as there you can see the characters jump from 25 to 50 and from 50 to 75.

One hears of many aspirants to pigeon fame, the interest, no doubt, in some cases, having arisen through the use of pigeons in the war. Probably several of these beginners are quite new to the business of even pigeon keeping, and it is said that it is best to buy experience; but, compared with years ago, they can escape a great deal of painful and expensive groping, as now there are means available from which directions for effective management can be obtained.

The pioneers of the sport in England have, from the inexorable law of nature, passed their last milestone; but the present generation of pigeon fanciers owe a heavy debt to their immediate successors. Perhaps there is no such thing as a generation of pigeon fanciers, and this word hardly seems to meet the case, as though the distemper may be escaped in puppyhood, age appears to be no safeguard against the development of the germ in an acute form.

From about 1880 the sport began to make headway, and a name that will always be held in honoured remembrance in connection with it is that of Mr J W Logan.

From about 1890 the sport of pigeon racing began to expand at a rapid rate, and round about that date I think it is safe to say that the best-known (to English fanciers) Belgian strains were those of Gits, Hansenne, Barker, Delmotte and Jurion, followed by Pletinckx, Janssens, Rey, Servais, Thirionet, Offermans, Debue, Wegge, Bovyn, and a few others.

Old fanciers, and new ones who read their pigeon paper, including the advertisements, may wonder why the great name of Grooters is omitted.

What's in a name? – a lot sometimes. The birds for which 'pure Grooters' or 'Grooters' blood' is claimed, have been, and, sad to say, still are, legion. The Belgian custom of calling a bird by the best-known strain in its pedigree, though there might be only one cross of the blood generations back, has been partly responsible, and this in itself is evidence of the fame of the Grooters. It is human nature to make the show window attractive, and things are not always what they seem, whether applied to pigeons or something else.

A few years ago there was a very instructive and illuminating correspondence on this subject in *The Racing Pigeon*, though the effects, which could be seen for a time, appear now to have worn off.

At a low estimate, fifty per cent of the British racing pigeons of today probably contain some of the blood of the five strains mentioned first, and I expect that pure birds could be obtained of any of these five strains. What constitutes pure is open to considerable difference of opinion, but mine is that for all practical purposes a bird cannot be called pure Gits unless it descends without a cross from birds bred by M Gits. Even this very moderate working definition of purity – commercially pure – which would be anathema to scientists, would, I feel sure, knock out a vast number of birds which are described pure this or that. Numbers preclude anything in the nature of a general stud book, and it must be admitted that for various reasons, one being that very few Belgians seem to have kept breeding records, exact pedigrees of Belgian importations were difficult to obtain. Mr H W Doll no doubt has a lively recollection of what research in M Vanderhaeghen's old pocketbooks meant.

This carelessness about the accuracy of pedigrees is regrettable, and it is a subject I feel rather sore about, as more has been claimed for birds I have sold than I myself claimed for them. Birds sold as containing some Grooters have, in other hands, become pure Grooters. It is a state of affairs that seems to be tolerated with pigeons that would not be tolerated with any other livestock. Fancy calling a horse a pure St Simon; it would be perfectly ridiculous. Referring back to a remark about the best strain in the pedigree, it may be of interest to give a few instances. I bought a hen that was in the late M Carpentier's sale list as pure Grooters. No details could be found out at the time, but later, when Mr Doll was in Belgium, he kindly made some enquiries and found that the sire of this hen was bred by Mons E Grooters and that the dam was bred by M Delmotte, this hen always being known in the loft as 'la mere Delmotte'. To carry it further, some few years later I bought another hen described as 'pur-pur, pur Grooters', and by a simple method that I had discovered of reaching a Belgian pigeon owner's heart, succeeded in finding out exactly how she was bred. Her sire was a mixture of Grooters, Coenen and Carpentier, while, oddly enough, her dam turned out to be an own sister to the Carpentier hen (half Grooters, half Delmotte) previously mentioned. So much for the 'pur, pur, pur Grooters'; but I do not suggest that there was any intention to deceive. It was custom – Belgian – but I can't see why it should be ours too, who pride ourselves on our studbooks and the purity of our livestock.

There was another very shadowy claimant to the title – a Grooters cock purchased at a sale. I had not bought him expecting him to prove one, but for other reasons. Through the kindness of M Durieux, then editor of the *Martinet*, who must have taken a lot of trouble, we traced his pedigree back by means of old sale catalogues till we got to 64 ancestors, and the best we could do for him, though even then it was not quite definite, was 6-64th Grooters. A typical Grooters instance was Thirionet's famous Julienne (bred by Jurion). Her sire was the Fameux Gros Bleu (Jurion), and his sire was a son of the Vieux Boeuf de Grooters. The dam of the Gros Bleu was a daughter of Diable (Jurion). The dam of Julienne also was a daughter of Diable, so Julienne was an inbred Jurion with a Grooters cross two generations back. Eloquent testimony to the reputation of the Grooters, but not satisfactory to admirers of our British stud books.

Of the Barker strain I have had no experience, and personal experience with the Delmottes and Jurions has been very slight; but I have had birds from M Hansenne and M Gits.

The Donkeren craze with those owning birds of the latter fancier's strain seems to have dropped out, and in connection with this it may be interesting to mention that 27 years ago – in 1893 – I bought a cock from M Gits, and the sire of this bird was the youngest son of the Donkeren who was hatched in 1875!

Another popular card was the old red Soffle Gits cock, which one not specified, as there were two of them – brothers, 893 and 927, both winners in the National races. Ten or twelve years ago 'his' children were nearly as common as blackberries; yet, 19 years ago – in 1901 – Mr Doll bought one of 893's three remaining daughters for me, and 927 was dead.

The last birds I had bred by M Gits were bought in 1907, two cocks and a hen. Mr Romer came to spend the day shortly after they arrived, and on entering the loft immediately spotted the blue chequer cock as an 'old Gits'. This blue chequer had not struck me as being typical of the earlier Gits birds; but the other cock, a red chequer, had; in fact, I should have picked him out anywhere as Gits. Mr Romer did not consider this red chequer typical of the strain; but the blue chequer must have been so, though I couldn't see it, as Mr Romer had not been told what the bird was.

M Gits was one of the few Antwerp fanciers who used to take prominent positions in the Grand Nationals. He must be an extraordinarily clever breeder to have maintained the type of his birds in spite of the number of crosses introduced.

The type of the late M Hansenne's birds appealed to me very much, and I think he would be well in the running for the honour if it could be proved who was the greatest fancier that there has been. One noteworthy feature of his birds is that at the time of his death (I think it was 1904) there was no red chequer or mealy in the loft. Beginners should bear this in mind when buying pure Hansennes.

I should like to have some of my old Hansennes now. Among others there were an own son (1885) of the mighty St Vincent, 'the best pigeon in

Europe', given to me by Mr Doll; a daughter (1889) of Calvi; a sister (1891) to Bruxelles 2nd, beaten by one second, in the Grand National from St Jean-de-Luz, 5,543 competing. The St Vincent cock was not quite of the usual Hansenne pattern; this was the bird mentioned by Mr Doll, in an article he wrote about M Pirlot's loft, that refused to trap in a National race, losing the pool money. As he was bred by M Pirlot, he may have been from one of his own hens, and not from one of those purchased with St Vincent from M Hansenne, which would account for the difference in type; but this is only supposition.

The Verviers birds are most attractive, the Delongs are similar in many ways to the Hansennes, and the Delrez are pleasing, with a fine record of long-distance ancestry. M Delrez has used the Delong cross; a notable Delong in his loft was a bird called Mairlot (named after a loftman of M Delong's), that flew 25,000 miles during his career, and won no end of prizes, including 1st St Benoit, 1st Bordeaux, 1st Biarritz, 1st St Jean-de-Luz, 10th Vittoria (700 miles), and many others, from St Sebastien, St Vincent, etc – a champion if ever there were one, and evidently no believer in an eight-hour day.

The Barcelona record of M Delrez's wonderful pigeon (1896) of that name is interesting. – 1901, 43rd; 1902, 72nd; 1093, 10th; 1904, timed, and the same year 151st Dax in the National, 5,000 birds; 1905, 6th; 1906, 2nd (Liege), 5th Brussels; 1907, 1st; 1908, 2nd (Brussels), 8th (Hirondelle Liege), 9th (Cornet Liege). This old champion must have been made of iron, and an extraordinary fact about him is that he was bred from a brother and sister. The sire of Barcelona had also done a bit more than his share; he was called Le Bayonne and one of his achievements in 1895 was winning 11th La Couronne, and the following week 1st Bayonne and 11,000 francs in pools. M Delrez did not keep his birds in cotton wool.

Several years ago I used to spend an occasional day at Lancing, and a more wonderful collection of pigeons than that owned by Mr Doll it would be impossible to see. It was a collection rather than a strain, as many of the birds had been presented to their owner by his various Belgian friends; but the individual pigeons were bad to beat. Mr Doll was generosity itself, and many a good bird came my way as a present or loan. I think his red cock, the last son of M Gits's 927 mentioned earlier, one of the most nearly perfect pigeons I have seen. Mr Doll always seemed to be particularly sweet on his Pletinckx, especially if they were a dirty blue with some white about them.

However, these rambling reminiscences refer rather to milestones that have been passed, and after suggesting that as the late N Barker was an Englishman, his strain probably is more largely represented in this country than any other, I must pass on to the present and future. Mr J W Logan, Mr H W Doll, Mr A Darbyshire and the late Mr H W J Ince have all played an important part in introducing to this country the parent stock that our racing pigeons of the present time descend from.

There is no doubt that the war has made a large number of people who, prior to 1914, either were not aware or hardly aware of the fact that there was such a sport as pigeon racing, take an interest in it, and though the interest may never become active, it must be all to the good that many more than formerly now know that there is such a sport.

Newcomers may be of all ages, shapes and sizes; but though this is of no moment for the part they have to play, it won't do if it applies to the birds with which they are going to start their lofts. Those who happen to have a friend with a loft that they have taken an interest in will probably have assimilated his type of pigeon as the type they intend to keep. The difference in type of the birds in the Army lofts was remarkable. A loft of 150 birds might have half of that number, or more, owners represented, and it was interesting when a strain happened to have a few representatives, as most of them could be picked out.

Those who have eyes to see – and unless they have, real success as fanciers is unlikely – will find a very marked difference in the appearance of the birds as a whole in the lofts of successful fanciers.

Having had a fairly lengthy experience, some of the following suggestions may be useful to beginners: First and foremost, in founding a loft I would place importance on strain – one that is alive and going strong. I would rather by far buy birds from a successful loft, with what for all practical purposes could be called a strain of its own, than buy chance-bred birds from what I will call a 'made-up' loft, even though the parents might be of individual merit. There is a saying that blood will tell, and I would rather go to a loft with a strain and take what the owner (jealous of its reputation) offered me, though in their own loft the birds might not have been actual winners, than buy individual good performers without the breeder's and family record behind them.

There is a considerable difference of opinion as to the respective merits of in-breeding and cross-breeding, and I think it is a question that resolves itself into the ability or otherwise of the owner. I think that every loft whose owner has established a strain will be found to have practised in-breeding to a certain extent; in fact, it is difficult to see how one can establish a strain without employing this method.

I agree entirely with Colonel Osman's advice to beginners not to buy promiscuously, but to obtain birds from two or three lofts only; but would add, that in my own case, if I were doing so, I should want to satisfy myself that the types were similar. I am sure that no lasting satisfactory results are to be secured from mating extremes together; it is possible that the first cross might succeed, but what is going to happen later?

What is of almost equal important to starting with the right stock birds is, when you have got them, to give them time. Success may not come at once, but if it doesn't, do not blame the birds. Wait to see what the second generation is capable of when they will have become more acclimatised to their new environment. Another point is that when the treatment is found out that suits the particular birds you may have, do not change it because it happens to differ from that of someone else.

The late M Vanderhaeghen was one of Belgium's greatest fanciers; his record from all distances was most remarkable. As I had many of his birds, he, through Mr Doll, very kindly sent me his system of feeding, which was varied and elaborate – very different to my own. Under these circumstances, as they had to fall into line with a new system, it is not to be wondered at that I never succeeded in getting a pure Vanderhaeghen (that is, a bird bred from two of the original importations) to fly 300 miles; but I should have done a very foolish thing if they had been cleared out, knowing what their breeder's reputation was. The birds were all right, but required time to adjust themselves to new conditions, and, later on, the majority of mine, if the pedigrees were taken far enough back, would work out to be anything up to half Vanderhaeghen. The Bekemans, another of the original strains, had probably been used to very similar treatment to my own, as the change of home seemed to have very little effect on them.

Beginners now will not be going to Belgium, but buying their stock from home lofts, so the change of climate, etc, for the birds will not be so marked; but if – an important if – they are sure that the birds they have are of good enough strain, my advice to them is not to throw up the sponge in disgust should immediate success not come their way.

If the initial keenness is retained and the birds are well reared and treated on commonsense lines, some success is bound to follow, but the measure of it is a doubtful quantity and not one that can be forecasted. After a start has been made, what is to follow is largely a personal question, and if half-a-dozen men were started on equal conditions, one of the six is going to be on top in a few years' time.

Concentrate on the longer races as the ultimate object in view; young bird races and the shorter old bird races are merely the means to an end and milestones in a future champion's career. Real champions are very scarce indeed; calling a bird champion something won't make him or her one.

I am afraid what I have written will be stale news to old fanciers, but I hope if some beginners happen to read this article they may find something to interest them, and also feel encouraged to persevere if their interest in the sport is beginning to slacken.

How my winners have been Conditioned

by J T CLARK

Great Northern Marennes Winner

Having been asked to contribute an article to SQUILLS' DIARY, I will endeavour to describe my methods as nearly as possible.

I have been keeping racing pigeons for about 25 years, and it has always been my ambition to win the cross-Channel races.

I generally keep about 25 to 30 pairs, but do not breed a lot of young from them.

Few fanciers realise what a great strain it is for a bird to race 500 to 600 miles, and to be on the wing 15 or 16 hours at a stretch is a long day's work, and nothing but a sound bird in the best of condition can accomplish it.

I separate my birds from September to March, and during the winter months I like to carefully watch the birds and choose which I shall mate together in the breeding season, and which I will send to the longer races.

I mate my birds up the first week in March. They are allowed to rear the first round. Those I intend for the cross-Channel races are not allowed to rear any more until racing is over; they are then allowed to rear one. I think it does them good. I let the yearlings that are mated together rear one during racing to see how they shape feeding. Some birds race best feeding; others do better sitting.

My Great Northern Marennes winner was sent sitting 12 days, and was due to hatch day of toss. She was sent to Nantes, 1920, in the same condition, when she won me £80.

To get the most out of any bird, you must study such bird's peculiarities; there are no hard and fast rules that apply to all of them.

The training I give my birds is as follows:– Youngsters, 4, 6, 10, 15, 20, 35, 47 miles, first race 68 miles. I stop a few after doing Stafford (112 miles). I do not care to send them further than Worcester (152 miles). Yearlings have the same tosses. I stop a few at Bath (208 miles), the others at Bournemouth (255 miles). I very seldom send yearlings across the Channel. Third season birds have the same tosses as the youngsters up to the 35 miles stage, then 68 miles, 112 miles, 255 miles; they are then fit for the cross-Channel races.

As my employment does not allow me to attend to my birds at regular times they are hopper fed. I always give them as much as they can eat. I fill the hopper at 9 am, 12 noon, and 6.30 pm. I also put some in the hopper at dusk ready for the early morning. I feed on mixture during the breeding season. When racing commences I feed on two parts maples, one part tic beans, with a few handfuls of Velo after the evening fly. I also give a little Velo after a training toss or race.

After the moult is over I feed on small maize until the middle of February. I always have a supply of good grit before them, a constant running supply of water, also a bath outside the loft all the year round.

I am very particular in keeping the loft clean. I do not use sand on the floor, but scrape and sweep the floor three times a day during the breeding and racing season; once a day is sufficient during the off season.

I do not force my birds to take exercise until racing commences; they are then made to fly for half an hour once a day in the evening. This is increased to one hour as the longer races come on. They are allowed full liberty up to 12 noon. This enables the hens to get out for exercise. I close the trap at 12 noon until 6.30 pm. It is then left open until 12 noon next day. For the forced exercise I turn out all birds half an hour before sunset.

I have not changed my method of feeding and training for 15 years, and I can claim a fair amount of success in the cross-Channel races.

My birds are fed on Hindhaugh's No 1 mixture both in the breeding season and during the moult, that will be complete about the middle of December. I also give them a little linseed every other day. They are hopper fed until the end of November, then once a day until the middle of February. I always give my birds as much as they will eat. I feed on maize from the middle of December to February.

I am not very keen on showing, although I like to send to a few shows during the off season; it adds interest and helps to get over a rather dull time. I do not give my birds any special preparation to fit them for the shows, but just pick out what I think will suit the judge. I do not turn them out for exercise the day before sending away or on wet days the week previous to the show.

When sending young birds for training I generally basket them the night previous. This teaches them to sleep in the baskets. I also put on the water trough for a drink, but I do not send them until the morning by rail. If sent overnight there is the risk of a bad morning, and as the train service is good up to the 90 miles stage I can get them through in time to be tossed by 12.30 am, that is 11.30 am by the sun, which gives them ample time. This is also about the time they are tossed for a race. The old birds are basketed the same morning as toss, up to the 90 miles stage. They then go over night to a liberator, half cocks, half hens, separated.

I am very fond of a good hen for racing, especially in the long distance races. I have also won good prizes with cocks. My best performances this season has been done by hens. Hens are rather more difficult to get fit, through having to lay, but they are more reliable than cocks when you catch them fit. I always like my best birds to have the same nest boxes and the same mate each season if possible. My two birds that won 5th and 7th Lancs Combine have won between them £250 this season, have been mated together two seasons, had the same nest box, and homed together from Nantes (498 miles). My Great Northern Marennes winner also had the same nest box and the same cock as she had the previous season when she went to Marennes.

The type of bird I prefer in the show pen is a good medium-sized bird, long cast, must come into the hand nicely, good length of keel, close in the vent, well feathered with good flights and secondaries; bright eye, not particular to colour of bird, but must be sound in colour. A pale coloured mealy or a blue with narrow bars, deep keeled, and lobby birds I don't care for; but above all they must be in good condition.

POOLING

I am very fond of pooling my birds. I like to see my choice come up first, even if beaten in the races. I do not handle my birds. I pick out my pool birds by the way they are going about the loft. I only handled my Great Northern Marennes winner once – about ten days before she went to the race. It is a great mistake to be continually handling your birds; they don't like it, and it also takes all the bloom off them.

I hope these few hints may help readers of SQUILLS' DIARY to gain success, but can assure them it needs plenty of hard work and care.

Methods that have gained my Success

by J W TOFT

In reply to your request I am pleased to write you my methods. I have kept pigeons nearly all my life, and commenced seriously with racing homers in 1893. I watched to see the winners in the long distance races, and bought squeakers from the winners. The first year I lost every bird I bought but had bred youngsters from them. Mr W H Bell got the fever, and I took him round the various lofts and he paid colossal prices in those days for birds that had won, 'Liverpool Pilot', 'Maritana', etc etc. He gave Barker carte blanche to send him four of the best pairs to be got in Belgium. I had youngsters from the Barkers, and all the birds he bought in Liverpool from Wilson, Gibson, etc, passed into my hands, and it was these birds and the Logans I got from J Matthews, Wadsworth Wilson, Mr T H Harrison, and from W C Moore that were the foundation of my loft. I have never kept any prisoners, and never kept a bird to look at, but have always raced them year after year, and bred from the best long distance birds. I think I was the first in this district to send the best birds in loft to Bordeaux. This was really the result of a bet with Mr Romer, whom I met at the meeting we had at Mr Logan's. I nominated two birds to beat his previous yearling winner. This was months before the race. Result: I sent five birds to Bordeaux in the National. Won 1st, 5th and 8th. The two I nominated never came, but they went to the race and I did not lose my bet.

The first few years I nursed my birds till their third season before sending them across the Channel. I have changed my methods somewhat, and now send yearling across. But if I have any yearlings that have flown well as youngsters and as yearlings in England I invariably keep them till the following year before sending them across. My object is to win the longest club races. I had six birds fly Marennes this year, five on the day won the race last year. I also won the Gold Cup for best nominated two-bird average in the four Championship Continental races this year. One has got to know the birds and have them in condition for this very wide radius – Southport, Cheshire, etc.

As regards training I pair up in March, let all the birds intended for Channel races rear one youngster in the first nest, and after that pot eggs, unless I have a cock bird that will fly better to a youngster. My experience is hens always do best when sitting about ten to twelve days, and I have had more success with hens than cocks. I am a believer in plenty of exercise round home, early morning and late evening. My birds have had very few tosses of late years. I like to send them a few miles, single up, the morning before putting them in the baskets for marking. Feeding: I always give them as much as they will eat; prefer beans and a little canary seed for racing on. Always separate Autumn to early Spring; and feed on anything so long as it is sound. This season, good English wheat and best Tasmanian peas. Always Squills and de Lacy grit. Rock salt in loft till paired up, and then taken away. During racing season a good strong dose of Epsom Salts is often beneficial, especially if the birds are not up to concert pitch. I am no believer in long pedigrees. What I want is work. There is no room in my loft for anything that will not work.

I have found long distance pigeon racing a good healthy hobby, as it keeps one at home, and there is always something to learn by watching the birds.

The mating needs to be done with care if you wish to breed long distance winners, but if you only breed from strains that are known to be of the best long distance blood you cannot go far wrong.

I often go round the Garston working men's lofts, and always find them at home with their birds. This is good for the home and good for the family. They get their birds tame, and their birds feed out of their hands outside. These men are very keen, and their lofts are kept scrupulously clean. I have given them dozens of eggs to encourage them, and they do not fail to rub it in when they beat me, but this is all in the game.

I used to think 50 miles far enough for young birds, but do not today. I send all my young birds through to Bournemouth, over 200 miles, and this year they were all given single up. With plenty of food this work does not stop their growth, as some may think, but it adds to their knowledge, and you get rid of the duffers early instead of wintering birds that will let you down later.

I think you can breed good young stock from yearlings, or even latebred birds in their yearling stage. In fact, I am very partial to the birds bred from maiden hens, that is to say, the young hatched from eggs they first lay. Some good birds I have owned have been bred from yearlings.

Early in the morning my birds are kept flying for half an hour, but I attach most importance to the night fly, when they are put on the wing and kept flying a good forty minutes towards the longer stages.

Success depends upon a good strain of pigeons, but fitness counts, and unless birds are absolutely sound and fit they will not win.

It is much more difficult for me to write about my methods than to have a chat with a fancier on the subject and talk pigeons.

My loft is a wooden structure on the ground, with plenty of air running through it; cool and well ventilated; also plenty of room for the number of birds I keep.

I have never kept a large flock of hundreds of birds. You cannot keep so many good ones.

I do not show my birds, as they are not bred for that part of the game, but for racing; and nothing delights me more than to see a real good one land from a 400 or 500 miles race. Many of my winners that have flown Nantes, very shortly after arrival have so soon recovered that they were difficult to spot from other birds that had not been racing.

Make a note of those that can stand a long journey without distress. These are the ones to pool next time.

How I Won from San Sebastian

by ERNEST A TURNER

I am a good reader, but a rotten writer, especially on this occasion. But I hope to live long enough to be a really good pigeon fancier, and I hope my fancier friends will forgive me if this article lacks interest.

My father kept pigeons when I was born in 1883, in Kennington. I caught the complaint from my eldest brother, S J Turner; he, being a very close friend of Alf Donaldson, was able to secure some of Donaldson's best birds at that time, the Red Listowel hen blood. My brother secured also two squeakers from T W Thorougood, and two from the sale of the late Schrieber's birds. These three fanciers' birds are the foundation of my present-day loft.

To the fancier who has kept racers, and has never had his share of winnings, I say this. Start again. Clear out every bird; never mind who gave you the old cock or the old hen, scrap them all. If they had been any good, surely they would have bred you a winner, or their young would have bred you a winner or two. If they have bred winners for others, or supposed to have done, are you at fault? Have you been spending your spare time at other sports and pastimes in the moulting time and the breeding time? If so, you must not expect to win pigeon races in the company of real good pigeon enthusiasts.

To be on the job, race after race, you must leave nothing to chance; you have got to be with the birds every moment at your disposal, for there is always plenty to be done. Always be on the look out for the birds that do not look healthy. Pick out the mopy ones; these have got trouble in front of them, they will not breed you a champion. Birds that are sick and sorry will not fight out a finish on a hard day.

To the beginner I advise him to go to a good fancier of some standing and buy some of his latebreds. Carefully nurse them during the autumn and winter, pair them up in May of the following year, and don't forget to feed them well. Let them have as much food as they can eat, for they have not finished growing yet. Exercise is of great importance at all times, especially on fine days, not forgetting the bath. Take my advice and thoroughly digest *Squills'* 'Food for Novices' each week, and you cannot go wrong.

My own loft management is simple.

I separate the sexes in September, except the latebred ones. These I put in the young bird loft, and let them have this loft to themselves until I think they are getting too forward. I then part the sexes of these, and let all the young cocks have the YB loft. I think they require more food than the old cocks. I let out both sexes on Sundays together, and let them have a bath, and part them again in the afternoon.

I like to cut down the corn ration 14 days prior to pairing up. You want the hens on the thin side when you pair up, or you will not get a good hatching. Fat hens mean a bad breeding season, soft shell eggs, thin shell eggs, young-sters dead in shell, etc. A dose of salts will help matters considerably. I think this important, especially with prisoners; they may look all right, but they may be fat inside.

I have no fixed date for pairing up. If the weather is mild in March I shall pair up my stock birds the first week and my racers 14 days later. My best racers will not rear their first pairs of eggs. I generally put them under the yearlings I am not going to race out, such as later bred birds.

As to my feeding, I use a mixture of beans, peas, tares, wheat, as supplied by my corn dealer. I never use small seed, not even for trapping.

As to training the birds on the road, I like leaving this till the last moment, for I think the birds come on, whereas if you give them a lot of tosses before racing I think they tire, and come much slower. Two weeks before the first race I generally start at two miles, then five and ten and 20, then 45 miles; first race 77 miles. I like them to do the 20 miles toss twice.

I think it is a mistake to give the birds rock salt. At any rate, I never give mine any. I have never yet had rock salt in my loft unless there is any in the grit I use, but to put it in the loft I never have. Look at the youngsters and the old birds that do have it; they look soft and flabby, loose inside, dirty look-ing, etc. They cannot be in top hole condition, and the water they drink must be enormous. At any rate, I think rock salt is too salt, far too strong, but a lit-tle table salt sprinkled on some lettuce or watercress is good for them, and also helps to keep them away from the fields.

I like to keep the loft clean, and usually, if not always, clean them out before I feed them, taking out the drinkers beforehand, and put my fingers round the inside of them to get out any filth, for clean water is law with me.

I use sand, for it keeps their feet clean, and limewash loft twice a year, once after the moult and once when putting in the nest boxes, and use Keating's powder freely in and around the nests. I use sawdust for the nest pans. I always wash my youngsters when they leave the nest. I wait for a nice warm day, and put them in the trap to dry. I think this gives them a good start in life, and healthy birds should not be seen dirty again after this.

Just a few words about the champion might be of interest.

I used to work with Mr Brett, the breeder of the good bird. We worked side by side from 1909 till 1920. Brett saw me reading *The Racing Pigeon* the first week he came on the firm to show us his merits as a cabinet maker. He had been a member of a Dartford FC.

We soon got going on pigeons, and I gave him a re-start. He raced young birds in 1914, when the war put a stop to his and my little plans. In 1920 he had only three pairs of birds and about six or seven squeakers. He asked me to try one of his youngsters. I, knowing the family from A to Z, soon spotted the one I wanted, and he has turned out to be the cause of me writing this article, but I am very lucky to possess this good pigeon. I will tell you about 'luck' in this pigeon.

I had only had him 24 hours when I let him out with some other youngsters, and they got on the wing and flew till dark. I lost two out of four that took off, and the champion was one of them. About four or five days after I was taking my clock to the clubhouse, a gent in the saloon bar told me a railway platelayer had picked up a bird on the line, and would I send round and get the bird and try to find the owner? Now we were racing from Rennes, and I went upstairs to hand in my clock. I found the race was going to be a very tight one, and the excitment of the race, as I turned out to be the winner. I was winning nearly every race I entered for that year. I quite forgot to mention about the bird the gent had spoken to me about when I entered the clubhouse (a pub, of course).

About a week after I received a PC from RP telling me Mr Sumpter, then president of the Croydon HS, had in his possession one of my birds. I went to see Mr Sumpter, who was very sore at me not going to get the bird from the railway man. I asked him how he came by it, and he told me the man had taken it to him to find the owner. Having thanked Mr Sumpter and greased the hand of the platelayer, I thought then, what a good youngster not to go into another loft, and, believe me, reader, there was only skin and bone left of this game youngster. I never had any more trouble with him, and he flew very well as a young bird, and he dropped with my winner from Exmouth that year, they being 1st and 2nd Fed, a very hard race, vel 928. He should have had the win, for he was first in the loft, but the other one who was with him was my fancy, and was also in three clubs, and the champion was only in two clubs, I not having him in time for nominating in the other club, for he was very late bred for YB racing, late in May, if I remember right.

Shall I send him next year?

I don't know. There is a lot that might happen before next July. Do I think he has done his bit. Yes, I do. For his age I know of no other bird which has flown so many races in three years and been in the clock every time.

But in my opinion these are the sort to send to long races, birds that will race very time out. If I thought I could do without him I would not chance him any more, for he may go once too often to the well. That I have in mind.

Did I see him arrive from San Sebastian?

No, I did not. I will explain.

I had been training my youngsters with a view to giving them a toss of 45 miles the day I expected to time in from San Sebastian.

Now, reader, I did not want to be caught napping, so rung up The RP Office asking if they had heard if the birds were liberated, and they said they had not heard anything yet. Time 2.30 pm. I went to business, and arrived home again at 8.30 pm, went straight to the loft. I never went in the house for quite half an hour. The other folks were not at home.

On going in the house I find on the table my telegram of liberation. I say to myself, "What about the timing clock?" It has been running eight days. I will get it reset first thing in the morning". Just then a local fancier called in, who was also competing in the race, and said he could find a volunteer to take our clocks to be reset in the morning. The morning arrives, the clocks go, the young man assuring me he will be back in an hour or two, and he really thought he would arrive back in that time. I then sent the young birds for a toss to Petersfield, and told the wife I would be back to lunch at 1 o'clock. I then go to business (Croydon), having done what I wanted to do. The time was then 1.25 pm, a tram ride, 15 minutes, and I am home again. I give the birds clean water. I go in to lunch, putting my chair in a position so that I can see the first home, naturally. Half way through my meal what do I see?

Not the Petersfield youngsters, but the San Sebastian cock standing on the old bird trap! Down went knife and fork. I shout to the wife, "I've got the red!" I need not say how pleased I was.

The time was then about 1.50 pm, I should think. I let him in. He flew in his nest box. I picked him up and kissed him, stroked him down and kissed him again. He had a drink before I did this , for I coaxed him in by putting the drinker in the trap, and he drank heartily.

Now the trouble started, for I had no clock. The young man had not returned with my clock. The wire I had prepared for verification was useless without the clock time, so I had to hunt for more telegraph forms. These had got mixed up with various books and papers, and, what with the excitement, they wanted some finding. Having found them, what shall I do for the best? Rules, NFC rules, I had seen in my RP three or four weeks ago. Out come a bundle of RPs.

I read the rules, and treat my clock as if it had stopped. I write rubber ring and bird's ring numbers, forgetting to put the bird's wing mark on the wire. Away goes friend with all speed, then it dawns on me there's no wing mark on that wire. I rush for all I am worth to catch up the runner. He had gone about three hundred yards, and a whistle brought him back. I write another wire, and add W29. This done, I ring up RP, and learn there is an arrival at Exeter. I ring off and hope for the best.

In conclusion, I need hardly say how hard I have worked to obtain the blue ribbon of this fascinating sport.

How Records are Made

by L GILBERT

Records are made by birds sent to the races which fly home faster than other competitors' birds. Yes, this is undoubtedly a fact, but when one has been asked to write an article with the above title, one cannot expect to receive the blessings of *'Squills'* for such a bald answer, so one must resort to the time honoured phrase "that many roads lead to Rome", and endeavour to show the safest and easiest road that leads one to that most desirable place, the 'Mecca' of the racing pigeon world, and record a few of the most interesting performances.

What are records?

The governing body of our sport is silent on the matter, and does not record any outstanding successes, as do the governing bodies of all other sports. The following are just a few which must stand as records, and undoubtedly my readers can come forward with many more.

J T Clark of Windermere, when he won 1st Marennes Great Northern in 1921, received £403, the largest amount won in a pigeon race in this country by one pigeon.

S P Griffiths of Northwich, holds two records. First, he has won 1st Marennes Great Northern on three occasions, viz 1913, 1914 and 1920, these being in succession, the war intervening, and two of them, viz 1913 and 1914, were won with the same pigeon, the famous blue chequer racing hen 459.

J Wones of Womborne, surely holds another long distance record, for he has won three premier events of the fancy in 1st San Sebastian in the open race of 1907, promoted by the Market Harboro' Club, 1st Mirande 1908, National Flying Club, and 1st Pons, 1914, National Flying Club.

F W Marriott of Saltley, the winner of four premier races in 1st Marennes, Midland Counties Combine, 1911; 1st Lerwick, North Road CC, 1920; 1st Lerwick, North Road CC, 1921; and 1st San Sebastian National Flying Club, 1924, holds a record any man can be proud of.

F Mattock of Old Southgate, in 1907 established a record which will stand for many years, for he won 1st Thurso and 1st Lerwick on the same day in the North Road CC, one a 500 and the other a 600 mile race.

Lt Col A H Osman of Leytonstone, made the earliest records on the North

Road into London from Arbroath in 1891, Aberdeen 1892, and from Thurso in 1894, winning the latter race by nine hours, since when many 1st prizes have been won from Banff, Thurso and Lerwick, no other London loft showing so many firsts from these places.

J W Toft of Liverpool, has many records to his credit, such as: 1898, 1st and 2nd Marennes, 526 miles, only two birds home in race time. 1st Bordeaux, 1899, National Flying Club, and in 1900 clocked four birds from Bordeaux day of toss, 599 miles. 1921, 1st Marennes, Liverpool Federation, only bird clocked day of toss.

J W Logan of East Langton, following many splendid performances, set the seal to his fame with 1st San Sebastian, National Flying Club, 1922, with his 1826, a red chequer hen bred in 1918 by himself.

In 1902 J L Baker of Sedgley, matched a hen he called Little Wonder against a cock surnamed Victor, flown by E Reynolds of Kensal Rise, in the Grand National from Bordeaux, and they finished 1st and 2nd in the race in the order named, 1,601 birds competing.

His Majesty The King, by winning 1st Banff National, 1923 and 1924, with the same pigeon, a fine blue chequer cock, holds a unique record.

E E Jackson, whey flying to Wheelton, Lancs, put up some astonishing records. He flew two youngsters from Marennes, 548 miles, known as Excelsior and Peter the Great, and also flew a younster called Marvel from Bordeaux, 600-odd miles. In 1914, flying with the Manchester FC from Dol, he won the first ten positions, club and open, 1,001 birds competing.

W H B Peters of Plumstead, won 1st San Sebastian, National Flying Club, 1921, with Royal Blue, a big blue cock, and in 1923 he won 1st Bordeaux, London Federation open race, while another London fancier in E A Turner of Norwood, won 1921, 1st Marennes, London South Road Combine, and in 1923 he won 1st San Sebastian, National Flying Club, with the same pigeon, a red chequer cock surnamed The Champion Cock.

T H Burton of Liverpool, will always be remembered as the winner of the first race ever flown from San Sebastian, he winning the event from there in 1907 promoted by the Manchester FC with a blue chequer cock known as Alfonso.

Dr M E Tressider of Westcombe Park, holds the speed record for a 600 mile race, he winning 1st Lerwick with a velocity of 1,684 yards per minute in 1913 with a red chequer cock known as Champion Pollution.

So much for a few of the records.

Now for the most difficult task of all – that of how to make them.

First – your stock must be of a family of pigeons that has won and is still winning from the 400, 500 and 600 mile race points. Families which won years ago and are a back number today are of no use. Two and three hundred mile winning families do not produce the long-distance winners, but the latter do and can win the shorter events.

By reading 'Food for Novices' weekly much can be learnt as to the training and conditioning of a bird, so that it is capable of flying the long race points, and no good purpose will be served by my detailing that which appears weekly therein.

I favour the old Verviers type of pigeon, which is apple bodied, close vented, broad shouldered, shallow in keel and medium in size.

I watch my pigeons flying, and find that the best fly at the head of the pack or on the extreme outside, and these I put my confidence in when sending to the races. A good pigeon never drops excreta on its perch, for a pigeon that does that is a lazy pigeon, and you will usually find it is one that is too lazy to race.

In February of each year examine your pigeons' tail feathers. If the middle two have commenced to fray at the end, this indicates weakness. These two, being in line with the centre of the bird, are a sure guide to a bird's vitality. When your bird is in the height of condition you will usually find on blowing back the small feathers along the keel that the skin is clear and free from scurf, and on pressing the skin tight across the keel you will notice the little blue veins.

The successful pooler must study all these points, and the man who does not study these and many other little things which happen daily in his loft can never hope to be a consistent and successful racer and a breaker of records.

How to Obtain Condition for Exhibiting

by R J WORTON

It gave me great pleasure to receive an invitation from *Squills* to write an article for his Diary on Showing, and the methods adopted by me for conditioning my birds.

It is nearly a quarter of a century ago since I exhibited my first bird, and still cherish that 2nd prize card, but by what my parents have told me I was a very keen fancier before then, as at the age of 3 my late father had a nice lot of tumblers, and when he was away in trying to catch them I pulled all their tails out! What happened is best left unwritten.

The great art of showing successfully is exhibiting your birds in the best possible condition, and studying the different judges' whims and fancies.

To obtain the former you must have birds with sound constitutions, then, with good management, perfect health will ensue; but must admit those of us who live in the country have a great advantage over fanciers who reside in large towns and manufacturing districts, where the air is not so clear and pure, and the chimneys belch forth clouds of black smoke.

My lofts are all wooden structures with wire fronts, which face south, but on the north side there are sliding shutters, so that during bad weather either side can be closed up, thus keeping out all rain. The floor is always dry, which is most essential, as pigeons with wet feet will never carry any bloom. During the summer months both shutters are open, with a current of air passing through the lofts; they are always cool. Pigeons love fresh air, but the thing to guard against is dampness.

In the showing season my birds are fed once a day, at about 11 am, choosing this time because they can then be despatched to the shows with their day's rations, and are quite all right until they are fed at the show. They are given as much as they can eat, but never leave any food lying about, as it only becomes soiled, and does them more harm than good.

I feed on five-eighths Tasmanian Maples, one-eighth Australian wheat, one-eighth small maize, and one-eighth beans, buying the grain separately and mixing it myself. The only time canary seed is used is when birds have been delayed in transit or return from long distance shows, thinking it just aids digestion, as it does after a hard fly.

I consider Squills' and De Lacy's grit invaluable for getting birds in show condition.

On fine mornings I like my birds to have plenty of exercise, and during the fine weather experienced during October and November they would readilly fly for an hour each day. I call them in directly they drop. Their appetite is always better after a fly, and it also keeps their wings strong and body muscles firm.

Some fanciers do not agree with much flying during the show season. Providing the weather is right, I strongly advocate such exercise, as I think fat, lifeless birds fail to appeal to keen racing judges, and always remember it is the judge you have to please to win.

Birds sent out in perfect condition often have their chances marred by not being trained to show themselves in the pen, as wild birds try a judge's patience. When they are squeakers is the best time to give them a few lessons. A method adopted by myself is to use show pens for breeding cages, putting them on a shelf about 4 inches wider, so the old birds can get access easily. When the hen is about to lay again place a nest-pan in the next pen and close down the other, as they will finish feeding the youngster through the bars. Always have a few pens rigged up in your young bird loft, and you will find those reared as described will make for them, and encourage others to do likewise, so saving a lot of time and unnecessary handling later on, as when birds are ready for the shows it is a great mistake to keep handling them.

Separation of the sexes is most advisable, and do so early in August, giving plenty of baths during the moult, as this helps the growing feathers. During the show season give them a bath first fine day after a show. Then they are ready for their next outing when they have regained their bloom.

In my opinion judging was never better than it is today, as we have most of the best racing judges officiating, who select real racing pigeons for their awards. Years ago it was somewhat different, but glad to say the alien type of racers, big and gawky, have now found their level in classes of their own in the Exhibiton Flying Homer classes, etc.

My experience has been that most judges like a medium-sized bird of good balance, with a nice even keel (not deep), strong in shoulders, wing and back, close in vent, clothed in good rich feather, and possessing an intelligent head, with a good coloured eye.

In conclusion let me add that if this rambling article helps any young fancier to be successful in the show pen it will have answered its purpose. I know showing is not favoured by all; at the same time feel certain the sport of racing our birds, which we all love, would be the poorer without it.

Some Interesting Methods of Preparation

by E J SPARE

In response to a further request for another article for publication in your 1927 Diary, it is with much pleasure I comply, and will endeavour to make it as interesting and educational as I can, by giving the young fancier advice and hints.

I shall write only from experienced gained, and not theory. I have made a life's study of the breeding and racing of pigeons, and the many little things I shall relate are those that have brought me success, especially in hard races.

If this article seems a little disconnected, pardon me, for I do not profess to be a journalist by any means. First of all, I am not going to tell you the old, old story of fresh water, clean corn, and a clean loft; this is very good and important in the conditioning of racing pigeons, and a part of the programme. Now, I shall relate the facts how I cheat my birds. I hope my readers will pardon me in mentioning my own pigeons. I would rather have refrained from doing so. This is how I cheated my Combine birds in 1923. I may mention it is a well-known fact, and advocated by many fanciers, not to pair two birds together destined for a long race. The reason for this being, the cock bird may want to be got ready in one way, and the hen in quite the contrary manner. This is how I scored with my two Combine birds. I am a believer in pairing my best together, providing blood, structure and temperament suit. I have had as many as seven pairs of 500-milers paired together when the long races come along. I put so many of the pairs for Marennes races, and so many for the Bordeaux races, and the rest for San Sebastian, and by following this method one does not get half the trouble and strife in one's loft. The birds might want a little more cheating.

As I mentioned previously, the pair of birds have to be got ready in different ways to get the best out of them; but that is nothing to a good fancier who understands his work. My two Combine birds were paired together (mother and son). The hen bird was a keen sitter, and would do best for me on eggs, but the cock bird I noticed was rather jumpy, he could not rest himself on eggs owing to the condition I had strung him up. As every fancier knows, pigeons, horse or man when put into training and are getting very fit and highly strung, it naturally plays on the nervous system. I thought to myself, I will cool his ardour and, knowing that his best condition for racing was when feeding young, I obtained a young one from another nest ten days old, and the next morning I waited for the hen to leave her nest as usual, and before the cock could get settled on the eggs I had placed the young one in his nest, and straight away he fondled and caressed it. I placed a tot of water and corn he would require for feeding, and when the time came for the hen to go to her eggs I withdrew the young one, and fastened up the nest box. I proceeded with the same routine every morning and night for one week until the birds were sent to the race. This pair of birds won the following positions, and both dropped together on my loft:

The hen bird won 1st BHS pools and special, 2nd Central Counties, also 3rd Midland Combine.

The cock bird won 2nd BHS, 3rd Central Counties, also 4th Midland Combine.

You will note that each bird of the pair was prepared in two different ways for the race.

I will now relate how I prepared and cheated my good cock, 'Defiance', the bird that has won for me many prizes, including 1st Nantes, 1st Birmingham Federation, and 1sts in my club. In 1925 he won for me 1st Marennes, Birmingham Federation, and 1st Club; this race was flown on the Sunday and, being on the look out, I observed a bird nearly two fields away from my loft racing with extraordinary vigour. Little did I think at the moment it was my bird 'Defiance', till he was nearing the loft. At last he dropped after giving his usual fly round. I clocked him and thought how fresh he had finished; he impressed me greatly. However, his hen had given her eggs up while he was away; he was so vigorous he started running her up again for four or five days, his vitality was wonderful. I made up my mind at once to send him to San Sebastian, providing I could cheat him and his hen, and, believe me, a bird to be sent to San Sebastian race wants some stomach in him. I had not so many days to play with him and get my usual weight on a bird for such a distance. My inducement for him to fly to was a youngster, 14 days old. So, late at night, I placed the young one in a bowl, and put it in the nest box. I caught his hen, and she is a perfect mother, I placed her on the young one, easing my hand over her shoulders for quite twenty minutes; but as soon as I removed my hand she gently came off the young one. I placed her back again, and had to repeat it several times till at last her affection had come to her for the young one. I said to myself, all's well with the hen, and fastened the nest box up.

My mind then turned to 'Defiance'. I caught him, put him in a basket until next morning. I came up from business just at the time when the hen bird is ready to leave the nest. I reached 'Defiance' out of the basket, and, after dropping the tab of the nest box and putting him through, he started to brush, the hen came off the young one, brushing to him. 'Defiance' then walked quietly on to the young one, making a great fuss of his new addition to his home. I also placed one egg in the bowl to bluff him from running his hen. I then gave him light exercise, using my brains and tact to get weight on him, as he had not so many days before he would be dispatched to the San Sebastian race. I clocked him from San Sebastian, winning for me nearly £40. Some fanciers would think this too much trouble, but you see the pleasure is mine, and I reap the benefit for my cheating and trouble. So you see 'Defiance' flew for me over 1,100 miles in a little over a fortnight, and finished as fresh as paint.

How I cheated and bluffed 'Combine Gwen' – the race she flew previous to the Combine was the Nantes race – an awful smash, which cleared out many lofts in my district. I flew her from Nantes, sitting. When she returned I got her to take to her eggs again until my time was ripe to work a change. I then adopted a week-old youngster, and she fondled it, and the cock bird followed and took up his domestic affairs. I then had to calculate my time to get her down on eggs for the Bordeaux race. When the youngster was a fortnight old, I noticed the cock paying attention to 'Combine Gwen'. I thought to myself, I must save her from laying and draining her system in doing so; therefore, late at night, I placed a pair of eggs under her which, of course, were quite warm. I lifted 'Gwen' off the young one and put her on the eggs. My method was to make her defend them, so I kept touching her under the beak to make her feel annoyed, until I could bring that nature and instinct to her to defend her eggs, as hens will do. That night everything was satisfactory. I then closed the nest box. I repeated the same methods with the cock bird. I was more than pleased, for he covered the eggs, also the young one in the natural way. My method was that after I could see everything was in order and they were sitting keenly, in a few days I withdrew the young one. My readers will see that by studying the temperament of 'Gwen' and her nature, I saved the strain of her laying a pair of eggs, which enabled me to put her down for the race in the pink of condition. At 6 o'clock in the morning, before sending her to the race, I placed a young one, about two hours old, under her. I turned her out about 7 o'clock for light exercise, and, behold, she came down over my head and through the bob wires like lightning, straight to her nest. I may say, I would not apply this method had she been sitting any length of time. I have come to the conclusion that pigeons like novelties, or, we will say, surprises; in my opinion it seems to nerve them up and, more so, it encourages them to do better in a race and keener to regain their lofts. It is sure to be the truth in every kind of sport. There is not any good done without the school master, and our particular kind of sport alone can make or mar his birds.

Particulars of how I cheated my good hen, 3872 ('Doreen II'), a latebred bird, 1923. This bird was racing well 1924, and I had great confidence in her. Before the Rennes race, however, in which she was going to compete, her mate went wrong, and I had to destroy him, at the time they were sitting eggs. 'Doreen II' being left alone, my plan was to get her attached to a young one. I therefore gave her a ten-day-old young one; in a few days I sent her to the Rennes race, and she won 1st special and pools VHS, also 2nd Birmingham Federation (2,062 birds). On her return I had her going again on a week-old young one, having changed the youngster. I then acted the part of a good Samaritan in helping her to feed it, to save her strength before I sent her to

Bordeaux, in which race I timed her in flying Bordeaux when under twelve months old. My readers will see by my cheating methods I got two more races out of her.

I would like to mention to young fanciers they must never think their hopes are blighted when they are forced into such circumstances as I was with 'Doreen II'. There is not a difficulty made that cannot be overcome, and it is surprising what one can do in the bluffing and cheating of one's birds. I have always gloried in it, and I know of no fancier more clever than the late N Barker. He was an adept at cheating hens, and he would rather race hens than cocks.

I will now relate to my readers about one more bird, my 'Village Queen'. By doing so I shall have given in this article a few of the many ways a fancier can cheat and bluff his birds.

'Village Queen' required handling very differently to my other birds, the reason for this is one cannot keep the weight off her. I have to feed her accordingly by this bird's nature. I have to give her plenty of work on the road. This hen is at her best about chipping, or with a young one a week old. At the time I was sending her to a Salisbury race she was due for chipping, and she was sitting dummy eggs at the time, so on the night before being basketed for the race I blew a new laid egg and then inserted a worm in the egg from the garden, and sealed up the ends of the egg with stamp paper. Therefore, I sent her to this race in her favourite condition, to be flown on the Saturday, and she confirmed my opinion of her by winning for me 1st prize, special and pools, also 9th Birmingham Federation, over 3,000 birds competing. On her return from the race I had removed the supposed chipping egg and substituted a week-old youngster. My reason for doing this was to stop all soft food and get her fit again to be put in the basket for the 500-mile race from Marennes on the following Saturday. On the Tuesday afternoon, when basketing my four birds for the race, one of my competitors was a blue hen I had prepared and rather fancied. I was bringing her out of my loft and, on looking her wing through, found it was not to my liking, so I called to my wife to cross all pools off her and put them on 'Village Queen'. There happened to be a friend of mine with me at the time, and he remarked, "What a conscience". He thought I was mad, having had the 'Village Queen' in the loft only three days, and raced her so hard on the Saturday. I remarked, "I know my pigeon", and she confirmed my opinion and confidence in her by winning in one of the worst smashes ever flown from Marennes. 'Village Queen' and my 'Favourite II' were the only birds clocked in two clubs at close of race, winning 1st and 2nd in each club; 'Village Queen' winning 1st in the Birmingham Federation, 867 birds competing. She put up the finest performance of any bird tossed at Marennes that day, winning four 1st prizes in a week. Fanciers can see what I had at the back of my mind. If I had let her chip her own eggs she would have been in soft food for the Salisbury race, and on her return she would have been out of gear for me to race again from Marennes. So the young fancier will see by scheming and using tact it brought me success.

I am now going to write a few words on the art of conditioning racing pigeons – it is a most vital and essential part in putting birds down fit for the races. I will admit we have improved these last thirty years; but have we improved enough, considering the number of fanciers there are in the British Isles? For my part I say no, far from it, we can certainly improve a great deal yet.

The methods employed years ago are completely out of date: one must move with the times to deal with the conditioning of racing pigeons, and a more scientific way to secure success in the keen competition of today.

The large convoys that we have to fly against make competition harder every year, and that old method of thirty or forty years ago, turning every Dick, Tom and Harry out of the loft and making them fly a certain time, in my opinion, is the wrong way of conditioning. Suppose a fancier has a team of birds to get ready for racing, there may be sixty different temperaments to deal with. The fact which the fancier must not forget is that what may be good for one pigoen may not be the slighest good for another. There are some pigeons one must train for stamina, others have to train for speed, and those that require hard work to keep the weight off, and the light pigeons that require light exercise with special food to put weight on – a compound of ground maize, egg and a little milk, and a small portion of sugar candy. This is very helpful for a light bird. It is essential that birds should be handled scientifically and individually, because all pigeons are not endowed the same physically. The above is my method of handling my team of birds but not being a great user of hard corn, I am not a believer in wearing out the digestive organs by using too many beans, etc.

I have a greatly different opinion to many fanciers as regards putting the birds on a short diet. After the moult finishes, instead of keeping them on the lean side I put all the weight on I can. The method of keeping them on the lean side has its advantages certainly, especially for short races up to 300 miles, and also the laying hen. Why I differ and do oppose it, the reason for doing so I will outline and enlighten my brother fanciers. I presume the birds have gone through a trying season, which will have a certain amount of undermining effect on their constitutions. Any fancier will admit the birds have had the stamina taken from them to a certain degree, dependent upon the rearing and feeding they have undergone according to the fancier himself.

Following the end of the racing season, birds start to moult, as we all know. This is natural, but still they are again sapping the blood and vitality from their system in renewing their plumage. So much for that. And now the birds are facing a cold winter, and at that time the vitality of the birds is at the lowest ebb, and with a depleted store of energy. We must now reflect and reason with ourselves whether it is wise that we should make further calls on that depleted energy of the birds. I say, no. If one does so, a fancier's birds being kept on the lean side and no weight to work off in conditioning them before the season has advanced far, fanciers will have what I term the skeleton in the cupboard, that is, the birds have gone light and will lose their form and, in all probability, a most disastrous season may follow.

To prove my idea is correct take, for instance, on the break up of the football season, the trainer will pat his boys on the back, and by giving them sound advice, will say, "Now, boys, come back with plenty of weight on, so that I may have something to work on to get you back to form, and not be worked out by Christmas". Take also the trainer of a racehorse. When his horses have finished the flat season they are mostly wintered out, and when the horses are brought up for training purposes their trainer will remark the horses have wintered well and grown, and naturally he is all smiles. But let the horses come up as poor as rooks, it is a sad omen, and he has only the frame to train, which means a loss to their owner. No matter whether it be man, horse, dog or pigeon, there must be weight to train off in order to get them into condition.

There is just one more little hint I would like to give the young fancier. When going through your birds at night in the winter months give to them occasionally, say twice a week, several small pieces of mutton suet from round the kidney, with a couple or three chilli pods. This is a splendid thing, and does one's birds good, and will improve their condition. In concluding this article, let me say to the young fancier I have not written to tell them what they should do, or should not do. I have written as my inclination leads me to do, and in a good spirit to my brother fanciers. I always believe that interchanging ideas is beneficial and helpful to our sport, and by writing these articles, young fanciers should sift the chaff from the wheat, and reap the benefit from the experience of older fanciers.

I have tried to give details of my life's experience to the younger fancier by giving hints and advice, and will continue to do so until the end of my allotted span.

Breaking Pigeons
by F W MARRIOTT

My first experience in breaking pigeons was when I got married. The loft that I flew to when I was living with my parents was built against the house, and the pigeons always alighted on the house. The new loft was right away from houses and about 500 yards from the old loft. I started the removal in November, took the whole of the pigeons to the new loft and let the cocks and hens run together for about a week. They could see out of the loft from the front, which was quite open. I then opened the door, with the pigeons hungry, but as soon as they put their feet on the floor outside the loft, up they went and home to the old loft, which had been left standing, but stripped inside of all perches and nest boxes. I immediately brought them back, put them in the loft and gave them a little seed, and immediately opened the door and coaxed them out again with seed. This time they stayed on the ground for a few minutes, then up they went again and to the old loft. Again I brought them back and put them in the new loft for about 15 minutes. Then I opened the door and coaxed them out again with some more seed. This time I left my brother at the old loft with a flag to frighten them away. After flying about, backwards and forwards, for about two hours, they started dropping on the new loft, one or two at a time, until dark. That night, all but three came to the new loft and I made those three stay out all night. Next day, they came to their new home and were very little trouble. Although they had been used to dropping on a house, they never touched a house at the new home. The following season they were quite successful racers, and raced right through the programme to the NFC race, which I believe was Marennes.

One of these broken birds won 6th prize, San Sebastian, in 1907. That was the first year San Sebastian was ever tried. The late John Wones won the race and held the record for speed until I broke it in 1924, with my good pigeon 'Triumph'. I had numerous successes to this new loft, including 1st Marennes Midland Counties Combine, and generally was premier prizewinner in my local club.

Now in 1913, I had to remove again to my Alum Rock Lofts. This time I removed at the end of July. Some of my friends told me that my successes would finish as I was going to one of the worst of places for flying pigeons – down in a hole and closed in by buildings. However, I removed my main loft, that is, the breeding and racing loft, and left the young bird loft a week, until the birds were broken. This time I removed about a mile from the old loft. I put up nest boxes and let them rear a late nest. I had the birds broken inside a week. I repeated what I did previously. I only brought the birds back from the old loft three times, and had no further trouble. But it is very necessary that they are not allowed to stay in the old loft, and are made as uncomfortable as possible there, with nice seeds and everything to make them happy at the new loft. This time the pigeons alighted on the house first. I had no chance to see how the broken pigeons worked in after years, as the War broke out and stopped racing for four years.

When we got settled down again to racing, I had numerous successes with this loft, including four National races, winning 1st prize and HM The King's Cup twice from Lerwick (two years in succession), then again, two years in succession, from San Sebastian, which entirely confounded my friends' prediction. It does not matter what the position is the pigeons fly to so long as you have good pigeons that are managed properly. This year I decided on another removal, and took my pigeons nearly two miles east of the old loft. I built an entirely new loft, designed altogether differently – two storeys, each loft 6ft 6in high inside 17ft long by 8ft wide. I took the birds to the new loft in March, and mated them up before trying to settle them. The old loft is still standing intact, so is a greater inducement for the pigeons to return. The yearlings settled quicker than pigeons that had bred youngsters at the old loft.

What I did this time was altogether different, as flying to a two-storey loft, where pigeons do not come out on to the ground, requires the loft door to be left open and the pigeons allowed to come out at will. The first time out they nearly all cleared off home and were immediately basketed again. I took them back and put them in the new loft, and after about three journeys in the basket, they settled down. Some of the older pigeons were very obstinate and would not settle on the new loft. I therefore got a show cage and placed it on top of the loft, then put the pigeons in for about two hours. Then I replaced them in the loft and left the door open, so that they could come out at their own will. In every case that did the trick. The pigeons seemed nervy to drop on the new loft, but being put on top in the show cage gave them courage.

It is an altogether different proposition to break pigeons when the old loft is standing with birds still there, as at the least upset at the new loft, they will clear off to their old home. I had one particularly good hen which every time I sent for a training toss would return to the old loft and stay there until I either drove or brought her back. Often she would not be driven away, so I stopped training her to let her settle down for 1928. Even now, most of the pigeons that were two years old and upwards, and who had bred youngsters at the Alum Rock Lofts, took it into the heads to return to the old lofts occasionally, especially after the separating of the sexes. During racing, some of my best pigeons returned to the old lofts before coming home, and often stopped there hours.

After the experience I have gained, my advice to anyone who has to remove a loft any distance above 500 yards and fly to an entirely new loft, is:

Do not race the older pigeons the first year. By older pigeons I mean any pigeon two years and upwards. Yearlings are quite safe. I consider where a pigeon breeds its first nest of young, that place is its home.

If you can take the old loft away and put it up anywhere, that is quite a different matter, especially if you arrange the interior exactly the same. The pigeons fly to the loft and not the surroundings. They will get used to the surroundings in a very short time and fly just as well to the new ones. There are exceptions to the rule; I had one pigeon that raced just as well to the new loft as it did to the old, but eventually I lost it at 533 miles, although she had previously flown that distance twice. I thought, then: Did the new loft take away that 'Do or Die' spirit that we try to instil into our pigeons? My opinion is that it does take away that bit of extra energy that is required for such a very hard race as that turned out to be. I believe a pigeon not only wants to be fit to win, but must have the will to win; it will then strain every nerve in its body to get home.

In my first two experiences of breaking pigeons I took away the old loft entirely, and flew the pigeons to the same one. But this last time the old loft is standing intact, with pigeons still in it, and took the birds and raced them to new surroundings and a new loft. I mention this again because I want to drive home how different the two circumstances are. The cock that I consider my best pigeon, I would not train at all. I was not prepared to take the risk with him, immediately after breaking. He has now bred youngsters at the new loft and seems quite settled, so in 1928 I shall see how he shapes. He will be my candidate for San Sebastian, 1928; he has flown Bordeaux, Marennes twice, and several shorter Channel tosses; has won at the shortest race point and also the longest races. That is the pigeon to send to San Sebastian, and not a pigeon that just comes home; every fancier should send their best if they wish to win.

I mentioned previously that a pigeon not only wants to be fit to win, but must have the will to win. I will try to explain how to get both. Some of my friends say I have some little trick or dope for conditioning pigeons. I can assure my readers that I use nothing in the shape of dope, nor have I any tricks. I use nothing but good sound corn and water. I get the pigeons out for a fly early in the morning, and give them a light feed as soon as they come in. It is generally too early for them to want a meal, so I give a pinch of English wheat in each nest box. The wheat must be very dry (dry it in the oven if you can, but don't bake it sufficiently to discolour it). Then at 8 o'clock I give the morning meal; peas one day, beans another, or both each day, but each grain fed separately. At dinner time, or mid-day, I give a mixture of beans, peas, tares and maize, and about 6.30 another feed of mixture; but always give beans first as they will always eat them last unless you see that they eat up the beans first. Then last thing at night, I feed them on something they like such as whole lentils, tares, and a very little bit of good white canary seed, and fill right up with peas. It is much better for pigeons to have more meals than to be gorged up or over fed, which happens when only fed three times a day.

The oftener they are out for exercise without worrying them the better. As long as they do 15 to 30 minutes twice a day, that is all that is necessary, and if they go for a training toss any day, flying round home is unnecessary that day. I give a pinch of linseed about three times a week, first thing in the morning. If pigeons are fed as described you will find pick-me-ups and dope unnecessary. You will get them on their toes all right and keep them there. You want to get a lasting condition on your pigeons. That is the reason I can win at the shortest and longest distance with the same pigeon. You will find pigeons fed as above eat less than pigeons fed three times or twice a day. Feeding, in my opinion, is one of the most important items of pigeon management.

I am often asked: "In what condition does a pigeon race best, sitting, feeding, driving, or chipping out a youngster?" There is no hard and fast rule; different pigeons have different temperaments, but the safest conditon is sitting from six to 14 days. I had one cock that took years to find out what suited it best, and eventually found it was driving, but driving is the most unsafe condition for a cock (for which reason I did not find out earlier). Some hens are very keen chipping, but I have found the best time for a hen is sitting, eight days.

Don't give too many baths while racing – one a week being plenty. I always let my long-distance racers rear a youngster first nest, but get it away at 20 days old. If it is a good youngster it will be ready to leave at that age or earlier. I often have them away at 17 days old; they certainly thrive better than when left with their parents.

I never use rice, having found it had a scouring effect. A bit of good white canary seed is what I use for trapping and as a tit-bit.

I am most particular with regard to cleanliness in the nests, each nest bowl is scrubbed twice during the rearing of young.

Having had numerous enquiries from fanciers in this country and abroad, which have been mainly dealt with in this article, my two South African and American friends may take this as an answer to their letters.

Now, about the will to win. What takes away the will to win of a pigeon, is, in my opinion, over-anxiousness of the fancier. He has his pigeons out early morning, during the day and again at night, flagging them and probably making them do in all two or three hours' flying each day. I believe this tends to sicken a pigeon of home. You must use discretion, some days a pigeon will fly that time and think nothing of it; another day the atmosphere is quite different, and pigeons fly with difficulty and do not enjoy it, that is the time to let them drop and enjoy their home duties. A cock that is fit will continually be clapping off the loft and showing the merest novice he is fit. A hen will have her tail trembling, and when she blinks the eye it should hardly be discernible, almost like a snapshot shutter. If a pigeon blinks its eyes slowly it is not well; the faster they blink the better the condition. How often do you find that when there has been a few days holdover we get an absolute novice come up and win, because his pigeons have had some rest in the basket and come into condition during the holdover, entirely due to not being worried so much with the flying round home.

Over 56 Years' Experience

by REGINALD SLACK, Southport

I have been honoured by my old friend *'Squills'* in being invited to give my experiences as above. There may be some with a longer record of active work as a racer of pigeons. From 1872 to 1928 I have raced every year, even during the Great War. A few of the registered locals were granted facilities to train a few birds. They were sent to Fleetwood, and after examination were rung and sent some 50 to 300 miles by gunboat, whither we knew not, still we had these races, and Frank Battersby, Brereton, E Wilkinson, Hardman, Rutter, etc, competed. We all had birds on service for a long time. Those terrible days are over, let us hope, for ever.

My first race was on Christmas Day, 1871. About 20 fanciers within an area of three-quarters of a mile sent a bird each to Ormskirk in charge of John Littler. They were liberated at three minutes' interval. I was a proud man as I won first with £3 prize. Many little races were flown during 1872, generally about 20 miles, always a single bird per man, and again on Christmas Day, 1872, we had a 'big do', viz from Liverpool, 18 miles, 5/- each. Ball Keen organised the race and gave a splendid copper kettle as a special. This I won along with £5. Mr Littler was again in charge with an assistant, Billie Lowe, who is still alive, well over 80 years of age. It was a case of single up five minutes apart. I may mention that after 56 years I had my tea Christmas Day, 1928, using this kettle, as good as when new. They were real copper then. It has outlived every competitor bar your humble.

We now had a club, and I was elected president and Mr Farrington, treasurer. We made good progress and flew three races 50 miles, 100 miles and 150 miles. There was to be no mob flying as the birdage was 1/- each 1st and 2nd birds, 2/- the 3rd bird and 4/- the 4th. Each competitor usually flew three birds 4/-, and year by year we extended our programme till at last we flew Cherbourg. No special trains or convoyers. We booked through Southampton, where the birds were fed and watered, food being sent with them, and after several days' detention through fog, the liberation was effected with results quite equal to present days.

This is all more or less personal. In the middle eighties great strides were made, and by chance I got to know J O Allen of Lytham. He had left Manchester expecting to pass on very shortly. He was very nervous about himself. He had established a big loft and one day sent by steamer from Lytham Pier to Southport Pier about 30 young birds, distance as the crow flies about seven miles. There was half a gale against them and only a few crossed. Half a dozen entered my loft. They were stamped with dogs, horses, etc. I found out the owner and secured most of the wanderers from the market, where they had been sold as 'strays'. I paid 1/- to 1/6 each and sent them on to Mr Allen. He suggested rewarding me, but of course I refused anything, and for what I had paid my reward came in having gained a friendship which lasted till he passed on. I believe most of these youngsters were from birds Mr Doll got from Madam Gilson when she sold out. Mr Doll kept some and J O Allen and John Wright divided the others, and what history these birds made. One of the youngsters that entered my loft and never saw the outside whilst I collected the others as above, two years later again entered my loft. I thought I recognised it, but it slipped out when I opened the door and within 15 minutes Mr Allen found it in his loft. It had been liberated at Nantes the previous day and had as a yearling crossed the Channel three times and as a two-year-old twice before going to Nantes. What a memory this bird had. He was, I believe, sire of Sumner's great winner from La Rochelle. T W Thorougood's loft was founded chiefly on Allen's champions and John Wright's blood probably the Gilson's.

As the sport grew clubs sprang up all over, and most of them had one rule, 'The decision of the committee to be final'. What injustices were done under this rule. A favourite could break any club rule and his position allowed, whilst another less popular was disqualified at once for the same action. There was a cry out for a tribunal, disinterested, to see fair play, but for years we went on, but in 1896 a change came. Three determined men from the North, Allen, Stabler and Wilkin, said we *must* have a Union, and called a meeting at Leeds in March, 1896. There were not many present, but they were the right sort; they saw that the sport could not prosper whilst injustices continued. So far as I remember the following were present: L R Halstead, Rushton, Galloway, Holdsworth, O I Wood (a mere youth), and Chas Plackett (who engaged the room), a few from Lancashire, Roberts (secretary of the Pigeon Protection Society), etc. I was voted to the chair, and soon found that I was in the midst of 'honest questioners'. Were we going to be ruled? was asked by each visitor. My answer was always the same, "You would make your own rules to suit your own requirements and the Union would see you kept those rules without fear or favour". I remember so well Mr Rushton cried out, "This is what we want", and so came the first applause. Those present then showed they were determined to act even if they were to have to start a Union with only themselves. I suggested we invite 21 well-known fanciers from all parts to offer some ideas for the agenda for a meeting to be held in Manchester a month hence. Twenty out of the twenty-one showed some sympathy for the movement, but the simple mention of this man's name who had refused to assist was 'fat in the fire'. John Wright very quietly said don't trouble about this man, we will try and go on without him or anyone else who does not want a Union. Mr Wright gained the confidence of all present, and it seemed as if most present were of one mind, and the tiny *acorn* planted at Leeds was nurtured and has grown to the *mighty oak*, the National Homing Union, with its thousands of members. It has gone through many storms, but 'justice' has ever been its watchword, seen of course through different spectacles. It would be a sad day when we all saw alike, but the same principle has, I believe, always been present, viz "Justice for all, favours for none".

In February, 1927, Mr Logan was elected president, and rendered great assistance, his name alone gave it 'go' and confidence. I will not mention other presidents; all have been inspired with the desire to keep the sport honest and straight so that it may be enjoyed by all classes of society. I am rightly proud of the Unon and jealous of its good name. Regulations for measurements on a common basis and timing clocks naturally came up, and such difficulty has been honestly faced and conquered. We have much to thank Dr Tresidder and Geo Yates for our measurements. For years we had various systems from 1d per mile charged for railway tickets, but nothing that could be checked second hand in case of dispute till we got the 'Great Circle', which under the conditions laid down by the Union can be checked by all authorised calculators to practically a yard or two. This has removed much heart-burning, particularly when hundreds of pounds, depended on a decimal point and the competing lofts were scores of miles apart. Yes, I repeat, if for nothing else we have the Union to thank for this God-send.

Practically all the above was written from memory. I now have a list of those who attended the meeting at Leeds, which contained in addition to those I gave W Rhodes (convoyer), Haley, Broughton, Clegg, Hardace, Senior and Fletcher, and it was definitely decided to *form a Union*. C C Plackett was appointed secretary pro tem. On April 18th, 1896, a committee was formed to formulate rules. Many well-known names were down, and some few are still alive, A B Taft, W C Moore, A H Osman, J Armstrong, Dr Garlick, Pointer, Wormald, John Wright, and Geo Yates, etc, etc. C C Plackett for about 28 years rendered most excellent service, when the Union had grown so strong as to require an 'all time' secretary, and Mr Selby-Thomas was elected. May his run be as happy, as long and as useful for the Union as his predecessor's.

In our early days I frequently sent a bird to London in a double paper bag well ventilated and seldom lost any. Our training in those days was more thorough than today. Youngsters were sent in all directions, perhaps 25 miles, twice single up, and they were then rested a week before being put on the arranged route. Losses were few as if over carried they knew their way back. Year by year there has been a tendency to draw in a club's area in English races and to extend competition over the water. This is all to the good, but do we not all keep too many birds? I fully believe shortly our big races will be with Championship conditions, limited numbers, such numbers declining as the distance increased. Men will then have to select and the competition will be fairer. All other races will be with the object of finding the best and then pooling according to your purse. Single up at 200 miles has put more than one fancier at the top over the water and mob flying cannot pull them down.

Be very careful who you take into your club. Most men are honest, but above all things don't put temptations in their way. Let there be no looseness in ringing or before panniers are handed over to convoyers, but where you find deliberate and planned fraud or conspiracy to fraud prosecute criminally as a warning.

I have seen the sport grow from pasteboard prizes to £5,000 distributed at a dinner. I have seen one pannier carry a club's consignment, and I have seen two special trains each with two engines carry a Combine's 9,000 birds for a race of 450 miles over the water. Such progress is wonderful. The sport has come to stay, but only so long as the aim is, as it was at the time that little band of determined fanciers in 1896 formed the National Homing Union with the watchword *'Justice for all, favours for none'*.

Pulling Together

Things that went to the making of 'Pride o' the East'

by W B REEVE,
Holder of the San Sebastian Record

No one man and no one pigeon ever pulled off a great performance such as a San Sebastian record, unaided by the work and influence of others. Particularly is this the case in areas which are unfavourably situated either geographically or by other circumstances. The chief credit of course belongs, and rightly, to the actual pigeon and the ones who trained and bred it (I may be as 'bashful' as 'Quintinian' says in his notes, but there are limits to my modesty!) But there are many influences which, though they may seem remote from the actual performance, have nevertheless a direct bearing upon it.

For example, we in the east owe the *possibility* of winning the National race to the late J W Logan, whose investigation of the 'Drag' problem resulted in his urging the NFC to adopt San Sebastian as the National race point, with a late toss. Again, had East Anglia continued upon the old Channel Isles, Rennes, etc, route, I am convinced that I should not be trying to write an article for Squills today, as the winner of the National.

The adoption of a route leading straight for East Anglia from San Sebastian, through Cholet, Laval and Caen, and the gradual building up of a considerable team of good birds experienced on that route were the prime factors which led to the East Anglian successes this year. It had been proved in the course of a series of more or less unsuccessful attempts on the National that the sending of a mere handful of good pigeons from an area so remotely situated as East Anglia was fruitless, even under most favourable conditions.

In 1929 we in the east, encouraged by a localised race in conjunction with the National promoted by the Angouleme-San Sebastian Club, sent a much stronger entry of well-tried, winning pigeons in their prime than we had ever done before. The result: 1st, 7th, 13th, 16th, 25th, 44th, etc, Open; and 1st, 5th, 8th, 9th, 11th, 14th, 17th and 18th, 600 mile section prizes won by East Anglians (who had had it dinned into them for years that they would *never* score from San Sebastian) was the reward of 'pulling together' on logical lines on a logical route. That is why my King's Cup winner is called 'Pride o' the East' rather than 'Reeve's Pride'. Only the united efforts of my fellow enthusiasts in East Anglia made it possible. With this acknowledgement to local concerted effort and to those who despite opposition and discouragement have worked for this result since 'Quintinian's' first route articles in *The Racing Pigeon* in 1921, to Mr Cruickshank's final 'push' in 1929, I will turn to the more personal factors which went to make the present San Sebastian record-holder.

I started with racing pigeons in 1908, and like most people I had my ups and downs – especially downs – in the course of which experience I came to the conclusion that most of us waste a good deal of valuable air space, corn, and time upon pigeons which we do not send on, either because they are 'too valuable to risk' or because we hope they will improve with age. I came to the conclusion that the quickest way, the surest way, and in the long run the cheapest way of finding your best pigeons was to put them to severe tests early in life. It means, particuarly at first, the loss of a number of apparently good pigeons, and sometimes the loss of birds you subsequently discover would have been successful stock birds. I do not mean that a novice who buys expensive youngsters should race them before he has bred from them – but I do advise him not to be too merciful with those he breeds himself, either as youngsters or as yearlings. He will get some heavy knocks and disappointments probably, but at any rate he will not be burdened with a crowd of untried pigeons, the respective merits of which he certainly is not competent to judge.

Another opinion I formed in course of time was that it was safer to obtain stock from the best teams in one's own district than to buy haphazardly from remote districts; for however successful the breeder might be in his own locality it by no means followed this his strain would succeed in a locality in which the climate or racing conditions differed considerably. I am well aware that there are strains which score anywhere and everywhere, but I give it as my opinion, for what it is worth, that, other considerations being equal, it is best to buy from successful lofts in your own district contending with the same conditions in racing. Other advantages are that you know the personality of the vendor, can see what you are buying, and can probably have the benefit of the vendor's advice as to matings.

The maternal grand-dam of 'Pride o' the East' was the first really outstanding pigeon I discovered by the adoption of the method of 'ruthlessness'. She was 3663, a blue hen bred in 1920. As a youngster she flew March, Spalding (4th Ipswich Flying Club), Gainsborough, and York, on August 21st. I then lifted her straight off the north road to Bournemouth, 170 miles south-west, on August 28th (as she carried an NFC ring), and she won 31st Section without any South Road training whatever. I then sent her on with the North Road programme again: Durham on September 4th (18th IFC), the Berwick, 286 miles, in which hard race she won 1st prize and all pools by half-an-hour. In 1921 I stopped her at 35 miles because I wanted her stock, and thought she would be the better for an easy season. In 1922 she went all stages to Thurso, 490 miles, and won 11th IFC from there. The following year I sent her to Thurso again, but she homed badly shot. Her sire, 369, was bred in 1919 by Mr Harry Salmon of Lowestoft, who won 1st Section Pons, on day of toss, in the 1914 National Flying Club series of races. He was about 'top-dog' in East Anglia at that time and is still one of the 'warm' ones from 'across the pond'. The dam of 3663 was bred from brother and sister paired together. They were bred from O Rigby's (Winsford) 9623 x 799, both of which were inbred to his very successful Goossens, 'Miss Bouchier' 1st Gt Northern Marennes, 1910, 'Old Slaty', 'Alert', 'Blue Tick' etc, all Marennes or Bordeaux birds, mostly several times.

As my birds were getting closely inbred I looked round for some fresh blood, but with the intention of making 3663 the base of my future breeding operations. In 1922 in a fairly strong class at Ipswich Show, Capt Lea Rayner gave 3663 a 2nd prize, and it occurred to me that if she suited him, his own birds would probably suit mine as to type for mating. As I also liked their breeding and the way they were already shaping in the local races, I paid a visit to his loft (part of the arrangement of which I afterwards copied), and booked a couple of pairs of eggs suitable for intermating. In 1923 I hatched a black from each pair of eggs, and as they were cock and hen I mated them together, and I have no reason to regret it. A black pied 210, bred from them in 1924, sired 'Pride o' the East' in 1925, and several of my other winners descend from them. Old 3663, the Thurso hen, when mated to 7913, another of the good pigeons I had from Mr Harry Salmon, bred the dam of 'Pride o' the East', a plum red 291. The combination of the 'Plums' and 'Slatys' of the Goossens on the one side, and the blacks, and barless mealies of the Lea Rayner blood on the other side, has given me some peculiar variations of colour, but they suit me all right as they thrive on hard work, are easy to condition and I don't have to wait three years before sending them to the far end with hopes of success.

'Pride o' the East' himself is a sample. He is neither red, chequer, nor strawberry, nor mealy, but something between them all. As a youngster in 1925 he flew Guildford and made a bad blunder, homing badly knocked after a night out. He recovered so well, however, that – fretmarks notwithstanding – I decided to give him a chance to redeem himself. He struck me as being a 'trier' with the will to punish himself getting home and the stamina to recover afterwards. As a yearling in 1926 he went to Laval and Marennes, homing 2 pm second day (very ahrd race, none home on day, winning velocity EA Fed, 714 yards per minute). In 1927 I sent him to Caen, Laval, and Marennes again. This time he was my second arrival and won 3rd Ipswich Championship Club at 8.40 am second day (north wind). 1928, Caen, very hard race in east wind, homed 5.15 am second day; Laval, 3rd Ipswich CC, 12th EACC, velocity 860; and Bordeaux, another very hard race indeed, 1st Ipswich CC, 12th EACC and £1 single nomination, velocity 613. In 1929 I found myself with a strong team of Marennes and Bordeaux pigeons, and decided to concentrate on an attempt on San Sebastian. We had in the Angouleme-San Sebastian Club a single nomination competition (April 30th), and of the five birds I intended for the race, 1592 ('Pride o' the East') was my single nomination, because he had almost always been my first arrival under difficult conditions, and always finished strongly on the second day when unable to home on the day. I decided also to depart from my usual practice and to save him for the big effort this time by reducing the amount of preliminary work. I gave him one toss at Manningtree, 25 miles, three at Romford, 64 miles; Laval, 300 miles (an easy fast race as it happened), and then San Sebastian, sitting eleven days and fondling a big hen youngster (his own) which I had hand-fed every morning before the parents were fed.

He was mated in the first week in March. I mate all at one time; it is less trouble, and one can still arrange the different conditions for the various races. When basketed he carried 3⅓ new flights and six old ones. He homed, with the fifth flight still safely held, at 1.04 pm on the day following liberation, to win the National with the record velocity of 1024 yards per minute. This is the highest velocity at which he has ever won a prize, by the way. I have other birds which score on the 1,300 days, but it is the hard-day birds like 'Pride o' the East' that I value most.

I got three home out of the four I sent – not bad for 623 miles, though no better than many others locally and elsewhere, I know. It was a wonderful race and a credit to the organisers, and to the convoyer – the condition of the winner on arrival was astonishing. He is rather below medium size and never carries a lot of reserve weight, but he handled almost as heavy on arrival as when sent away.

At the beginning of this article I referred to the results of 'pulling together' as regards fanciers in an isolated district. It is even more important that the principle of 'pulling together' should be applied to the fancier and his pigeons. A fancier may have the best birds in the world, but unless the fancier himself does his best for the pigeons they will be unable to give him of their best. This means work – hard work, if you keep many – and regular attention to detail the whole year round. It is also essential for the fancier to gain the *confidence* of his pigeons. No fancier whose birds are *afraid* of him can reasonably expect them to break records to get home to him, and still less can he expect them to trap on arrival. It is even more true to say that the cleverest fancier in the world could not win a single, medium, or long-distance race unless his pigeons had the necessary breeding, physique, courage, and stamina to respond to his treatment.

No patent food ever invented (and some of them are excellent in their way) will ever compensate for irregularity, neglect of detail, foolishness, or impatience in the loft, on the part of the fancier; or for softness, poor physique, or lack of quality on the part of the pigeons.

So much has been written on the subject of management that I can add nothing new – I owe as much as most of my generation to the excellent books and articles which have been published from time to time. The reader may have noticed that the condition in which 'Pride o' the East' was sent to the race – sitting and fondling a big hand-fed hen squeaker – is one which 'Squills' himself has often recommended, and mentions in his book 'Secrets of Long Distance Racing'.

There is nothing mysterious or exceptional in my loft management. I am away at work all day, but my wife helps me by changing the water, feeding, etc, when necessary, and sometimes times in for me – as she did in the case of the National.

In the racing season I exercise at 5.30 every morning, again at noon, and in the evening, feeding in each time. Between those hours I give the birds a free hole, as long as they keep to their own boundaries and do not trouble the neighbours' gardens, etc.

I feed, for breeding and racing, on a mixture of one part small maize, one of tares, two of maples, and one of tic beans – all of the best quality I can obtain. As the longer races approach I make the last feed of the day all beans. As a tit-bit I use a little rice occasionally, and also for trapping from training or racing. A lump of rock salt, fresh grit, and some sweet old mortar is always before them. I do not use a hopper for feeding. Pairs intended for racing are only allowed to rear one youngster in the first nest. The one 'Pride o' the East' was feeding when sent in July was his second of that year.

To the aspiring novice I can only reiterate the old slogans – clean, fresh water, good food, fresh air, a clean loft, good grit, regular and systematic attention, patient study of the individual pigeon, commonsense, and perseverance.

The path to the King's Cup is no secret or mysterious one. It is simply one of steady application of these old-fashioned, oft-written maxims, to pigeons worthy of the work entailed.

Deserve well of your pigeons, they will then repay you, *if they have it in them to do so.*

The Problem of Breeding Long Distance Winners

by Dr M EVERARD TRESIDDER

The training of long distance winners is a subject different from breeding.

Of course, all long distance winners are bred by somebody, and if we could examine the breeding of all these winners, we might find some law governing their production; it is a task however, beyond my reach – so we are led to observations in a more limited sphere – where the detailed knowledge is greater, and try and draw upon such knowledge as a general guide.

There are tens of thousands of pigeon fliers; there are thousands of long distance racers; there are hundreds of long distance winners, but the numbers of breeders of winners from long distance are comparatively limited. The first difficulty that confronts me, is the question: What is a long distance race? The second is: What is a winner? Velocity in any race cannot compare with velocity in another race, and all experienced fanciers realise that velocity alone is not a sure guide in determining the value of a performance. Time taken in any race governs the velocity. Therefore it seems to me that these two factors judged conjointly may be a guide to classify a race.

If, therefore, we state that a velocity of 900 yards and not less than nine hours, should constitute a long distance race, we find that such a race will be no longer than 276 miles.

But there are races of longer distance which take less than nine hours – these I would exclude – which will bring us to a fly of nine hours for a 400 miles race at a velocity of about 1,300 yards.

Some arrive at a minimum standard of a nine hours fly, and a velocity of 900 yards, as a long distance event.

Every fancier knows that a 900 velocity from 300 miles, is a greater test of a pigeon than 500 miles in ten hours.

These points are difficult for a beginner to realise and to appreciate accurately in estimating the value of his stock. It is quite certain that the long distance (as defined above) racing pigeon, has now become an established breed. The parts which go to make up their breed is not obvious, nor is it easy to describe and define.

To the careful observer of pigeons exhibited at shows confined to 400-500 milers, it will at once be evident that colour has little to do with the subject, the ordinary physical appearance is not so marked as to be easily distinguished, except, perhaps, that generally the size is uniform, neither extreme of size being common, the big ones are rare, as also are the very small ones. In hand, again, there is a general uniformity and this is made more obvious if the handler is blindfolded. These physical points do not lend themselves to a detailed description and can only really be learnt by practice in the appreciation of minor differences of conformation, and even then are not conclusive.

There must therefore be other factors in this breed, which undoubtedly exist. Such as constitution, internal physical development, courage, and a homing sense. Just as the external physical characteristics of breeds are hereditary, so also are these invisible features, which go to make up a breed.

Experience has taught over and over again, that given a breed of true racing pigeons, unless the practice of racing is continued year after year, resulting in the keeping only of those that succeed, the breed will diminish.

Fanciers have tried to evolve a pure coloured strain by selecting an admired type and keeping these for reproduction, and then have discovered that the racing ability has diminished greatly. Therefore the keeping of racing pigeons for type-production, or colour, or breeding from birds of a strain that is no longer raced, is a matter attended with many disappointments, and if continued for long, I am certain will lead to the deterioration of the breed.

How can we find out a bird's constitution unless he is physically tested, and exhausted? Courage is a feature of perfect fitness, as well as perfect physical development. An unfit pigeon lacks courage. Internal physical development is largely a question of the heart and lungs. Good pigeons like good racehorses must have large hearts and sound lungs to meet the requirements of great events.

The increase in the size of the heart is an accompaniment of early training, and so long as their early exertion is not excessive, the heart will grow to meet the demands made upon it; if the exertion imposed is greater on the heart in the early stages of life, it will dilate and will consequently be injured and will fail to hypertrophy or increase above the normal so too in a less measure with the lungs. This fact is one which should be borne in mind by the advocates of early long distance races.

I am prepared to admit that although I consider it a most extravagant method to adopt, it may be that the half a dozen that remain would have for the long races a better chance of success than a selected half dozen not so treated. But beyond that they would have a lesser chance than the *best* treated with great caution. It is obvious that parents not raced for several generations may not possess these attributes of development and hence the advice to go to successful racing studs for purchasing.

Homing sense, we all believe, is an hereditary quality, but its exposition is not always evident because it is so intimately dependent on these other features; it is however a sense which is extremely necessary to cultivate and breed for always.

The second consideration of What is a winner? again places us in difficulties. In England, I am glad to say we do not express winners as the 220th prize, as is done in Belgium. A prize is a prize to a Belgian fancier; to us in England we want to be on top and devil take the hindmost.

The classification of prizes in England leaves much to be done – the sub-division of the money, the decision on decimal points, the areas of competi-tion – all lead to a hopeless confusion and in many cases is not pigeon rac-ing at all. Winners vary in respect of the numbers engaged; they also vary in accordance withe the calibre of the competitors.

A two-bird competition among giant fanciers is a harder race to win with a small number, than a race in which thousands compete, because among those thousands are many which can never fly the distance. Many sent by novices which are not fit to race, but may get home late.

A first prize in a small club cannot compare with a first prize in a Federation, although sometimes the two prizes coincide. The race cannot compare with another. The variability of our climate is such that races from the same place on the same day, to the same locality with different times of liberation with equal birds, vary beyond all explanation. Comparisons of long races may therefore be made by the time taken, the velocity, the distance, and the numbers competing. Quite a nice mathematical problem to reduce to a race-winning factor which, one day may be solved.

If we believe that there is such a breed as a long distance race of pigeons, we should then satisfy ourselves that the performances upon which our val-uation of them is based, is that of long distance. We must separate winners of short races or medium races from long distance races.

In order to do this, we must know the distances of the races: Bournemouth to London is 95 miles; Bournemouth to N Lancashire is about 200 miles; Bournemouth to Aberdeen is 450 miles; therefore the name of a race point means nothing.

It has become the fashion of Belgian writers in recent years to make use of the word 'Ace' in describing a constant and regular performer.

Then we find that this 'ace' has won prizes from Creil, Corbeil, St Quentin, Quiverain and St Denis. What do these names convey to us? Nothing at all unless you know the distances. What is the value of 220th prize worth? Nothing if you know the distances. All these races are for short distance pigeons. St Denis, which is Paris, is only about 190 miles, and all the other places are considerably less. They may be 'aces', but they are not in the trump suit, and a deuce in a trump suit is better than these 'aces'. If you are going to breed long distance winners, this point is most necessary to keep in mind. I would rather estimate a fancier's ability by his long-distance birds, whether they be winners or not, than by a fancier's prizewinners.

A man who has five birds flown 500 miles is more of a fancier in my eyes than a man who has won 50 prizes and not got any 500-milers. Wins, unless they are long distance wins, to my mind, do not count. Of one thing I am quite certain is that these *'pigeons de vitesse'* – sprinters – which have under-gone all kinds of special preparations, are valueless as reproducers of long distance winners. The crossing of these two breeds does not blend to make improvement in either one or the other variety. It has been tried over and over again and failed. These fast sprinters up to 200 miles are wonderfully regular in special hands; but when they are lifted 100 to 150 miles for a moderate distance race, they just fail to arrive.

Why? is not an easy question to solve. Their training is so different. Many tosses at short stages. Repeated mid-weekly regularly and prepared under a special regime. Treat these same birds as you would train long distance racers in preparation for 300-400 or 500 mile races, they are no better, and often worse than long distance birds in these short races, and when sent to the long races, well, they 'stay'.

The fancy in Belgium since the war, has undergone a radical change; the number of fanciers may have increased, but the number of long distance pigeons has decreased, simply because there are not so many sent as in previous times.

The Belgian Grand National from Dax or St Jean de Luz used to provide regularly 5,000 or 6,000 birds. The numbers are now only about one-fifth and so the total has dininished.

Long distance fanciers require to keep more birds and often have fewer races. Fewer birds and more races, when corn is too expensive and labour so unbalanced, is one cause.

The pioneers who founded the great strains are rarer today. Probably the best long distance racers today in Belgium are as great as their forebears, but they are given less credit, because their sport is less popular and requires greater energy, patience, and forethought. A general historical survey of the production of long distance pigeons in England, may help us.

The late N Barker of Brussels (a Yorkshireman), was a great fancier; a consistent winner from all the great races of long distance. The late J W Logan purchased his whole loft of old birds; to these he added, as occasion arose, winners of long races, from every source he could. He was the pioneer to teach fanciers that long distance birds were a breed of their own, and that nothing short of winners in long races was worth while. Then came Mr Alfred Darbyshire, who scoured Belgium for real long distance pigeons. He, too, was a pioneer.

The Fancy Press at that time devoted its space to collecting accounts of long races from France to Belgium, and invited the lofts of the great fanciers of that time. Mr Walter Jones of Sandringham, may be considered responsible for introducing into England the pigeons of Jurion, Duchateau and Delmotte, all members, as also was N Barker of the select (25) National Flying Club; M Alex Hansenne, the victor of Calvi in Corsica; Gits of Antwerp, Delrez of Verviers; Dardenne of Liege; Pitvil of Antwerp; are all strains which have left their mark in establishing the England long distance pigeon.

None of these great fanciers ever worried about winning races from Creil, Corbeil, Quieverain or St Quentin. They probably jumped over these stages and only began to feel serious when reaching Vendôme, a distance of about 240 miles, which nowadays is the termination of the *'pigeons de vitesse'*. These fanciers discarded all but *the trump* suit. Their *aces* were aces in the trump suit. The Blue d'Hippolite of Plétinckx, who flew till 18 years old the GN from Dax, La Vieille Bayonne of Jurion, Le Barcelone of Delrez – the Donkerin of Gits – La Voliere of Delmotte.

From such pigeons as these and others which I don't call to mind has built up the strain of English long distance racers.

It will be a sorry day for the English fancy if we forget these lessons of the past. Distance and more distance is our road to progress. Successes in short races will not lead us onward – the fancy as in Belgium may divide itself into two sections. I believe it already is tending that way, and it behoves breeders of long distance racers to beware. Don't be like the greedy dog, lose the substance for the shadow.

Apply the lessons of the past, which are general, and have operated successfully over years, to your own colouring.

Classify your birds, those that fly 10-12 hours in a day are rare birds; concentrate on their young ones. Breed such birds if you are fortunate enough to have more than one; breed these together. Don't worry about a 1st prize from Dorchester.

Those families that won't and don't stand the big jumps are better away. Faked sprinters are not lasters and rarely breed great long distance winners. If you want to have both sorts, keep two lofts or two separate strains. Don't mix them or you may spoil both.

To breed winners of long races, you must have a strain that can be relied on to do the journey – whose parents and grandparents have raced and done the journey. Not birds kept for stock descendants of a great strain. Not birds kept because they win short sprints.

My own feelings in regard to the various systems of racing today. 'Averages' and so on, is that those who practice it, are not out to produce a long distance strain so much as a long sequence of success from any distance. If this be so, then I am certain it is bad in the long run for long distance production.

The natural practice, adopted by the pioneers in creating the great strains, is probably a better criterion of a bird, than the performances of doped and maddened pigeons, who may be injured by these sexual stimuli, so much that they will fail to breed at all.

I can imagine I hear the modern fancier say of this article: Why worry? Pigeons are cheap and easily reared and bought. My retort is: that his sort may be, but I know that the great pigeons which can fly 15 or 16 hours, two or three times in a day in their whole lives, are neither cheap nor common, and are only to be obtained by careful breeding over many generations.

A Tynesider's Tips

by F GUTHRIE
Winner of six Silver Cups in 1931

I am certainly honoured by the request to write an article for Squills' Diary – all that I am troubled about is as to my capabilities of writing something that will arouse the interest that is usually created each year by the pioneers and experts who have adorned its pages.

How long have I kept and raced pigeons? With a very slight break I have raced pigeon since 1890.

During this period I have had many and varied experiences of pigeons and those that keep them – some very happy and others aggravating.

This I can truthfully say; that in my early days when my purse was limited, I got the maximum amount of pleasure with a minimum amount of effort, and naturally expense.

Years have brought about many changes which, whilst they have allowed me to obtain exactly what I desired in the way of stock has not in my opinion improved the quality, but simply the advance made by the sport has brought out these latent qualities that were simply waiting development.

New strains have flashed across the pigeon world and are now very seldom heard of.

I have had at one time or another practically every newfangled record-breaking strain, but they have not stood my management.

I have had a few good pigeons during my time, but since the War I think my best have been practically confined to the old strains of Barker, Gits and Hansenne.

I had a good hen 'True Blue', that I lost in the Troyes smash of 1922, which won three Federation races out of four prior to going down, and my 'Angouleme' blue cock is own brother to her.

'Angouleme' won 1926, 3rd Amiens; 1927, 2nd Nevers in two clubs; 1928, Nevers; 1929, 2nd Angouleme, 1st Single Nom; 1930, 6th Angouleme, whilst his daughter won in 1930, 2nd Northern Section from Angouleme, distance 625 miles.

Both birds have bred winners from inland and cross-Channel races. These are the Barker family, whilst the Gits family since the War have produced me splendid birds, notably 'Lion', 'Pilot' and 'Lump', who cleared the deck from the Channel races two years in succession; in fact, in 1922 they won 1st and

3rd Fed Troyes, 500 miles, one being the only bird home on the day.

'No fads', is a general expression used by writers. Clean lofts, water, and corn. I wonder if this is absolutely true.

My ideas have in the years undergone a radical change, and whilst men would say they have no fad, shall we call it system or method?

I have proved it up to the hilt that given fit birds you can take almost any liberty so far as jumping them 3, 4 and 500 miles is concerned, and this was driven home to my mind when I purchased 'Mons Hero', who won 1st Up North Combine Mons and 1st Up North Combine Troyes, 500 miles. This bird was jumped from Doncaster to Hastings each time winning from the latter point and then to the respective races that he won each year, viz Mons and Troyes.

My Angouleme winners were trained privately to Bournemouth, 300 miles, then jumped into Angouleme 625 miles. My San Sebastian hen last year was trained Bournemouth, then San Sebastian, 784 miles. My Barcelona hen this year was trained exactly similar being jumped from Bournemouth, 300, to Barcelona, 930 miles, and my this year winners from Melun 465 and Nevers, 567 miles, were jumped from the 190 and 278 miles race points.

We are generally apt to take too much out of our birds prior to the final test, and that is why so many pigeons fail to score.

If a bird lives to three years old with me and fails to score its chances with me are nil. I did not always think so, but I am not prepared to keep them longer now for serious racing.

I have become firmly convinced that a yearling will beat old pigeons up to 500 miles every time, unless vile weather conditions are experienced and even then my fancy is the two- or three-year-old.

I am a convinced believer in the survival of the fittest and I will not excuse more than one mistake – in fact I will not breed from the bird that has made one mistake.

Breeding is not a game of chance, and I have proved over and over again that birds of a certain family will reproduce winners as has been done by similar pairings years ago.

I like good hens for both racing and breeding – they are more difficult to work as racers, but I am confident they are more reliable and worth the extra trouble.

I test my youngsters through to the bitter end and so long as there is a race and they are fit they must go with the exception of five or six hens at the most, which I stop at the 100-miles stage.

I believe in a thorough training for youngsters – at least twenty tosses prior to the first race and early and late exercise – mark this *early* exercise, not 8 and 9am, but before 6am.

I used to be very half-hearted in training youngsters, but experience has taught me that to make successful and reliable old birds you cannot give them too much training as young birds.

I have read 'Nor' West's' many notes on young birds being raced from Guernsey into Lancashire, and then their work as yearlings.

In 1929 I experimented and bred some early youngsters putting four on racing with the old birds. I got one black chequer cock, a grandson of 'Mons Hero' to do Amiens, 365 miles, after ten days holdover and in 1930 he was in my first six from Nevers, 567 miles, whilst this year Amiens in the first six, and then in the day homing during a terrific thunderstorm. Nothing wrong with his constitution!

The Up North Combine race a tremendous crowd of birds, and my only regret is that whilst we have an Up North CC we have not a Two-Bird Specialist Club liberated on their own on a different date to the Combine, similar to the two or three that Lancashire boasts.

As to lofts and management: My lofts are open half-way down, and the drinking water (which is changed twice daily), is on the outside of the bars, thus preventing dust and feathers getting into and fouling the water.

I use for racing right through a sound mixture containing 90% beans and peas. I do not believe in what is described as winter feed, I used to do, but this last three years I have, on completion of the moult (during which they get a liberal supply of linseed), put them on to beans only. A good dose of salts then before pairing, and I have no trouble as to bad eggs. Not a single bad egg.

Exercise twice a day during racing, with Sunday a rest day when they can have their full freedom. It is peculiar that my birds, although the bath is available, reserve such for Sunday – it must be a habit.

I hopper-feed old birds, the corn always being available, but I hand-feed the youngsters.

I have little difficulty trapping, as I drive the birds in with a long cane which they obey every time, being trained to same.

Once let a bird alight on the verandah I guarantee to have it in under five seconds. We cannot afford in the Up North area to give seconds, as very many trap through the open door.

I pair up middle of March and allow each pair to rear one nest, and after that nest it depends on the temperament of the pigeons as to wheather I allow them to rear any more. You cannot rear and race both at the same time, and successfully.

I am very partial to the first round from a new pairing, and again let me say I have experimented with out-crosses and inbreeding.

I have at the present time two different birds bred from own brother and sister. One is a blue cock, my first from Le Puy, 702 miles, 1930, and another blue white flighted that has crossed the Channel four times and was my first up from Amiens in 1929, a very hard race.

My ambition is to head the Up North Combine, a very difficult proposition. I have managed 2nd, and twice 3rd, whilst I have been in the first dozen a few times. I really thought I had won from Nevers, 567 miles, this year when I timed in at five hours before any previous season's time in the area.

The last ten years shows tremendous advancement over previous pigeon history, and I am certain we have not reached the limit of the possibilities of our birds.

I am convinced in my own mind that 1,000 miles will be accomplished in the next three or four years, that will break all records.

Some people call it freak racing, but there is not the slightest doubt that from the family that produce these pigeons we get the very best birds in the classic races.

I am looking forward to cheap aeroplane transport for our birds to the cross-Channel race points, as the boat and French rail facilities are all against the condition of our birds.

The novice will ask: "Is there a Royal road to success in pigeon racing?" My answer is "No".

Certain generalities are necessary, such as good stock, clean lofts and corn and water, but when it comes to method or system I have known men win under all kinds of methods as to feeding, training, etc.

Try and win the confidence of your pigeons, and you will have gone a long way towards success. A dog is only very sensible when its master or mistress try by gaining the dog's confidence to make it understand whilst it is trained – unless you gain the dog's confidence you will never make anything of it.

Apply that to your pigeons and you will find your reward sooner or later.

My greatest pleasure is to see my birds home from the races and the last arrival is just as important to me as the first.

In conclusion, I must ask your readers to bear with my short article as I am aware it is perhaps a little disjointed – I am not a journalist. If I have interested your readers in the slighest degree I shall be delighted. Just one word more: "A bad loser should never go in for pigeon racing, as it is built on future hopes."

Interview with Mons G Stassart

by 'NOR WEST'

A visit to Brussels would not be complete without calling upon Mons Stassart, and the members of the Lancashire Social Circle party of 1932 are indebted to him for his kindness to them and the time he willingly and unreservedly placed at their disposal.

I took a note of many of his observations on the sport, an incomplete record I am afraid, but even so, readers of Squills' Diary will, I am sure, find much to interest them in M Stassart's observations, and the practical words of wisdom he let fall from time to time will, I hope, prove useful to fanciers in this country.

Mons Stassart's loft occupies the top floor of a large restaurant or café in the Park d'Anderlecht in the suburbs of Brussels, and his home address is 2 Rue de la Democratie, Anderlecht, Brussels. Not only did M Stassart show us many of his racers, but he gave us a whole evening in the Falstaff Restaurant in Brussels and submitted to a bombardment of inquisitive questions both from Mr R Dunn, Wigan, myself, and other members of the party. A formidable ordeal and it put M Stassart's knowledge of our English language to a test from which he emerged triumphant. Only when we made the pace too hot did he call upon our interpreter, M Malsenaere.

To attempt to give even a remote idea of the big wins and the money won by M Stassart's birds is a task beyond me, and I think it is even beyond that of M Stassart himself. In the 1932 Grand National race from Pau, nearly 600 miles, flown July 21, the birds numbering 2,998, were liberated at 10.45, wind SW, so that no birds could get home on the day. He sent 15 or 16 birds and won 12 prizes, viz: 7th, 10th, 33rd, 35th, 36th, 43rd, 46th, 73rd, 198th, 335th, 457th, and 576th. His winnings in this one race totalled 63,000 francs, or just over £500.

M Stassart is great on the Angouleme Young Bird National race, which is roughly 410 miles. Very nearly the distance of our Nantes to Lancashire OB race. In this long-distance YB race he has in five years won the Gold Medal four times for best average, or rather best loft, with the most prize-winning pigeons clocked in. The year he missed was because he did not send (1931), as there was a disease epidemic among young birds in Belgium, and he called in the vet, and had all his youngsters innoculated against contagion. In the

1932 Angouleme YB race he won 2nd, 5th, 6th, 18th, 19th, 27th, 41st, 49th, 63rd & 94th prizes.

M Stassart showed to us, and we handled these twelve Angouleme youngsters which flew the race on September 10. His 2nd prize winner a blue chequer hen, was a very nice, good-ribbed, well-built hen.

There were several other well-built youngsters among them, but to be candid they were what we would call a mixed lot, some on the biggest side, some on the small side, some awkwardly shaped.

M Stassart made no secret of the class of young pigeon he sends to Angouleme. If a bird did not please him in the handling and was awkwardly built or not his type it was sent to Angouleme. If a youngster was irregular in its training or race tosses it went to Angouleme. You are quite at liberty to surmise from this and M Stassart would not deny it, that his Angouleme youngsters are not exactly rejects, but birds that did not please him. Wait for the sequel!

Next we handled the 'saved' youngsters, birds which this year were stopped after flying 200 miles. Here, indeed, were some beauties, much nicer handling birds than the Angouleme youngsters. Oh, yes! I grant you, some gems among them. For instance, the nestmate to the 2nd prize Angouleme YB 1932, is a beautiful blue cock, and this was 'saved'.

M Stassart thoroughly and heartily believes in this long-distance YB stunt, to try out finally and severely and test the sort of pigeon that does not please him. I am hiding nothing, and neither does M Stassart; he just gave us the plain, blunt, unvarnished truth.

I cannot wait any longer to give you the sequel. It was in the nature of a bombshell I cast at M Stassart. I said: "Tell me, sir, how many champions or real good big-winning pigeons in later life do you get: (1) from the hard-tested Angouleme YB's; and (2) from the 'saved' 200-mile youngsters."

After short consideration M Stassart replied: "I certainly get in later life quite an equal number of champions or real good long-distance winners from my Angouleme long-distance youngsters as I do from my 'saved' 200-miles youngsters. They are equally as good in both categories." In plain English he means it is a case of 'fifty-fifty'.

The conclusion of this great authority on long-distance old and young bird racing is particularly valuable, because he has been in some awful Angouleme YB smashes, and has stuck it through.

This proves, as I have so often written in *The Racing Pigeon*, that the more 'leathering' you give a good pigeon, you cannot kill. This is what I have so often written about long-distance YBs and yearlings. But M Stassart does not agree with me about yearlings, as you will read later.

These Angouleme youngsters are not flown a yard as yearlings. The 'saved' youngsters are only trained 100 miles as yearlings. Latebreds are given many tosses of from 10 to 25 miles as youngsters, and yearlings, but none of the youngsters are raced or punished as yearlings.

M Stassart told us, "I think a *yearling* is in its most critical condition – it is developing and should be carefully watched. It is like a youth of 15 years old or so. The three-year-old pigeon is the best for the long-distance races."

He mates his Angouleme youngsters and his 'saved' youngsters, when they reach the yearling stage, to his old bird racers and thus ensures the latter having a mate always at home waiting for them. As two-year-olds, both the Angouleme youngsters and the 'saved' youngsters go right through to Angouleme, 410 miles, and no farther, unless it happens to be an odd bird or two he wants to stretch out and test.

It is as two-year-olds, up to and including Angouleme, that he gets some very good performances out of both the Angouleme youngsters and the 'saved' youngsters.

For instance, take the 1930 Angouleme Young Bird National, a disastrous race with only nine youngsters on the day. Six of them were M Stassart's, and the won 1st, 2nd, 4th, 5th, 7th and 8th approximately, these are the positions. To complete our education M Stassart let us handle all his 1930 Angouleme YBs, all hens mark you! With their rest as yearlings these 1930 long-distance youngsters had, in 1932 as two-year-olds, all flown exceptionally well up to Angouleme, and they handled well. His 1930 Angouleme YB winner is a lovely red chequer hen and she has this year won three extra good prizes in races up to Angouleme. All these twelve 1930 Angouleme youngsters that came through the 1930 YB disastrous race have won prizes in 1932 as two-year-olds up to and including Angouleme.

It is interesting to note that M Stassart races his old birds, perfectly Natural. We also handled a few of the champion old bird long-distance racers.

We did not see M Stassart's champion: 'Baladin', a red chequer cock which in six years' racing flew in the long races 12 times and was nine times in the first nine in big and important races, winning big money and lots of good prizes. He has now finished racing.

Practically all the birds in M Stassart's loft are from medium to just a bit above medium size, thick-set and powerful, but not big or clumsy. They are 'all pigeon', and the type the late Col Osman used to call 'wiry' or 'Corky'.

They have a kind, quiet, intelligent look, and generally the eye seems to be a combination of purple-brown and violet, with occasionally a much-prized one with a pearl ring.

M Stassart told us he firmly believes in hens for racing and the best time is when they have been sitting three or four days, and again when they have been sitting thirteen, fourteen, or fifteen days.

For the cocks their best time is sitting eggs over the time-limit. A hen, racing in its favourite condition, is the best bird always. if a racer casts a flight in a natural way, it does not bother him. The bird has to go to the race, the fact that it has naturally cast its feathers proving it is in a good natural condition.

M Stassart separates his birds from October 1 to March 1; the sexes are always separate during this period and are not allowed to run together at any time. At 8am the birds get their first feed and half-an-hour afterwards the cocks or hens as the case may be are turned out and shut out, the loft being closed until evening, when they are re-admitted to the loft and fed for the evening.

This enforced absence from the interior of the loft prevents the old birds from nesting in corners, and also induces the birds to take to the fields, a practice which is becoming increasingly popular in Belgium during the winter months.

In the first nest the birds are allowed to rear double youngsters if good and healthy. If any youngsters are poor or not doing well one of a pair is killed.

After the second lot of eggs are laid and been sat on ten days they are taken away. Then when the third lot of eggs are laid, training and racing is commenced. He can generally get in three races whilst the old birds are sitting the third round of eggs. These eggs are allowed to hatch and one youngster is allowed to be reared quite natural.

M Stassart finds in his experience that both hens and cocks generally race best to eggs.

The first training toss is 15 miles, then 22 miles and 40 miles, and after that the first race 100 miles.

All his birds must then take forced exercise both morning and evening, about one hour's fly on each occasion. No training down the road between races, and he does not believe in single up tosses.

The health and conditon of the birds when you first commence to train is most important, and on this all depends whether you will have a good or a bad year. Give a bird a bad smash early in training and it is often finished for good racing that season.

M Stassart feeds on a mixture as follows: Beans 60 per cent, the remaining 40 per cent is made up of fairly equal parts of Maple Peas, Maize, and Wheat. This is the feed all the year round, up to the last ten days of December, when Linseed and Barley are gradually introduced, and the mixture is cut out.

It is Linseed and Barley during January and February, this being gradually knocked off towards the end of February, and the beans, peas, etc mixture are re-introduced.

Half-an-hour after the last fly at night three grammes per pigeon of a mixture of Hemp, Rape and Rice are given during the race season. This roughly, is M Stassart's system, and feeding for 30 years, and it has given him great success and he still keeps at the head with the champions of Brussels and Belgium.

M Stassart wound up by advising us as follows: "See a fancier's young birds first; then his hens, and you can then judge him and his team through these. Never mind so much about his cocks."

Successful Competitive Racing with Three Pigeons

by HUGH CROPPER

It came as a great surprise to be asked to write an article for this most excellent annual Squills' Diary. I must say that I have read this annual for a few years now and have picked up a lot of useful information from same, that I must give credit for part of our success to the gentlemen who have contributed to the past editions.

My brother gives a hand in the management and this is a great help, as I am away from home at noon.

Our birds are separated about the end of September or as soon as they begin to forsake their pot-eggs. We have tried keeping the sexes together; although some fanciers manage it, we could not keep the hens from laying. I put this down to the way we feed during the moult. We feed three times a day during this period: 8am with a light feed, 12 noon, and 4pm with as much as they will eat without leaving anything lying about. Up to the end of November we feed on Racing Mixture with a little linseed three times a week after the morning feed. In fact, up to the moult being completely finished, we feed just as well as during the racing season. A regular supply of grit is always available. This is put in a hopper and changed once a week.

The hens are given the open hole every morning up to 12 noon (providing the weather is fit); they are then fed in and the cocks turned out up to 3pm. During the moult they are given a bath twice a week. A keen watch is kept on them, and should a bird show signs of distress during the moult it is not raced and youngsters are not rung from it during the next season. We are very keen to see that they have always clean water. This is changed three times a day. Every Sunday morning right through the year a pinch of ordinary table salt is put in the water. This we think helps to keep them in condition.

From the first week in December to three weeks before mating we feed on a Winter Mixture. To this we add a little American Flat Corn, and if the weather is very cold we add beans to the afternoon feed, these being fed separately. We have never tried Barley for winter feeding.

A fortnight before mating the lofts are limewashed. This makes them cleaner and seems to give more light. The nest pans, which are removed when we separate, are now scrubbed and given a good coat of Creosote, and as we put them back into the boxes a good sprinkling of an insecticide is put under each pot. This keeps down insect life, and prevention is much better than cure. A week before mating the birds are given a small dose of Epsom Salts, and if the weather is anyways cold they are kept inside for a couple of days to prevent them catching cold.

This season we mated the racers at the middle of March, the stock birds a week before this. Until they are settled in their boxes we spend a lot of time with them and do our utmost to prevent fighting, as a few broken flights would ruin the winter's work. We always try to put them together on a Friday afternoon so that we can have the weekend watching them.

When they are well used to their boxes they are given the open hole and allowed to build their own nest. Having them in the farmyard there is plenty of building material handy. They are now being fed morning and afternoon on Racing Mixture. Midday we give them a few handfuls of small seeds which we gather from under the threshing machine when it comes to the farm. We also get some from a neighbouring farmer whose land grows a lot of Ketlock. (This may not be the name used in all parts of the country – it grows among the corn and comes with a yellow flower.) This is stored in a spare room all winter and kept perfectly dry. There is something in these seeds which the birds seem to relish, and strange to say since we started to use them we have not been bothered to the same extent with fielding.

When the youngsters are about eight days old they are rung with RP rings and transferred to clean nest pans. They are not touched again until they are transferred to the young bird loft. When transferring to the clean pans we always take great care that there is no catching cold. This we think very important as it would check their growth and ruin them. We always chop some straw and put same in a bag, and place it in the house near the fire, and leave it there overnight to make sure that it is perfectly dry and warm.

No forced exercise is given until the young birds are removed to the young bird loft. They are given the open loft and allowed to do as they wish. Being egg racers we only allow the racers to rear one youngster in the first nest. They are then put on pot-eggs right to the end of the season. We have tried substituting a youngster just before sending to a race, but we have only had one bird perform well like this; in fact we have had disastrous results with same. The best results have been obtained when sitting six to 12 days. We give three training tosses, about 10, 25, and 40 miles. After the first race and up to a fortnight before the first cross-Channel race we exercise twice a day, 6am and 7pm, and give them 20 minutes each time. They are turned out at noon for a go-as-you-please fly.

After this the exercise is increased to 45 minutes twice a day, and they are turned out an hour earlier in the morning, viz 5am. We now give a little best Spanish Giant Canary Seed and Scotch Groats after the morning and afternoon feeds, midday we give the small seeds from under the thresher. After the last fly the loft fronts are opened and they are allowed to do just as they wish. A piece of rock-salt is put outside the loft, and it is very interesting to watch them. Nearly every bird visits this before retiring.

Before sending to a race every bird is dusted with an insecticide, and we never allow them to drink cold water when they return. We always have the chill taken off it: Water is now changed as often as possible, and the lofts cleaned out three times a day. We never give our birds greenstuff although lots of good fanciers do so.

We exercise the young birds twice a day about 30 minutes morning and night, and feed them twice a day with Mixture, and give them as much as they will eat. We never allow them open lofts. We try to send all of them to the last young bird race, 206 miles, and those that are in good feather and condition we send to Guernsey, as we believe that Guernsey youngsters make good Channel racers.

Although we only competed with three birds which had flown Guernsey as young birds, they won the following for us in this season's cross-Channel races: 1st and 5th Club Vitré; 1st North West Championship Club; 5th Liverpool CC; 3rd Federation Vitré, and also 1st Club and 4th Federation Nantes, winning between them about £80.

We firmly believe that to succeed nowadays a fancier must put in all his spare time with the birds. Although it is hard work one is amply repaid when he sees his name on top in the cross-Channel races. Do not trust to memory but book everything down, and when a bird makes a good win try to get it the same way the following season, and more often than not it will do well again.

When the birds are sitting or just hatched, walk about the loft as if you were walking on glass and disturb them as little as possible. We never handle or mess our birds about more than is absolutely necessary.

I have stated as accurately as I am able the methods on which we manage our loft, so if these notes prove to be of any assistance to those who have not had the pleasure of winning in the cross-Channel races, I shall be amply rewarded for my effort in try to inform them of our methods.

The Easy Road to Success

by H MARSDEN FLINT

Having again been honoured by being asked to write for Squills' Diary, I feel rather helpless about it, as today everybody knows everything about the sport of pigeon racing. I started in the usual way, swopping rabbits for the first pair of pigeons I ever owned. That was about 50 years ago. The madness, or pigeon fever, which I caught with those pigeons has never left me. I prevailed upon the headmaster of Trent College to allow me to keep the pigeons and at the end of each term usually got a letter from my father saying "do not bring back any pigeons". However, my mother always put that right. Nothing in the sport line has so fascinated me as pigeon breeding, racing and all the enjoyable, and sometimes annoying, things that go with it. I have shot grouse on the Derbyshire Moors until my gun got so hot that I could hardly hold it. I have killed 126 trout in one day, played cricket and football on some of the most famous grounds in England, but all this gradually took second place to the pigeons.

No doubt some fanciers will be wondering why it was I was so confident this year in pooling my good hen 'Sanfariann' in every pool she could possibly be entered in. Well, that is easily answered, for the previous year she won all pools to £2 from Libourne in the Combine and was entered in £5 pools, but at the last moment, as the car was waiting to take the birds to the marking station, I altered my mind and crossed her out of the two £5 pools and put these on a cock which I now call the 'Pocket Emptier'. I sent four birds. Two did not carry a penny in pool money, but the other two had me guessing and, of course, I did the wrong thing as the hen arrived in time to win both £5 pools, etc. I am aware other fanciers do the same thing, still the loss of over £100 did not really annoy me, for I had gone against what I knew, and that is, if you have a hen or cock about as good as each other, put your money on the hen always. The fact that I did not do this served me right and was very bad judgment on my part. However, the winning or losing of pools on pigeons has never troubled me, and yet if I have a bet on a horse and lose, as frequently happens, it leaves a nasty feeling. Hence the name of the hen, 'Sanfariann', which I suppose means 'What does it matter?' Bar accidents I made up my mind that nothing would stop me pooling this hen to the limit in 1934, from Libourne, 533 miles, and as she pleased me, I did, with the result that she cleared well over £200 in Combine and Central Counties FC. The only bet I had with this hen, made months before the race, was with one of the best working-men fanciers I know and it was for one packet of cigarettes, the winner to buy it. I had the pleasure of buying it, but only just.

With regard to pooling I have always been a heavy pooler. It is a game that requires very keen observation to even hold your own today, and if you happen to have your birds just ordinary, without that bit of 'extra' fitness, on a fit pigeon, you might just as well throw your money in the gutter. Now, I had a little mealy cock that had won 1st Nantes and 7th Birmingham Federation in 1933, a performance I looked on as a fluke, so I sent him to Nantes again same week and he won 3rd Central Counties and pools. I altered my opinion (he had beaten my big pool winner by 20 minutes in the first Nantes race) and kept him back for season 1934. He disappointed me all through the inland races, perching in his box-perch and opening and shutting his eyes, a sure and certain sign of a bird being off colour. I thoroughly examined him, could see nothing wrong whatever, but I was terribly worried, as I have lost too many birds of this 'blinking the eye' type not to realise that here was another pick of the loft gone west unless I was careful, so I thought the best thing to liven him up was to give him much less food, shut his nest-box and take his hen out of the loft, which I did. The effect was amazing; in four days he was no longer sitting, blink-eyed, in his perch; he was all over the loft looking for his hen (I never keep odd hens in loft). I saw then I could put his hen back and she would be sitting right for him to go to Dol with the Central Counites FC. He was A1 when basketed for this event, which proved an awful smash, about twenty members not timing on day. He won 2nd prize, £5 Nom, and all pools to 10/- (the 20/- pool I won with a yearling).

I sent the mealy next to Rennes. He picked up the £2 pool in the open Rennes race, and then Nantes, my first up, and he finished up by winning a prize and pool money at Libourne, four times across in 1934.

Now, I firmly believe that had I left him just to carry on lazily with his hen he would have done nothing for me in 1934. 'Starvation brightens the intellect' and it really means to say that we nearly all feed too well. You can learn something from a mean man, the sort that has a pressing engagement just about the time it is his turn to pay for drinks. Generally very successful fanciers, as they dole out the corn to the pigeons, think how costly the corn is.

As regards feeding and corn used, personally I favour maple peas, Tasmanian for preference as they are sounder than English, but I was very successful with English peas in 1922. If your birds don't please you, change the corn; it may appear A1, but change and watch for the improvement.

There has been much interesting matter in the good old *Racing Pigeon* about the part played by the man and the bird. I say straight out that I do really believe that the birds owned by a man who is always kind to them will make a difference of yards a minute in his favour, and will stick it to the very last to get to him and their loft. Many of us, I feel sure, do not give them credit for the intelligence they are always showing, if we have eyes to see it. They get kindness from me and honest grub, no fads, no pills, no fancy food with cod liver oil; just clean water, good attention all the year round and, as I have told so many, I have nothing whatever secret that I give them in the food line. It amounts simply to this, that the man who understands how to feed is the one who does the winning, irrespective of what kind of grain he uses.

An old friend of mine, a very successful fancier, once had a very wealthy client who purchased most of his winning birds. Just as he was going, he said, "Well, I have bought most of your best birds; you will be an also-ran for the next few years". My friend replied "Yes, but you haven't bought the man with them". I am sure many of us really race against the crack man of our club.

It has always been my ambition to beat the cracks and I tell all youngsters who come to me for advice to aim high. There has never been a crack yet at this game who does not have his colours lowered now and again. If you have a bird that has done a really good performance, breed round about it. That is, put it, if a cock, to several different hens and then pair the progeny together and also with the best of your other birds. I fortunately have had such birds as 'Newsboy', 'Bonny Chick', 'Gipsy', 'Marennes King', the 'Ch White 'Un', to blend up together, and I can assure you, you never know, with all your care, when you are going to get an ace. My old friend, the late James Baxter of Lostock Gralam, who used to race with me in 1897, 1898, etc, in the Manchester FC, offered his champion yearling at £2. No one bought this gem, but he went on to win over £500 for Jim.

The wing of a pigeon today seems to be little talked of, or dismissed as of no real importance. I regard it in a very different light. Years ago we used to cut the end of a flight in the hope that this would prevent its dropping in the basket; it was a vain hope. The best condition for cocks for holding feather and flights is, of course, feeding. I maintain that the nearer a pigeon is to having a full wing, the greater advantage it has. Many fanciers stop breeding after the first round and fly to pot eggs, and I know that many fanciers, just when they do not wish it, find their birds, or bird, has dropped two flights in each wing. Now this is serious, but what about it dropping another in the basket? Is there anything more annoying in pigeon racing than to have your champion all ready, fit as possible, and then to find it has dropped two flights in one wing, or both? 'Gipsy' once did this on me and I am glad to say I refused to send her to the big race though she would probably, but for this, have won me a considerable sum. I never forgot the disappointment and ever since I have tried to find a remedy, and I believe I have succeeded, as for the last three seasons the birds have gone through a special process, and all the six birds I clocked in from Libourne last year went through the process, and not one of them lost two flights together. At any rate, it has given me more confidence in pooling and helped me to win not far short of £400 last year, and against the views of many celebrated fanciers who say that the flights matter little as to how many of them have been moulted.

I wish, however, still to experiment for the one or two years I may be spared to race pigeons, and will state here quite candidly that I firmly believe if I gave to the Fancy my views, every one of them would be rushing to try 'Flint's Process'. I can hear them all saying "What a nerve this chap has, and what a swanker".

Well, I am sorry, Mr Editor, I have sent you such a poor effort. Your letter, received this morning, hurried me up and I just sat down and wrote what came into my thick head. I wonder if everyone knows how easy it is when you get a bird with a broken leg to look after it? When the leg is set, simply put the bird in an old stocking, leaving its head free and swing it up in an exhibition pen at such a height from the floor that it can eat and drink.

Racing to a Backyard Loft

by W WASDEN

It gives me great pleasure to write these few notes for this year's Squills' Diary. The backyard is not large enough to house a loft on the ground floor. The loft is housed on a flat concrete roof over the WC and when you climb the ladder you will find not much room left around the loft. Stand at the door of the loft and cast your eyes around and you see plenty of overhead wires, a scrap yard, works, and a busy street a few yards away.

My loft houses 14 pairs of old birds and 18 to 20 youngsters, and these are only possible by thorough cleanliness. I use sand on the floor, nest boxes, and nest bowls, and the loft is limewashed five or six times each year. Pigeon racing is a 365 days' job and there is always something to be done.

The birds are so tame that I can fly to a trap and beat the open door experts. In fact the way to win races is "to get your birds to come home quickly, get them in at once, and no messing about hiding behind loft or other places".

A good start is necessary, and a pair of birds that were presented to me, on owners giving up the sport, are the base of my loft. The cock, 'Bryant Grooters', flown Marennes five times, 500 miles, winning at all stages and only bird home in club, one race; the hen, 'Spanish Princess' and 'Alfonso' blood, from Mr Burton of Liverpool. This pair bred me a hen to win 14 races, another one 1st Fed Poitiers, 500 miles about.

I keep buying each season three or four good birds from good parents, from good fanciers that race their birds. I am always on the look-out for something better and prefer pigeons bred from actual winners.

One or two actual cases. A hen from a friend had never done the Channel on the day I found her, and she won 2nd Club, 4th Fed Dol, 1st Club, 7th Fed Nantes. Another, from a real good fancier that could not do anything with it, and he won me a lot of races and pool money in plenty and was a topper.

Several of my best birds have been bred from birds brought in and mated to own birds. Buy the best, they are cheapest. You must be kind to your birds and they will repay you. One of my best birds was paired to me and would not have a cock all one season and she has won me 14 races including Marennes, 500 miles. I managed to pair her the next year to a good cock that had won me 7 races. What a pair: twenty-one 1st prizes between them, and they bred me a bird to win a NHU gold medal, and they, other youngsters, won from 70 to 500 miles.

I have several birds that will meet me at the kitchen door on my shoulders, head, and hands if I am a bit late. They want to know where I have been and have hard work to get them to take exercise, for they help me to clean out and have a regular game. But they win races, they hurry home to see me, as they love me, for they know I love them, wife and children first and then pigeons.

I feed on the best mixed corn all the year round and fresh water at every possible chance. I send my birds for a training toss, and when they come home they have a clean loft and a bit of rice and they are pleased to see me, and it is the same on race days. They never go away anywhere hungry; in fact I race them hard and feed them well, and if they could eat gold they would have it, for they would win more back for me.

My birds have open loft until noon and then exercised for 20 to 30 minutes at night. This is enough to get them ready, any bird, for any race, and leave plenty of energy in them for the races. Early morning is the game, and I am writing this at 5am and have had a good night's sleep. I am two miles from the station and my birds are basketed, taken to station, and home again before 7.30am. It suits me; and they win; so it agrees with them too.

Since racing finished I have added new wheat to my mixed corn and linseed and they could have plenty of baths, for the feathers they have grown will be of vital importance in next season's races.

I pair up the nearest weekend to February 14th, Valentine's Day, and let all pairs rear two youngsters, and when these are taken away the parents are allowed to go in and feed them for about a week. Be kind to your youngsters. The old birds are then basketed and put in the kitchen, and Mrs Wasden finds shavings a bit of trouble for they do blow about. The loft is limewashed. This is the second time, for it was done before pairing up; no green pastures here! The birds are put back again and training will soon commence and no more youngsters are reared. You cannot rear youngsters and race at the same time.

The training spot fixed has a good train service, about 20 miles away, and they have this two or three times a week. I am always waiting at this end for them.

Flying in three clubs and having 28 old birds they all have to toe the line, which means hard racing, and this means plenty of good corn. One goes with the other. Work never killed anyone but a lot rust up.

I spend all my spare time with them now, and you should see first one and then the other tell me they can win on Saturday. They tell me lots of little things that have helped me to climb the ladder. I find that cocks and hens can have all manner of things tried on them with success after sitting about 10 days. I try all possible to find something they like and make a note of it, and when they tell me they are fit they get it.

The youngsters are now flying freely and they are put in the basket and then given two or three local tosses, and then a big basket is required for they now go to the training base with the old ones to speed them up a bit ready for the Channel. The youngsters also soon get the idea of 'to come home quickly and get in'. This is the great 'will to win'.

We are now once more in trouble, for Mrs Wasden keeps pigeons in the kitchen for the loft to be limewashed again. Different pigeons, different ideas. Some want eggs 3 days, 10 days, or up to 17 days; a newly-hatched youngster, 7 days' youngster, all will help and give that extra urge. About 10 days on eggs is about the safest, and I have won 1st Federation positions with two different hens from 500 miles in this condition, but the best of all is the 'love of home'. Breed for this; your tamest pigeons are your best pigeons. Race to the corn tin at 300 to 500 miles on a fair flying day and you will have cold feet and not much chance of that thrill when they drop out of the sky like a stone on the trap, and seeing me awaiting in they come. What a joy!

And under this system I had a loft of 20 birds, all flown Poitiers, just under 500 miles, in 1934, including yearlings, and they had done it in either 1933 or 1934. In this year, 1934, I had 17 different pigeons out of my 46 (total old and young) that had won club positions and had won £160 in club and Federation prizes and pools. And note that Sheffield is not Manchester, Liverpool, or Central Counties.

The youngsters are now ready and know the business of racing, and they have all the corn they want, and I find new pals in these, and at the end of racing my troubles once more begin.

Getting ready for next season, picking out the old racers that will have a go, a dreadful business this for me to make room for the new pals that have pushed their way in and youngsters that win at from 70 to 200 miles and will fly 500 miles as yearlings and two year olds, and pick up their prizes and pools at all stages to the 500 miles. This is not meant for the old hands, for I can still 'listen on how to win' and admit I don't know it all.

I tell my pals to beat me and they will be near the top in club and Federation. And to the young fanciers I say get the best birds possible, from winners preferred, and buy from a man that wins out of his turn. Feed on the best of corn and give plenty of clean water. This is cheap and, finally, make pals with your pigeons and make them understand that your method is 'kindness' and you will succeed.

The Necessity of Training Old and Young Birds

by M EVERARD TRESIDDER

It would seem at first sight that such a title was absolutely elementary. It may be so, and yet often the most elementary things which the beginner starts to learn are neglected by the more experienced.

Recently, it has become quite the vogue to deprecate '*so much* training', and from this has arisen a school that the less training the better, providing your birds are fit. We have all been taught that training is necessary. Why? Chiefly as a result of years of experience. But let us think a little of this ability to return home. I will offend nobody, if I can help it, by giving it a definite label.

We know that if we were to send 100 fit young birds, their first toss at 100 miles, under ideal conditions by themselves, the results over a given number of cases would be disastrous. At a venture one might say that 90 per cent would be lost and 10 per cent only return. If, however, given favourable conditions generally, the young birds have a number of graduated tosses up to 100 miles, then we should expect over a series of experiments to get 90 per cent home, and a loss of 10 per cent only. Why is this? It must, therefore, be that experience has taught the pigeons something.

Therefore, whatever it may be, it is dependent on education. Which means that it is an educated faculty. There is undoubtedly a strong predisposed hereditary tendency to learn readily. Such as the foxhounds selected for their scenting powers, the pointer for its pointing ability. These characteristics of learning and acquiring proficiency have been handed down through many generations. This power of learning to reach home has equally been handed down through many generations of selected stock.

So certain has this power become, that fanciers are inclined to confuse the result with the hereditary ability to learn. They learn so easily that non-thinking fanciers think they can cut out one step in reaching the result. This most definitely is not so. Let those who disagree form a club to fly by itself, and not with a Federation, preferably in midweek. To fly young birds, yearlings, and old birds races separately, with separate liberation, at separate places, and this without any training at all, I feel confident such a club would not continue to exist very long, and the number of candidates it sent to the classic races would be greatly diminished.

To consider the results obtained by untrained birds, in conjunction with others which have been trained, is a fallacious conclusion. Is it not often said that the first two Federation races are very confusing to the birds by reason of so many improperly trained birds being engaged? Let those who carry out this practice enter into a competition with trained birds, not for his first arrival, but for his tenth arrival in each race. I venture to think the trained team will win nine times out of ten.

This education, this training admittedly is necessary. Then we wonder what do they learn?

A complete answer to this question would solve the mystery. Nobody can give a complete answer. Hence, the many theories on the homing faculty.

Does this education teach them a sense of direction? Practice and tradition has proved that pigeons which undergo a graduated system of training, succeed in flying 500 miles – when that training is carried out in a definite direction. But by some it is concluded that the same direction is not essential, because individual, isolated results have been obtained by *'turned round'* birds, sent with a club flying from another direction.

Here again confusion of thought arises. These experimental birds are sent with others properly trained. Is there a club, say, in Birmingham, near the centre of England, which would carry out races? Say the first race from Lowestoft, the second race from Haverfordwest, the third race from Newcastle, the fourth race from Rennes, the fifth race from Tralee, and the sixth race from Lerwick. If not, why not?

Would they practise what they preach? It might succeed, but I greatly doubt it, because I believe that graduated training in a given direction teaches the birds a sense of direction. I don't tell you how it teaches them; I wish I could. Therefore, we arrive at the conclusion that the homing faculty is an educated faculty, dependent upon graduated training from a given direction, whereby they learn or acquire a sense of direction.

The experiments arrived at by *The Racing Pigeon* recently gave results in accordance with the opinion expressed above, by a high percentage of losses.

I give to you an example of how history repeats itself. More than 30 years ago the Greenwich Club sent 16 birds to Whitchurch, but omitted to put 'Hants' on the label. They were liberated at Whitchurch, Salop, and only one reached home.

I was in the Isle of Wight this September and was told that the Freshwater Club sent 54 birds to Whitchurch. At this stage I nearly gave the answer. These birds were liberated at Whitchurch, Salop, instead of Whitchurch, Hants. Only five returned home in reasonable time. A case of ancient and modern history. This year, in conjunction with an invalid friend of mine, I selected a dozen birds to be sent to San Sebastian. Eight of them were birds which had flown well from San Sebastian, Marennes, and other cross-Channel races. Four of them had never been across Channel. All had a similar private training in England from the coastal race points. The result was that seven out of the eight trained birds returned, but not one of the four uneducated pigeons.

A novice fancier – so he styled himself – told me I was wrong in my conclusions about the necessity of training, because he had sent three birds to San Sebastian which had never crossed the Channel, and one of these with a jump of 500 miles had won £40 in the race.

I congratulated him, and hoped he might continue his success in the future. If all such birds had been gathered together and tossed separately, on a different day, what then, would be the result? I wonder. It is not only a question of speed, but it is also one of the percentage of losses.

Why do young birds occasionally – and it is an experience common to all old fanciers – from a distance of 30 to 40 miles, on a perfect day, take 3, 4, or 5 hours to return, and then arrive together? One has met with the experience that, on the same day, from the same place, one basket of pigeons liberated 15 minutes apart, home in an entirely different manner. One lot may be quick, the other take a long time. One lot may return and the other get lost. Why? Old trained birds rarely behave like this.

Examine this subject a little closer. Every old bird team probably exists of yearlings, two-year-old birds, and older. Suppose, for the sake of illustration, we say there are 15 yearlings, 10 two-year-old, and five older. From which lot do we expect the greatest losses at the end of the season? What does the term 'reliable-consistent' mean when applied to our racers? They are proved birds, proved by experience over many distances under variable conditions, while the inexperienced, the uneducated fail. Therefore it would appear that training and education lead to efficiency.

Lack of training leads to losses.

The remarkable exceptions which exist all over the world of birds returning from long and short distances, even after a great period of time, without any training are known to us all. But we also know that if the bulk of birds were submitted to this experience the losses would be excessively high. There is no doubt, whatever, that some strains require more education than others. Some men have better memories than others, and so with some strains once educated – all that is necessary is to ensure their physical fitness to engage in any race with very little training another year.

To the beginner in our sport I would warn him against accepting the policy and practice of jumping his pigeons without adequate training and education. As I have pointed out in previous articles, a trained and educated bird is an entirely different proposition to a young bird or a yearling.

When we come to consider the subject of long-distance young bird racing from 400 miles, one is bound to admit there are very few, if any, fanciers in England who have any actual experience. Their advice may be good or indifferent, whatever it is, it is unquestionably a theoretic opinion, based on hearsay. So we must look elsewhere than in the British Isles for information based on actual results.

In Belgium, the 400 miles young birds race, organised by Brussels fanciers from Angouleme, is an established annual event. These birds are often hatched in February or very early March, and so are frequently six months old in September, when the race takes place. Together with these birds are sent old birds. It is supposed that these are properly trained, but on that score I have no definite information. But if they are not trained properly, then I see no possible reason for their presence; rather would I regard them as an objection.

This last practice, I think, would be useless in England, because, I ask, what good fancier will engage good experienced birds to fly from 400 miles in August? The inclusion of birds which had failed in the ordinary old bird races, were improperly trained and belonged to a class of indifferent performers would be far from helpful, and should be excluded.

Then we come to America – the United States – where long-distance young bird racing has been carried out for many years from 400 to 500 and 600 miles. No old birds are mixed with these young ones.

Without further elaboration, and in proof of what I have written, I give you the complete programme of the Buffalo Flying Club. It will be seen that this club evidently believes in training.

GREATER BUFFALO RACING PIGEON COMBINE

Young Bird Schedule for 1936

1.	75 miles	North-east Penn	Aug 7 for Aug 8
2.	100 miles	North Givard, Penn	Aug 14 for Aug 15
3.	130 miles	Kinsman, Ohio	Aug 21 for Aug 22
4.	150 miles	Warren, Ohio	Aug 28 for Aug 29
5.	200 miles	Canton, Ohio	Sept 4 for Sept 5
6.	200 miles	Canton, Ohio	Sept 11 for Sept 12

Buffalo Centre Races

7.	355 miles	Portsmouth, Ohio	Sept 16 for Sept 19
8.	300 miles	Lancaster, Ohio	Sept 24 for Sept 26
9.	300 miles	Lancaster, Ohio	Oct 1 for Oct 3
10.	500 miles	Louisville, Ky	Oct 1 for Oct 3
11.	600 miles	Galliton, Tenn	Oct 7 for Oct 10
12.	300 miles	Lancaster, Ohio	Oct 8 for Oct 10

In addition to this, my friend, Mr Lang Millen, writes as follows:

"Here is my method of training young birds, embracing a period of one month and a half. On September 16th we ship our birds to the National Convention Race, distance, 350 miles from Buffalo.

"On August 1st we flew our first race from 75 miles, and each week thereafter we flew the following races: 100-130, 150-200, and 200 miles.

"I have given the birds that flew in these races, and that are to fly in the 350 miles race, the following tosses:

40 miles–65 miles–75 miles race.
115 miles training.   100 miles race.
75 miles training.   130 miles race.
75 miles training.   150 miles race.
75 miles training.   200 miles race.
100 miles training.   200 miles race.
65 miles training.

"You will note that the birds get all this experience within a month and a half before they went to the 350 miles race.

"There is one other thought which might be helpful at this time: I think young birds, for long journeys, have to be kept after it rigidly, and cannot be prepared over too long a length of time, and it seems the harder you give it them, a month or so before the race, the better are your results, although I admit the losses in this preparation are somewhat heavy."

Mr Lang Miller has flown these long-distance 500- and 600-mile races for many years, and it is evident he believes in training. I wonder how many strains in England would stand this severe treatment?

This, too, is accomplished by a strain which had no out-cross for 23 years. Admitted, as I pointed out last year, the climate in the States makes these performances more possible than in Europe, it, however, does not detract from the practice of systematic training. And if in the future there be another 400-mile young bird race, it appears to me essential that the pigeons should be adequately raced from at least 300 miles before being engaged.

Why Keep a Mob?

by J CHANNING, Newport, Mon

I was very pleased when asked to write an article for this year's 'DIARY' as I believe this is the first time a Monmouthshire fancier has had this opportunity.

I have kept pigeons for a number of years, but never had a large team, and start the season with 20 birds, of which 17 are put into training. The remaining three are old stock birds. I never intend to keep more than I can properly attend to, nor than I can afford to keep and race, and being an ordinary working man these things have to be considered. I am certain the failing with most fanciers is that they keep too may birds, and cannot treat them as individuals, but must treat them all alike.

Good birds will always respond to good management, and what success I have had, I consider is due to keeping my birds clean and happy, and also watching their actions in the loft. My loft, which is an ordinary wooden structure, is 16ft 8 in long and 6ft wide, and divided into three compartments; is scrubbed out every day during the summer and many times during the winter, and in addition is also scraped out twice a day. This may seem a lot of work, but I am satisfied that it has amply repaid me. I cannot see how birds can be expected to give of their best if the loft is not kept clean.

I would add that the centre compartment of loft is 4ft wide, and is used for trapping, the open door being used. Traps or drop-holes are of no use in the district in which I reside, and if you lose any time at the home end on race days you can reckon you are among the 'also-rans'. Birds are never allowed to go on the house, or pitch on top of the loft.

With the 17 birds mentioned I fly in two of the strongest weekly clubs in Monmouthshire, namely, Newport Championship Club – 45 members with a 10 bird limit – and Pill Homing Society – 25 members with a 6-bird limit. I am also a member of the Welsh Championship Club, which comprises the cream of Monmouthshire fanciers, who fly their only Old Bird race from Thurso. Duplication of birds through clubs is not allowed.

My family of pigeons, which is an inbred one, is based on Osman, Logan, Thorougood and Hansenne, and as they win from Ludlow (56 miles), to Thurso (488 miles), for myself and others, I do not want anything better. Although inbreeding, I do not go too near, and think that grandsire to granddaughter is sufficient. From this mating I have bred some real good birds, and although my birds are not wild, and allow people to walk among them, the tamest birds have proved my best racers, while the highly-strung birds have proved excellent producers. I do not believe in chopping and changing with strains and types of pigeons, and never introduce a cross, however good the breeding may be, until it has proved its worth on the road. I am not keen on pedigree, preferring the basket to find the right ones to keep for me.

Only flying a small team, my object is to get the most out of my birds and win whenever possible. The birds are therefore mated in groups, yearlings and birds for races up to 300 miles on March 1st, and those for the longer races on March 17th. By doing this I find I have a few birds in good condition for each race on the programme. I mate my best racers together, provided they are not too closely related.

I only breed one round of youngsters from the racers, and do not mind how old a cock or hen is so long as they produce the goods. My old stock hen, now 8 years old, flew Thurso three times, has bred numerous winners up to Thurso, and a youngster from her won 3rd Stockport (NCC), this year.

Old bird training is started at the end of April, the first toss being 17 miles, second, and last toss prior to racing, at 56 miles. Until the 300-mile stage is reached birds are exercised 30 minutes morning and evening, and for the longer races are exercised 60 minutes morning and evening. Should it be found necessary they have a midweek toss of about 50 miles. Peas are the staple food used during the racing season, and when the longer races are coming along the birds are given canary seed and meat extract mixed – one tablespoon to each pair of birds after the evening meal.

I have found the medium-type pigeon does best for me, with good shoulders, strength of back, tight in vent, and not too shallow in keel. My favourite condition both for cocks and hens is sitting 8 to 10 days, although I have won 1st Banff with a hen sitting 21 days, 14 days of which she sat on her own as I lost the cock to which she was mated. This hen also won 1st Stockport, 2nd North Berwick, 2nd Perth, sitting the last two days prior to basketing on each occasion on her own, as I removed the cock. She was also 21st Lerwick Welsh Championship Club. This particular hen was very keen when sitting, and always wanted to be on her eggs. Through observing this I made good use of it.

I am very severe in my selection of young birds, and only rear the strongest – 20 in number. I wean them when 4 to 5 weeks old, and when ready for training they receive a few tosses of 2 miles, one at 17 miles, and then 56 miles, ready for first race. They are raced naturally right through to Ripon (188 miles), which is the longest YB race on Federation programme. This year, however, I sent five with the Welsh Championship Club to Tynemouth (247 miles) and had four on the day, a hard race.

As yearlings I stop some at Perth, and send some to Thurso. I have flown some excellent yearlings from Thurso, and find this does not stop them winning in later years. For instance, my light chequer cock, bred 1931, flew 247 miles as a youngster, Tynemouth smash, and Thurso as a yearling; 1933 – Thurso, and in 1934 won 2nd Thurso, Welsh Championship Club.

While the birds are moulting they are only exercised twice a week, and I do not mind if they do not fly at all. During this period they are fed on a mixture, with an addition of linseed. All the year round they are given plenty of green stuff, and fresh grit is always available for them. They also like once a week some homemade cake, and know when this is being served up. I am a heavy feeder summer and winter.

During the racing season my birds get a bath twice a week, and I put plenty of ordinary table salt in the water. I consider this hardens and improves the feather considerably. During the winter they only get a bath once a fortnight, and then the weather has to be good.

The methods I have outlined have enabled me to win my share in the three clubs mentioned, and since joining the Welsh Championship Club in 1924, and never sending more than three pigeons each year to their race from Thurso, I have won the following positions, three 2nds, 4th, 5th (sent one bird), 7th (sent one bird), 9th, 10th, 17th and 19th, winning in these clubs and Federation from Thurso alone, approximately £320.

I have not kept a complete record of all my winnings, but during the past four years have won: *OB's* – four 1sts, one 2nd from Shrewsbury to Carlisle, including 1st Mon Federation, Lancaster, and with a yearling red hen won 2nd Thurso Open South Wales Combine (velocity 1020 yards), 2nd Monmouthshire Federation, 1st Newport CC and 1st Pill HS. I was also 9th Mon Federation in same race. – *OB's*, 3rd North Berwick, and in the Perth smash with two different pigeons won 1st Perth, Newport CC, only bird home in club on day (a yearling hen), and 1st Perth, Pill HS. Only three birds in Newport day of toss, the other belonging to a clubmate in the Pill Club. I was also 9th Thurso, Welsh Championship Club, smash. A record number of birds were sent to Thurso last year, and I think this one of my best performances. – *YB's,* four 1sts, two 2nds, two 3rds from Ludlow to Preston, including 1st and 2nd Mon Federation, Ludlow. – *OB's,* three 1sts from Ludlow to Stockport, 3rd North Berwick Pill HS, 3rd Perth, Newport CC, 1st and 9th Thurso Monmouthshire Open, 1st and 9th Thurso Monmouthshire Federation, 1st, 2nd and 3rd Thurso Pill HS, 1st, 3rd and 5th Thurso Newport Championship Club, only 17 birds home on day of toss in Open and Federation Welsh Championship Club race, 4th Thurso (smash). Also winning Combined Average, Newport Championship Club.

My Thurso winner last year is a blue chequer cock, bred 1933, and was sent sitting 10 days. He has previously won prizes from Shrewsbury, Stockport and Perth. My second pigeon is a blue chequer hen, bred 1935, and she is also the dam of my third pigeon – a yearling flying 15 hours 40 minutes.

Although as previously stated I only started 17 birds in training, I was able to send 10 birds to Thurso, 8 with the Federation and 2 with the Welsh Championship Club, and had 8 arrivals. This is the largest number of birds I have ever sent to Thurso.

I do not send my birds away to shows, but only show them locally and with this same team have won since 1935: fifteen 1sts, fifteen 2nds, ten 3rds.

My birds, therefore, give me a lot of pleasure, in addition to work, and I am sure they do all they can to do me a good turn when it is required of them, as I know each one of them through not keeping a mob.

IMPORTANT NEWS TO BREEDERS

AND

FLYERS of PIGEONS, POULTRY and
AVIARY KEEPERS and DOG FANCIERS.

All Breeders and Flyers should know that Hall's Sanitary Distemper is the best thing known for ensuring perfect cleanliness in breeding and store cages, show-pens, fowl houses, kennels, etc.

It is sold in Tins ready to mix with water, and is supplied with a whitewash brush.

One coating effectually destroys all the insect life and fever germs, etc., which collect in cages and kennels.

At the same time, Hall's Distemper is non-poisonous to bird and animal life.

It is made in two qualities ; the outside quality is the one specially recommended for the above use, and suitable colours are—

No. **17** Cream.   No. **66** Pale Green.   No. **47** Pale Blue.

HALL'S DISTEMPER

is sold by all Oil and Colourmen, Stores, etc.
Remember to ask for outside quality.

Sole Manufacturers

SISSONS BROTHERS & CO., LTD., HULL,
199ᴮ, Boro' High Street, London, S.E.

Full Particulars with sample and shade card will be sent post free on application.

The One-Man Loft

by H SMITH, Darfield

Three times winner of Barnsley Federation
Cross-Channel Average

My small loft is known as 'The One-Man Loft'. Why? Well! I do not believe
in partnerships. There sometimes comes a time when partners fail to see eye
to eye.

The loft is divided into three sections. The Old Bird section is 7 feet 6 inches long. The centre section, in which my corn and utensils are stored and in
which I sit, is 3 feet long. The Young Bird section is 6 feet long. At each side
of the centre section is a sliding door so that one can enter either Young or
Old bird section without having to go outside.

The loft is well ventilated at the front and there is an air escape at the top
of each section.

The methods of trapping are very simple. Just a small landing-board and a
two feet clear entrance.

I can truthfully say that I have never started a season with more than seven
or eight pairs of birds in my loft. And they have to serve the dual purpose of
Breeding and Racing.

The strains are Gits x Delmottes; Logans x Osmans. Only the very best of
these strains have been introduced into my crosses.

As is well known, I have more than held my own both in the Darfield and
District HS, and the strong Barnsley Federation of over 500 members. To
hold one's own in the Darfield and District Club nothing must be left to
chance. It consists of some of the best and keenest fanciers it is possible to
find in any local club. Barnsley Federation results prove this.

I am classed as a 5 per cent Fancier. Why? In the first place one must have
the right class of bird to handle. Hence the pigeon gets the 95 per cent credit. For no matter how clever a fancier may be, he cannot successfully compete with second or third rate birds against the A1 class. Given the 95 per cent
pigeon a 5 per cent man will use the remaining 5 per cent to build it up for
racing and breeding in a way that usually spells the difference between success and a moderate performance.

My 'Pigeon-year' starts the first week in September. A vital month for the
birds. For then they drop into 'The Big Moult', and this is the time to make
or break them.

Even with the best of birds, a bad moult will certainly be followed by a poor breeding and racing season, the following year.

So study the comfort and general health of your birds in much the same way as you should study your own health and comforts. You will then be amply repaid for your trouble, as undoubtedly health is the main key to success.

During September and October I feed my birds on good, sound new wheat in the morning. Twice a week at morning feed I give them a feed of three-parts sound linseed and one-part hempseed, mixed together. About two handfulls for twenty birds. For evening feed the following mixture: three parts best peas, two parts good, plate cinq maize; one part good tares. I emphasise that all food must be the best as birds require all the nourishment they can get during 'The Big Moult'.

In November, when the days are short and I can only feed once a day I add to this mixture 50 per cent last year's tic beans.

During the moult, a piece of common brick salt is placed in a dry spot in the loft for the birds to pick at. Twice a week they get fresh grit; greenstuff, and a tablespoonful of Parrishes Chemical Food mixed in a quart of drinking water.

They get as many baths as they will take. Into the bath water goes a thimbleful of permanganate of potash. It helps to keep the birds free from vermin and does not injure the feather.

I do not use Epsom Salts, but the following for physicking:– Two tablespoonsful of Glauber Salts and twelve senna pods scalded together in a quart of boiling water and allowed to stand overnight before using.

During racing season my staple food consists of four-parts best Tasmanian maples; two-parts best goa tares; one-part best plate cinq maize. Every other morning the birds get a little mixture of three-parts Australian dried wheat; three-parts linseed; one-part hempseed.

Tic beans are used from November until the first week in February only. After that date beans do not find a place in my feed.

To win the Barnsley Federation Cross Channel Average in 1934, 1937, and again in 1938, also two NHU gold medals in 1934 and 1937 with my small workingman's loft, never utilising more than eight birds in all three water races, proves beyond doubt that my birds are 95 per cent, and my methods of feeding and management not far wrong.

There are eight nest-boxes in the Old Bird section. In each nest-box there is a half-pound jam jar for the drinking water and a small home-made box, four inches long, two inches wide and two inches high for the corn. This entails extra work, I know, but experience has proved that it pays, as each pigeon or each pair can be fed and watered and physicked if necessary to suit the individual. It also has another great advantage. The youngsters, seeing their parents eat and drink soon learn to do likewise, and can thus be weaned much earlier.

My birds are well fed. In return they have to work for me. I can afford to keep only racers. Not homers. The youngsters are sent right through to the last race of 202 miles, if fit. Yearlings to the first two water races. A black mark is placed against any that do not perform satisfactorily. And out they go. This, in my opinion, is the only way to maintain a select, high-class loft.

All my best birds have been worked hard as youngsters and again as year-lings. To fly the 200 miles stage is easy. To fly the 500 stage year after year is a different proposition, and only the best and most strongly constitutioned birds can hope to accomplish this task.

Therefore, breed from only the best and soundest of your stock and let the pannier be your guide. Pedigrees are often misleading. The best of stock does not breed good, sound youngsters every time. They have to be tested before their worth as Racers or Breeders can be ascertained.

Now for a word about the youngsters.

In the Young Bird section is an old race pannier into which the youngsters are placed when taken away from their parents. There is a water trough attached to it into which I let them see me pour water. They are also fed in this pannier, which, by the way, is left open. And when I enter the loft they scutter back into it.

I consider this schooling a great advantage to them in later life, especially during holdovers, as they are 'at home' in the basket.

Around five weeks old the youngsters learn to leave this pannier and fly up to the perches. When they do this I get them acquainted with the entrance to the loft by putting them out on to the landing board and letting them run in again.

At eight or nine weeks old, when they start to fly strongly round, I put them into the training basket for the night and next morning liberate them at a point two or three miles away, in the direction of the first racepoint. At twelve weeks old, when I am satisfied they are leaving the basket all right, their seri-ous training begins.

Four or five tosses at 8 miles; repeat at 15 and 20 miles. Then on to the 40 miles stage, which they get as often as possible before the first race to which the whole team go, so as to obtain the experience of mass liberation.

Then I divide them up. Selecting so many for each race. Those that do not race one week get a mid-weekly toss at the 40-mile point. All that are fit go to the last race.

Being on 'shift work' in a mine, that is, morning work one week, after-noons the next, I have to depend on the capable assistance of my wife, who not only cleans, feeds and exercises strictly to my time, but also clocks in when I am away convoying. The birds know her as well as they know me. In fact, when my good hen 71 homed and won the Fed Paris race, I had to leave the loft as she would not trap for me, but directly my wife went inside the loft she followed her in.

During my recent ill-health she has flown the youngsters to win, amongst others prizes, a lovely tea service from the last race at Hastings, 202 miles. Any fancier blessed with such a wife has a lot for which to be thankful. I know I am grateful.

I will close with a few hints to those just starting in this great sport.

When stocking loft, buy from fanciers who win year after year. Pay the price and get a few 95 per cent pigeons instead of a mob of second and third raters. Feed on the best food. Never allow your loft to be master of your pocket, otherwise the sport will cease to be a pleasure to you.

When in the loft talk to your birds and gain their confidence. In time they will learn to know you as their friend and master. Then by patience and perseverance you may become the lucky owner of such birds as my 'Inkerman King', 'Harvest Queen', 'Lightning', and many more that I could mention that are and have been a delight to handle.

And I hope to handle many more of their calibre before Anno Domini forces me to say farewell to this great sport that I love so well.

In conclusion, I may say, "Where there's a will, there's a way." "If at first you don't succeed, try, try, try again."

CAN YOU SPOT A CHAMPION?

by A W GARRATT

Ever since its inception I have taken the greatest interest and derived much pleasure from *Squills' Diary* and when on a winter's evening I have read over and over again the different articles by the great lights of the Fancy I would build castles in the air and wonder if ever I should do anything to make me worthy of writing an article for the old *Diary*, and now I am asked to say a few words I feel absolutely at a loss. There is one thing I am not going to do, however, and that is try to teach the present day keen young fancier how to win inland races, because they can invariably whack my head off in races up to 200 miles, but what I should like to help them in, if I can, is to recognise and be able to pick out a good individual and outstanding bird. Remember, please, he is a lucky man who breeds one really good bird every other year.

In my younger days we used to go miles to see 500-mile pigeons, now they are as common as flies, but I often wonder how many really good pigeons are ruined? I will try to explain what I mean by taking my Bordeaux Section winner this year as an example. He was bred 1937 and flew Guernsey (236 miles), as a youngster. Nothing in that, as I have too many pigeons and hate selling, and always try youngsters to the bitter end to see which are worth wintering. As a yearling he flew Guernsey again and I liked him and stopped him. Last year he flew some good races and the harder the race the better he seemed to like it. In the first week of racing last year he scored 3rd prize, beaten by decimals, from Gloucester in a bitterly cold Nor'Easter. After that I worked him through to Guernsey. He was most consistent and both my son and a fancier friend said he would win from Rennes. They both thought I had taken leave of my senses when I said I was saving him for Bordeaux. "But he has never been farther than Guernsey in his life", they protested. "I do not mind that", I replied, "he is bred right, has never had a night out and if he has a sporting chance he will get home from Bordeaux in the day".

I had a month to wait and my trouble was to keep him fit. I had promised a cup for a race organised amongst local fanciers in aid of the Burton Infirmary and they decided it was to be raced for from Bath (104 miles), so I entered 15 birds and my Bordeaux was 1st home for me, but the clocks went wrong or something of the kind, and the secretary and chairman of committee came to see me on Sunday morning (the race was flown on the Saturday), and said it had got to be re-flown on the following Tuesday from Bath (one member – one bird), to be liberated at 5.0 pm, any weather.

I sent my Bordeaux bird and the Tuesday evening turned out a 'fair stinker', visibility about 200 yards with light rain, and I timed him in at 8.0 pm, and I was never so pleased to see a bird drop in my life.

We were sheltering in the greenhouse and my son was chaffing me about my prospective Bordeaux bird when he came through the mist and rain, hit the loft and ran in like a lion. I said, "He will do".

After that I got him sitting nicely and gave him a private toss with a few others to Weymouth (157 miles) just as a kind of trial and he came bouncing home first batch. Well, although this pigeon was having a jump of 320 to Bordeaux, 555 over country he had never been in his life, I pooled and nominated him and he was 4th Open and 1st Section.

Now I come to the point I want to impress on young fanciers.

If you have a good bird, bred to stay, and it has raced well up to 200 or 250 miles, do not break its heart by sending it to the 350- or 400-mile stage before the final event, but give it basket work up to 100 miles, and then jump it. You will find it pays, for after all it is in the big final events that the money is to be won.

Of course, my methods will not appeal to young fanciers who are inclined to be impatient and mate their birds in January and breed or race their old birds to death before the final event. One cannot burn the candle at both ends. I still mate my birds about March 15th, the same date my late and esteemed friend, Col A H Osman used to fix on, and I have done pretty well with young birds in my time. One little incident which happened about 35 years ago regarding young birds I must tell. I went to Altrincham Show (where *The RP* used to give a £5 Special for Best Young Bird), with Col Osman, 'Nor' West' (I believe his name was Vickers), and H J Longton (all of whom have joined the great majority), and we met all the big fanciers in Lancs. I remember the main topic was the wonderful successes of one, Gordon Clarke, with Soffles. But I transgress. An argument arose as to the best age for young birds, and a home-and-home match was arranged for the following week between 20 from Col Osman's loft and 10 from anywhere in Lancashire. Col Osman gave his 20 a toss at Cambridge because he thought Burton, where I had offered to liberate for him, too far, and won the race handsomely.

I have no hesitation in saying that he was the cleverest man that ever flew a pigeon and the finest tutor the sport has ever known.

HOW I MANAGE MY OWN LOFT

by MRS R JOHNSON
Winner of Osman Memorial Trophy 1940

It is a great honour to be invited to write a short article for *Squills*, and also a bigger honour to be the first lady to win an Osman Memorial Trophy. I have been a fancier for about six years on my own, having three lofts and 60 pigeons. This is how I started pigeon flying. My husband, a keen fancier and big prize-winner, also a business man, was unable to look after his pigeons so left me in charge, and it was so interesting that it just got into my blood. I was left to clock in many winners. From Thurso I clocked in the only bird home on that day, winning a great number of prizes. This gave me the cue. I would have some birds of my own, and started right away, winning several prizes, but not a first until this year (1940), when I have won four Firsts and several other prizes. Old Birds – Harrogate 2nd and 4th, all pools; Northallerton, 1st and 3rd, all pools; Morpeth, 1st and 4th and Special; Berwick, 2nd and 3rd; Perth, 3rd, all pools; Banff, 1st and 3rd, all pools, topped Fed; Berwick Yearling Race, 1st in Club, 4th in Section, won Combine Average Cup, Skevington Cup for the best three longest races, Fed Cup, and Gold Medal, money prize for two longest races – old birds and young.

The breeding and training of my birds is entirely my own responsibility.

Management of Loft – My pigeons get plenty of fresh air, living out of town is nice and open. Cleanliness in the loft is one of the main factors to success; I do not believe in sawdust or sand on the floor, but use slack lime. This I dust on perches, nest-boxes, scrape floors well, then brush well in and sweep up. This is done morning and evening in summer, and every morning in winter. Water cannot be changed too often, they love a nice fresh drink in hot weather. Every week I give them two teaspoonsful of Epsom's Salts four days before race, after their feed in morning, changing their drink at mid-day.

For racing I feed them on Tasmanian Peas and Beans entirely, and a good mixture for winter, with Linseed for moultings. If possible, I exercise three times a day in summer, all weathers bar fog, and once a day in winter. I always feed them at the same time every day. This they look forward to, I give them plenty, but not to leave any, as I like to see them a little hungry. I do not feed them on the floor but use wooden trays. These I can always scrub out. They also have a fresh box of grit every day. They have a bath twice a week with a handful of common salt in water. This hardens the feathers and keeps them clean after the bath. I hang up a bunch of lettuce or watercress sprinkled with a little salt. I like them to have plenty of green food.

Training – Old birds, as youngsters, I train them 40 miles, south before starting to train north, then I give them 5 miles, 12 miles, 22 miles, 36 miles, 44 miles, 60 miles, three times. I like plenty of training. Young birds I give much the same, but before starting to train I give them three or four lessons in the basket and always teach them to eat and drink in it. They then never get basket fright. I always give them a nice little surprise when they arrive home from training, getting them in as soon as possible. They enjoy this little mixture containing baked bread, crushed very fine, millet, rice and a little linseed – this they look forward to. I never exercise my youngsters until early evening, until they have had their first training. This is good advice, as I know many fanciers lose birds due to their being unable to find their way back until trained.

Breeding – I mate my birds the first week in March, completed about the middle of April. I think they do better if they are clear of the first nest before training. One youngster is reared in the first nest, and nothing after. Some of these are reared with foster parents.

My Osman Trophy winner did not feed any. I think they are best flown dry, as some of my birds eat very little when sitting and are very dull and are also first in from exercising. I like them best when starting to call the hen to nest. Mine are very keen, then; when racing, I separate them overnight and put together again half-an-hour before going in basket, but I always make sure they are fit before sending them. A word to young fanciers: do all you can for your birds, and if your birds are good enough they will pay you for your trouble. Good pigeons are very intelligent and I put a lot of time in with my birds, getting them very tame and making them happy at home, so that they will strain every nerve to regain their loft. The blue bar pigeon I like best.

In concluding this article, I would like to say that if it helps anyone to win a good race, and bring them pleasure, I shall be repaid for what little trouble it has been to write.

Come on ladies – you don't know what you are missing.

HOW I LOOK AT THINGS

by J KENYON (Latham nr Ormskirk)

Having the great honour to be invited once again by Squills to write an article for this popular and interesting Annual, I will endeavour to put before readers my views and methods which have gained for me this highly-valued distinction.

I will commence with what to my mind is the most important point of all – 'The Value of Strain' – which cannot be over-stressed. Whenever I am thinking of introducing fresh blood, it is my first and deciding consideration. Size, type, colour, etc, are of insignificant value in comparison, and here is where pedigree is of great assistance. Should I go to some of the leading shows, and decide as to what I would like, from amongst the many beautiful birds in the pens, for here the type of bird, and quality of feather, is everything one could desire. Decidedly no! The most important points are missing, from a racing fancier's standpoint.

What races has the bird won? What good performances have the parents, grandparents and the family for generations back got to their credit?

These are the questions, or the answers to them, which influence my decision. For a starter in the sport, especially if the strain happened to be a local one, I should say a satisfactory answer to these questions would be all that was necessary and the best way for him to start.

But, for the purpose of introducing a cross, it is a recognised fact that some strains, however good in themselves each may be, definately will not blend successfully, and knowledge or advice on this question is of valuable help before making any decision.

I have often advised fanciers, who have written or visited me with the idea of purchasing one of my birds for a cross, when I have happened to know what strains they kept, which I thought were unsuitable as a blend with mine, not to do so, as the result would probably be failure. To further impress upon readers the importance of Strain, what better illustraton can I give than that which was published some few years ago by 'Nor-West' in his 'Jottings' for *Racing Pigeon* readers, in connection with the Grand National Steeplechase for racehorses, run at Aintree. Many readers will remember how he tipped, as his fancy to win the race, in different years, horses that were by a noted sire – 'My Prince'.

I am not interested in horse racing, but when 'Nor-West' stated that it was sufficient for him to know that when he looked down the list of runners and found one by 'My Prince', to select it to win, and his selections did repeatedly win, I thought how well the same method applies to pigeon racing. The importance of strain alone guided him in his selections.

As another example of confidence and trust in strain, let me give you the history of the greatest stock bird I ever possessed – 1018-25.

1018 had a noble head, with a rare violet-coloured eye, beautiful feather, and, exceedingly well-developed wings, but when handled she was very open-vented, and had also a rather badly crooked keel, two faults, each of which, alone, would have been sufficient for perhaps 95 out of every 100 fanciers to ring her neck as a youngster. Moreover, as she was bred from a father and grand-daughter mating, I thought an outcross as a mate would suit her best; this was done, I bred one from them to win a 1st prize at 76 miles, but in her first four years at stock, with three different mates, this was the only bird she bred worth more than a pinch of salt. Did I regard her as a failure? No!

In the following year – 1930, I mated her to a 3-year-old – 1107, off my 1894 and 1886. Both of these were good cross-Channel winners, and each were bred from a long line of winners. 1107 was a complete outcross with 1018.

I only kept two youngsters from them, the first, hatched on May 1, from the 2nd round was 'What's Wanted', the other, 2469, was hatched on June 9, and owing to carrying nest flights in 1931, was not raced until 1932, they were my first two birds home from Dol that year. 'What's Wanted' winning the Ormskirk Continental Club open race. But in his next race from Nantes, 2469 failed to return, and to make matters worse, I had lost 1107, their sire, in 1930, the year they were bred, from Dol.

So in 1931, I mated 1018 to 1097, an inbred black pied, and right from the start until 1936 they bred me a continuous flow of my best racers, and this blood is the base of my loft.

I ask: How many fanciers, remembering her supposedly structural faults, and her almost complete failure throughout her first four years at stock, would expect to produce anything of promise, or allowed her to live any longer? Yet I had a feeling that she should breed champions. To confirm this, Mr Charlie Burton of W & C Burton, formerly of Wigan, came to my loft in 1929, just after 'Valiant' had won the Manchester FC open race from Marennes. He wished to borrow either 'Valiant' or a good black pied, to mate with one of his champions for a nest, the suggested arrangements being that we should each have a youngster. I thereupon said to him, that I had in the loft what ought to be a better breeder than 'Valiant', but that so far she had not bred anything good. He asked which bird did I refer to. I replied, 1018. So he took her and mated her to 'Viking', we each had a youngster from them, mine proved to be a barren hen, I believe his got killed.

But 1018 eventually justified my faith in her. Some fanciers say they never keep a bird more than 5 years of age, asserting that after that age, they are not of much value, either as racers or breeders. I am glad that I did not incline to their way of thinking. To me, it seems, one of the quickest ways of destroying a strain.

As war-time racing only permits of races of up to about 250 miles on the south route for many of us, the tendency will probably become stronger in some fanciers to breed for more and more speed in their birds by introducing birds of this class. My advice is, do not sacrifice the future for the present. When this German menace is destroyed forever, which we all hope be at no distant date, the sound, game, intelligent birds of a good long-distance strain will be essential. The thrill of seeing our 450- and 500-mile winners arriving at the edge of darkness will be enjoyed again, but not by those who have changed over to the fast sprint racers.

Nothing has happened to make me alter my previously expressed opinion, regarding which type of bird I like best. I still prefer a bird of over the average in size. They have proved to be the best for me, and what is more, nearly all the best winners which I have handled in Lancashire are of this class. Perhaps a somewhat smaller type may suit other parts of the country better, but that is for some other fancier to find out.

I only remember once having weighed one of my birds. It was 1018. When I killed her off in August 1938, I thought it would be nice to know her weight for reference in later years; she was 16¾ ozs when dead, at the age of 13 years.

But to give readers a better idea of the size of some of my best birds, I have this mid-November day, weighed half-a-dozen, before feeding, so their crops were empty.

Here are their respective weights:

'What's Wanted', ring no 63.30	18 ozs
'Blue Star', ring no 633.32	18 ozs
933.33	19 ozs
1097.27	19½ ozs
1817.34, 'Ruffec Hen'	17½ ozs
869.39, 'National Effort'	19 ozs

I frankly state that the type of some of my black pied strain does not suit some fanciers.

As a rule, they are bold-headed birds, mostly with rich-coloured eyes, beautiful feather, and extra well-developed wings, but what most of the critics do not like is that many of them, 'What's Wanted' included, have somewhat round-shaped backs, which makes them feel rather lean on the under part of the body. But generally speaking, I think the old craziness of condemning a bird if when in the hand, its back would not hold an eggcupful of water, has disappeared. I have always found the hollow-backed birds to be extra well developed in flesh on the under part of the body, handling nice and plump, but I do not consider them as strong in the back as the former.

I well remember, a good many years ago, when the late Mr C Thorougood was at the height of his fame, and was a member of our club, a fellow member remarking to me that he did not care for the type of his birds; whereupon

I said: "It is not his type that is wrong, it is ours". His birds were flying to the most north-westerly position of any in the club, yet he was beating us sometimes in races, even with north-west winds blowing; but what I had especially in mind at the time was that his birds were putting up wonderful performances, from all cross-Channel racepoints, in the strongest possible competition. So I said, if the birds could do this, and in races which sometimes compelled them to fly a full day, from early morn to late at night, there could not be much wrong with the type.

To sum it up in a few words, "Performance is the Best Pedigree".

I race on the Natural system, the favourite condition, for either cocks or hens is, when they are sitting, from 8 to 13 days. Some of the cocks race just as well when feeding a youngster and noticing his hen, or beginning to drive without having any youngster.

In a previous article, I stated which condition of wing I liked best, namely, the bird to have nine and a quarter flights in each wing to race with. This refers, of course, to the ordinary ten-flighted birds, and not to those odd birds, which have eleven flights or more. To be in this condition, the bird must cast the two old flights (one each wing) about 12 days before date of race.

I don't mind if the preceding two flights are only half, or three parts grown.

Let me relate how I, in 1941, profited by sticking to this opinion. My best racer that year was a blue pied cock, 869, 'National Effort'. I had prepared him specially for the Wigan (2-bird) Club's Penzance race, on July 5th, he was sitting 14 days, but on June 29th he cast a flight in each wing, which is about the worst time, so I kept him at home. Anyhow, I thought it might turn out to be all for the best, and so it did.

I took his eggs, separated his hen from him, until July 9th, then on July 10th I sent him to the National Flying Club's Penzance race, flown July 12th. I knew his wing would be to my liking this time, so I pooled him in all pools in Section, up to 10s in open; also in all pools Liverpool Championship Pool Club race, flown in conjunction. He won 7th Section, 22nd Open, and 3rd Liverpool Championship, winning in all £86, 3,575 birds competing.

A good many fanciers take no notice of the wing, so long as the bird does not throw two flights in each wing they consider them all right, maintaining that it is perfectly natural for the bird, which is moulting without any check.

Is it not also natural for some birds to cast two flights in each wing at the same time, yet these fanciers will admit such birds are not fit to send.

During the moult I like to give my birds extra seed, but linseed is at present scarce, so I visit some of the farms nearby, when they are threshing corn. Underneath some machines can be found a nice pile of small wild seeds, of many kinds, which have been harvested with the corn, and being so small, they trickle through the grain riddle. The birds enjoy them. As they are a natural food for birds, they should be beneficial.

If what I have written will help some fanciers on the Road to Success, or help to prevent others from giving up in despair, I shall feel rewarded.

MY EXPERIENCES AS A NIGHT FLYER

by H J ANDREW, Ruddington

From time to time we have heard of the strides the American novices have made with night-flying, yet I believe there are many fanciers in the old country that have been black-out flying for years. Personally, I came to the conclusion that to win races to the extreme South Coast from such points as Perth, Banff or Thurso, with invariably headwinds, the 10-12 or even 14 hour bird was not quite good enough. Being a farmer, I had the space, with no risks of wires, etc, to attempt to produce the 16 or 18 hour pigeon. At first, losses were heavy, especially with hens, and the type of bird which I later discovered was useless when I moved to my farm in the South (Bexhill-on-Sea). I found the Federation covered a very wide radius with some members flying 20 miles less than me. With headwinds generally south-west, and my loft west of the majority, overfly on a 400 to 500 miles distance needed a bird that would keep it up even after the light had failed. As I had to travel a lot exhibiting my farm stock, I spent a considerable amount of time studying the visibility over London just where our birds would pass over after already having flown 400 to 500 miles, with another 50 miles to complete their journey.

From this I came to the conclusion that the inability of my birds to beat the rest was entirely due to the failure of the light around London. Possibly my experiences as a night-flyer will be of interest.

I selected 20 birds all two years old. These were given single-up tosses all along the route, including Perth and Banff, approximately 416 and 468 miles – of course, tossed early morning. None of these birds were raced that year; 16 survived this test. They were then given dozens of short tosses (1 to 4 miles) in the evening. During the winter these birds were liberated in fogs (sea fogs) and a few minutes before dark. All hens failed to stand the test. The same thing applied to birds lacking in eye colour, only the rich ruby-red survived. My difficulty then was not so much the dark, but I discovered the birds could not locate their loft. I overcame this by erecting a white-painted sliding shutter over the prop-hole.

Personally, after my experience, I am perfectly certain night-flying is something just lying dormant, waiting to be cultivated. Something my birds well repaid me for my trouble was by their extra early arrivals in the morning following an impossible day, and we had many down South.

Due to my success with OB, I selected 20 YBs. These were trained short journeys, always sent on the last train with arrangements for tossing. Returns were rather erratic, or the lot would arrive next morning. This I defeated by including tested OBs – by the way, YBs were trained South and North. That season my YB attempt was ruined, as by an oversight on my part I omitted to get in touch with stationmaster, and, worse still, young birds went to Berwick-on-Tweed (300 miles) instead of Berwick, a matter of 13 miles. Incidentally, 8 returned together.

The question is: does night-flying pay or is it practicable? I say "Yes". The training my birds had, produced the following: 1302 – 1st Newcastle (wind SE, vel 700 yds, 300 miles). Perth – 1318, 4492, 4505 (vel 800-750-740) – 1st, 2nd, 3rd (wind West, distance 416 miles), nearest competitor vel 513. Thurso – 1301 (vel 801), 1st Thurso (565 miles, wind East); 1318 (vel 799), 2nd Club; 2485 – 3rd Club (vel 699); 4489 – 4th Club. No other competitor timed in. Perth race (416 miles) – 1301 (vel 799); 4505 – 2nd (vel 765), nearest competitor 622 vel, wind East. Newcastle race (300 miles, wind SE), 6411 – 1st Club (vel 896) nearest competitor 781. Perth race (416 miles), 1305 – 1st (vel 873); 4505 – 2nd (vel 698) nearest competitor 633. Berwick (vel 700) 1st, no other competitor timed in. Thurso (565 miles), 1303 (vel 890), this bird could not locate loft, fell in field. Together with a fellow fancier, a Mr Bristow, we picked her up in the field. Birds were tossed late, 6419, 4.20 next morning, 2nd Club, nearest competitor vel 405.

On another occasion I had left the loft. A Mr Clarke waited long after dark to clock three birds in one shuttle; before doing this he had to clear all birds from the racing loft and time in with a lantern. If he reads this, he will remember the occasion when, in my excitement, I asked him what three birds he timed in. His reply was: "Don't know; don't care; they are in". When I examined my clock and found but one puncture, I thought: "What a joke"; but little knew the three had been placed in one thimble to win 1st, 2nd, 3rd, by hours.

Just one more outstanding instance; a fellow club member (Mr Ashburnham) clocked my winning bird after I had to leave the loft at 10.37pm. This was from Thurso (565 miles). How he managed this I could never understand, as the light had gone before I left. His nestmate I found on loft to win 2nd Club long before daybreak next morning. The performances of my birds enabled me to win the big cup presented by the late Lord Dewar (a competitor) for best combined average in the South Coast Federation, which I won outright three years in succession. No night-flying has been carried out here, as Ruddington, being in the Midlands, nothing could be gained, as 99 per cent of the races are won on the day.

Given the right class of pigeon, I am certain night-flying is not a difficult matter, and when the long races are with us again the fanciers in the extreme North or South can reap a rich reward with that little bit of over-time flying. By the way, strains play a major part, as I found the same family that bred our old pal Caldwell's 'After Dark' had that bit extra. I had at least 50 birds bred by Jim. The late Billy Pearson once got a shock with my latecomers. I think the great advantage in late flying is the morning bird after an impossible day out.

The above is only a brief outline of my methods. Feeding by artificial light, generally about 11pm, was one of my golden rules, summer and winter.

I had almost forgotten possibly the outstanding performances of my night-flying; 1302 was picked up on day of toss by a late arrival in a nearby street to my loft. This was after turning-out time. Liberated next morning with three fields to cover he won 1st Hastings Club from Banff (465 miles). Incidentally, this was my only entry, and he homed with a broken leg and five flights; never flew again, but came here with me at the age of ten years – a memory of the past.

The open door system is not nearly as successful for this purpose as flying to a trap; the birds seem to fear the open door. I do not think a special diet (carrots, etc) is required to enable them to see. As a farmer I have noticed wild rabbits can see at night, tame ones too, with difficulty. While geese and ducks can see almost as well as by day, other bird life, except partridges and gulls, seem helpless. There is one thing certain – the progeny from night-flyers are most difficult to beat. I believe they won about 200 money prizes for me during my years down South. I could fill The RP with other features such as colour, etc. Sufficient to say I have never bred a good mealy or light-red, as these always seem to fail in eye colour, a point that meets the eye to one like the writer, who has bred and exhibited with a certain amount of success most species of livestock.

WHY I HAVE BEEN CONSISTENT IN LONG AND SHORT RACES

by W WASDEN, Sheffield

It is a great honour to be asked by the editor to write another article for Squills.

My advice to beginners in our sport of racing pigeons, who I hope will increase when the war is over and the boys come home again, is to go to a good fancier who is winning races from start to finish, as pedigree is no good at all if the pigeons are not winning races. Buy two or three pairs of youngsters, train them about 20 miles, then the year following ask the good fancier how to pair them up. He will know best which birds to put together. Later on in the season train the old birds about 100 miles, youngsters about 20 miles, then, for the third season, you should have a good team for club racing. You must go steady at first till you get a team together, or else you might throw them away, and have to start again. Give the birds a chance. Feed on the best mixed corn all the year round, and plenty of it, as starved birds won't win hard races; clean water twice or three times a day, more in the summer if possible, also fresh grit and linseed, and baths twice a week. I use sand in my nest pots as this keeps them warm. Some birds will fetch straw and twigs to nest with – all the better, it will keep them fit and happy. I also use sand on the floor. It is easy to riddle when cleaning the loft out twice or three times a day while birds are exercising. Don't throw at the birds to make them fly, and so fall out with your neighbours. Make the birds happy, as they fly for the love of home.

This is a 365 days a year job, and it will repay you when you go up for your prizes in club or Federation. When training, have a good, clean basket with clean chippings. This will give a good impression of the sport to outsiders. Don't overcrowd. Train birds 20 miles twice or three times a week, pick your days. If in doubt as to the weather, wait another day, as bad days will come soon enough. Exercise birds 20 to 30 minutes twice a day. This is sufficient to keep them fit for races up to 500 miles, and leave plenty of energy in them for Saturday, when the money is on. If you should win a few races, don't think you know it all, as it is quite easy to get a good beating when you think you have got a grand team. This is the glorious uncertainty in our game. Let birds take it easy on the Sunday after the race, and give them a bath. Treat your last bird the same as the first. He might be first next week.

When training be ready to get your birds into the loft straight away with a bit of rice, and talk to them. They will soon take notice. Don't handle them rough, keep your temper or you might spoil a good bird. Don't keep too many birds, a few good ones are better than 40 also rans.

My loft is 11ft long, 8ft for old birds and 3ft for young birds, and I keep 12 pairs all told, sometimes less. No stock birds, they all have to race, and if they don't shape well, they soon are killed. Take notice of your birds which are keenest when sitting or with youngsters. I like birds sitting 10 days as you can't really race and breed with a small loft. I have had some winners with birds sitting 10 days which, I think, is the best possible condition. My old hen 'Kindness' won me 14 races, including 1st Federation Poitiers, nearly 500 miles, 458 members. 'Starling' won me three races, including 1st Federation Nevers, nearly 500 miles, different routes. 'Rockfellar' won me three races and a lot of money, including top cash winner (£72 1s 2d) in Manchester Flying Club's open race from Laval, 1,365 birds; I sent 4 birds only. My good hen 'Lovely', won 1st Club and 3rd Federation Truro, hard race, 1st Stockport Federation North Section; 2nd cash winner in whole race (£34 4s 6d), sent 3 birds only, was the only bird in Sheffield on day. Last year she won 1st Club and 3rd Federation Penzance, pooled to £1, winning nearly £100 on day, also about £30 in other races that year. I tell her she is lovely every time I go in the loft and she knows when cash is knocking about. I bred her off 'Kindness' and a latebred cock off Jim Kenyon's 'What's Wanted', winner of £600 racing. All these winnings were with birds sitting 10 days. I keep improving with black pieds from Kenyon, also from Westwood Bros of Yorkshire. They sent me a latebred which, in his first race, won 1st both clubs and 1st Federation. I hope fanciers will excuse me if I am rambling a bit, but this is harder for me than winning races. I fly to a trap. Open doors have advantages and disadvantages, but if I can see my old bird coming 20 yards away it will be in the clock before any man can close one open door. I have proved this many times by winning 217 races, as many 2nd and 3rd and £2,028 in money. I have made good old fanciers sit up and cough, who have sent a lot more birds than I have.

I am still willing to learn and always pick something up when I visit a winning loft. Pigeon racing is a grand game. It is nice to win, and it's nice to see the other man win as it makes competition keener. Look at the good chaps you meet, old friends and new, all smiling, saying "he has got a better team than ever". When you visit the shows, what a pleasure to be in good company. Better than all doctor's medicine.

SUCCESS AND FAILURE - WHICH IS WHICH?

by Dr W E BARKER

For over forty years the pages of Squills have teemed with invaluable information on pigeon management from the pens of men who have told the story of how they reached, if only temporarily, the Olympian heights of pigeondom.

As readers already know, Mount Olympus is in Greece, as was Marathon, where one of the world's decisive battles was fought, and this name is still perpetuated in the title of our greatest long distance foot race. The Ancient Greeks were great athletes and Mount Olympus was their highest mountain, and no doubt, in those far-off days, it appeared to them as the very roof of the world. Many were the times, during the last war, that I gazed upon its summit across the bay at Salonika, as it stood, stern and aloof, and still snow-capped, throughout the sweltering heat of a Mediterranean summer.

Now neither the summits of mountains, nor the summits of success in pigeon racing, are reached without resolute endeavour, for there will be disappointments to face, mistakes to remedy, and faulty footsteps to be retraced.

How fanciers have overcome these difficulties has been told by writers in Squills, but as we often learn more from our failures than from our successes – if we are wise that is – I want to try and point out some of the probable causes of those failures, and how they may be avoided, and later I want to try and show how even failures may be associated with enjoyment. I want to cheer up the underdog who is perhaps contending against a more powerful opponent.

Now it is probable that amongst the most frequent causes of failure are:
1. *Unsuitable lofts.*
2. *Inferior birds.*
3. *Too many birds.*
4. *Incorrect management.*
5. *Too frequent changes in either birds or methods of management.*

As regards lofts, nothing elaborate is necessary. No doubt it is a satisfaction to an owner if he can afford to erect a substantial building of brick or stone, with a number of sub-compartments, and conveniently arranged perches and nesting boxes in profusion, but to a pigeon, home is home, however humble, and in it, with reasonable precautions, they will be happy, healthy, and successful. To ensure this it must be damp proof, draught proof, though well ventilated, and kept in a cleanly state, and these requirements are within reach of the most moderate purse.

As to the inhabitants, that is a more serious matter, and particularly nowadays when prices of pigeons have reached an almost fantastic level. This, however, is the fault of purchasers rather than of vendors, and the prudent beginner will hold his hand until matters right themselves, which in due course they undoubtedly will.

When purchases are decided upon one should select not merely pedigrees, but sound pigeons of good pedigree, and by pedigree I mean good ancestry, and again by good ancestry is meant birds whose parents and grandparents, for several generations, have raced and won.

One swallow does not make a summer and one outstanding pigeon in a loft may be a mere sport, an accident as it were, and is never likely to reproduce its own merits in its offspring.

Buy then the close relations of the most successful birds from lofts that have been turning out winners for a number of years, as such pigeons are almost sure in their turn to produce winners if reasonable patience and care are exercised in their management.

As to the system of management to be adopted; it is a good plan to find out from the vendor what his methods in general have been, and how he has fed and exercised his birds, and if these methods can be conveniently continued by the purchaser it is advisable to do so. This, of course, may not always be possible for a man away from home at work all day or who may not be able to continue the management of a loft in which a full-time man has been employed, but the point is worth keeping in mind. One thing is certain, however, that frequent changes in management, such as constant alterations in feeding and exercising, are fatal to success, and indeed are perhaps one of the most prevalent causes of failure.

Another frequent error is the keeping of too many birds, for it is essential that a man should know his pigeons individually, and treat them so far as possible as individuals.

The ideal way of managing a loft of pigeons would be on the same lines on which a racing establishment, containing forty or fifty horses, is managed. There, by careful observation and trial, the capabilities of each particular horse are ascertained.

One may be at its best at five or six furlongs, whilst another may excel at one or two miles, and their training is arranged accordingly. I do not suggest that such ideal methods can be followed in a loft of pigeons in their entirety, but the general principle is there, and should be borne in mind, and if careful records are kept of every race it will be found that certain birds distinguish themselves at certain distances, or at particular times of the year, or when sitting or feeding a youngster, as the case may be, and this knowledge may be put to useful account in mapping out a season's programme.

Sometimes, for some reason or other, a plan may have to be modified or even abandoned, and it is a fundamental rule never to send an unfit pigeon to a race under any circumstances.

Careful observation and experience will enable one to tell when a bird is in good form.

Sound food and clean water in abundance are essentials, as is regularity at all times in all things.

And now a word of cheer for the underdog.

There is no greater mistake than to imagine that in this world success, so called, is synonymous with happiness. It all depends on one's disposition. One man, a misanthrope, may win second prize, and £100 in cash in a big race, and be disappointed because another competitor has gained first position and perhaps double his own emoluments, whilst a tail-ender, of cheery disposition, picks up a 'couple of quid' and goes off home as happy as a clam at high water. A real sportsman this. One of the world's happy warriors most to be envied.

Many are the men, who plod along year after year, obtaining the greatest of pleasure from the mere tending of their birds, watching the youngsters grow, noting how they resemble or differ from their parents, and, crowning triumph of all, seeing an old favourite drop in from a long hard race which they almost feared might have been his downfall. What care they that he has failed to add to their banking account. He has come home, and they are spared that heartbreaking spectacle; 'the empty perch'. There is nothing of failure here.

Doubtless a little prize now and then provides a cheering interlude, and the less frequent these are the more they are valued, but these men know that great wealth never bought happiness, and that unsatisfied ambition is one of the curses of existence.

Fortitude in failure and modesty in success are the attributes of really great men, who have schooled themselves, in Kipling's words:

> To meet triumph and disaster,
> And treat those two imposters just the same.

There is as much pleasure to be gained from a quiet stroll along the cool sequestered vale of life as there is from a scramble on Everest, and it is for each man to choose for himself which route he will follow.

It is an important choice. A choice, almost, between peace and war.

MY EXPERIENCES IN ESTABLISHING A SUCCESSFUL RACING LOFT

by R BRAMMER

It is indeed a pleasure to contribute an article to the ever-popular *Squills*.

My first connection with the sport dates back to 1908, when at the age of 17 my father condescended to my wish – the keeping of a few racing pigeons. I joined a local Society and to my delight, won a prize or two in my first season. However, my father was in business in the butchering trade, and my full assistance was naturally required in that direction. Incidentally it was not long before I was reminded by father that the business was priority No 1 over pigeon racing, and so the pigeons eventually had to go, bringing thereby my early fanciership to an end. Nevertheless I had always expectations of the day when I would be my own boss, and be at liberty to re-enter the sport in earnest. Time passed on however and it was not until 1928 that I renewed my association with the racing pigeon fraternity.

From this up to the present time (a matter of 17 years), I think I can claim, without wishing to be boastful, to have created a family of racing pigeons capable of winning and breeding winners at practically any distance from 100 miles to 550 miles.

For the benefit of *Squills'* readers (and especially those fanciers who have recently embarked on this great adventure, pigeon racing), I cannot do better than give my own reflections in relation to my experiences in establishing a successful racing loft.

Foundation Stock – Being in business, and well and truly tutored by my father, I ever have in mind his words: "The best lad is always the cheapest". I was therefore determined to procure the best pigeons that money could buy. It was the famous Osman strain that became the base of my loft, and I have never regretted it. By the way, I must confess that I was most fortunate in having two personal friends and fanciers of repute in Mr Harold Jarvis of Middlestown, and Mr William Berry of Macclesfield. These two Osman specialists have never refused me any birds that I have wished to introduce. It goes without saying that while the Osman strain still remains the backbone of my colony other strains have been used as a cross, those that have crossed admirably with my Osmans being Toft, and Putmans, Logans.

Loft Position – At the outset I had no alternative but to convert a hay loft over the stables into a pigeon loft. This was suitable in some respects, but time lost in trapping was inevitable. Notwithstanding this, I enjoyed the beginner's luck in my first year, winning with young birds – 2nd Leicester; 3rd Rugby; 2nd Banbury; 1st Didcot; 3rd Gosport; and 4th Federation. However, some two years later I was racing my birds to a loft which I had situated in a field a few minutes' walk from my residence. Here, apart from other advantages, I was able to trap through the open door . . . a most essential practice if one is to compete with the majority of other fanciers, especially on fast or good racing days.

The Best is Always the Cheapest – In making that statement one, of course, is quite aware that one could spend fabulous sums of money on thoroughbred stock, yet never attain any really high degree of success. Racing pigeons, like anything else, needs every attention, all the year round, and of course there are all the other hundred and one things that spell success or failure. Under this heading I cannot do better than quote some of the successes that have come my way since founding my loft on what one might term thoroughbred stock. 1930, 1st Rennes SGCC; 1931, same club, won Silver Cup for best average Rennes-Marennes old birds, and Gosport young birds. 1932, my best performance was from Marennes, 500 miles, when only 28 birds were clocked in Doncaster Federation, of these seven were mine, their positions being 4th, 6th, 10th, 13th, 14th and 16th; 10th and 16th Federation were two yearlings, red chequer hens, and the only ones timed in this race. Previous to going to Marennes my birds had done well from Rennes – eleven birds were entered and all home in good time, the first two birds taking 2nd and 13th Federation.

I naturally fancied my Channel birds more than ever in 1933, for they had given ample proof of their gameness and ability to win through under the most trying conditions. They did not let me down, for after clearing the deck, 1, 2, 3, from Jersey, the winner my good blue hen, 'Sunrise', went on to win 1st Club, 3rd Federation, and 5th South Yorkshire Combine, Marennes. She was entered in all pools, Club, Federation and Combine, winning over £40. Had this success been registered in recent years, with money more plentiful, 'Sunrise' would have won at least £200.

In 1934 another gruelling race was flown from Nantes. I had sent a few good experienced Channel birds. Liberating had taken place at 5.45am, but at 9pm I had not seen a feather. However, at 9.25pm I clocked my first bird, a son of 'Sunrise'. Out came the car and off I went with clock to Hemsworth, the Federation Headquarters, leaving a spare clock at the loft (I might add here that I always believe in having two clocks set for Championship events – better to be sure than sorry, that is if you have two clocks available).

I never dreamt that I should have any further arrivals that day, for it was very dull, and almost dark when I left. I arrived home rather late after meeting a few old fancier friends to learn that three other racers had dropped in together at 9.45pm. Butterscotch of The RP and my loftman Bill had clocked them in after striking matches to help see the birds in the darkness. These three birds, all hens, were on the wing 16 hours, they were 'Sunrise', 'Shades of Night', and a sister to her. Only 14 birds arrived home in the Federation day of toss, four of these were mine.

In 1936 I had the good fortune to win 1st Rennes Hemsworth Federation, and was rather optimistic about my chance in the Yorkshire Continental Championship Club's Marennes race. The convoyer lad wired 'No toss' on the Saturday, so about 2pm Sunday I rang my old friend Harold Jarvis at Middlestown for time of liberation. I got this, and he forecast that on such a lovely day, with a helping wind at the home end I should have a few birds in by teatime. Don't be late at Huddersfield, he said, then we can have an hour together. That was not to be, so it transpired, for the race resulted in a disaster, only two birds getting through up to close of race, Monday. It was my good fortune to own one of these birds. I well remember how it had rained Monday night, and that I was about to ring down the curtain for a black weekend, when out of blue, so to speak, came my game hen 'Eventide', to be clocked at 9.8pm. The winning bird belonged to another Doncaster fancier and was clocked at 4pm.

It would be quite easy to go on indefinitely renumerating my many successes up to the present time, but suffice to say that I have provided adequate evidence that the best stock is always the cheapest if your goal is to become a successful racing fancier.

Coupling of Birds – I generally mate up in three separate batches – the first on the 1st of March, the second on the 15th of March, and the last batch, comprising birds I want for the longest races, about the 1st of April. By this method it is possible to have a few pairs sitting nine to sixteen days for any race I may be desirous of entering them in, and incidentally this is my favourite condition for getting the best results from any racepoint.

As Regards Training – The procedure I have pursued and found satisfactory is as follows: Old birds, two to three tosses at 20 miles, twice 40 miles, and then into the first race, which is about 60 miles. Birds thereafter that miss the race on a Saturday are generally given a midweek spin at 40 miles. When it comes to the big events, I think nothing about jumping my selected candidates 200 or 250 miles, so long as they have had good cross-Channel experience. For instance, 4367, red chequer hen, when she won 1st Federation Le Mans 1939, was jumped from Basingstoke, 150 miles, to Le Mans, 385 miles. My red chequer cock, 1329, was jumped from Didcot, 131 miles, to Rennes, 374 miles, to win 2nd Doncaster Championship Club.

Young bird racing has never really fascinated me although I invariably compete in the YB events. I like to give the youngsters plenty of training up to 30 miles, but seldom send the most promising ones to the far end races. With yearlings it is different, they are given a fair amount of work, and at least one cross-Channel race. In fact, one is sometimes tempted to send odd yearlings to the far end. This was the case with my 1329, he simply asked for more work, and he flew Le Mans, Rennes and Tours, to be clocked each race. He is still in the loft and has yet to fail to home on the day of toss.

Feeding and Exercise – During the racing season my birds are exercised regularly at 7am each morning. At this time of the day they they are not too hungry, so they are given a light feed of peas, tares and hard wheat. Round about 9am the birds are ready for a good feed of mixed corn. Then at 2pm I turn out the hens for half an hour, and again feed them in. In the evenings I like my birds to be on the wing for half an hour, increasing this of course as the longer races draw near. At this stage I always add a few extra beans to the diet. Actually I am rather a heavy feeder, and perhaps my racers are at somewhat of a disadvantage on very easy racing days. But not so when the going is hard, and all shoulder work, on such occasions I can generally rely on my racers giving a very good account of themselves.

The Closed Season – Some fanciers advocate a very open loft at any season of the year, and are not adverse to turning their pigeons out during the severe winter months. They contend that such treatment tends to make the birds tough, and therefore more capable of standing up on a 'sticky wicket' – in other words, when flying conditions are hard and difficult. On the contrary, I part my birds about the end of August and, whenever damp and cold wintry weather arrives, I apply glass shutters to the open front of the loft. Then again I seldom let my pigeons out at this time of the year, in fact as I pen this article, it is two months since they were out of the loft. They have been fed mainly on horse beans. Incidentally, the birds have enjoyed an excellent moult and look pictures in their new 'clothes'. By this I am not suggesting that pigeons should be treated like 'hot-house plants', but for my part I do believe that racing pigeons do appreciate or benefit by being kept light and dry and sheltered from driving winds and rain.

In conclusion, I trust that the less experienced fanciers and those just entering the sport may be assisted through these rambling notes in establishing a successful racing loft.

TWO SIDES OF A PICTURE
by Dr TOM HARE

I never learned the name of my fellow-traveller, nor he mine. He paid me the courtesy of lending me his lighter when mine failed to work; I returned it with a cigarette. Then we settled back in our respective corners and chatted of this and that. Before many miles had gone by our talk turned to pigeons; I enjoyed his enthusiasm. As he warmed to his subject he gave me a non-stop account of all his doings since his return from the Army. He was glad to be back at his hobby; he was repairing his damaged lofts and collecting the nucleus of a racing stud with which he hoped to do big things. But there were far too many difficulties for ex-Service men who were building up for a fresh start in the sport. I gathered that he had had a long tussle with local bureaucrats about the position of his lofts; materials for roofing were scarce; paint was poor quality and not fit for high-class lofts; feed was a nightmare and not the stuff for building racers. His grousing was good-humoured and to the point; and it seemed to me that he had prepared for and was overcoming every difficulty but one – his pigeons. So when he paused to light up another smoke I asked him if he was satisfied with the birds he had bought from various studs. Satisfied? From his expression I realised that I had dropped a brick. Did I not realise that in buying a pair here and a pair there he himself had selected each bird? After all, he was not exactly a novice at the game; he knew a good bird when he saw it. What ever made me ask such a question?

The train began to slow down and it was time for me to collect my traps; so I could only reply that I was wondering if he had left room in his plans for the vetting of his pigeons. Had he considered the possibility of disease? Disease? No, by gad, he hadn't; and what was more, none of the chaps in his club ever mentioned such a thing. And as I bade him good luck he asked what had disease to do with racing anyway; and had I had a packet of it among my birds; and what did I do about it; did I ring their necks?

The train moved off and I had forgotten to ask my companion if he read Squills; I hope he will do so this year because it is my only way of answering his questions? Moreover, there may be other ex-Service men who have planned with minute care every detail about lofts, equipment, food and so forth; and are scrutinising the quality and durability of everything but the health of their birds. To the art of 'spotting their fancy' racing pigeon men bring a wealth of ideas about performance, pedigree, conformation, colour, etc, but very few of them appear to ask themselves if the bird they fancy is likely to be diseased; and still few of them would think of having it vetted.

I suppose the thought of disease would rob us of the pleasure of seeking our fancy and buying it. I have done it myself too much in my fancier days to preach wrath and damnation at the foolhardiness of the optimistic buyer. So I will not spoil anyone's fun in spotting his fancy; but I do ask for thought and care once we have brought our bargain home. Then is the time to wake up.

There is one essential thing to be done with a newly purchased pigeon, whatever its origin – it must be separately isolated – put into quarantine. The prevention of trouble is the justification for quarantine. If the new purchase is diseased it should not be put with healthy birds. If it is healthy we should not endanger its health by putting it with diseased birds. Is our bargain healthy? Are our pigeons, to which we wish to add our new purchase, free from spreadable disease? These are significant questions; significant for all members of the club and of the sport as a whole; they concern others than the owner. And the answer to these questions cannot be found in just looking at and handling the bird. The experienced examiner may be able to recognise that a pigeon is not healthy; but in order to identify the particular cause of its trouble he will require technical methods and equipment and the knowledge of how to use them and the information obtained through them. There are numerous diseases of pigeons; some are spreadable, others are relatively unimportant so far as other pigeons are concerned. Quarantine is designed to put a check upon spreadable diseases and the fancier in quarantining his new purchase is taking out a sound insurance policy against spreading trouble in his own and fellow-fancier's birds.

According to my records there are two particular diseases which account for over 90 per cent of the total of spreadable disease among the racing pigeons of Great Britain. Thus if the fanciers of this country would but take reasonable measures of quarantine against these two diseases, they would make a profound change in the performances of the British birds and effect a big saving in money. With the names and main characters of these two diseases every fancier should try to make himself familiar.

One is a germ infection of the bowels and is known as Coccidiosis (cok-sidi-osis); the other, Capillariasis (cap-il-ar-iasis) is a threadworm infection of the bowels. Both diseases are spread through contamination of food, drinking water, sand-grit, perches, floor, etc, by the droppings of the diseased bird. The living parasites of these two diseases as passed in the droppings, are capable of remaining alive in dust, litter, soil or manure-heap for eighteen or more months. Moreover, they are so minute as to be invisible to the eye without the aid of high-power microscopes; many thousands of them would occupy no more space than that of a pin's head. Consequently they are readily conveyed on the feet and plumage of birds which are fouled by infected droppings, on the cleaning gear, on utensils and on the boots, clothing and hands of the fancier himself. Therein lies the danger of the racing basket as the means of spreading disease about a club's birds.

Pigeons of any age can contract either disease. Squabs and immature birds tend to be more rapidly and more gravely affected so that the death rate among them is higher than among adult birds. Though adults usually withstand Coccidiosis and Capillariasis much better than youngsters; and though for many months they may appear to other than the expert to be unaffected; they progressively loose strength. As racers they rarely come in among the winners. As breeders they tend to fail; some of them fail to fertilise the egg; some produce fertile eggs which fail to hatch out; some hatch out their squabs and then convey their disease to them during feeding. The infected squabs either die or become the weaklings of the young bird loft. And so the disease goes round the full year, becoming more widespread and more securely entrenched in the stock until its ravages are obvious to the most unobservant.

Now let us turn to the bright side of the picture. Both Coccidiosis and Capillariasis are capable of being diagnosed by analysis of the droppings and both are curable and preventable. These are big gains, thanks to the progress of veterinary science. Both cure and prevention take time and trouble since they necessitate working under a veterinary surgeon's supervision on a long-term policy. On this occasion I have not the space to enter into discussion of the many variants in the mode of curative and preventive treatment; so much depends upon the number of birds involved, the duration of the infection, the layout and resources of the owner, etc, etc. But I can emphasise the point that by quarantining new purchases and vetting the stock before the breeding season opens a fancier can prevent these two diseases from making a mess of this breeding programme.

The diagnosis of Coccidiosis or Capillariasis involves the analysis of the droppings of each pigeon. For this purpose each bird should be separately caged or basketed; the cage floor being covered with a sheet of clean paper. The droppings should be collected and wrapped in grease-proof paper and labelled with the bird's ring number. The samples should be posted to the veterinary surgeon along with the fee for the analysis and a note giving a list of the birds sampled with their ring numbers. Unless the fancier's records and history of his loft are well known to the veterinary surgeon, the initial letter should also include a note of the fancier's observations, method of management and layout of his lofts. This information will assist the veterinary surgeon in detailing the curative and preventive measures which are advisable for the particular case.

Had my train journey permitted it, this article would have been the pith of my answer to my fellow-traveller's questions. I should have felt my journey had been worthwhile had I succeeded in persuading him to have his birds tested for Coccidiosis and Capillariasis. I should have hoped that he would have grasped the meaning of the risk run from these two diseases and the benefit of insuring against the risk by having the droppings test of all his birds before mating up was decided upon, and of any newly purchased bird, while it was in his quarantine cage. After the 1914-18 war we were not equipped with present-day knowledge for helping and guiding the ex-Service fancier. Today the returned soldier has the service available to him for starting with birds clear of the two principal plagues of the racing pigeon and of keeping a firm control over their possible appearance as a result of the unavoidable conditions of racing. The sport has an opportunity to clean up much of its loss owing to disease and to save its birds from much unnecessary suffering. Will our fanciers take it?

WIDOWHOOD

by HENRY MARTENS
Brussels, President of the Pigeon Olympiad

In response to the request of my old friend, Major W H Osman, it gives me great pleasure to give a description of a method of Widowhood racing which has been proved a good one by its results.

Before I go into detail, I wish to calm the apprehensions of readers, and declare that, contrary to some statements which have been made, the Widowhood system does not present any great difficulty. This system, simpler than the natural game, needs less time and produces wonderful results. A number of reasons are regularly given to excuse failures of pigeons flying naturally, and, because of this fact, it is very difficult to judge their true sporting value.

CONSTITUENTS OF THE LOFT

The constituents of a loft for flying the Widowhood system does not necessitate anything special. The front of the nest-box is divided into two panels, fitted from top to bottom with thin wire bars. One panel is immovable (see diagram 1), usually the left hand one; it is a little to the rear in order to fix the right-hand panel, which is movable and serves as a door through which the bird enters the nest-box.

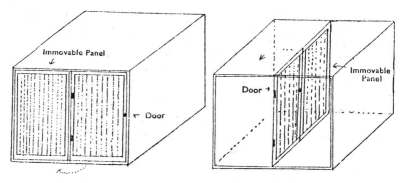

The interior of the nest-box, in its depth, will also be divided into two panels made of thin wire bars reaching from top to bottom of the box (see diagram 2). The rear half of the dividing panel will be immovable. The front half will be movable; it will serve as a door, and will separate the box into two parts. The interior panel serving as a door will, when opened, back up against the immovable panel in the front of the box. Therefore, when the nest-box is completely open, the two doors come into place one on either side of the immovable portion of the front of the box. The two fronts of the box should be painted in different colours.

The dimensions of the nest-boxes may vary. Here are those which are most often used: total width of the front should be approximately 24 inches; total height, 16 inches; total depth, 20 inches.

THE ENTRANCE TO THE LOFT

A window with double openings is preferable as an entrance to the loft; being open, it allows the pigeons to fly directly into their respective nest-boxes. Very often the pigeons waste precious moments in resting on the board at the entrance of a loft. Birds flying under the Widowhood system are accustomed to re-entering through a window.

THE PIGEONS

Pigeons are better suited for flying under the Widowhood system when above the age of two years than at an earlier age. In general, all cocks lend themselves to this system easily, and good birds learn very quickly. Under the Widowhood system pigeons will quickly find their true form, and, if they are well looked after, will keep this form for the whole of the season. There is, therefore, a minimum of strain during the racing period. From distances of over 200 miles, the cocks should not be kept in baskets away from their lofts for more than a fortnight at the most.

WIDOWHOOD

After the first cold weather has passed, put the cocks which you intend to race in the Widowhood nest-boxes, as well as the hens which are definitely earmarked as the mates of the respective cocks for the whole of the coming season. When all the pairs have taken possession of their boxes, close all other boxes which are not being used. If the moult is finished, that is to say, if the 10th feather is almost entirely up, begin the winter diet on October 15, with light food. In very cold weather give a handful of large grain at the second meal. The first meal should be given at about 8.30am, the second at about 2.30pm. Give the birds full liberty from 9am to 2pm, except on very bad days. Give them fresh water at every meal.

Winter diet should be maintained until February 10. The hens should then be taken away for a few days pending the definite reunion of the couples. As the days pass, gradually increase the daily ration of food.

WIDOWHOOD PUT INTO PRACTICE

I expect that you will begin flying on or about April 17 with a toss of about 100 miles. Work in the following manner. On February 18 pair off the birds definitely. About February 28 will come the first laying. About March 18 throw away the first round of eggs. The pairs will then carry on to the second laying. This will be about March 27. As soon as the pair have their second round start training so as to be ready for a 100-mile toss about April 10, with the pigeons still sitting. Above all, remember do not put the hens into training, for if you lose one her cock will lose all his sporting value. When the cocks have returned from their flight on April 10 (a Saturday), and have gone back to their nests so as to hatch out their eggs, you should put the hens in the compartments specially reserved for them. The cock will probably sit on the eggs until the following Tuesday. If some of the cocks are still sitting on Wednesday morning, you should take away their eggs, and see to the chips in their nests. They will be as fierce as lions for a few days. You must make them fly around the loft twice a day, whatever the weather, after the first laying, leaving the hens in the nests. All the time during the two laying and during the driving period birds will remain at liberty, both windows being open. Flying time is increased each day until the target of one hour's flying both morning and evening is renewed. It is not necessary to adhere too rigidly to set times, however. In bad weather fifteen, twenty or twenty-five minutes will suffice. During the flying periods, both morning and evening, clean the loft and renew the water.

When the birds return from the flight, close the windows and let the cocks take charge of their nests. A few minutes after give them their usual meal of large grain, preferably in a little terra-cotta pot which will allow you to control the amount they eat. Cocks will sometimes be difficult to feed; in this case, go away for a moment; if they are still obstinate in their refusal of food, add some small grain which they like. When they have finished, take away the little eating pannikins and make sure that no food whatsoever remains after your departure. They must get to know that when you go they will receive nothing more. Do not visit them except at meal times. Always make certain that another hen cannot enter the loft, for she will create a pretty bad disturbance between the pairs. Continue two flights a day and two meals.

BASKETING DAY

Flight and a feed in the morning.

About 3, 4 or 5 o'clock in the afternoon, if the marking is in the evening, allow the hen to go to the cock. Take one hen at a time, for if you try several at a time, the cocks will evade your watch and tread the hens. Leave the couples for four or five minutes together, making sure that they don't cheat your supervision. Then put the cocks in baskets fitted with separate compartments. Have them taken say, two or three miles and released at intervals of three minutes. On their return, allow the cocks two minutes with their respective hens in the nest-boxes. Allow one minute for taking him away and placing him in the basket and await the arrival of the following bird. The time taken to complete the short distance, the way the bird enters through the window and the way in which he finds his nest-box are precious indications, and should be taken a note of, the best and speediest performances especially.

Experience will teach this. On the return of the cocks on the race day, allow them a few minutes in which to coo around the hens (a short time suffices, especially if the race has been over a short distance), then put the hens back by themselves. After a short while give the cocks some small grain as a mid-day dinner.

SOME ADVICE

As I have remarked above, the Widowhood system has great advantages. In general, all good fanciers of Belgium attribute all their success to it. In expounding the advantages of the system, there are certain considerations to be borne in mind. For example, it does not do to couple good racing hens with 'Widowhood' cocks. The keeping aside of the hens for a whole season detracts greatly from their quality and very often, after a season spent entirely in the loft, they will not give good results, either from the racing point of view or from the point of view of breeding. It is, therefore, preferable to pair the cocks with hens which are 'not so good', who will serve purely the role of mates. At the end of the season, when the races are all finished, it is a good thing to pair the cocks which have proved their worth with good hens in order to breed some good late youngsters. Thus you maintain at full strength the strain which has been so carefully cultivated.

Also, it is a good thing to have in addition to the Widowhood loft, another loft in which the good hens are paired with cocks of under two years of age. This allows for the flying of the good hens under the Natural system and thus they maintain all their good qualities and their values can be judged with a view to mating. As a consequence of this, also, the yearling cocks fly under the Natural system, and value can also be assessed.

This organised system offers very desirable advantages:

(1) The Widowhood system is honestly applied.
(2) The good hens maintain their quality.
(3) The good yearling cocks become known.
(4) The education of yearling cocks is perfect from all points of view.

CONCLUSION

There are different methods of Widowhood, all of which give satisfaction. I advise you, however, to keep to one system, and not to change from one year to another. What must be done, above all, is to apply a method and to keep to that method for as long as you can.

In conclusion, I earnestly advise readers to attempt the Widowhood system and to study the particulars which I have given. I am convinced that not only will Widowhood bring great success, but in addition, it will provide a new pleasure to members of the Fancy.

THE RED CHEQUER HEN . . . AND SO ON!

by TOM MILLAR

When the editor of *Squills* invited me to write another article for his annual he said he did so because someone had told him that I had a hen which had won some £600 in the last two Scottish National races, and he felt sure that his readers would be interested to know the details of her prowess on the road, her breeding and how she was managed.

That was easy enough to do as there is nothing mysterious about either of these aspects. She was treated exactly in the same way as the hundred or so loft companions which have passed through the loft during her lifetime were treated; but as she is unfortunately the only one amongst them which has won me £600 I have concluded that the method of her management hasn't much to do with it at all. Raced out to 185 miles as a youngster, winning club pools in three races, she homed from her last race in time to win open money, but because of a thrashing she suffered in error the previous evening she sat out for four hours. No immaculate lady she, you see.

There was, of course, a reason for her action. There had been a young red almost similar in colour and build straying in from a local fancier during the week, and as dad didn't want to be bothered with it during the clocking time next day, he got permission to give the little red a really good hiding to cure her of her wayward ways. After administering the medicine in the good old measure of the pioneers, however, he discovered our little visitor sitting smugly on a perch, and our own intended pooler sulking after her mistaken pasting.

Perhaps naturally enough, I didn't hear of the incident until I tried to reason out our hen's peculiar actions on homing next day. That being the last race of the season, she was carefully handled during the winter, and apparently forgot all about it in her preoccupation with the new maternal duties she assumed in the spring. For she gave us no bother for years and always was a model trapper.

But hey presto! She arrived home from Worcester this year in time to win the race, but dad appeared round the loft side in what must have been the attitude he had adopted when thrashing her five years ago and that little photograph stored in her memory was uncovered. Off she went to a neighbouring field as she did that day and stayed some twenty-five minutes, the first time she had missed the trap first go in years.

This gave me some thoughts when arranging the pools for this year's National from Guernsey, her next race. But banking on the fact that she would probably be too tired to bother if she got back in time to win anything. I entered her in all pools and nominations again. Even so, dad was leaning against the loft-side in the early morning grey when she appeared and back she went to the chimney-stack, giving me visions of treasury notes flying off a pile one by one as the seconds became five minutes. Luckily though, only a few earlier Scottish arrivals were pooled as heavily and she collected her feed-money once again. She'll probably get another chance from Dinant National in July if all else is normal, which will make her third change of National race-point in as many years, an added test for the best of birds.

She was lifted from Lancaster (148 miles), straight into Rennes (565 miles) in 1947, and had no special training between and no flagging at home; and was sent to Worcester in 1948 only because I wanted to test her condition, and the weather appeared to be settled enough to risk an even fly.

Like all my best hens, she is a very free layer. Old 'Merrywings' would lay regularly within five or six days of mating and would be on eggs again before her young were sixteen days old. The red hen is so difficult to time that she went to Rennes only two days after laying her second egg of the fourth pair of the season; this year she was down eight days, and I shall try to plan for twelve days next time!

I don't claim that there was any special science behind the mating which produced her, becuase none of the other offspring from the pair stood up to the same treatment; nor did the pair themselves do any record breaking, being from the same breeding as other pairs in the loft, none of which produced another red hen. But one of her grandparents was old 'Merrywings', probably the best thing in wings ever to race to this district; and one great-grandparent was Henry Michie of Devonside's old red Grooters hen which had an amazingly successful racing career over some nine years of open competition, continued breeding till she was sixteen despite these strenuous years and was closely related to Peter Lawrie of Dunfermline's 'Highlander', which was second from Rennes in 1924. Another great-grandparent was our old mealy hen, a great racer in her day and such a grand stock bird that she went to the 1926 smash with some seven of her direct children. She was half Van Cutsem of the same family which won the National for Dan MacAuley of Dunfermline in 1926. And as 'Merrywings' went back on one side to a long line of Barker stars, and on the other to a sister of Paul Sion's famous 'Red Mascot' that I had from Dr Anderson, it will be seen that with all these winning germs swimming around there could hardly be a miss at some point, which in this case was lucky for me. Luckier still that that fancier didn't instruct dad to 'neck' the red stray in 1943!

You may have gathered that I am no believer in what is known as the pure strain. I don't think there is any such thing. I am, however, keen on a bird which comes from a family which has been carefully nurtured by a good fancier over a long period, and have had my best successes as a breeder from a first cross between an importation and an inbred member of my own colony; this without prejudice to the fact that I bred all four grandparents of the red hen, which I don't regard yet as the best bird I have bred although she has won me most money. Intending purchasers should look beyond the cash value of winnings to the more important asset of much hard work performed with distinction plus a long line of hard-worked ancestors. I do not know the answer, but the unions should compel advertisers to indicate distances and the number of competitors in wins claimed as a reason for the higher range of prices charged to purchasers unfamiliar with the geography of the route on which the wins are made.

Our unions do not do much to initiate the novice into the easiest way to become a successful fancier, nor have they so far laid down a code of loft management which, though optional, would still, if practised, remove much of the criticism levelled at us by non-fanciers. The manner in which we conduct our loft hygiene – the silly license given some birds when the man next door has just laid out his garden – the noise made when chasing birds from some-other-body's housetop – the grinding of the loftscraper early on Sunday mornings when the neighbour has his long lie-in; trifles all, but added together they tot up a wealth of distaste which seems to grow with time. And all that's needed to prevent the deterioration of relations is a rubber ball for the housetoppers, no open trap while the greens are taking root, no scraping till a reasonable time on Sundays, and a box to collect the scrapings for the next-door-man's leeks.

While I personally favour a daily clean-out, or at least once a week, this is not fundamental to success, as is evidenced by the phenomenal success of men like Hugh Cropper, the Liverpool crack, who has his loft cleaned out about three times a year! But even so, being a hard taskmaster, no man puts in more time on his birds than Hugh does in the racing season – up at four-ish in the morning grilling his birds into condition which is maintained during the season. Many men do this, but I must either have had too much to do, too many irons in the fire, or been just plain lazy, for although getting up to exercise the birds next season is one of my perennial New Year resolutions, the succeeding season sees the birds on the old easy method of the open hole at night for an hour or two, and big jumps without previous hard training for the long races.

Which won't stop me having the same old intention for 1949!

Many fanciers are thorough faddists over food and would no more think of changing the type of grain supplied to their birds in mid-season than they would of sending their favourite away with a flight feather bursting. I don't think either factor would prevent a good bird from winning or that it matters much whether the second or fourth flight is being moulted during the distance events. Just look at the number of photographs taken of winning birds immediately after they have won, and see how many show primary feathers missing. This suggests that such birds were at least on their fourth flight when the test was on, and indicates that the trouble some men take to retard this natural development is not justified by results.

I am well aware that in the vast cross-section of the Fancy through which it is my good fortune to move, many of my best friends differ strongly from these views, as they do on the more political aspects of the administration of the sport.

For instance, I have long been an advocate of the one big union for Britain, suitably linked with our friends in other countries through an efficient International organisation.

This is imperative on many grounds, notably on feeding, race facilities, and new legislation affecting the keeping of pigeons. More generally, the repeal of safeguarding clauses operating in favour of many factors which create injury and death to many of our birds *en route* – hawks, electric cable and plain telephone wires included. Then we need the erection of a system of weather information calculated to minimise our more serious disasters, almost always caused by faulty forecasting or inadequately trained liberating personnel; and a wider diffusion of publicity material in the non-Fancy Press so as to encourage more recruits to our sport.

Talks have recently taken place between the Scottish Union and the National Homing Union, and I believe that some contact has been established between the latter and the Welsh organisation. The NHU has suggested in the Scottish case that we could become a centre of the bigger body, retain our present secretary and generally operate as now, except that lost birds would be dealt with from Gloucester. I see no danger in such a process, which, as the NHU points out, is the simplest form of resolving the question. But old customs die hard, and it would be idle to pretend that the majority of Scottish men are willing to become little Jonahs inside the English whale, and even less likely that the North East would follow their example.

I have little knowledge of the background quarrels which induced the smaller unions to sever connexion with the parent body; and am painfully conscious that I leave myself open to the charge of being an interfering busybody, with already so many pre-occupations that my personal correspondence has fallen woefully behind, for which my friends please forgive me!

But, whether it would initially be a popular move or not, I am fully convinced that one union for Britain has become an imperative necessity. No one with experience of even the most benevolent officialdom will gainsay the fact that numbers count, and unit perhaps even more so; and an over-burdened rank-and-file can hardly be expected to wait until all the antagonists disappear from the current scene. The unions should, therefore, meet in conference and continue their labours until a suitable scheme for unification emerges. I do not rule out complete amalgamation, with workable local autonomy in matters not involving changes in rules or principles, which obviously must be universal. It is possible, though not so desirable in my view, for some scheme of confederation to operate. But no matter how, a scheme there should be, put as attractively before the members as may be in an endeavour to secure the requisite majority for the fusion.

I hope no one will advocate waiting until everyone agrees, for that is impossible in any matter; and I expect to be labelled in some Scottish quarters as a Judas of the worst kind. But since I conceive the duty of those who are privileged to act as leaders for the time being to arrive at what they think right and proceed to practise accordingly, I rest content in the knowledge that even if the majority of Scots do not accept my thesis, they will still give me the credit of being sincere about it while they elect to proceed on different lines.

With a single membership, a universal ring, and a similar code of rules and bye-laws, the membership could even out the anomalies inherent in any such amalgamation. A suitable fee would cover the administration in such a way as is consistent with the dignity of a great sport; it could canalize information so as to enable the enquirer to decide his policy on loft management and kindred matters after a comprehensive study of comparative results in every sphere; it could create a suitable system of dealing with wrongdoers and of compensating friendly outsiders who help the sport by detecting malpractices against us and our birds; and it could set up a special department for arranging campaigns for the alteration of laws militating against us, or even more useful still, secure amendments to draft legislation before it becomes law. I hesitate to mention lost birds or the feeding shortages which will be with us for some years yet.

We might even have sufficient data to determine whether there is anything in the current eye theory or in similar fads which interest discussion groups from time to time. For myself, I have yet to handle a really good performer on the road which had not some distinguishing feature about it – either a knowing look in its eye, a marvellous build or stance, a cheery disposition called 'personality' in the human, or some other asset which lifted it above an occasional bent keel or soft ventspace, and spotlighted it to men with the 'fanciers' eye'. Yet I do not believe that pigeons, or any other species for that matter, can be sorted out for racing or breeding by any physical standard applicable to all. The forces of genetics and herediatary influences are all against it, and thank goodness, in humans it doesn't apply at all! (If it did, Mr Editor, where would you and I be?)

Please do not think I place no value on arranged breeding. Far from it, I am convinced that careful selection of tried stock over a period will enhance the number of capable performers produced from a given loft; but I emphatically reject the idea conveyed in many advertisements that some fanciers can produce an unlimited number of top-quality squeakers. The number of fools will always outweigh the stars, else the performance of the red hen, which started all this, would not have been outstanding. I shall not tresspass further, as I am sure you will be convinced that I am no believer in sanctity either of loft management or sport administration simply on the grounds that such and such has previously been the practice. I hope to see continuous change after mature but speedy consideration, for only thus can the Fancy hope to keep abreast of evolutionary change and maintain control of natural forces on behalf of our birds.

Successful Racing

by T A WARRINGTON,

The Gold Cup Winner, Manchester FC, 1949

I am honoured by the editor of *Squills* asking me to contribute another article to his Annual, and I will do my best to make it interesting. It is forty years since I took up racing pigeons and I can honestly say that I have never had a bad season.

I have often been asked by friends and visitors how I manage to keep up winning season after season, and I will try to answer this question. I attribute most of my success to the fact that each season, before I start training my young birds I spend a full day picking out the best six youngsters from my winners; I take these and put them out into a loft of their own. They are fed well and allowed to take what exercise they like, and generally make up into fine birds. The difference between two nest mates – one raced, and the other kept like this – has to be seen to be believed, both as to size and feather. I try to pick three cocks and three hens each season, and these, as far as possible, are perfect birds.

These special youngsters never go into a basket, and have nothing taken out of them. As a result of this selection I have always young stock on one side or the other of a mating, and the majority of my winners are bred from these unflown birds.

I do get winners direct out of actual winners, but you can take it as certain that the bulk of my winners are bred from the direct sons and daughters of my best racers. You must have youth and vigour to pull out that extra bit of speed in these days of keen competition, and I feel sure from my own experience, that this is the only way to keep up the standard of a loft. Providing that you do not overbreed from these unflown stock birds they will continue to breed winners for years, but I always give a younger mate after the third season. Cocks will produce winners with a yearling or a two-year-old hen for a long time, but I find that hens fail to produce racers, after the fourth season. Their youngsters come all right, but they just do not have the strength to get in front in races.

Please note that all my four 1st prizes this season, 1949, from Guernsey, Charleroi, Belgium (two 1sts), and Virton, Belgium, are all bred from parents kept for stock alone and not trained.

First of all let me say that pigeon racing is hard work, and if one is to be successful it means making many sacrifices. Above all do not be in too great a hurry. It takes years to found a decent loft of pigeons. I know the temptation of wanting to have a go in all the races, but it does not pay when you are founding a loft. A very successful fancier gave me this advice when I started to go in for racing . . . "get your stock from a fancier who is winning consistently and do not quibble about the price – ask his advice and make a friend of him at the same time, if you can, and he will be ready to help you. Make your loft a real home to your birds and do your best to make it worth while to the birds to come home quickly, and they will respond if they are any good. You will get out of good birds just as much as you put in them . . ." Better advice was never given, and I acted up to it. If this good fancier were alive now I think he would agree that I kept to his words of wisdom – and it paid me!

The above then, is the answer, and I am sure that it is responsible for my success.

When I started to keep racing pigeons I bought them here and there at 2s 6d each, and expected to win races with them, but I soon found out that they were homers and not racers. I asked an old successful fancier why he could send one or two birds and win, and I sent twenty and got nowhere at all. His answer was very brief and to the point, and it was this . . . "Your pigeons are no b——— good", and he advised me to clear the lot out and start afresh – which I did! He advised me to write to four successful fanciers, so I took his advice, and paid what to me was an awful price – but they were worth it.

I bought four youngsters which I never trained at all, and all four birds bred me winners, and further, every good winner that I have ever had since contains the blood of these four birds. By the way, just a little story about one of these youngsters that I bought. When I went for this bird I did not like it at all, and this good old fancier could see that I was disappointed, and I shall never forget what he said to me . . . "Laddie, that youngster is a good one, never mind what you think, take great care of it, and it will help you found a loft of winners". How right he was, as the bird (which turned out to be a cock) bred me winners with every hen I mated with him, including 500-mile winners. I leave the moral of this story to my readers, and hope they see the point of it. Pigeons to me are a 365-day-a-year job, and they get the same attention all the time. As a matter of fact I think I watch them more closely during the moulting season than at any other time, since every good fancier knows that pigeons are made or marred at this season.

My birds are separated the day after racing is over, and never go near one another again until mating time. I feel very strongly on this point of separation, because it is, in my opinion, the only way to ensure a perfect moult. It is impossible to feed the birds as they should be fed during moulting time if they are left together, because some pairs will always be playing together, even if half naked in moult, and you will get eggs laid. I have tried leaving the birds together until Christmas, and the result was that I had to curtail the food to prevent egg laying. I am convinced that separation is the only way to make sure, as far as possible, a satisfactory moult, since they can be given plenty of food, which in turn will give the birds every ounce of health and strength to grow good feather, as well as giving the hens a complete rest during the winter months.

As long as the weather is fairly decent with no fog about, my hens are given open loft until noon, and the cocks have the same in the afternoon. They are fed light during the day, and at the last meal are given just as much as they can eat. Any bird which shows signs of weakness at this season cannot be sound, and is soon spotted and suppressed. During the moulting season the bath is before them every day, always cleaned and refilled each day, with a little common salt added to keep down insects.

When you are satisfied that your birds have had a satisfactory moult you have won half the battle towards having a successful season the following year.

TRAINING AND RACING

This interesting part of our sport must be dealt with separately. *ie* Inland and Continental long-distance racing. There is a vast difference between them, and if possible the birds for each should be kept in different sections of the loft, because the treatment and training are so opposite. For the inland races one has to have the birds lean and nippy, always 'on their toes' so to speak, and always ready for a bit more food. They must be flown hard and trained often, fed light, and treated always like sprinters.

The long-distance birds must be treated in just the opposite way, since it is staying power, NOT sprinting, that is wanted for this game – but I shall deal with this later on.

I do not say that one needs a different strain of birds for each type of race, since the same birds will win in either short or long races, but with different treatment and in other seasons. Perhaps you will pardon me if I quote my old champion 'Dolly' to prove my point. 'Dolly' won 1st prize at Worcester – 87 miles, and 1st prize Bournemouth– 190 miles, in her first racing season, and in later seasons under different treatment she won 1st prize Guernsey – 277 miles, and four 1st prizes from Marennes – 530 miles. Any fancier who thinks he is going to mate early, rear youngsters, and train early in the cold weather we usually get at this season, in order to get his birds fit and in form for the early races, and then carry on with the same birds in the long races is bound to have failures and disappointments. There are, I know, odd exceptions, but they are very few, for in my opinion, you cannot have it both ways. The form and condition of birds got ready for these inland races is gone for the season by the time these long races come on.

INLAND RACING

My birds are raced on the Natural system, and I think I can safely claim to have won my full share of inland races, whether they have been fast or slow. I pick out during the winter months, the birds which I am prepared to finish racing with after the first Channel race of 300 miles. These are chiefly yearlings which have had a good testing up to 180 miles as young birds, and have had plenty of experience. I mate them about February 14 to 16, and they are allowed to hatch and rear one youngster only right up. During this first nest they are allowed an open loft and get themselves fit for the work that is coming. When the second round of eggs are laid I substitute pot eggs, as I want no hatching now under any circumstances. As soon as I am satisfied that the hens have got over the egg laying I start to train them, all cocks and hens together. Then go to ten miles which is repeated four times, this to make quite sure that all fatness and stiffness is worked off and to tune them up after the winter's rest. I am sure this method does this job far better than flying them at home. If the weather is good, they get these four ten-mile tosses on successive days. After these tosses I let them have a day's rest, then I send them to 25 miles twice, followed by 40 miles twice, which by the way is the limit of my training. I never send either old or young birds any farther in training.

Once they have had this 40-mile (twice) toss I never send them training above 25 miles, and they get this toss every other day except weekends, right up to the Thursday before the first race. On this day I send them to a friend at 30 miles, who very kindly spends hours giving them a single toss, after which they are ready for the race. I keep up this toss for 25 miles every other day right up to the time when the birds have to be in baskets above one night, after this they get one toss between races. In these short, and usually fast, races, quick trapping is very important, for if the bird flies round three times you can have lost the race by it, so that all the time you must be making the birds trap like lightning. Contrary to most fanciers I do not force my birds to fly at home until later on in the programme, because it makes for bad trappers, and in the earlier races when seconds count it makes all the difference between winning and losing. Although I do not give them forced exercise on the off day between tosses, they are all turned out three times a day, if only for a few minutes, and then called in. Any bird which does not come at once, get no grub until it does come in quickly. On these off days I often give this lesson. I put out one bird only, watch it fly for a minute or two, which serves a double purpose, since it tells you what sort of condition it is in, which you cannot notice if they are all out together. Then I call the bird in, and if it comes at once I give a little special food, but if not it gets nothing at all, but it has to come in on its own just as if returning from a race. This lesson teaches the bird confidence and I am certain has helped me to win many prizes, and made many bad trappers into good trappers.

Every one of these sprinters has this lesson many times, and whilst you may say it is a lot of trouble, I have proved that it is time well spent. Other than the trapping lessons the birds have open door on the day between tosses. The cocks race best sitting seven to ten days, but if overdue to hatch, and you see him trying to get hen to leave nest, take him away until a few minutes before sending to race, and he will come quickly enough and maybe win. Hens race best for me when sitting over 14 days when you can dodge them in many ways, *ie* give a chipping egg the night before basketing day, or give a two-day-old youngster three hours before basketing for race, even if the hen has only been sitting 14 days, and they will usually go on sitting again when they come back from race. These inland racers should not be fed on heavy foods for they will put on weight which you do not want for this kind of racing. Tares, dried wheat and groats during the day with a few peas at night after all exercise is finished.

The above is my way of training and racing inland birds, and it has given me many successes in races up to 200 miles.

CHANNEL OR LONG-DISTANCE RACING

I feel certain that it is the ambition of every good fancier to win in this class of race. Personally, I would rather win one good long-distance race than a dozen inland races. It is the real test of the birds and the owner, and puts him into that class that can be called 'A good fancier'. To win a long-distance race gives the owner a thrill which is beyond any words of mine, and the satisfaction of knowing that all the hard work he has put in preparing the bird has had its reward. Success in this class of racing means good birds with the blood of long-distance winners in them (not above two generations back) mated at the right time, little or no feeding of youngsters right up, perfect timing in all details, and finally storing up in the birds a great reserve of energy and strength to stand up to the task of flying 15 or 16 hours with the possible chance of getting home on the day of liberation. If these birds are going to have a real chance in these long races it means months of care and preparation. You must take no risks of bad tosses, and everything must be done to conserve energy and vigour of the birds, using patience in every degree, or in other words *make haste slowly*. You can have the best bred pigeons in the world, but they will not succeed in long races unless you have done your full share in fitting them for the big task you are asking them to do. I spend months in selecting my candidates for the long races, and generally they are the birds which have done well in the past season's races, together with the yearlings which have had one Channel race, or 300 miles (these, of course, are now two year olds). First of all, I want to stress the point that I do not want these birds to come into their best form until the first week in July, and to do this I keep back the mating as long as I can. If the weather remains on the cold side I can usually manage to delay it until April 15 or even later, but never before the first week in April. This late mating is, of course, done with the object of delaying the start of moult, and at the same time holding back the peak condition of the birds. As a result of this my birds have usually got the first flight re-grown by the time the first Channel race is on. They are allowed to hatch and feed one youngster *only* for ten days, when it is taken away to be brought up by the stock birds. During this time they do just as they like with the open door (except when I am playing with the inland birds) and as I have open fields all round my loft they have a real happy time for a month. When the second lot of eggs are laid I substitute pot eggs which they sit as long as I can get them to stick it, and I have several times had birds sit for 30 days. I do not attempt to train these birds at all until about a week before the Bath race – 130 miles. Then I start work on them. I do not send these old trained birds short tosses, since they have had plenty of flying at home during their *easy month*. I send them all, cocks and hens, to the 30-mile toss twice, miss a day, and then 40 miles twice, being very careful to pick good days since a knock now would upset all my plans. From now on they get the 30-mile toss every other day until the second Bath race, and they all go to this race. I do not expect anything from them in this race, but I have had some of them actually win prizes in this, their first race, although please note that these birds are getting as much food as they can eat. Now that these birds have had their first race they have to start doing their daily work and have to fly at home three times per day, 20 minutes each time, which is gradually increased to 40 minutes each time.

By the time the first Channel race is reached they are having a 30-mile toss once per week, and the 90 minutes on other days, and are sent to the first Channel race after this preparation. I always try to keep these birds on their second lot of eggs for the first Channel race, and if they show signs of leaving the nest I can usually dodge them by giving a little youngster for a night or so, and then taking it away again again when they usually go on sitting for another period. If I get a stubborn pair then I have to give them a four-day-old youngster which never gets any bigger because I keep changing it for a smaller one. The main thing is to keep them steady and happy with no chasing or egg laying, and I do not hesitate to take the cock away for a day or two if there is any doubt about him keeping good. During this period the weekly toss of 30 miles is kept up, and the three periods of exercise of 30 minutes is kept up; the motto now is plenty of food and plenty of work. And so we come to the first Channel race, which, by the way, I do not expect to win with any of these older birds. If this race is a normal one the yearlings usually whack them easily, but if it is a hard race then the old birds do come on to win prizes.

The only thing that I am concerned about is to get all my selected birds safely home on the day without having too much taken out of them. When this first Channel race is safely over I am happy, and can now plan to get them into the condition they like best for the long races. Usually there is about a month between the first Channel race and the longest race, so that one has plenty of time to get the birds as you want them. I always try to get my birds for these long races with a four- or five-day-old youngster, which, in this case, is their very own. These, by the way, are the third round of eggs, and I do not hesitate to separate these birds for a few days, if there is any doubt about the hens laying too soon. When these eggs are laid I take them away after carefully marking them with number and date laid, and put them under stock pairs to be sure of them hatching on the right day. I do this, of course, because the continued flying and occasional tosses would spoil the eggs otherwise. The exercise is now increased to 45 minutes three times a day, and all of them (cocks and hens) have to do it except on the day when they have a toss, once a week to 30 miles.

Four days before going to the final race I try to get in some way a single up toss at 30 miles, and after this I let them have a happy time for the last day or two. The eggs which I took away to stock birds are due to hatch the day after I have this single up toss, and I put the exact eggs back to each pair on this day. The birds, therefore, hatch their own eggs and have their very own youngster until the moment of basketing. Needless to say that the youngster is there when the birds come back from the race and I can assure you that in the case of hens, the first thing they want to do is to see that the baby is there before either eating or drinking. This condition for racing long races may not suit some birds, but it certainly suits mine and I have kept to it for many years, with a full measure of success. For the last month these birds are fed often and well. Peas, tares, dried wheat (if I can get it) during the day, and the main meal after the last exercise, when I feed beans with a little Cinc corn and they get as much as they can eat. At noon they always have a little tit-bit, which is a special cake made by my wife, this is baked very hard and then broken up into pieces the size of peas, and the birds go mad for it. If I have any in my pocket (which I usually have) they will be all over me, so they must be able to smell it. Do not ask me what is in it, because I do not know all the ingredients, that is my wife's secret.

Many of my friends have tried to get her to give them the recipe for this food, but she will not give it away. Anyway, it has helped me in no small way to win many good Continental races. I would, at this stage, like to pay a tribute to my wife, who is as keen with the birds as I am, and can race them and handle them with any fancier. She has contributed her full share in all my successes and rightly claims (and gets) an equal share of all prizes!

I advise all fanciers to get the wife interested in your birds. It will surely pay you, even if you have to share the spoils! I have faithfully given my readers the methods and treatments my long-distance birds get, and can only hope that it will help young fanciers to obtain success in these races.

SECRETS IN PIGEON RACING

A lot of ink has been spilled on this subject and it is well worth a word or two. Every fancier is (or should be) trying to get *a step ahead* of his club-mates. I am always experimenting one way or another with different kinds of food, etc, to see the reaction on the birds. Let me say, at once, that I do not mean dope of any kind, but genuine foods, which *every fancier can get* at any grocer's shop. I maintain that once a fancier has mastered the art of feeding correctly, either for short or long races, he has won *half the battle*. We have all heard the old maxims, *ie* good birds, clean lofts, good corn, clean water and regular exercise. That might have been good enough in the old days, but not now, at least not in my opinion. Let's be candid, and admit that we all have our little secrets which we believe will bring out that extra energy, vigour and speed in our birds. Perhaps it has taken years to prove that certain foods and tonics will help to do this. Surely we are entitled to keep these secrets to ourselves and reap the benefits which we may get for using our own ingenuity and common sense.

LUCK IN PIGEON RACING

We have all heard the expression 'lucky beggar' used in connection with our sport. Luck is the lazy man's word, and the fancier who pins his faith on it, is in for a very lean time. If it is luck then many good fanciers must be chock full of good luck, since one sees the same good fanciers' names at the top or near it season after season. Every fancier today has good birds, and if he does his share in getting the best out of them, his turn is bound to come. There may be odd flukes in blow-home races, but in nine cases out of ten the prizes go to the fanciers who earn them by the care and attention they give the birds. No, the fancier who uses this expression is merely making excuses for his own inability to get the best out of his birds.

The funny part of this is, that if these fanciers (who are ready to use this expression) do happen to get a winner, they suddenly change their opinion and are ready to explain why they have won.

If I write any more I am afraid the editor will think I want to fill the book, so I will close with the hope that my humble effort has at least given the novices something to read and study.

The Lofts of Louis Schietecatte

by HENRY MARTENS

It gives me great pleasure to write an account of the loft of the Belgian champion, Louis Schietecatte for the readers of *Squills*. There are, each season, some amateur fanciers who achieve remarkable results; a few even maintain their success over a period of years, but there are very few indeed who remain at the head of our national sport for very long. One can appreciate the enormous difficulty of becoming the long-distance champion of Belgium, when one realises that there are more than 250,000 fanciers in the country.

In order to succeed in the national races, one must have, not only a variety of pigeons, all in perfect condition, but also the experience, care and loft-planning that come only after many years. M Schietecatte started in his father's loft in 1896, when he was most interested in short-distance races, in which he was very successful. In 1910 he introduced some of the late Guillaume Stassarts' birds (another champion racer, one should note) into his loft. The effect of this was postponed by the 1914-18 War.

Immediately after the war, he moved his loft, and the Schietecatte 'Stassarts' achieved resounding successes in short and middle-distance races. Stassart was considered, at that period, to be the Grand Champion of the national long-distance races, and it was in 1930 that Schietecatte followed his example, by attempting their distances, with quite extraordinary results. The two were always great friends, exchanging young birds year after year.

SOME RESULTS IN THE NATIONALS

With a view to showing how successful is the combination of Stassart birds and the Schietecatte loftmanship and technique, here are some typical results.

1937: Angoulême, June 12 (412 miles), National eight prizes; Brabant, six. Bordeaux, June 26 (450 miles); National 7th and 167th; Brabant 1st, 9th, 18th, 19th: Flying 1st and 18th.

Angoulême, June 26, 1st, 10th, 29th, 30th.

Derby (412 miles), National six prizes.

Dax, July 14 (550 miles), 18th, 37th, 38th, 39th.

1938: Angoulême, June 11, National 55th, 111th, 142nd, 158th; Brabant 3rd, 11th, 14th, 19th, 42nd, 51st: Flying, 2nd, 4th, 6th, 9th.

Bordeaux, June 25, National 8th, 208th, 219th; Brabant, 1st, 7th, 8th:

Flying, 2nd, 18th, 19th.

Libourne, July 9, National, 2nd prizewinner and pools; 1,000 francs. Dax, July 16, 2nd, 12th, 30th, 34th.

1939: Angoulême, June 10, National 43rd, 67th, 238th, 253rd, 287th; Brabant, 4th, 9th, 24th, 26th, 31st: Flying, 7th, 11th, 21st and 23rd.

Bordeaux, June 24, National 15th, 45th, 46th; Brabant, 2nd, 5th, 6th: Flying, 2nd, 6th, 8th.

Libourne, July 1, National, two prizes.

Angoulême, Derby, July 1, National, eight prizes.

Dax, July 16, six prizes.

St Vincent, July 21, National 23rd, 41st, 47th, 243rd, Doublage National 13th, 26th, 30th, 132nd; Brabant, 3rd, 6th, 7th: Flying, 4th, 7th, 8th.

The list of other results recorded each year would be much too long for publication. But one that must be mentioned was the Chartres YB race (218 miles) of August 20, 1939, where despite hard weather and a velocity of 950 m or less, M Schietecatte took 29 prizes, namely: 2nd, 6th, 7th, 11th, 25th, 32nd, 40th, 43rd, 45th, 46th, 53rd, 57th, 58th, 59th, 66th, 70th, 71st, 84th, 88th, 89th, 92nd, 101st, 106th, 108th, 112th and 116th.

During the disastrous years between 1940-45, M Schietecatte succeeded in preserving most of his best birds, and he re-established his position as soon as racing started again. In 1949 he won 1st from Brabant at 660 km, and 1st at 720 km, and he repeated these performances in 1950. One of his best post-war birds was the Ace Pigeon 'Heron', which won the 1st prize from Libourne two years running at 720 km, and has won more than 150,000 francs as prize money during its lifetime.

Note: eight kilometres = about five miles.

TRAINING AND RACING

Widowhood

M Schietecatte follows, for the most part, the Widowhood system with his cocks. The birds are paired off towards the end of February. The widowers rear one young bird per nest, and then, at the second egg, on about the tenth day, the hens are removed, and the cocks are widowed until the end of the main long-distance races. The cocks are re-paired after the end of the season, and rear two young birds, a procedure that assists in achieving a good moult.

The widowers are trained during the first rearing and become widowers when the second round of eggs is removed. Every morning and evening an enforced flight permits regular exercise; it starts with 20-30 minutes working up to 50-60 minutes. Before being put into the Widowhood sections for rac-ing, some of the cocks are presented with hens, but this is a matter for care-ful observation of the character of the birds.

The widowers receive a light meal each morning, consisting of wheat with three grammes (about $^1/_5$ of an ounce) of a mixture of small corn, hemp, colza, millet, for instance. After the evening flight, large corn is fed to them, 20 per cent of beans, 20 per cent Marle peas, 30 per cent of maize, 20 per cent dari and 10 per cent barley. This diet is maintained until the end of August.

From the beginning of September, the racing foods should gradually be replaced by wheat and barley, so as to arrive at a figure of some 60 per cent of these grains by the end of the moult. The other 40 per cent should be 10 per cent beans, 10 per cent Marle peas, 15 per cent of maize and 5 per cent of dari.

A fortnight before pairing, the winter food should gradually be replaced so as to arrive at the breeding and racing mixture.

THE HENS

The hens destined for racing are paired with the year-old cocks. Both the cock and the hen fly 'naturally', that is to say at the time of sitting, rearing, etc. Young cocks who are successful are transferred to the Widowhood loft.

At two years old they will race 660 km as widowers, and at three years they will fly at distances from 170 km to 900 km. They must, from this age on, fly in at least three or four 660 km races.

The Lofts

The Widowhood lofts normally hold 28 cocks. The lofts inhabited by the hens and their year-old partners hold 16 nest boxes, while the breeding lofts can hold 12 pairs.

The Breeders

Breeding cocks are chosen from those with extremely good results, and they are paired with the best of the young hens, in appearance and sporting qualities. A good breeding bird among the old hens may be paired with a young cock.

Young Birds

Young birds are trained without being paired. They must fly various distances up to at least 350 km. Before the 1939 war, the young birds had to fly a 660 km race in their first year, and in some years as many as 40 young birds would be flying this distance. Many of these birds would qualify as 'extra pigeons', and M Schietecatte has won a number of gold medals for the largest number of prizes won by young birds at 660 km.

Selection

At the end of each season a severe weeding-out is essential. Any bird that has not shown any ability is removed. IN ORDER TO BREED GOOD BIRDS, ONE SHOULD ONLY RETAIN GOOD BIRDS.

Pairing

M Schietecatte, attaches great importance to pairing, considering that the future rests on good pairing-off. He aims at retaining three main traits, the strain, the results and physical form.

Strain — The birds must have a well-defined pedigree, showing the origins of the strain, the results, even those of the parents, and the physical qualities. One should aim at always uniting birds with a maximum number of similar qualities.

When cross-breeding, one should bring together birds of well-established strains, able to fly the same types of race. It is inadvisable to pair a short-distance bird with a long-distance one, or a large one with a small one. In general, mixtures should always be avoided.

Conclusions

It is essential, for success, to have pigeons of a good, well-established strain, and to preserve this strain as long as possible.

Make all your birds fly as much as possible.

Weed-out your birds with the maximum severity. The main source of success is HEALTH.

NO HEALTH, NO SUCCESS, NO PRIZES.

Sound food, fresh water, cleanliness and exercise.

I should like to thank M Schietecatte for the wise advice he has given us. For the last 50 years his success and his experience have guaranteed his becoming a champion.

(Translated by WAO.)

The Secret of Consistent Winning

by ATWELL BROS (Newport),

Winners of Welsh Grand National Lerwick Race two years in succession, 1950 & 1951, also 1st Welsh Young Bird National, Carlisle, 1950, and 1st, 3rd & 6th, Thurso Welsh Grand National 1951

It is a great honour to write an article for *Squills*, writing alongside men of international repute in the pigeon fancy,

We were born and brought up with pigeons, Father being one of the pioneers of long-distance pigeon racing, winning the Arbroath race to Newport as far back at 1895. We have been racing as Atwell Bros. for thirty years, and found pigeon racing a 365-day job. A system of management must be maintained throughout the whole year to gain any measure of consistent success.

We write for the novice, outlining our system of management, but take this opportunity to say you benefit from your own mistakes, and you learn by experience. Our loft, facing south, has separate compartments for cocks and hens, and youngsters; nest-boxes have hinged doors to shut up when mating and when birds are away in races or tosses. These main points we recommend, ventilation, design and the other details you can make to suit yourself.

The main thing that makes for success is to get the best pigeons obtainable and manage them properly, always remembering that the best man in the world cannot make a bad bird into a good one, but bad management can make a good bird a bad performer. We attach so much importance to breeding that we confine pairs to their nest-boxes until eggs are laid. They only have their liberty when we are in the loft; thus we are sure of parentage.

We are also very careful when introducing a cross. We only put new blood in once; that is to say, pigeons from a first cross must prove themselves in long races before they are bred from, so that if the cross is not successful we only have to get rid of one generation, and not a lot of pigeons that have eaten plenty of good corn. The working-man fancier cannot afford to keep a lot of coddled pigeons.

All our best pigeons worked hard as youngsters; as yearlings they should fly 300 to 350 miles at least, and we are not afraid to send a yearling to Thurso, flying 487 miles, if flying in form. We had the pleasure of clocking a yearling from Thurso in the 1951 Welsh National, winning 6th prize. Fed well as youngsters and throughout their lives without stinting, to trap in short distance and young bird events, the birds, if bred right, will take a lot of los-ing and last some years as long-distance racers. Our best hen, 'Mary Ann', has been raced hard for ten seasons. She has flown Lerwick, 598 miles, five times, each time a prizewinner, including 1st Welsh Grand National, Lerwick, 1950 and 1951, at eight and nine years old.

The birds for Lerwick are mated together, and the birds for Thurso mated together where possible, to have them both ready at the same time with the same treatment. Only one youngster is reared in the first nest, the eggs or youngsters having been transferred to other pairs put down for the purpose. When birds have youngsters in the nest they want plenty of food and we keep jam jars full of food in the nest-boxes. At this time they look for the green-stuff – cabbage, chickweed and dandelion, etc – that we always supply in abundance at all times. Baked salt and a little salt in the drinking water will stop your birds from going to the river banks and fields.

We never change the nest until youngsters are weaned and birds down on eggs again. We use insect powder to keep down pests, and do not handle the youngsters, even to ring them; we simply raise the right leg backwards to slip ring on, and they are then undisturbed until weaned, which at the latest is at twenty-four days old. They are better parted off, for, besides the strain in the old birds, there is a danger of the youngsters being scalped by coming out of their own boxes and going back into the wrong box.

Birds are trained and raced to eggs, some being allowed to rear one in nest to eight days if weather is fine and warm, for we notice they work well to youngsters if weather is good. But if it is cold and wet they do best to eggs, and although we have won good races with a cock flying to a hand-fed big youngster, our best results have been with pigeons sitting from seven to four-teen days at basketing, and it is foolishness to race a hen to a youngster over seven days old, as the danger of laying is too great a risk to take.

Care should also be taken after the race, as many good pigeons arc ruined by bad management after a hard fly. Tired birds should have a light feed of baked bread, baked wheat or canary seed; no hard corn at all until bird is well rested, usually next day. This is fed in the nest-box, with warm water and sugar or oatmeal water to drink (also supplied in the box), and bird shut in box to rest. When birds are racing and rearing, cabbage water once or twice a week will keep them healthy. We feed on beans, peas and wheat, all of the best quality obtainable, and feed separately, as we find they won't eat the beans if they can have peas or wheat. The water, of course, must be absolute-ly clean at all times.

Birds fit and healthy should fly twenty minutes or half an hour night and morning, but hens, if turned off their eggs, will drop to the loft and spoil the swing of the batch, so they should exercise on their own between 11am and 12 noon and between 4pm and 5pm at night. We never exercise late at night, being afraid of their birds striking the wires in the half-light, for wires are plentiful in the vicinity of the loft. Nevertheless, we have clocked in very late and on point of darkness several times.

We believe in plenty of short tosses for youngsters, but old trained pigeons, if any good, will stand big jumps. 'Mary Ann' went from loft to Shrewsbury (80 miles), then Stockport (131 miles), and jumped from there to Lerwick (598 miles), to win Welsh National in 1951. We fly the Natural System, having tried the Widowhood System and found it to be no good over 250 miles. Immediately racing is finished we separate the sexes and they are not allowed together until mating-time, which is from February 14 for yearlings to March 14 for the long-distance candidates. They are well fed at the moulting period. Give them some linseed every day at this time, for this is when next year's races are won. They need the best of attention now, for the feather they grow will be carried to next year's trials.

After the moult we curtail the hens' food when necessary, as we have had trouble with hens mating together. The cocks are allowed as much as they like; driving and rearing youngsters will take all their fat off next year before the racing starts.

The best birds are given two mates, and the youngsters most like the bird you are breeding back to, are mated together. This method of line breeding has produced our best racers. When you get pigeons to fly fifteen hours to win on the day, you know you have the right stuff, and care must be taken to see they are not spoilt through bad management or indiscreet crossing.

If the novice follows these lines, success will surely follow, but the measure of success will be governed by the work and perseverance they are prepared to put into it. The distances we work the pigeons should be modified a little by the novice fancier, as he cannot take the same liberties as established fanciers with teams of tried pigeons. We hope what we have written will be of value to the novice fanciers and of interest to the veterans.

My Way of Success

by ALF BAKER

Winner of Osman Memorial Trophy for performance in London NR Combine Thurso, 1952

It was indeed a great honour for me to be asked to write an article for *Squills*. I have been keeping pigeons for 30 years, since I was ten. I used to run errands to buy pigeon corn and my mother used to say pigeons would be the death of me, standing out in the rain, but I love them. Wherever there were pigeon lectures or talks, I was there, wet or fine. I think successful pigeon fanciers are born, with a gift for the knowledge in them. I used to spend hours watching every movement and their ways. You will learn more that way, than reading about them. I never see if a pigeon is fit for a race by handling, as I've found this will let you down. Some of my best birds never handle hard and firm when ready for a race. My champion 'Plum' mealy cock never carried any flesh, but he won 17 firsts, eight of them in succession.

I shall never forget the last race he won out of the eight from Berwick (300 miles). The late Wally Dimmock said, "Sending 'Plum' again? You'll lose him," to which I replied "I bet I don't. Never stop a winning bird." It was only a figure of speech, but Wally being a betting man put a £1 on the table. "Cover that," he said and before I could say a word, they all wanted to have a bet. Well, I took over £5 on 'Plum'.

The Saturday morning opened up fine, birds were up 7.30, wind SW. I knew it was going to be a hard fly. At 10am it came over black – started to rain and never left off all day. I was sitting down having tea when my daughter said "'Plum', dad." There he was, shaking the rain off. I clocked him in 5.32, 1st Club, 2nd Fed; what a relief! I made a fuss of him, not because he had won £5; but because I could still hold my head high, and he proved what a great pigeon he was.

You may ask how he kept on winning. I had him paired to his mother; she went barren the year I bred 'Plum' and he was always after her, so I paired him to her and sent him calling to nest first race. When I came back from the club I put a pair of eggs under her. He used to come home and go straight on the eggs till I took them away on a Wednesday morning to get him calling to nest again. He won the LNR Fed five times and all his races were won in this condition.

I find that if a cock races that way, so will his sons and grandsons. My second open LNR Combine from Thurso, 502 miles, was sent the same way;

there is not much that will beat a semi-Widowhood bird on good days.

My young birds are raced semi-Widowhood with wonderful results; it is not only clean water, clean food, plus good birds that make a successful pigeon fancier, it is the tricks you play with them. Two cocks to one hen, or vice versa. You will find you will have great losses at first, but those that work on it, as I have previously stated, will breed the same. My losses were great when I first tried them, but I hardly lose a bird, young or old, now.

When you are buying birds, go to a successful fancier in your club; if he won't sell, go to any successful fancier who is winning today, not those who won years ago, selling grandsons of champions that have been dead years. You have very little chance of reproducing them.

I must repeat what I've often heard, "when you have mastered the art of feeding you will be on the road to success."

My birds are hopper fed summer and winter. I never cut their corn in winter, for that is the time they need it most, I give them plenty of exercise in the early spring to get down surplus fat before breeding. I feed on peas only, with a little maize when I can get it. A lot of fanciers don't like maize, but I fed on maize only in 1947 and I had one of my best years.

There are no hard and fast rules to the game. A lot of fanciers don't like sending yearlings driving; I have won some good races this way, but you must know your pigeons. Birds raced on the Natural system will win, but most successful fanciers have their little tricks. There are no certain types that will win.

Personally I like a good eye. I know this is a tricky subject with most fanciers, but there is a breeding eye and a racing eye. Most of the fanciers who don't believe it, don't know what they are looking for. I am not a great believer in the circle of correlation, but I am in the colour of the iris. My best producers have the violet eye, and breed the red pearl eye, or light violet eye. I think pigeons with a good eye break from the batch quicker when they are nearing home.

I like to train my youngsters with plenty of single ups, from about 15 miles, picking a certain landmark, making that their main training point. I don't believe in stopping old birds up to 300 miles. I like to keep them at it. If they can't stand the stick up till then, they will never fly 500 miles and win. I am not a wing faddist; I like my birds to moult freely then they are in good health.

I remember one of my good cocks 1729, the year he won £58 from Fraserburgh, 428 miles; moulted two flights at once in both wings. When I saw that I fancied him all the more as they were half grown. I like youngsters for the long races completely to finish the body moult, that is to have moulted new bars. I never worry about the wing or tail as I have won good races with youngsters with only six tail feathers and a gap in both wings, but they have finished the moult. It is obvious that with the covering feathers missing the wing has not got a good downward movement, and the air can get through.

My birds have an open loft during the racing season. When I let my birds out in the morning, they only fly for about 10 minutes, and go straight down in the garden. I often come home and find only a few of them there; the others have been out running.

I like to keep them as near nature as I can. I use about 2 cwts of acorns every year and my birds love them. I don't know if there is any goodness in them, but the birds like them and that is the main thing. It also helps the rations out.

I never go in the loft without some sort of titbit, rice, Quaker Oats, baked bread crusts, anything they like. The main thing is to keep them happy. I never shut pairs up in a nest box to pair them up. I let them out all the morning together, then get them in and let all the other birds out except the pair. The cock will take the hen to his nest box. I let my race birds pair how they like as I believe in love matches, for they race better. Another one of my best cocks 1733 was paired to a tippler; it was his fancy and he won seven firsts. After a couple of years I paired him to another hen to breed from, but he did not race as well.

Three days before the Morpeth race, which is the Gold Cup race, owing to the wrong liberation time, my birds were all out. When they all clapped off, 1733 was dropping in, he didn't stop but went with the others. I dropped him when they came round and he was 1st Club, 3rd LNR Fed, missed the Gold Cup by 3 yards. There was a love match for you.

This was only one instance. As I have already stated, you must know your pigeons. I'm sure my space is limited and hope my methods will prove helpful. You can only try. We are never too old to learn. To those who have learnt it all, I trust it has been good reading.

WHAT ABOUT THAT SPLIT SECOND?

As a Mathematical Measurer of Pigeon Flying Distances I found it necessary for Secretaries to adopt an easy way of working out velocities and variations to save time and for the purpose of accuracy. For this reason **JACKSON'S "SIMPLEX" Velocity and Variation Chart** has been contrived, and is now more or less adopted universally by CLUBS. It is hopeless for Secretaries to work by rule of thumb, besides being unfair to the members.

JACKSON'S | VELOCITY | VARIATION
"SIMPLEX" | CHART | INDICATOR

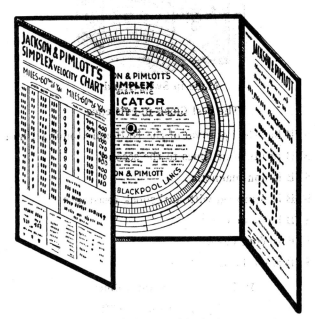

An extension of the Great Circle Logarithmic System and very easy to read and work as the figuring has already been done beforehand.

See this: " I have just taken over the post of Secretary and the B Federation advises me to get your 'SIMPLEX' Variation and Velocity Chart."

Read this: " I would like to be supplied with a Velocity and Variation Chart. I believe it is called the 'SIMPLEX,' and I have heard a lot of praise about it."

SUBSTANTIAL ARTICLE, BOUND LIKE A BOOK, TWELVE INCHES SQUARE.   PACKED IN BOX AND POSTED—32s. 6d.

Jackson & Pimlott | LARGEST PIGEON | 189 HORNBY ROAD | MILLION
** | GOODS MAKERS | BLACKPOOL, ENG. | PARCELS POSTED**

Winning the Scottish National

by MONTGOMERY BROS, Catrine, Ayrshire

King's Cup Winners Scottish NFC Rennes 1953

It is the ambition of every Scots fancier to win the National, this National of ours has everything – magnificent and coveted trophies, honour and glory, and hard cash. Next to winning the National we appreciatde very much being asked by *Squills* to write an article for his Year Book, and thank him very much for the honour he has conferred on us to be numbered among the long list of fanciers who have contributed to its pages down through the years.

For us to try and tell the really big-time fanciers anything about pigeons would, in our opinion, be presumptuous, but, on the other hand, we sincerely trust that our success may serve as a source of encouragement to other fanciers who, like ourselves, keep eight to ten pairs of pigeons and who, year after year, endure hiding after hiding but still come up for more.

Some years ago, after having sent several birds to the National with not the vestige of encouragement, sitting with my brother weighing things up, I asked him to say frankly what he thought our chances of winning the National were, and he replied, without hesitation, "We have the same chances of winning the National as we have of being struck with lightning."

My advice to fanciers, for what it is worth, is to get as big a loft as posible but not too many pigeons. Our loft is a small one of two compartments, but its lack of space is made up for in the fact that the intmates have a free trap and can come and go as they like. Every bird is treated as a separate individual; its habits and inclinations are closely studied and the bird treated accordingly.

Whilst our birds live a wild, free life they are exceedingly tame and easily managed. Keeping so few pigeons, it is a difficult business trying to have a bird or two in tip-top condition for every race week after week during the racing season. We try, however, to have a good bird or two for the final races of the Federation and, of course, the National.

On the subject of breeding, pigeons with long detailed pedigrees, with all their T's crossed and I's dotted don't necessarily produce the goods, but on the other hand, the really good ones have actually a background of worthwhile performance and breeding. We have found our best guide is to breed to the bird which comes up when the velocities are low (because of the wind in their faces) and returns are few. It is the long hours on the wing under hard conditions which finds them out, make no mistake about that.

Feeding. We are no authorities on dietetics, pigeon or otherwise. Our method has been, after mating up, to give birds complete liberty, and each pair to have a pot of mixture and a jar of water in their nest box and to be left to it.

Somewhere or another we have read that pigeon feeding is deficient in fat, so we have made it a practice to buy a carton of suet and give a little of it to our birds, and they have become very fond of it, and gobble it up immediately it is offered to them. Regarding the new scientific approach: Eye theory, wing shape and length, white marks on tail flights, short keels, long keels and what have you, well, we are no scientists and just wouldn't know.

In the final analysis, the two measuring sticks we understand and find most accurate are the basket and the print on the timing clock. As for 'Local Girl', our winner, she is, to look at just an ordinary little blue chequer four-year-old hen with a peculiar colour tint of bronze showing on her checks and bars of her wing. She was bred by our near neighbours Gordon Allen and Sandy Blackwood of Mauchline, from a family of pigeons which have rendered yeoman service to those fanciers, and also to John Cook, a good National racer of that village. How we came to possess her is worth relating in detail.

In 1947 Messrs Allen and Blackwood won the Ballochmyle Federation race from Guernsey (480 miles) and we were second in the same race the following year (1948). After this race Messrs Allen and Blackwood suggested we should exchange youngsters in 1949 from our respective winners. We agreed to the idea, but requested that the youngsters we have from them should be from the parents of their winner rather than the winner itself. In 1949 we duly received two youngsters, 1344 and 1345, and it is of more than passing interest to note that while 1344 'Local Girl' won the National, her nestmate blue cock 1345 was timed in in the same race as our second bird, and to remember that Allen and Blackwood's Guernsey winner in 1948 is a full brother to 'Local Girl' and they also won with a son and daughter of the sire of 'Local Girl' the most coveted trophy in the Fed – The MacRae Memorial, awarded for the best average over the three longest races in the Ballochmyle Federation.

What a wonderful pair of stock birds these two pigeons have been. But outside our immediate locality they have never been heard about. The National winner has been a very consistent performer, particularly so when the going has been tough; she won 169th Section, 296th Open in the Rennes National 1952, in 1951 she had a good position in the difficult Dorchester race of the Ballochmyle Fed, thus repeating her yearling performance from the same race point in 1950, when velocities were low and returns few. She is dam of our blue chequer cock, winner of 1st Cheltenham 1952.

In conclusion, to all you good fanciers who are perhaps losing heart, please member that if it can hapen to two chaps like us, it can happen to you.

Third Time Lucky

by T BETSON

Winner 1st London NR Combine, Thurso 1954

It is with a feeling of pride and satisfaction that I accept the Editor's offer to contribute to *Squills*, and although I feel much more efficient with a garden fork than a pen, I will do my best to make this little article interesting to the fancier.

We had won the Thurso race (517 miles) for the past two seasons in our club, and naturally we were very keen to complete the hat-trick with a win in 1954 and it was fitting that we should accomplish this feat and win the London North Road Combine with the best bird we have ever owned or bred, Chocolate Soldier, NURP51WC816.

Chocolate Soldier is a deep red pied cock, an honest racing pigeon, who takes the weather conditions as they come, and always flies a good race. Naturally, all birds have their top condition, and Chocolate Soldier is no exception. His favourite condition is 13 days sitting when put in the race basket; he will fly a very good race at two days or nine days sitting, but he 'tells' you by his actions in the loft and the way he leaves the basket on a training toss that his top condition is 13 days, so he was sent to Thurso in this condition. Probably we were a trifle lucky to get this exact state but then (in my

'Chocolate Soldier', 1st
London NR Combine
Thurso NURP51WC816,
deep red pied cock,
owned by T Betson &
son, West Croydon.

opinion) luck is always needed to pull off a big win. By this remark I mean that everything connected with the bird has to go right all the time.

To keep him happy all the season we decided to mate him to a hen that we would not be racing, so that she was always there when he homed from a training toss or race. As his top state approached we would take the hen out of the loft for fairly long periods, so that he could sit the eggs himself, as he really enjoys this. We decided to let him rear a single squab in his first and second nest, which he did; he also had to fly the race programme up to Alnwick (286 miles) which he did with ease, always coming well and never far behind the club winners. He then missed Berwick, flew a drop-back race, missed Fraserburgh, then flew two drop-back races before going to Thurso.

It was just before the Fraserburgh race that we decided to let him mate up to get the desired condition of 13 days sitting. Our Old Bird programme contains 12 races, so by the foregoing it will be seen that he flew ten of these. As to his feeding and management, this was entirely the same as any other bird in the loft; the best maple peas obtainable, clean water three times a day, *every* day, and a clean loft. At a desired time before the long races come along, we give the birds a few tic beans, say half a dozen each to start with, then working more in as we think, as the selected race approaches. The main feed is always maple peas, but we do use a little canary seed as a tit-bit, and for trapping from races, but only a very little.

I think that here we should inspect the record of Chocolate Soldier, which to my mind is very impressive.

1951: Six Young Bird races to Northallerton, winning Northallerton (1), Newark (2), Northallerton (2), in successive weeks, being awarded NHU Gold Medal for Meritorious Performances.

1952: 1st Club, 21st No. 3 Section, 82nd Open London North Road Combine Fraserburgh.

1953: 1st Club, 1st Crystal Palace Fed Northallerton, 1st Club, Morpeth. And minor positions in Club racing.

It was at this time that we thought he had a great chance from Berwick, but, unfortunately, he failed to home on the day, but arrived next morning in a very sorry state, his right eye just a blob of dried blood, his back bare of feathers and most of his secondaries missing from his right wing. It was obvious that he had been attacked by a hawk and that we were lucky to get him home.

Chocolate Soldier is not only a racer, for he has bred us a prizewinner each year, mated to a different hen each time. In 1952 he bred NURP52THO745 which won 2nd Club Northallerton as a baby, 4th Club Berwick, was the only yearling to be clocked in from Thurso, and was 2nd Club Berwick 1954.

In 1953 he bred NURP53THO779, and she won 1st Club, 1st Crystal Palace Fed, Newark (2).

In 1954 he bred NURP54THO749, and he won 6th Club, 2nd Produce, Selby and 2nd Club (beaten by loftmate) Northallerton.

Chocolate Soldier has also won two 1sts in shows under two different judges, so to sum up, we think he is a really good bird and entitled to the label

'A Champion', and we are very proud to own him.

It may interest beginners to know a little of our general management. Always buy the very best blood you can possibly afford and stick to one family. Always use the best corn obtainable all the year round. Always keep the drinking water clean, and use proper drinkers made especially for the purpose. Give plenty of baths all the year round, in water that has had the chill taken off of it; keep the loft clean and well ventilated and if the birds are any good they will respond. We always give our birds plenty of basket work, old and young, with single-ups in most weathers, at all points of the compass. We are great believers in this item.

In conclusion we would like to say that pigeon management is a 366-day-a-year hobby, and the man who works and thinks is the one who is most likely to get the best results.

I would say here that my son and my wife are very able helpers in my routine, and that without their help I would never have had the opportunity to write 'Third Time Lucky'.

The Meaning of Skin Irritation

by Dr TOM HARE

The amount of disease in British racing pigeons due to insects is very small. Such has been my experience during some thirty years of investigating the causes and treatments of their illnesses.

Before 1950, apart from occasional inquiries about lice or red mites, owners made no mention of insects when consulting me about disease in their lofts. A change has come about during the year 1950-1955. During these six years twenty-eight owners have asked me to identify the insects to which they attributed the illness of their pigeons. Most of them told me that they had been treating their pigeons with various liquid or powder insecticides before they consulted me. All of them expressed some measure of surprise when I told them that their birds were suffering from internal disease which had nothing whatever to do with insects. I then learned that they had suspected insects as the cause of trouble because their pigeons showed one or more of the following signs of skin irritation – scratching, nibbling, pecking, plucking, or stamping.

That 28 pigeon men living in different parts of the British Isles should have such strong suspicions of insects as the cause of disease in the lofts was to me a new experience. I have been wondering how they came to think of insects as the cause of behaviours with which pigeon fanciers are very familiar.

Possibly some have been influenced by the article on page 32 of Squills for 1948, which was written by Monsieur Charles Bondois of Lille, in France. It is a well illustrated and most informative description of various aspects of insects which injure the skin or plumage of racing pigeons. Monsieur Bondois gives particulars of the scabies mites, feather mites, red mites, lice, ticks and biting flies; but he omits any indication of their prevalence among the racing pigeons of France and Belgium.

Propaganda since 1950 in press and radio about the insect pests of agriculture may have influenced racing pigeon owners. The advertisements of the chemical industry have informed us of the various insecticides and the precautions to be adopted in their use.

In my experience scabies mites, often described either as 'scaly leg' or 'depluming itch', are very rarely met with in British racing pigeons. I have found scabies in only three cases; but none of these came from lofts in active racing. On the few occasions of my finding red mites in lofts the insects had migrated from adjacent poultry houses, the cleaning of which had been neglected. Red mites can be a serious pest particularly for squabs in the nest and we do well to keep them in mind when choosing the site for a new loft. Pigeons which are weakened by internal disease are liable to develop lice and feather mites in their plumage. Healthy pigeons give short shrift to a louse which gets on to their body. Thus if an owner finds lice on his pigeon he should not waste time by applying insecticides; the bird will be ill from other causes and should be examined without delay.

It is true that pigeons which are suffering from insects show signs of skin irritation. It is also true that of the pigeons showing signs of skin irritation the vast majority are unaffected by insects and their behaviour is due to internal disease. What is the explanation? First of all I want to say something about each of the signs.

Scratching means scraping of the head or neck with one or more claws of one foot. Usually the scraping is done by the claw of the second or digit toe. The frequency and intensity of the action are very variable. Scratching may result in lacerations of the skin and dislodgement of feathers. Lacerations due to scratching are more commonly seen on the eyelids, angles of the mouth and on the nape of the neck. Occasionally the third eyelid and even the eyeball itself are wounded. Scratch wounds may become septic and result in abscess formation. Another expression of irritation of the skin of the head is rubbing of the forehead on the perch or the cheek against the butt of the wing.

Nibbling means scraping the skin with the tip of the beak. It is a relatively gentle action which does not result in laceration of the skin or dislodgement of feathers. It is the most common expression of a pigeon's irritation. A pigeon can nibble any part of its skin from the lower neck to the tip of its toes. Usually as in preening the bird nibbles now one part, now another; but the action differs from preening in that it is expressed at any time, even during feeding or walking about the loft floor; also feathers are not 'combed'. In the more advanced stages of anaemia the pigeon may persistently nibble the skin of the 'frill area' over the crop. In so doing it may compress its windpipe so as mementarily to asphyxiate itself; when it falls or flutters to the floor. Such cases have been brought to me on the owner's assumption that his bird 'takes fits'.

Pecking means biting either skin or feathers. The injuries to the skin produced by pecking vary from a reddening of the feather follicles to bruises, lacerations, and bleeding. Small feathers are bitten off and the quill feathers of wings and tail are frayed, bent, or snipped through. If the action persists at one part of the skin damage to the underlying flesh may result. For instance, toe-pecking, which some fanciers appear to misinterpret as a particular disease, soon results in septic conditions of the foot and shank. Persistent toe-pecking may also result in the ends of one or more toes being bitten off. The various forms of egg-binding and disease of the lower bowel induce vent-pecking and the production of a septic cloaca, which complicates operative treatment.

Plucking means the extraction of firmly rooted feathers. Their own feathers are rarely plucked by diseased pigeons; more frequently they pluck the sprouting feathers of their squabs. Plucking has to be distinguished from the much more frequently observed 'false-moult', which is the intermittent shedding of feathers other than during the true moult at the end of summer. The number of feathers which are shed over a given period of days may be few or many. The shedding of a feather begins with the stopping of the blood supply to its root. This is followed by the gradual extrusion of the quill from the skin. The arrest of its blood supply owing to disease may occur at any stage of a feather's growth, from the sprouting 'pen' to the fully developed feather. Not infrequently while nibbling or pecking itself a pigeon will remove one of these extruded dead feathers with its beak; such an action is not the violent action of plucking. In case of doubt an owner can examine the quill of the feather under a magnifying lens; the squill of an extruded feather is smooth and dry; that of a plucked feather is rough and its tip is raw and spotted with blood.

Stamping means the sudden extension of one leg while the pigeon is standing on the other leg or while it is walking about the floor of the loft. A mild stamp is a downward thrust of the foot on to the perch or floor. A more viollent stamp is a kick backwards by which the foot scrapes, or is lifted off, the perch or floor.

The actions of scratching, nibbling, pecking, and plucking are more obviously directed to the skin and plumage than stamping, which often comes at the end of a bout or attack of one of the other actions. Stamping, while an expression of skin irritation, is also an expression of the pigeon's irritability.

The word irritability conveys the notion of a generalised upset rather than an irritation confined to the skin. This leads me to say that, except possibly in its mildest forms, skin irritation is associated with some degree of irritability. For instance, the irritated pigeon exhibits its irritability by aggressive displays towards its companions and attacking its mate; or conversely by perching well away from others, fluffed out, dull, lethargic, indifferent to food. These moods of excitement or depression come and go without rhyme or reason and the owner recognises that the bird's normal demeanour has changed and wonders what can be amiss.

All the signs which I have described are evidence of disease. The disease may be one of the various insect pests, but, as I said earlier on, it has rarely proved to be so in British racing pigeons. In my opinion the standard of cleanliness and general management in the racing lofts is far superior to that of any other bird or animal industry or sport in Great Britain. In the seventy years of organised pigeon racing British owners have recognised that success, especially in long distance flying, demands regular, systematic and thorough attention to the details of general management.

Despite the high standards of owners as a whole there is a heavy toll from disease of the internal organs, of which most are due to germs and worms. I have written more than enough of these depressing matters in previous numbers of Squills. I mention them in this article in order to remind readers that whatever the nature of the disease, if it damages the bowels and the liver, it will cause the pigeon to exhibit signs of skin irritation and general irritability.

That skin irritation can be caused by upset of the bowels and liver is not unfamiliar to many of us. After an injudicious evening's celebration think of the following morning when we scratch various parts of our anatomy, though we have neither lice nor scabies. A scratching dog may have mange, but nine times out of ten there is something amiss with its internal organs. The skin is the barometer of the liver and bowels in man, dog, and pigeons. If no insect is recognised don't waste precious time on dressing your birds with insecticides; have them examined by a veterinary surgeon.

A Visit to Some French Lofts

by S & D CALKIN of Barnet

On July 13, 1956, we left Barnet to catch the Boat Train at Victoria for France, eventually arriving at Dover about 12.30, and embarking on the *Invicta*, we sailed half an hour later on a sea that was as smooth as a duck pond, docking at Calais about 2.40. Here we took our seats on the Paris-bound express for Amiens.

We were met by the President of Sous-Groupement De La Somme, M Rene Boizard, of Salouel, whose guests we were to be for the next 14 days. After the usual introductions we left by car for the home of our host, which is in a beautiful wooded valley three miles outside Amiens, being met and welcomed at the door by Madame Lucienne Boizard, the wife of our host, and three of his four charming children (the baby still being asleep).

The family soon put us at our ease, and Madame Lucienne Boizard was soon showing us over the chalet-type house which they had recently had

S Calkin of Barnet (right) and M Boizard at the latter's loft

built.

We then had a quick look at the garden and loft, which is still under construction, while a meal was being prepared for us.

The loft was built of concrete blocks with a wooden floor and asbestos roof, four compartments having been completed and two others still in the course of construction (see photograph). Trapping is done by the usual bolt wires, with an asbestos drop board. Here we were shown a beautiful team of old birds that had flown and won at all distances to Tolosa (Spain), Barcelona, Biarritz, Dax, Limoges, etc, and in the young bird loft there were about 30 nice youngsters, all of which were not being raced this year, but only being trained ten miles.

M Boizard is also the President of La Concorde d' Amiens, which has a total of 300 members, and in 1953 he was the Champion of France for both long distance and short distance races, his wonderful pigeon 'Vedette' winning the Championship for all the long races, and his bird 'Lucrece' the Championship for all the short races.

'Vedette' also represented France in the 1955 Olympic Barcelona race, in which France finished second winner, and among her many other wins are several 1st prizes in the French National Shows.

'Netta', another champion in M Boizard's loft, has won over 50 prizes, winning from Chantilly, Morcenx, Tours, Libourne, Poitiers, Dax, Tolosa and Limoges, and is dam and grand-dam of many winners, but some of M Boizard's best wins are 1st Tolosa (Spain), 1st Exeter (England) and 1st Limoges, and we were shown and handled many other birds that had won races at all distances to Barcelona.

It was pointed out to us that under French rules fanciers are only allowed to train up to 20 kilometres privately. After that all training is done by the club – twice 10 kilos, twice 30 kilos, twice 60, and afterwards during the season, on Wednesdays, 50 kilos each week, both old and young bird season. The Old Bird season starts on April 22, at Bretigny, and finishes on July 22, at Poitiers, but in club racing Biarritz is the longest race. This does not include Combine or National races.

The young bird season, July 29, starts at Bretigny and finishes at Orleans, as they believe in stopping youngsters to give them a chance to mature.

When the pigeons are race rung they are displayed in a securitas, five at a time, so that everybody can see that they have the rubbers on the legs, then they are put into the big race panniers, the lids being sealed as in England. Twenty-five pigeons are put in each basket, size about 48 x 26 x 11½ in.

The birds are always liberated for every race – short or long – between 5 and 6 o'clock in the morning, and the liberations are broadcast on the wireless at approximate hour of liberating, ie, if birds are liberated at 6 o'clock, then it is broadcast that the pigeons have been liberated between 6 and 6.30, so that fanciers can get an idea of what time to expect the birds, but not know the definite time they were let go. This eliminates to a great degree the possibility of clock manipulation.

Incidentally, clocks are set for the race at 4 to 5.30 on the morning of the race.

While it is a known fact that the French have a much larger prize list than we do, it must be remembered that the number of birds sent to their races is much higher than in England; therefore a pigeon that gets into the first 12 in a club has put up a very good performance, and to win first prize in a club would be equivalent to winning a Federation in England.

We would like to extend our thanks to M Rene and Madame Lucienne Boizard for a most enjoyable and interesting holiday, and to the various French fanciers we met, our sincere thanks for the facilities given to us on our visits to them.

On Tuesday, July 17, we visited M Cyril Taecke, of Saleux. He is a retired electrician and now keeps himself amused with his pigeons and a large allotment. His loft is situated very high, approached by a stairway, and contains about 25 old birds and ten young birds, one of which I noticed was paired up and on eggs. I would have selected this young cock as his most likely young pigeon. He has a very good record of wins over the last six years. He timed in on the Sunday (day of liberation) from Biarritz, 4th and 8th prizes, 768 kilometres, sending only two birds. He won: **1955** – 1st Rouen, 1st Libourne, 1st Angervile, 1st Blois, 1st Poitiers.

Blue Chequer Hen (30254.50). 1950 – 32nd Angerville (110 miles, 167 birds), 8th tours (190 miles, 82 birds). **1951** – 5th Blois (162 miles, 190 birds). **1952** – 24th Chantilly (50 miles, 131 birds), 44th Blois (162 miles, 211 birds), 60th Blois (361 birds), 21st Tours (190 miles, 512 birds), 99th Tours (2,288 birds), 38th Tours (314 birds), 81st Poitiers (250 miles, 396 birds), 23rd Poitiers (2,025 birds), 274th Angouleme (310 miles, 1,387 birds), 12th Libourne (375 miles, 201 birds), 43rd Libourne (718 birds), 23rd Dax (455 miles, 119 birds). **1954** – 183rd Limoges (287 miles, 1,031 birds), 423rd Limoges (2,191 birds), 47th Libourne (375 miles, 230 birds), 26th Dax (455 miles, 129 birds), 83rd Dax (409 birds), 5th Tolosa (515 miles, 66 birds), 12th Tolosa (236 birds), 26th Tolosa (1,192 birds). **1955** – 19th Tours (190 miles, 302 birds), 5th Poitiers (250 miles, 353 birds), 13th Poitiers (1,903 birds), 1st Libourne (375 miles, 204 birds), 14th Libourne (751 birds), 40th Libourne (1,687 birds).

Blue Cock (325023.49). 1950 – 45th Blois (162 miles, 271 birds). **1952** – 1st Blois (211 birds), 106th Tours (190 miles, 512 birds), 14th Tours (314 birds), 68th Blois (405 birds), 2nd Rouen (70 miles, 32 birds), 40th Le Mans (162 miles, 226 birds). **1953** – 74th Angerville (110 miles, 604 birds), 71st Bretigny (95 miles, 612 birds). **1954** – 1st Blois (162 miles, 575 birds), 1st Blois (895 birds), 2nd Rouen (70 miles, 129 birds), 2nd Poitiers (250 miles, 256 birds), 21st Poitiers (1,519 birds), 74th Libourne (375 miles, 836 birds), 25th Libourne (230 birds), 159th Libourne (1,744 birds), 17th Dax (455 miles, 129 birds), 53rd Dax (409 birds), 124th Dax (863 birds), 23rd Limoges (287 miles, 242 birds), 106th Limoges (970 birds). **1955** – 56th Poitiers (250 miles, 353 birds), 335th Poitiers (1,903 birds).

One pigeon won seven prizes in seven consecutive races, but it was killed by a gun a few days later.

M Cozette of Cardonnette, and a section of his loft.

July 21 saw us at the loft of M Andre Disma, of Rue due Chateau, near Salouel. This good fancier works in a factory and is home at 4.30. He is a Champion of Amiens La Concorde for the year 1956, and has about 15 pairs of old birds and breeds only 30 young birds, but, like nearly all the long distance fliers in France, he does not race young birds, and often gives the selected yearlings one race only.

The only strain in the loft is Dordin of Harnes and Portet x Dordin. M Portet is the Champion of Angouleme.

The pigeons were beautiful blues, blue pieds, light chequers and very dark chequers. I have never seen such beautiful blues anywhere, and I was very sorry to have to refuse some squeakers which this good fancier offered to me, as it is not permitted to bring live pigeons out of France, but I was very delighted to be presented with an egg from each of his three best pairs.

His loft, like so many French lofts, is in the top of the house, and is reached by a stairway from inside. It has a magnificent view across the country, and, of course, being high, has a good circulation for air throughout.

The training of the long distance pigeons are ten km. and 15 km. privately, and then by the usual club method to about 60 km.

M Disma likes his long distance candidates to be sitting about 14 to 21 days on pot eggs, but for a big cock bird he prefers to have a large youngster in the nest. He sometimes sends two or three-year-old cocks driving to short races, but never to long races.

Linseed is given the pigeons each Monday and Colza seed each evening.

I consider that the quality of the food given by the French fanciers is inferior to that used by English fanciers, but, on the other hand, I should also point out that their pigeons do fly these long races on the day.

On Monday, July 23, our visit was to the magnificent loft of M Cozette, of

Cardonnette.

At 3 o'Clock we were picked up by M Cozette and taken into Amiens for the usual glass of wine at a large hotel as a start to our trip. After a while we resumed our journey, stopping first of all at the very large business yard of M Cozette, who is the biggest metal merchant in Amiens, and from what we saw his equipment and stock was worth a considerable sum of money.

On resuming we were shown some of the lovely countryside around Amiens, arriving eventually at what to me was a most wonderful sight, a loft that was the best I have seen in France.

It is a brick building about 100 feet long, forming one side of a square, two other sides being taken up with the house at right angles to the loft, and, opposite, a number of outhouses, the fourth side being the entrance and drive of the estate. In the centre is a lawn about 80 feet square, with an unusual centrepiece of decorative stone and very attractive ornaments.

The interior of the loft was fitted up for both Widowhood and Natural flying. Each section of the loft was made to contain only nine nestboxes, pigeons entering by large sliding doors. To enter the loft one mounted a short flight of stairs which led to a passage about 6 feet wide and 7 feet high, which ran the complete length of the loft, ventilation shafts being fitted every 6 feet, and an electric stove fitted halfway down the corridor.

To enter the loft properly, one enters an office, complete with bookcase, table and chair, etc, where everything to do with the pigeon matter of the loft was kept in detail, even to the amount of food eaten by each section every day.

The pigeons we handled were very beautiful. There were Sions, Ghestams, Commienes by the dozen, and it made one's mouth water to see such pigeons.

M Taecke of Saleux, watching for racc arrivals from his lookout.

M Cozette kindly offered me some squeakers, out of his best birds. They were just ready for putting out, but once again I had to refuse. Still, I was presented with four eggs to bring home – two Sions, the sire and dam being a blue and blue pied direct from Sion, one egg Sion-Commienes and one egg from the most beautiful pair of pigeons it has been my fortune to handle.

After seeing the loft we were taken to a volerie where the older stock pigeons, which were prisoners, were allowed to breed in peace. Here again were Sions such as I have never seen in England. It is practically impossible to describe them, other than to say that every one of them was like an oil-painting.

My friend M Boizard was offered the pick of any two youngsters he liked, and selected a silver blue pied and a dark chequer, the eyes of which were nearly black, and as he, M Boizard, was the Champion of France in 1953, also second in the Olympic Pigeon race, he knows what is what.

After a long discussion on the Natural and Widowhood methods of racing, we adjourned to the house, where we were introduced to M Cozette's daughter, who gave us tea.

As M Cozette is a wealthy gentleman it would be useless to try to describe the interior of his house. Sufficient to say it was beautifully designed, decorated and furnished.

I think the biggest surprise I had was when a very charming young French girl, who was the maid of the house and came in to help attend to our wants, told us that part of her daily duties was to clean out this large loft. This, by the way, contained not only pigeons, but photographs of pigeons, horses and some real French 'Pin Ups', and notices to the pigeons telling them that it was forbidden to leave droppings in the nestboxes!

As we had spent rather a long time in handling the birds and discussing the Widowhood versus Natural system, plus being shown over the house and refreshing ourselves, there was not a lot of time for writing up his very long list of wins, but I managed to jot down just a few of them.

We then left M Cozette's house in his car, to make a call at the nearby loft of M Desplanques (Leon), who is a farmer. Once again the wine appeared and we all sat down for the usual discussion, after which we adjourned to the loft, which was situated at the very top of a farm building, being reached by a long ladder, which my wife climbed very gingerly. After we had all got in the loft (eight of us and two children – everybody comes in when one visits) it was rather crowded, but nevertheless we did manage to handle many winners, especially one pigeon which was a pure LaCoste. Mealies and chequers were the most predominant.

The pigeon I mentioned previously, a grand strawberry mealy cock, had won in 1955 2nd Dax (730 km), 1st Toloso (825 km), and in 1956 it won 1st Dax and 1st Biarritz (775 km).

M Desplanques is a very consistent fancier for all distances, but better on the really long races.

As time was getting on and we had quite a distance to cover, we took our leave of M and Madame Desplanques (after wine again) and resumed our journey in the car of M Cozette to the home of our friends M and Madame Boizard.

Our final visit, July 25, was to the loft of M Corbier, another of the Widowhood specialists.

First of all we went to Amiens by bus to meet at the headquarters of the Amiens La Concorde a friend of M Corbier who had come to collect us and take us the 20-mile journey to his mother's home. We were met at the gate by M Corbier's mother. M Corbier lives himself in Amiens, but keeps his pigeons with his mother.

After the usual introductions all round we were shown the house and gardens by Madame Corbier, while friend Corbier put a bottle of champagne on ice ready for our return. The house is a one-storey building, but about five feet from the ground, with a wide veranda, reached by a wide stone stairway, running right along the front of the house. It is divided into a number of large rooms, which are beautifully furnished in modern style.

Surrounding the house are lovely gardens with long arches of climbing roses and wistaria, the flower beds containing many varieties that I have never seen before.

On our return to the house we were invited to take champagne, but, as usual, I settled for lemonade while the rest of the party had the fizz.

We then adjourned to the loft, and once again we were treated to a spectacle of oil paintings. Somehow these French fanciers have a way of getting tip-top condition into their pigeons and, needless to say, I have brought home with me some very useful tips which I intend to try out.

First of all we visited the Widowhood loft, where every cock bird was a blue with the exception of one beautiful light chequer. The strain of pigeon was Roussell-Masserrell; both strains are made up mostly of Sion and the best strains of the Champion fanciers of France, and in looks they resemble the Sion pigeons mostly. We handled many good winners and one pigeon that had won 96 prizes.

We then went to the Natural system loft, about eight pairs of birds only being kept here. Once again most blues and mealies, but one grand chequer cock.

The squeakers were kept in a separate loft in an upstairs department but, unfortunately, time did not allow us to see them, as we still had to visit the stock loft and the volery.

M Corbier has not been flying many seasons, and he only sends to selected races, but each time he does send he wins several prizes.

Our visit was concluded and we returned to Amiens, calling at the hotel for another wine.

The hospitality we were shown by French fanciers will always be an inspiration to us, and the memory of the birds we saw will keep us on our mettle for many a year.

Let The Pigeon Tell You

by J MEADOWCROFT

1st Nantes Lancashire Nantes Club 1957

If the reader hopes to find in this article a short road to success, he will be disappointed. If he hopes to find out the tricks I use, and formulae of secret speed-cakes and tit-bits, he will finish these pages with his appetite unappeased. I have no tricks, no speed-cakes, nothing which is not available to every fancier in the world.

The essence of success is contained in very few words: Perseverance, Cleanliness, Good Food, Patience and Study. Then more study, and more. If your loft is as clean as it can be, and your food the best obtainable; if by your perseverance and patience you have got your birds to know you as well as you know them, then the only possible advance is by study – study of the birds, and of the methods of those fanciers who have, by virtue of great wins stretching over the years, become immortal in our sport.

In their writings, going back to the beginning of the Fancy, you will find over and over again that stress is laid on 'condition'. Regularly in *The Racing Pigeon* informative articles appear in which 'condition' is mentioned. Any bird which is well-fed and well-housed, exercised regularly and trained sensibly, will, at some time during the racing season, come to an absolute peak of fitness and eagerness, a pitch of perfection. So taking for granted that the majority of readers are sensible in their approach to a feeding, training and breeding programme, I know that nearly every decent bird of every fancier comes into condition at least once a season.

WHEN?

Yes, that is the question a fancier must be able to answer about each of his team before he can enter birds in long classic races, and hope to be near the top! What is the use of sending a bird to a five- or six-hundred mile race, with so much at stake, and hoping that by a fluke you can appear in the result? You stand to lose so much; not only the entry fees and any pool money you may have laid down, but the bird itself, and with it, the years of accumulated work, the breeding programme, and quite a bit of self-esteem!

Each bird, like each fancier, is an individual, and will be at its best as a result of a chain of circumstances which might well have no effect on the life of another bird. Some world champion professional boxers swear by a huge meal of steak and chips before entering the ring, whilst others win fights on a most Spartan diet. One man's meat is another man's poison; so it is with pigeons, too.

The fancier, bearing in mind the date of the big race, must set in motion the sequence of events which will result in the selected bird being sent to the race at a pitch of condition. For this reason I cannot, and will not lay down definite rules about sending birds to races. Each bird has its own rules, and by persistent study of the birds as individuals, the keen fancier will find that he won't have to pick his race candidates, for they will pick themselves. Watch the birds. See them in the loft, at feeding, and at exercise. You will learn (or rather, be taught) more about the game in your own loft than you can ever acquire outside it. Don't continually handle the birds, but let them know you as a gentle friend. If a bird shows you, by his prancing and gay manner, that he is ready for a long, hard fly, and will do you justice, then check that his flights are right, and send him.

By listening to good local fanciers you can sometimes pick up useful tips, but more often than not you can be bewildered by the many different winning systems propounded, and if you try to put bits of each of them into effect you may well be killed by kindness, or choked with cream.

Don't be misled. Adopt one system, and one only, in your management, and provided it embodies as cardinal points cleanliness, good food and regularity, then stick to the system through everything. Nothing will ruin your chances more than chopping and changing a loft regimen. It is important to remember that in this sport, probably the only one in the world, you have a lifetime in which to reach the top. The experience of a 75-year-old man who has been racing half a century is worth far more than the muscular achievement of a 30-year-old. But the old man has won his experience by a lifetime of study. He didn't get it all at once by reading one article in Squills! So please, don't read just a few articles by winners, and then hope to sweep the board next year. You may do it, true; but it will be more with the help of luck than anything else.

My first bird in the Nantes race in 1957 was a three-year-old, and he had never before won for me a single penny in prizes or pools since the day he was bred. But more important, he never failed to home from any race point in race time. Usually, he was behind the others, being my fourth or fifth bird, but whereas those that had beaten him went down, 448 could always be depended upon to return in fair time, whatever the conditions over the course.

This year he told me he was ready for the race; he was a picture of suppressed dynamic energy. He was sent in his favourite position, sitting eggs and with a big youngster in the nest.

My second bird, 175, was one bred by one of my fellow club members, which never once won for his breeder, but was also consistent in arriving regularly in race time. He, too, before the race, told me he was fit, though I had qualms, because as we say in the North, "he was full of blow". Still, he had a two-day-old youngster, which I knew was his best position, so I sent him. With both birds, my faith was justified.

So there it is – condition and position is the secret, granted that everything else in the programme contributes to good health and fitness. How many fanciers have never won much, yet could have won had they studied their birds and sent the right one? We can all learn a lot, and perhaps win a bit more, if we realise that a well-known quotation by the poet Gray can easily apply in the pigeon loft:

"Full many a gem of purest ray serene,
The dark unfathomed caves of ocean bear.
Full many a flower was born to blush unseen,
And waste its sweetness on the desert air."

Rainwater and Hard Flying

by STAN CURTIS

1st Nantes National FC

Having been asked to write an article for Squills I find I have got something to think about. What to write, where to begin. I guess the best way is to start from the beginning. It is over 70 years since I had my first pigeon. I think I caught the pigeon fever from my uncle, George Salsbury. I was born in Liverpool. I am not a Welshman as many fanciers think. I moved to S Wales 40 years ago.

I first started racing pigeons at Sunnyside, Pentre Broughton, near Wrexham, with success over 50 years ago. I topped the West Cheshire Fed from Nantes, considered a great performance in those days. I did not begin to take a deep interest until I moved to S Wales. My two sons and daughter and myself practically lived with the pigeons. Up at daylight and still there at dusk, we were winning out of our turn.

What a difference to-day. Instead of 365 days with the pigeons I put in about 365 hours. My loft is cleaned out occasionally, water changed now and again – they drink mostly rain water – fed any time of the day, exercised when I feel like it. Yet still I get just as good a result as I did before!

I don't keep stock birds, every bird in my loft has to race – and I mean race not walk home. I don't make favourites however good a pigeon has been. I am never satisfied on one good performance, I want some more before I have a good bird.

My Nantes National winner was 7th Open from the Bordeaux National. She is only two years old and I think I have a good one. But I am never satisfied with one good bird; I am always trying to get a better one as they don't last long, although my pied hen, grandam of my Nantes National winner, has flown Pau five times and San Sebastian once and in the money each time. All being well she goes again in 1959 as nothing is kept here just to look at.

Feeding; I used to go in for the very best, but to-day I feed them peas, beans, maize and plenty of linseed. People who visit me say what poor corn it is. It may look poor but must be good to produce the results it does. To me every pigeon I breed is a good one until it shows different as I am sure better youngsters are destroyed than some which are kept. Take Ivory Girl, a crooked keel chequer hen, very small and disregarded by all, even the late Col A H Osman; yet he had youngsters from her and she won 34th Open San

Dark chequer hen NURP56W5753. Up to 1957 this hen had been raced on the North Road (having flown Tynmouth and Edinburgh and Thurso into Wales as a yearling). The Thurso race was the very hard race of 1957. She was turned South and was flown from Weymouth and Guernsey before being sent into the Nantes NFC race. She was bred from well-flown stock; both sire and dam have raced from Thurso and on the North Road. Grandsire is his San Sebastian cock which won money in the NFC San Sebastian race, being the only bird into Wales on the second day, beating all other Welsh pigeons by a day and a half. The grandam is his well-known Pau hen flown Pau five times and San Sebastian once, being in the money every time. The dam of the Pau hen won the Welsh SR National from Nantes and was extremely successful from Bordeaux.

Sebastian, 8th Open San Sebastian, 9th Open San Sebastian and 41st Open San Sebastian four years in succession. As a youngster I could not give her away, but look what a loft of birds she has left me – Barcelona, San Sebastian, Pau, Mirande, Lerwick, Thurso birds.

I only fly South Road now, but when I raced north and south I used to fly the same birds on both routes with good results. My first and second birds from Nantes both flew north, and last year the hen flew Thurso and the cock flew Lerwick, so you see it does not make much difference. I also flew Lerwick and San Sebastian same year, same bird!

Regarding my birds' training, I give my old birds one training toss. Lavenock, 28 miles, then first race which I nearly always win. If any miss the training toss they get the first race as it is as cheap to race them at it is to train them. My youngsters get the same Lavenock, 28 miles, toss and one from Barry Island two days before the first race.

You are always learning something new about your birds. If I go to a loft I don't ask questions but I look learn and listen. Now, if I was starting a loft again I should go to a good loft and buy three pairs of youngsters, asking the owner to choose them for me. With this foundation, patience and perseverance you are sure to succeed at the start.

OUR METHODS OF LONG-DISTANCE RACING

by McCARTNEY BROS, Moira

Irish National King's Cup Winners 1959

If by giving a few particulars of our methods we are of benefit to the budding long-distance champions of the future, then we shall consider that our effort has not been in vain.

To found a long-distance strain of pigeons it is essential to give every possible moment to one's birds. First of all, irrespective of whether you have been fortunate enough to be the owner of a loft of considerable dimensions or not, it makes little difference to the racing potential of your team, so long as the loft ventilation is correct. We find that plenty of ventilation will do the birds no harm, in fact we believe that is is absolutely necessary in order to keep the birds healthy, provided one takes the necessary precautions to keep out dampness. Livestock of any description will not thrive under damp conditions, hence our advice on keeping the loft dry.

Another priority on our list is cleanliness. One could never overstress the importance of cleanliness in the loft. We clean our lofts scrupulously at least once a day during the racing season, and at least three times a week during the winter. After cleaning we use a light dressing of ground limestone on the floor as we believe it is excellent for keeping the floor dry and helps to remove moisture from the droppings until they are cleared out of the loft.

We would like to make it quite clear to the novice that when purchasing a supply of ground limestone be quite certain that one is supplied with unburnt limestone as we believe that it is the only class of lime suitable for a floor dressing, and even then should be only used sparingly in order to avoid it being blown about the loft and becoming injurious to the birds's health.

We are fortunate that we live in a district where limestone quarries are numerous, and a supply of very high quality is easily obtainable at competitive prices. A number of years ago we were in the habit of using a dressing of fine sand on the floor, and found it fairly satisfactory, but each summer without fail a number of our birds contracted a type of one-eyed cold, or at least we believed it to be so. It appeared to affect the birds in a similar manner. A very prominent Irish long-distance fancier visiting our colony remarked on the fact that we were using a sand dressing on the floor, and inquired if we ever had any birds in the loft affected by one-eyed colds. Assured that we had great difficulty with this complaint he immediately placed the blame on the use of sand in the loft, and told us that he had had the same experience.

We discontinued with sand and began using limestone as a floor dressing. Since then we never have had one single case of one-eyed cold.

One thing is certain in the pigeon Fancy, irrespective of what amount of knowledge one may have on breeding and racing long-distance pigeons, the opportunity of increasing knowledge is an ever-open book. The intelligent, progressive fancier can always add to his experience and make use of it. In fact, the fancier always on the look-out for knowledge is the fancier who will be sure to succeed. We strongly advise the novice who is contemplating entry into serious long-distance racing to visit as many reputable lofts as possible that are forever prominent in long-distance events. Do not be afraid to ask questions on any problem that may confront you.

We have always found pigeon fanciers to be most courteous towards novices, and always willing to discuss their birds and system of training and racing them.

Handle as many 500-mile champions as possible and so gain experience of a practical nature in the qulities that are required in champion stock. Note the type of body, wings, eyes, and feather that are prominent in champion pigeons. Make a mental note of all these necessary qualities and the novice will eventually discover that practical experience in this field is of far greater importance, and of more use in the end, than spending years studying text books of a theoretical nature.

The type of bird which we find most suitable for long-distance racing in Northern Ireland is a small to medium corky type with good strong wings, and carrying most of the weight well to the front of the body. We have cultivated this type of bird over the past ten years because we find that they can remain on the wing for a much greater period of time than the big strong bird with the deep keel.

Our King's Cup winner, 'Moiralona Queen', was on the wing for 16 hours 59 minutes in this year's National, yet she would be described by the experts as "a bit on the small side". Still, however, if we can remain among the winners of National races in the future with birds of this class, and of this type, we shall be quite happy to let the show experts continue to collect the red cards at the local shows with the beautiful specimens that are exhibited in the pens. After all, one would not expect an ordinary cart horse to have much chance of winning the Grand National at Aintree!

The purchase of foundation stock. We have often heard it said that "blood will tell", and from experience in livestock we have found this to be very true. Our advice to beginners is to go to a fancier of repute in classic long-distance racing, one who is always fairly prominent in the prize list in national events. Let him know your requirements, and accept his advice on possible pairs that are likely to make good foundation stock.

Do not worry about the price because a good quality article is always cheaper in the finish. Always remember that good pigeons are as easily fed as inferior pigeons. If you have not got sufficient cash available to purchase your requirements, then wait until you have. Do not be content with the purchase of cheaper stock for the sake of cheapness. This, in our opinion, is the first step on the road to success or failure. We can speak on this particular subject from experience gained in the foundation of our present family of six long-distance racers.

Twelve years ago we began to take an interest in the longer races, but found that our stock were just not good enough, so in desperation we began to search for long-distance blood. In the next two years we spent almost £300 on the purchase of stock. We bought birds at this sale and birds at that sale, all were supposed to be 500-mile blood, but it was not until we hit on a Van Cutsem Logan cross that we began to succeed at the longer distances.

In 1950 we clocked two birds well up in the prizes from Guernsey, and ten days later we again recorded the same two birds on the day from Guernsey. We were very fortunate that this pair of birds were a cock and a hen, so the following year we mated them together, and found that they were producing the type of bird which we believed to be ideal for national racing.

One of the first birds from this mating was our famous long-distance champion, 'Moiralona Billy', flown France on four occasions. His best performance was 4th Open Irish National from Rennes in 1956, winning £250.

Our King's Cup winner this year, 'Moiralona Queen', is a grand-daughter of this foundation pair also. So the novice will see that it is not so easy to reach the top as it may at first appear, but it is much easier when one has hit a winning lead. This, in our opinion, is the crossroads to success or failure. Every year new names appear on the horizon of national racing only to sink back into oblivion and are never heard of again. This we believe is because such fanciers being unable to master the art of breeding, and so the inherited characteristics are gradually destroyed by careless and unsuccessful outcrossing until the fancier again becomes a mere subscriber to racing funds in national events.

We are firm believers in a sound system of line-breeding and have built our present team of 500-mile racers on such a system. Every bird in our colony today is closely related to the foundation pair which we have mentioned earlier in this article. We do not intend to go into details on line-breeding, as the novice can find very good artiles on this subject in the Fancy Press from time to time, and by abler pens than ours. Sufficient to say that we not only believe in line-breeding as a means of preserving a strain, but if properly operated we believe it will actually improve a strain.

In the selection of mating pairs we are always on the look-out for super health, because without it, super fitness which can keep a bird on the wing for 15 or 16 hours would not be possible, and could never be achieved. We have yet to handle an outstanding pigeon which did not display all the signs of super health.

Therefore, our advice to the novice is to try and breed birds with abundance of health, then good food and healthy surroundings will assist in the creation of those outstanding performances, which are the ambition of all young fanciers.

Our system of training and racing is very similar to that adopted by most fanciers. We believe that birds should be raced on the Natural System, keeping the home conditions as near to nature as possible. Living in open country our birds enjoy the benefit of an open flight practically the whole year through, with the bath always available on the lawn in front of the loft. During the breeding season the birds are allowed to build their own nests from grass and twigs which they collect in the garden. We have practically no trouble from cats at any time as we have a good dog running loose in the garden with its kennel under one of the lofts, and woe unto the cat that happens to trespass in its domain.

The birds are fed on a good sound mixture of grains the whole year round, and are always given all they can eat. We have never used any magic potions or have no hidden secrets for success.

Good pigeons and commonsense are all we depend on for successful racing.

Our system of training candidates for National racing is quite simple and has been practised successfully by us for a number of years. We never send youngsters that are potential long-distance candidates further than 150 miles. As yearlings they are sent to the 150-mile stage again and sometimes are even given one Channel toss of 190 miles. Then at the two-year-old stage we send them to any distance on the race programme. We never hesitate to jump an experienced bird from 200 miles to 500 miles when fit. Our usual practice in preparing birds for the Irish National, if they have plenty of Channel experience, is to mate them in April and give them one race at 75 miles, then one at 150 miles, and complete their club racing with a further race on the Channel at 190 odd miles.

From this stage until they are despatched to the National they receive plenty of single up tossing to prepare them for the big event. When taking the birds for a training toss in the car we always go in the opposite direction to which the wind is blowing. We never worry if it is on the line of flight or not, so long as the birds get a nice stiff flight on the homeward journey.

Often we are asked for our opinion on that very controversial subject – eyesign. Whilst we admit that a good eye is important in a champion pigeon, we still keep an open mind on the subject, but do believe that there is something in eyesign. We have yet to handle a champion pigeon without eyesign, and can state that it is very prominent in our own family of birds.

Many fanciers believe that if a pigeon has super eyesign, then it should be a champion, but if it fails to home from a race eyesign theory is immediately condemned. What they forget is that eyesign alone is not the only characteristic required to make a champion, the bird must be a perfect specimen in health and fitness, and possess the desire to home. Without these qualities eyesign is a useless characteristic.

In conclusion, we wish fanciers everywhere successful racing in 1960, and hope that these facts will be of some assistance in building up successful lofts for others.

BREEDING THE WINNERS

by Dr G H T STOVIN

Basically, breeding methods have changed little during the past fifty or a hundred years. Darwin's writings, possibly because they were insufficiently studied, stimulated a wave of optimistic breeding. Everything seemed so simple. All that was necessary was to mate best to best, re-select and re-mate. Such a system – with its emphasis on the individual as opposed to the family – was foredoomed to failure because it results inevitably in a steady and persistent drift back towards the average of the breed. This fact, ultimately appreciated by the early enthusiasts led to a rapid evaporation of the prevous breeding enthusiasm and gave rise to the introduction of the pedigree system.

The subsequent history of breeding is punctuated by a whole series of schemes from selection on a pedigree basis through numerous plans based upon sundry external signs of type and conformation, leading up to such elaborate standards as the escutcheon theory in cattle or the eyesign theory in pigeons. The only common denominator in all these expositions is the complete absence in all of them of a scientific or factual basis. This alone accounts for the extreme variability in the description of the signs to be sought and the interpretation placed upon the various phenomena by the many devotees of the individual cults.

During the past twenty or thirty years economic necessity has led to a search for some more realistic method of assessing breeding merit, and one, which depends for its interpretation upon facts and figures, as opposed to human ingenuity. Various forms of progeny testing for cattle, poultry and, more recently, for sheep and pigs have eventuated. Whilst these have not yet met with universal approval, there seems little doubt but that progeny testing, with its emphasis on family merit rather than upon individual brilliance, will increase in importance as the years pass. It is this belief which has caused me to look for some simple method of applying this technique to pigeons. I feel that to be generally acceptable, any plan should be capable of simple application in the average pigeon loft.

Fallacy
The advice on breeding, given in the majority of pigeon books and articles, is to mate best to best. It will be observed that no indication is given either as

to how the best is to be found or even as to what the writer understands by such a phrase. The interpretation, which the racing pigeon fraternity appears, almost universally, to place upon this expression is that the most successful racers should automatically make the best breeders – a fallacy which I shall endeavour to expose in the course of this article.

When considering the recurrent, yearly, task of arranging his proposed matings the average fancier tends to head his list with his most successful racing cock and best winning hen, the list being completed in descending order of merit. In some cases, in which it is considered inadvisable, from the racing angle, to have a pairing of two pigeons, both earmarked for the longer races, one or other of the pair will be a stock bird. All too frequently such a one will have been selected, not on the basis of its known ability as a breeder of first-class racing progeny, but as a result of a study of its pedigree.

Pedigree
At this stage it might be advisable to consider the value of pedigree as a means of selecting either birds to purchase or to use as progenitors of racing stock.

The first point to remember is that, to be of value, a pedigree must confine its attention to the bird with which it purports to deal. Pedigrees can be, and often are, swollen by a mass of irrelevant detail, whcih has no direct bearing whatsoever on the subject matter in hand. For example, it is quite in order to point out that the parents or grandparents of the bird in question won such and such races with cash prizes of so many pounds. It is, however, completely redundant to state that some relative, such as a brother or a sire's sister – some bird not directly involved in the pedigree – won certain races or was sold for some unusually large sum of money.

All such statements are merely padding because they are based upon the erroneous assumption that near relatives are genetically alike. In actual fact two brothers or two sisters, unless they chance to be identical twins, are no more genetically similar than they are alike in appearance, in temperament or in ability!

Secondly, because of the well-known moral laxity of pigeons, it by no means follows that every pedigree is a statement of fact. Knowledge of genetics helps to confirm this statement. From time to time, in the advertisement columns of pigeon journals, a bird appears of a colour which could not possibly have been born to parents of the colour stated. It is for this reason that I always advise a fancier to introduce an outcross through a hen, never through a cock. Whereas the hen must be the mother of any eggs which she lays, a cock can easily be the sire of an unknown number of offspring. If the introduction should prove undesirable, it is easy to eradicate the young of a hen, whereas a cock may well leave in the loft a sample of his genetic composition, even though all his known sons and daughters have been eliminated.

It is a curiously interesting fact that many a fancier, who would never countenance inbreeding, claims to practise line-breeding. In point of fact these two methods of breeding, if really followed, are synonymous terms, differing only in degree. Usually, however, line-breeding, as it is carried out in practice, has degenerated into a form of ancestor worship; by this I mean that a fancier will be satisfied that he is line-breeding because he happens to have an ancestor common to both parents of a mating, perhaps four of five generations back. If it be remembered that the parents contribute each 50% genetically to their young, grandparents 25%, great-grandparents 12½%, it will be seen how small a genetic influence any ancestors still more remote can bring to bear upon the present generation.

On the whole the above is a somewhat severe indictment of the methods of breeding commonly pursued in this country. Summed up it amounts to the statement that multiplication replaces breeding.

In this connection the comments of Roger Mortimer, writing in the *Sunday Times* of August 9, 1959, under the title "Failure of our Classic Sires', is not inappropriate. His opening paragraph speaks for itself:

"There is one striking feature about this year's list of successful sires; there are only two winners of an English classic race among the leading twelve . . ." or again "Few Derby winners have done more indifferently at stud than the French horses, Pearl Diver and My Love; Galcador looks like rivalling them in mediocrity, whilst another undeniable failure is the American-bred St Leger winner, Black Tarquin."

The advent of artificial insemination in the cattle world speedily drew attention to the necessity for some form of progeny test for bulls, if the system were to lead to improvement in the milking ability of the dairy cattle sired by AI bulls in this country. It is interesting to note that Dr Alan Robertson, having investigated the results of the early years of AI, wrote that the average first lactation yields of heifers sired by AI bulls, was better than the comparable yields of heifers resulting from natural service – but only by one gallon! The obvious deduction must be made that the method of bull selection, followed until then, was faulty. Shortly after this the present system of selection on the RBV (Relative Breeding Value) standard was introduced. Unfortunately I have had to quote Dr Robertson from memory since I have been unable to find the actual reference.

The one lesson, which these observations would appear to emphasise, is the impossibility of picking suitable breeding stock by observation, by pedigree, by type and conformation, by eyesign or by any other man-devised system. Because man is mentally lazy, he is eternally looking for a sign where no sign can be given to him. The truth is that the only way of knowing how any creature will breed in the future is to know how it has bred in the past; in other words, some form of progeny test is, or should be, an integral part of any rational breeding system.

It is frequently claimed that the highest proportion of winners is produced as a result of some form of outcross. There is no doubt but that this statement almost certainly holds good today, not because it is fundamentally true, but simply because it is the system most generally favoured in the pigeon world. After all, if 75% of all racers are produced as a result of some form of outcross, such birds, by the mere weight of numbers and of chance, should provide three winners out of four, even though each such mating may throw up only a relatively small percentage of winners. It might well be illuminating if this journal set to work to collect data under the following headings:

1. The method of breeding employed in the production of the winners of the longer races.

2. The *percentage* of winners produced by each such mating. This must include every youngster born, lost, killed or used for stock.

3. The *percentage* of winners produced as a result of coupling together two successful racers.

It is because the results of the breeding methods, almost universally pursued in the case of all forms of livestock, are so unpredictably erratic, and, therefore, unsatisfactory, that I have been endeavouring to work out some form of progeny test of practicable application to pigeons without too much labour, and one which does not demand the utilisation of elaborate equipment, nor of specialised knowledge.

In cattle the task is relatively simple because, normally, one adult bull can look after a large number of cows. By the employment of AI, permitting, as it does, considerable dilution of semen, the number of cows can be increased very considerably.

In pigeons, the stumbling block lies in the necessity of providing 'pigeon's milk' for the newly hatched. It is not impossible, by employing a special design of nestbox, to arrange for ten or a dozen hens to be mated to one particular cock, which is kept to itself in a special compartment, away from the general loft. Because of the desire of all hens to lay again before their first young have been weaned, it becomes much more difficult to mate these same hens subsequently to the same, or to some other cock. Moreover, for a progeny test of this type to be really valuable, it is essential that all conditions, including rearing conditions, temperature, weather and age of young at testing time should be as comparable as possible.

It will be clear that the method above tests only males. This is the only practical method in such polygamous and slow breeding stock as cattle, but it fails completely to make full use of the monogamous habit of the pigeon, whereby it should be possible to progeny test both sexes at one and the same time.

For the past three season I have been experimenting with the following technique, the only specialised equipment for which consists of keeping of accurate breeding records together with systematic training and racing returns; in effect this demands the maintenance of three notebooks. It is essential in order that the owner should not practise upon himself any form of self-deception, that these books must be kept scrupulously accurate and that no entry should be made from memory.

During the initial stages of any progeny test both parents can be raced in the normal way. So soon, however, as the test points to any particular bird as a valuable sire or dam, such bird should be at once relegated to stock. There are two reasons for this:

1. A bird, which can breed reliable racing stock is far too valuable in the breeding pen to be risked in a pannier, at all events until a better bird has been located as a replacement.

2. Whilst an outstanding stock bird may be also a racer of note, such is by no means always the case. There is, of course, a genetic explanation for this, which I will endeavour to put forward in due course in language as simple as possible.

It is clear that initial calculations must be based upon the young bird racing season, that each season, which passes after this, must diminish or enhance the value of any conclusions then formed, and that, if a bird has performed creditably as a breeder for three seasons, preferably with three different mates, it should entitle such a bird to be classed as a progeny tested sire or dam. In my experience with cattle and poultry it is but seldom necessary to alter the assessment, based upon the production of the first dozen or twenty young, more particularly if it is possible to compare these figures with those of other males operating in the same herds or flocks over the same period. Owing to the comparatively solitary habit of pigeons, to the small number of offspring bred per season, and, therefore, to the greater influence of each individual female, it would be reasonable to expect that results might be somewhat less uniform.

My experience with pigeons is, as yet, too meagre to allow me to speak with any degree of authority on this point. Nevertheless, so far as time has yet permitted me to make my investigations, I am inclined to believe that a cock or a hen which produced satisfactory young with one mate will probably continue to do so with the majority of other mates. I should expect this statement to become truer the longer the line of progeny tested ancestors, possessed by both parents, and the closer the degree of inbreeding practised.

Both these systems lead gradually and steadily to a purification of gene pairs with increasing propotency. Provided that the above plan were pursued for twenty or thirty years, it would not be difficult to visualise a parent, whose gene pairs were so homozygous, or pure breeding, for desirable characters that, no matter what mate were chosen for it, the resultant offspring must all be performers of merit.

In evaluating results it should be remembered that one excellent youngster might well be the result of a fortunate distribution of good genes; two would make this less likely, whilst four would still further reduce the chances of such a fortuitous happening. Again one is looking, not for the occasional champion, but for the family, which is reliably consistent. For this reason no young bird is credited with having flown a training toss unless it returns home on the day, with having flown a race point unless it negotiates this in race time. The beauty of this plan is that, automatically, the standard by which future stock birds are selected must rise since existing stock birds are only replaced when something better has revealed itself.

This brings me to the point of offering an explanation of the basic theory upon which all progeny testing is based; this necessitates harking back to Mendelian theory. In this connection it is necessary to remind fanciers that, with the sole exception of sex-linked characters, all factors are controlled by genes, arranged in pairs. At mating time there is a halving of both the male and female autosomes (chromosomes other than the sex chromosomes) with the result that the genes on these chromosomes are also halved.

Thus each parent now contributes one gene of every kind to the fertilised ovum; in this way the total number of chromosomes and of genes, carried on these chromosomes, is brought back to the normal number for the species. It must be remembered furthermore that it is a matter of pure chance which member of the gene pairs is contributed by either parent to the resulting offspring.

A moment's reflecition will show that, so far as any individual parent is concerned, it is a matter of no moment to the resulting offspring which gene, of a like pair, it receives.

$$\frac{A}{A} \quad \text{at division} \quad \frac{A}{A} \quad \text{can only pass gene A}$$

$$\frac{a}{a} \quad \text{at division} \quad \frac{a}{a} \quad \text{can only pass gene a}$$

The capital is customarily used to denote the dominant gene, the small letter to indicate the recessive.

Where the parent carries both the dominant and recessive, it is a matter of chance which of the two any of its young will inherit.

$$\frac{A}{a} \quad \text{at division} \quad \frac{A}{a} \quad \text{can pass gene A or gene a}$$

It has been shown that practically all characters of economic importance – milk production, number of eggs laid per year, ability to rear lambs well, height in humans, etc – are under the control of a very large number of genes indeed. There is every indication that this applies equally to racing ability in both horses and pigeons. Herein lies the explanation of the well-known fact that so many creatures can be good performers themselves, yet are seldom or never able to beget young as good as they themselves have been.

Because production is controlled by a large number of gene pairs, and because, in the cases which we are at present considering, a large proportion of the gene pairs are heterozygous, or mixed (Aa above), these birds will possess a sufficiency of the valuable genes to perform well, but, at mating, will be as liable to transmit inferior genes as superior. Herein also lies the explanation, as seen is the case of height in the human race, for the tendency always to slide back to the average for the breed.

I trust that I have made my attempted explanation clear. For reasons of space it has been necessary to keep it as brief as possible.

I give next the detail of the progeny test of

Blue Bar Cock, NURP56HPN236

Season 1957–

Mated to brown chequer pied hen 56/1145. Four young resulted: blue chequer cock NURP57HA137, blue chequer hen NURP57HA138, blue chequer cock NURP57HA132, blue chequer hen NURP57HA140.

Of these 137 returned from a training toss so badly injured that it was destroyed. The remaining three were all lost as yearlings.

Season 1958–

Mated to black frill hen NURP57HA131. As a matter of passing interest, the dam of this hen was a brown chequer blondinette. From this pairing four young originated:

Black Hen	NURP58HHC60
	NU58WHE3388

This hen flew Essendine in 1958. In 1959 she was turned south and flew Rennes.

Blue Bar	NU58WHE3387
	NURP58HHC58

This bird was lost in training as a young bird.

Black cock NURP58L9765. This cock was bred too late to compete in the young bird races. He was trained to Kettering. In 1959 he flew Northallerton and Berwick.

Black hen NURP58L9766. Nestmate to 9765 and trained with him to Kettering in 1958. In 1959 was mated to 9765, was given several tosses from Kettering but was not raced.

Season 1959–

Was mated to red chequer hen NU58Y1951. This pairing produced three young birds:

Red chequer cock NURP59RA8762. This bird arrived from the Lymington race two days late with its rubber race ring intact.

Blue bar wf hen NURP59RA8784. this was my only bird, out of twelve sent, to arrive from Lymington on the day.

Blue Chequer Hen	NU59WHF3
	NURP59LL5

This bird has been trained to Kettering twice. Owing to illness I was able to compete in one young bird race only – Lymington.

It will be seen that, with the first hen to which he was mated, 236 was a complete failure. It was not until I observed how his offspring were shaping this year, in addition to the performances of those of the previous year, that my interest in him was again aroused. I judge him to be a mixed breeder, with a preponderance of good genes, which enable him to breed successfully with the majority of useful hens. Next year, in an attempt to collect together his better genes, I propose to mate him to a daughter. All their offspring would be raced in order to make sure that this line of breeding was proving successful. At the same time each season resulting cocks and hens would be progeny tested until one that was breeding better young than 236 could be found to replace him. So long as the system continued to produce stock birds of a quality superior to their predecessors, as shown by the progeny test, so long would it be unnecessary to consider any outcross.

In this way the standard upon which the choice of stock birds is based must be continually and automatically raised, resulting in better race returns, a higher percentage of winning pigeons and ever more valuable breeding stock.

SURVIVAL OF THE FITTEST

by Messrs HICKS, SHERWOOD &
HANSELL

1st UNC Cormeilles and Lille (2)

We would like to thank Squills for giving us the opportunity to write this article for his year book.

We started keeping pigeons in 1953 as Hicks & Sherwood, Hansell joining us in 1959. We had the wonderful experience of being highest prizewinners in club first year we flew. But we learned how to lose in 1958-1959. Although our birds flew badly during these two years we bred some of our best birds in these seasons.

We keep 12 pairs of old birds and breed 20 youngsters. The birds are all related to black pied hen 51X28607 and birds from local fanciers.

Here are details of some of the birds in the loft today. Blue chequer cock 57U6479, winner of 3rd Club, 3rd Fed Selby, 2nd Club Welwyn, 5th Club 15th Fed, 16th North of England Championship Club, 60th Up North Combine Welwyn Nat, as a young bird. Flew three races to Ashford 240 miles as a yearling. Flew through to Lille 318 miles as a two year old. 1960, 2nd Club Lille, 3rd Club 19th Fed Bourges, 2nd Club 5th Fed Lille Nat. 1961, 1st Club Lille, 9th East Cleveland Championship Club, 14th Fed 82nd NECC. 1st Club, 1st Fed, 1st NECC, 1st UNC Cormeilles. Winner of Danish Cup for the best 2-bird average along with 7101, 1st and 34th UNC Cormeilles. Lille Nat, 2nd Club, 6th ECC, 8th Fed, 40th NECC.

Red cock, 7101, winner of 1st Club, 3rd Fed Bourges, 536 miles, 1960; 2nd Club, 5th ECC, 6th Fed, 18th NECC, 34th UNC, 1961.

Blue chequer pied cock, 6371, winner of 4th Club Welwyn, 5th Club Newhaven Harbour as a YB. 1st Club, 3rd Fed Welwyn; 1st Club, 1st Fed, 1st ECC; 1st NECC, 1st UNC, 1st Vaux Gold Tankard Lille Nat, 1961.

Black cock, brother to 6371, 6th Club, 11th Fed Peterborough as YB, 2nd Club Peterborough, 3rd Club Welwyn as yearling.

Dark chequer cock, 7130, latebred 1958, untrained as a YB. Our first bird to loft five times as a yearling through to Ashford, 240 miles. 1960, 1st Club Selby, 5th Club, 13th Fed Grantham, 4th Club Lille flew Cormeilles, 6th Club Ashford, 5th Club Lille National. 1961, 2nd Club Welwyn flew Lille, Cormeilles, Bourges and Lille National.

We have three hens in the loft which are unraced. These are paired to our best cocks and the majority of our youngsters bred from these. We breed from three or four pairs of birds the others sharing the job of rearing.

If a pair of birds do not breed winners first year we put them together again, as winners are not hatched from every egg. Some birds will breed a winner in every nest but it is possible for a champion to be bred from a pair which has never bred a club winner.

We pair our birds to type irrespective of relationship. One should be careful when introducing new blood because it can dominate a loft in two or three seasons.

Our old bird loft is 12ft x 6ft, 7th high at front, 6ft at the back. It contains 15 nest boxes 2ft x 2ft x 2ft, no box perches, in the loft but a perch inside the nest box. These birds are fed in the pens with beans before them all the time.

As soon as old bird racing is finished the pens are locked up and the birds hopper fed. When young bird racing is over the sexes are separated. We pair the birds up on the first Saturday in March but our second and third round youngsters perform the best.

We have had many prizes with youngsters bred in May and June. Latebreds have put up some fine performances but it has been our experience that these are often the starters of one-eyed cold and other set-backs during the moulting season. There is no doubt that races are won from the end of July to the beginning of April.

We prefer a consistent bird, but we have birds which perform under special conditions. We have had little success with broken birds racing. We do not do any showing but it is a fine sport for the winter months it must help to pass the long winter months away, but one thing we should remember, show birds are not always successful racers. Some of the best racers do not come anywhere near show standard.

We never worry about wing condition unless they are on the last two flights. If youngsters are stopped racing they fall very heavy into the moult – it is fatal to send them in this condition. We have stopped youngsters but they have never proved to be good old birds. We have had youngsters fail in the early races then do quite well later on.

We wean our youngsters at 28 days. They are put on a mixture and fed twice a day. They are exercised in the evening until they start running, then they are exercised as much as possible. We only train about seven miles, possibly single tossed once or twice but we always train them at 7 o'clock in the evening. When our YBs are on form they clear from the loft and will fly for 60 to 90 minutes, we never force them to fly.

If a bird has arrived late from a race it is best to turn it out when the others have had about 60 minutes or it will tend to drag the others down.

We do not use any medicines or potions, just good sound corn, grit, minerals, linseed and a little plain canary. If a bird is off-colour it is shut away on its own and allowed to recover without the aid of pills, etc.

We advise beginners to buy a few feeders and try to buy some eggs from a successful local fancier whose birds are flying well today not 50 years ago. He should train his youngsters 100 miles first year. Then he could race the inland races the following year after breeding a strong team of young birds.

It is stupid to pool them until they show some form. If a pair are breeding strong healthy youngsters keep them together and breed half a dozen each season.

Too may fanciers try to walk before they can crawl and lose heart before they have tasted the fruits of success. When a loft has been built up of consistent racers; race them hard and breed from the survivors. If a bird is flying hard on week after week he will get a day made for him and win Club or even Combine.

Our ambition is to fly 500 miles on the day in a head wind and then we may have a crack at Barcelona. We are sure the birds which fly 700-1,000 miles are the stock birds of the future. We are trying to create a strain which will win at any distance in any weather. It's a long time to keep them all the year round just for one or two Channel races.

We think a lot of knowledge can be gained on marking nights taking notice of the birds of the old hands and then watch how they perform from the race. When you are beaten think how the winner beat you and be determined to beat him the following week. There can be rivalry without being enemies.

Club members should remember that the club is dependent on them to have a successful club. It is surprising how many members cannot read a clock. Every member should have knowledge of clocks, etc. Do not leave every job to the officials, there is plenty of work for everyone.

Our blue chequer cock, 6479, a very honest racer never far behind whether fast, slow, long or short. We paired him in 1960 to the best bred hen in the loft. We bred some poor specimens which were wicked performers but we paired them together again this season and have won 2nd Club, 7th Fed with dark hen, 3rd Club, 14th Fed, 5th Club, 6th Fed with dark white-flighted cock.

During the winter months we buy the corn and other necessities. We have paired our birds up a 1,000 times on paper before making a final decision.

Finally one cannot read too much nor know too much about pigeon racing. we have learnt some most interesting points from Squills. You can always learn something new. When visiting a loft, look and listen and don't believe everything you hear.

Confidence in the birds, a little bit of luck and plenty of patience and perseverance and you are sure to succeed.

HOW TO PRODUCE A FAMILY OF RACING PIGEONS

by RON MITCHEISON of Winchester

Winner of 1st and King's Cup NFC Pau 1962

My son Clive and I feel deeply honoured to be asked my the Editor of *The Racing Pigeon* to compile an article for Squills Year Book, 1963, and we hope our efforts at journalism will be the means of helping those of our fellow fanciers who are in a quandary as to what is meant by a 'Family of Pigeons'.

Quality

To begin with, you must acquire pigeons of the finest calibre in constitution, bone structure, quality of feather, intelligence and above all, great strength of back. These are what I term 'Quality Pigeons' and I can assure you that if you go in for anything but the very best you will only be wasting your time. These birds should be obtained from an honest fancier who has excelled consistently in the long races, furthermore, he should be a fancier who is actually winning today, not one who is still resting on the laurels of 30 years ago. Take your time finding this good fancier and get to know him personally, it will pay you well in the long run.

The Hard Way

Way back in the early 1940s I had some wonderful winning birds and I was beginning to think I knew all there was to know about racing pigeons but as the years passed, I realised I knew so little about producing and preserving a family. Now, after further years of thought, study and putting into practice my ideas gleaned from experience, here are a few conclusions:–

If you were a millionaire and bought all the champions in the country, when it came to pairing them up, you would know little or nothing to assure you of immediate results. How you paired them would probably be a matter of luck and the progeny would be of all shapes, sizes and colours. No doubt a good bird or two would be bred from them but not so many as if you had known your own family and studied the matings properly.

As we all know, a lot of champions come from a direct outcross and so through the years, one could go on outcrossing all the time in the hope of producing champions. No doubt some would be produced (IF you kept enough pairs of pigeons!)

'Sir Lancelot', dark chequer cock 60Z649. Bred and raced by
R Mitcheison & son, 9th Open NFC Pau.

We have read statistics as to what percentage of champions are bred by out-crosses, line-bred and inbred but every blood I have read on breeding, whether it be cattle, dogs, horses or pigeons, not one can come to a definite conclusion as to which method is best. I have now satisfied myself that *a family of pigeons* is definitely the only answer because nowadays when I pair my birds, I know to a very large extent just what they are going to produce as to shape, colour, constitution, intelligence and that elusive 'Quality'.

The Foundation

Having acquired the type of foundation birds I recommend, the youngsters from these will be raced and you will decide, from their performance, which pigeons are the ones which show the characteristics I mentioned earlier. Any youngsters which are weakly or show signs of distress must be put down, suppressed. (Not raced until they are lost for this is not only the easy way out but it is cruelty to the birds.)

In all probability some of your youngsters will conform to the type on which you wish to build your family. If, for instance, you found that only two pairs of birds produced your best youngsters, I would advise you thus: Pair the cock of one pair to the hen of the other pair and if these matings also produce good quality youngsters, pair the youngsters of the first year to the youngsters of the second year. These would be half-brothers and sisters. Should these pairings continue to produce good birds (by this time the first year youngsters would be two-year-olds and would have been raced to 400/500 miles) and they were continuing to do well, your task is made a lot easier.

The Intuition

You may find however, that you have lost what you thought were some good birds or some of them home and return too slowly. If this is the case, you first ask yourself 'Were my birds put down in first-class condition to negotiate such long journeys?' – if the answer is 'Yes', then you must start at the beginning again for these are not the 'Quality Pigeons' you are looking for, do not waste further time with such birds.

I have personally found that the birds I selected as the basis of my family were nearly all the right ones, and in this I consider I was very fortunate. At the same time, something inside me, which I cannot explain, guides me in my selection. Very few fanciers will understand what I am trying to explain but I am positive that there is a very strong link between the pigeon and the master and to me this is very staggering and extraordinary, yet I cannot go any deeper into this subject for I am not able to explain it in words.

A Step Further

I will assume that the two-year olds came up to expectations. I would then pair my youngsters from the half-brothers and sisters back to the grandparents and by this time your birds would be attaining the quality and type you set out to produce. I would never hesitate to pair dam to some or sire to daughter if I wanted to double up some good characteristics of quality, but the youngsters from such matings would not be raced they would be kept for stock. When you find you have a champion producer, ensure it being on both sire and dam's side in future pairings, no matter how many generations in between. You may well breed a duffer or two, do not let this deter you but when you have eliminated most of the bad characteristics (remembering that in line-breeding and inbreeding you are 'doubling up' bad points as well as good ones!) you will be on the road to success. It is when the good points 'double up' that you can look for your new champions.

Maintaining the Strain

By careful study, judicious pairing and extreme patience you will eventually reach your goal. Now comes the real testing time for you will not dare to sit back and rest on your laurels for this has been the downfall of many a famous loft. Having produced one champion, you must now continue to produce others, you must think ahead all the time, way out ahead, even ten years

'Queen Guinevere', dark chequer hen 60WFC99, bred and raced by G Mitcheison & son. Winner of Pau NFC.

ahead, for what is winning today will not win then and if you let up, even for a minute, deterioration is liable to ruin all your good work and study in a very short time. I have found, however, that as long as the physique is of a very high standard, the other qualities remain intact as well.

Always be looking for a good bird or two and bring it into the family bearing in mind that the new blood must be of the same type and characteristics as your own.

Sundry Points

A champion is very often produced when an outcross is introduced to an inbred family and I have found that my best racers this way bred are of a medium size with the hens a shade smaller than most. I prefer a little depth of keel but I insist on a very strong back at all times. Some of my pigeons have backs like little camels.

Returning to the subject of breeding, you will find that in the pedigree of my King's Cup winner 1962,'Queen Guinevere', two prepotent stud pigeons appear, namely 'Steele's Mealy' and 'The Southwell Hen'. Both these pigeons appear in the pedigree of both the sire and the dam of 'Queen Guinevere'.

In Closing

Experience is a very good and useful thing to have and if you have a retentive memory, so much the better. I recall to mind the story of the young fancier who asked the famous champion how best to succeed with pigeons, the champion replied: 'Well son, I have always found it very useful to start about forty years ago'! (He wasn't so very far wrong was he?) We hope you have enjoyed our brief journey into words and we wish you all happiness and good fortune in the future.

THE COMPLETE CIRCLE

by B WAUD of Waud Brothers

1st GYA Bourges 1963

We consider it a great honour having the opportunity of writing in SQUILLS 1964, and hoping it gives you interesting reading as much as we will enjoying writing it.

Well, we became interested in pigeons in the early fifties, having a few in the back yard. I suppose many a good fancier started this way. I well remember sending four birds to Grantham, a distance of about 65 miles, this being the farthest we had ever sent them. It was quite a day for us as you can well imagine. Well, it turned out to be a bit of a bad day and we waited with our noses pressed against the kitchen window. It got to 7 o'clock in the evening and one came. Well, there was so much fuss made of it you would have thought it had won the National. We got them all home in half an hour so it wasn't a bad day after all.

Time passed on and we were lucky enough to obtain the lofts and birds of the late Wally King, a local successful fancier. It was quite a big step from our pets in the yard to some fine racing pigeons. Of course, we were very fortunate to have with us Wally King in our first year of racing. He gave us some sound advice which every beginner needs to set him on the way to being a good fancier in the future.

In our first two seasons we learnt a lot about racing. Looking back, we realise we overfed and didn't train enough, but one thrives on one's mistakes, so all this was taken care of by trying various methods of feeding, exercise and training. We also understood that the birds which were tired and tested to the best of our knowledge were not up to the standard that we required for the longer races, and especially in the Castleford area where there are almost 150 members in the Federation and they are always most prominent in the GYA and take more prizes than any other Federation.

We wanted the best so we wrote to Hammond & Whittaker, who in our opinion have the best Grooters in the country. We got from them two pairs of late breds, just for breeding, of course. You must buy one particular strain and keep them together; this way you are sure of hitting a few good birds and probably an outstanding one somewhere along the line.

We feel it is a mistake to pick an odd one here and there because you can't establish a good sound family this way, but I'm not saying you can't breed a

'Madcap', blue cock, NU58/259, bred and raced by Waud Bros,
Castleford, 1st Leeds & District Fed, 1st GYA Bourges.

good pigeon this way. We have had many a good pigeon bred from odd crosses, but you must get one string and stick to it until you have tested them to
your satisfaction. But don't be over eager, it may take a few seasons. There's
a good saying, and a true one: "From the best you get the best", which should
answer all your questions. We went on to prove this in later years.

I remember a beautiful dark cock, pure Grooters, the first good Grooters
we flew. He won 3rd Club, 6th Fed and 12th GYA from Amiens, 350 miles.
He also flew the water several times. We had good results from Nantes, Caen,
Amiens, Le Mans the following year and we won 1st Club, 2nd Fed, 6th
GYA with a dark flighty cock.

As you can see we had improved the standard in our lofts with these
Grooters. If you have good birds all you have to do is get them into condi-

tion for racing. This is what I would like to mention next. We feed twice a day, in the morning after making them take half an hour's exercise, or maybe more, it just depends how they are framing.

May we say that we believe that the birds seem happier being flagged around home than going away on training tosses, so naturally we do quite a lot of it. After this we give them half feed of peas, maize and a little wheat; this is repeated at night. Perhaps a tasty bit is given two or three times weekly. We don't believe in fancy foods at all.

Well, there you have our simple method, but don't be taken in the words 'simple methods'. What I mean is it isn't complicated, but after all it is you who are feeding and all birds are different and so are the distances of races you are competing in. All this you have to take into consideration when feeding.

I would like now to tell you of famous blue cock 'Madcap', its life story in fact. He was a single reared youngster in the middle of April. That year we bred about 22 youngsters, our usual amount, losing a few from the loft and in training. We kicked off in the first race with 16. Up to the first race they had plenty of tosses up to ten miles and then we tossed them in twos, a thing we do every year.

We do this about four times from ten miles, then they are jumped to 17 miles altogether, then on to 30 miles twice, a day or so before the race. When we reach this stage we usually stop six or more for the following season. If fanciers did this every year with a team of youngsters they wouldn't lose many, but, of course, there's always later youngsters they keep putting in among earlier bred ones and then try to catch the others up by jumping them too far, which more than often ends up by losing them. Again we learnt by mistakes. This is a time when you have to be very patient, then I'm sure things will end better for you.

After two or three races we stop a few more, leaving only about four for the young bird races. You see we don't think young bird racing is of much importance to a pigeon you want in later years.

'Madcap' was christened that name as a youngster because he was so wild and unpredictable. This was so unusual as we had such an exceptionally tame set that year. He flew six stages to Welwyn GC, scoring 10th Club Huntington, and dropping in with another bird which took 5th Club from Welwyn GC. The following year, in 1959, he had several races inland before going to Amiens, 350 miles, sitting nine days, winning 1st Club, 1st Fed, 1st GYA, beating 1,800 birds. He took 8 hours 20 minutes.

One of our ambitions achieved, we looked forward to the following year's racing with this fine pigeon. This didn't turn out to our expectations; we had to move our lofts owing to the Council building houses. After winning the Old Bird Inland Average the previous year this really upset our rhythm, and to top it all 'Madcap' must have picked up some poison somewhere, making him miss all the water races.

In 1961 he flew Newark, 6th Club Huntingdon, 5th Club Welwyn and flew Dover and Lille, 320 miles, north-west wind, 10th Club. 1962: Newark,

Huntingdon 10th Club, Dover, Huntingdon 5th Club, Cormielles clocked in race time, Huntingdon 9th Club.

1963: In the close season we had built new lofts to our own design. They have ventilation in the back, which we can close or open in any type of weather, also in the front. We are big believers in fresh air. The old bird loft is 12ft long, 7ft high and 5ft deep. Young bird loft is 10ft x 7ft x 5ft. We have a corn and basket storage cabin on the end which is 8ft x 7ft x 5ft. There is a 4ft canopy which keeps out all the rain and assures the lofts being dry all the time.

There is also glass that goes on the front to keep out all the cold winds. Tom, my brother, being a joiner by trade, built these luxury lofts, giving our birds extreme comfort. We thought these new lofts would upset the birds but they seemed to be quite happy in them.

'Madcap' showed his approval with the following performances. After going to Huntingdon, 108 miles, we tried him driving from Dover, 211 miles; he came like a bomb, taking 3rd Club. This gave us the idea of trying him on the Widowhood system. He was sent this way to Huntingdon, being our first bird, and coming home with all the song and dance.

You can imagine this, we were very pleased with running into good form at this moment. Seven days later he was sent to Beauvais, 350 miles, winning 1st Club, 2nd Fed, 3rd GYA, pooled to 10/–, 1,400 birds competing. I would like to mention that I was on my own that day and one of the old birds was acting awkward. As it was getting near race time I had decided to lock it up, closing the door behind me. Of course, at this moment 'Madcap' came. After flying round I eventually clocked him, losing half a minute. This robbed him of 1st GYA. What a pity after the bird had done its work well. It took him 8 hours 35 minutes and when he dropped he looked superbly fit.

I was amazed at the condition he was in. We knew then that he had a good chance in his next race which was 500 miles, the longest he had ever been in. The system of Widowhood had paid us well, but we don't advise it to novices as it takes very careful feeding as some of the birds fret and won't eat. You must vary its feeding and encourage the bird's appetite. Some members say that pigeons will not race Widowhood over 400 miles, but we had decided to send 'Madcap' on his last race to Bourges, 500 miles, along with a grizzle hen and a black flighty cock, also on Widowhood. All three went very light in hand, carrying no excessive fat at all. Liberated at 5.35, it turned out to be a dull day with heavy showers. After many anxious moments we clocked 'Madcap' at 8.27 to record a velocity of 962, winning 1st Club, 1st Fed, 1st GYA. Well pooled. It was a bad race with only 11 birds on the day out of 892 of the best. I must say that after his long fly of 14 hours 52 minutes he was far from defeated; in fact, early the next morning, waiting for the other arrivals, he went clapping around on four different occasions. What a constitution this remarkable pigeon must have. We clocked the grizzle hen at 6.20am taking 6th Club, 17th Fed, 32nd GYA. Our last arrival was on the third day. We now have the winter months ahead and I am sure the birds hit it as much as we do. Of course, it's moulting time as well. Our birds never come out through the week, only at the weekends if the weather permits.

I have known our birds being in for two or three weeks at a time. This does not worry us at all as we are believers in rest at moult time. Through the winter we feed the same variety of food with a little linseed but giving them ample enough. We have to feed once a day in our dinner hour owing to our jobs of work being awkward hours for the winter darkness.

About a month before breeding begins we start rationing their corn and giving a little exercise. Of course, the hens are given more attention at this time to avoid any uneasiness when pairing up. Well, we seem to have gone a full circle telling you what goes on in our lofts and have enjoyed doing so. We only hope that you have enjoyed the same and possibly picked up a few hints which have certainly brought us our success.

CASUAL SEMI-WIDOWHOOD FOR SUCCESS

by PETER TITMUSS

1st Western Home Counties Thurso

My first reaction when asked to contribute an article to SQUILLS was to ask myself which aspect of pigeon racing would be most likely to interest my readers. After due consideration I decided to devote my thoughts to the breeding and training of long-distance racers. It is in this study that I find most interest, and so perhaps my decision was arrived at with some bias. Nevertheless, I know from experience, that the aim of most fanciers is to win long-distance races, and their dream is to capture one of the classics. You may win 50 first prizes in races under 200 miles, but within 12 months it has all been forgotten. On the other hand if you achieve something in a race around 500 to 600 miles, the mark is indelible. You remember for ever how the bird was bred, how it was trained, in what condition it was basketed, and above all the glorious fulfilment experienced when it arrived home.

It is necessary for anyone taking up our sport, to first condition his mind to the ups and downs that will certainly be encountered. I have noticed through the years that often a particularly successful season is followed by a rather lean period. This is the nature of things, and one must never become disheartened. With the same birds, the same loft, and the same management, two seasons can be so very different. When one's luck is out, one must carry on with the same unswerving determination, still giving attention to the smallest detail, then once more success will come your way. So much for the man, now for the pigeons! There are many ways of setting up a loft and acquiring birds, but most essential is to make certain that the stock you acquire is from birds that have achieved success from the long-distances. I would recommend either children or grandchildren of actual long-distance winners, not birds that are called 'pure' this or 'pure' that. The founders of these strains may have been dead for a quarter of a century or more. Providing your birds are not more than two generations removed from consistent long-distance winners, you can rest assured that at least a proportion of their progeny will be capable of the same successes. In short, my advice is to buy from reliable long-distance winning lofts.

The science of genetics is a study which I am for ever attempting to conquer. There are many very informative books on the subject, and my advice is to get hold of as many as you have time to read, and glean as much useful information as possible. All knowledge is useful, but I find it very difficult to harness my limited knowledge of this subject to any advantage. I have seldom raced an inbred bird with any success. I have found that they are the easiest to lose and the most difficult to win with. It does not matter how well their parents have performed, inbred pigeons are the most unlikely to succeed. I have found exceptions to this, but it is the exception that proves the rule. My most successful breeding method has been to inbreed to a champion and then outcross the result with a bird of proven stock. Here are details of an example of this practice. This season I was 1st Section NRCC Lerwick with a chequer cock 271. He has been a great racer since a youngster, in fact he is one of the best pigeons I have known. His dam is 2141, a stock bird that has bred many winners. In 1961, when 271 was a yearling, I paired him back to his mother (2141), and they bred a rather frail looking cock No 91. This pigeon, although nothing to look at, has the ability to reproduce birds of its father's calibre. In 1962, I paired 91 to an outcross hen of proven stock and they bred a blue chequer cock 2666 which has proved himself a champion. This year alone he has won two 1st prizes, one 2nd and a 3rd, including 1st Club, 1st Federation and 1st Combine Thurso. Last year, I paired 91 to another hen and they bred chequer cock 2875, who has scored twice this season. This year I paired 91 to yet another hen and they have produced two most promising young cocks. There may not be any scientific evidence to support this practice, but I would recommend it, as it has given good results both for me and other fanciers. As far back as 1925, I well remember my father bred his San Sebastian winner by using the same cross. I realise that most champions are the result of pairing two unrelated birds, but it seems obvious to me that if one can breed winners as a result of some inbreeding, then one's future as a consistent performer is more secure.

Last season I departed from some of my usual practices with satisfactory results. In fact, it was one of my best seasons. I am not saying that I have discovered some revolutionary system that will automatically bring success, but I do think that what I have to say is a system simple to operate and the results have been encouraging. No doubt, it is practised by many fanciers, but for those who have never tried semi-Widowhood, here are a few details of what I did. All my birds were paired on February 14 and one round was reared from each pair. Prior to pairing, the hens were given plenty of exercise; this I think is a good practice, as often during the winter months, hens are overfed and become rather fat. When the youngsters were ready to pick up, I took the hen away and left the cock to feed them alone for a few days. This, in some cases, prevented the hen from laying her second pair of eggs. As soon as the youngsters were weaned and the cock birds nicely settled in their Widowhood quarters, I started to train, by tossing at about 10 miles. It was very noticeable at this stage how their general condition improved. Within ten days of separating, they were ready for the first race, although the programme was not due to commence for another fortnight. I let them out for exercise three times each day, but did not force them to fly, and I gave them as many short tosses as was possible. I am a big believer in short training flights. When basketing night for the first race came round, I caught them in the normal way, and did not allow them to see their hens. On the Saturday, I let the hens enter the nest boxes and there await the arrival of their mates. After they had been together for about an hour I took the hen away until the next Saturday. The hens I raced in the same way when their mates were rested and it was then the turn of the cocks to wait patiently.

This method I practised until after our Northallerton race when my Lerwick and Fraserburgh candidates were re-paired. After the next race which was from Berwick, I re-paired the remainder of my birds, most of which were intended for Thurso.

On looking back at the race results I noticed that these un-paired cocks had taken prizes every week including Berwick, where one of them won 1st WHC Combine, over 2,000 birds competing. The hens on the other hand had only taken a few minor positions. From now on I was racing the Natural System and with equal success. I won 1st Section NRCC Lerwick, 1st St Albans Club, 3rd Harpenden Club Fraserburgh, 34th Combine and 1st, 2nd, 3rd and 4th Club, 1st Federation and 1st WHC Combine Thurso. These positions were won by cocks that had been re-paired. It is interesting to note that although my hens were competing in the same races the only success I had with them was from Perth in an Open race flown mid-July.

From my first season's experience of flying this rather casual method of semi-Widowhood I have arrived at certain conclusions. It is a simple practice, where cocks will race well each week, they are always fit. Cocks are keener as a general rule than when raced Naturally and are especially fit after re-pairing. The 'unpaired' period delays the moult, resulting in a good wing for the long races. The problem of hens pairing together is ever present. They often seem dull and are seldom at their best. They do not reach peak condition until sitting on their second pair of eggs after remating; this is rather late in the season.

Where it can be arranged a good idea would be to re-pair the best racing hens to stock birds, say about six weeks before the race for which they are intended. By this method these hens would reach peak condition at the right moment, and would, I feel, give a good account of themselves.

No matter how well bred a pigeon may be, it cannot be expected to win races unless it is healthy. In recent years respiratory troubles seem to break out where least expected and many of these attacks, I am sure, are owing to overcrowding in a stuffy atmosphere.

It was to combat this problem menace that about three years ago I built an aviary big enough to house all my birds. The main idea was to be able, for a period of four to six weeks to completely empty my main loft. This new construction runs the whole length of the loft. The frame-work is two-inch angle iron bolted together and covered with half-inch wire netting. The floor is also half-inch wire netting and is constructed three feet above the ground to enable the droppings to fall through and thus save a lot of work. It is built so that the back of the loft forms the back of the aviary and by doing so gives protection to the birds on this side. There are two perches running the whole length situated 18 inches and two feet from the back respectively. Two feet above the perches, and forming part of the top, are sheets of corrugated iron fixed long ways, these protect the birds from above when it is raining. The aviary is divided by a wire-netting partition and has two doors through the back of the loft.

To all intents and purposes my birds are living outside during these four to six weeks, apart from the overhead protection they get when sitting on the perches. Even during the rest of the year they are allowed to sit in the aviary if they so choose, but for a month or so during the winter they are made to live there while the loft is thoroughly cleaned and sprayed with a disinfectant.

I have not seen a sign of cold in either my old or young birds since the aviary was built. Plenty of fresh air is the finest preventive of cold and respiratory infections. Pigeons can stand cold weather but soon succumb to a warm stuffy atmosphere. It may surprise some fanciers to learn that since I adopted this practice I have been far more successful in our local club shows. I realise that limited space and other factors prevent many fanciers from erecting such a construction, but for those who have the room I would advise giving the idea serious consideration.

Finally I would like to ask the readers' indulgence for only touching on the fringe of our fascinating sport, but I have tried to expound some of my theories relevant to the breeding and training of long-distance racers.

WALK BEFORE
YOU RUN
by L A WENT

1st London North Road Combine Thurso 1965
1st London North Road Combine Thurso 1963

Having been asked by Squills to write an article for this famous annual relating to our methods of training winners, I do so with the hope that it will make interesting reading and that someone, somewhere will have gained a little knowledge.

Our association with pigeons began, like so many other fanciers within the London area, by purchasing a pair of youngsters from Petticoat Lane. This was the sowing of the seed and our interest began to grow. Our first loft was erected from war-damaged doors and I often chuckle to myself when I look at old photographs of it. It was while I was away on National Service in Egypt that my brother Lol joined the Enfield Wash Club and began YB flying in 1949. Success was hard to come by in those early days and on my return to civvy street the 1st prize had still eluded him.

It was in 1952 that the partnership of Went Bros began. Also in this year a new club was being formed in our area, this being the N London 5-Bird Club, and we were one of the founder members. With the purchase of new stock success in club flying began to follow. The most valuable in terms of breeding was a blue cock 10259 obtained from J Farrenden of Enfield and it is this cock's progeny that is well represented in our loft today. Very little Combine racing was attempted in those early days and it was in 1955 before the first open race success came our way, this was to trigger off a succession of open race wins. With a few years of experience behind us and having more knowledge of our birds Combine racing began to interest us. It was during 1955-60 period that our name began to appear on the Combine results with some regularity.

In 1959 a bird was purchased from Hollis & son of Tottenham, blue chequer 1212, and in a very short while he made his presence in the loft known by producing 'Steptoe' and 'Stubtoe' when paired to a blue hen 23512, this hen being the nestmate to our champion 'Jimmy's One'.

In 1963, after a long illness, my brother Lol died at the early age of 35 years. I shall always remember the many hours of enjoyment we shared together in this sport of racing pigeons. We are now beginning to be regular intruders in what you might describe as the 'top twenty' where LNR Combine racing is concerned. Some of our recent performances, 1962 – 2nd

'Steptoe' 1st London Championship Club, 1st London North Road
Combine, Thurso.

Open Northallerton, 22nd Fraserburgh; 1963 – 2nd Open Thurso, 9th Open
Northallerton, 14th Open Thurso; 1964 – 14th Open Fraserburgh, 22nd Open
Fraserburgh, 3rd Open Fraserburgh, 16th Open Berwick; 1965 – 1st Open
Thurso, 11th Open Northallerton.

I cannot conclude this section without paying tribute to my mother, who for
the past five years has attended to all the chores. Always an early riser the loft
is cleaned every morning at 7am, this 365 days a year. Exercising, feeding,
bath, all come under her wing. Not to forget my younger brother Doug who
trains our race team and does a lot more in the way of taking the birds to the
club, etc.

Method

Many are the methods of past masters of pigeon racing and to each have
brought success. The first thing to take into account, is the amount of time

one is able to spend with his birds, each and every fancier follows a different occupation or profession. What will suit one person will be impossible for another to operate. Therefore a method has to be one which can be operated to the best of his ability, and will be most beneficial to his birds. We are fortunate that our birds have an open loft during racing season, the open sky is their's to fly at will.

The season begins in earnest come March when the breeding season begins, in my opinion the most important of the year. It is from my carefully selected pairings, that we hope to breed worthwhile pigeons, that will replace lost and ageing favourites. For a fancier to stay on top, in this world of pigeon flying, this is a must. One will soon topple from his perch if he fails to breed at least one or two above average pigeons each and every year. I spend many hours during the long winter evenings matching my future pairings, these are often chopped and changed about, come the actual time for mating. It is in this field that our past records come in for scrutiny, these will give me a guide to what I think in my opinion will be our best pairs.

Our nest boxes are used as perches, having 18 in number this simplifies the task of mating. This means that one of the birds is already accustomed to the nest box. I have tried retaining the cocks in the breeding section during the winter months, but this was not highly successful and I reverted to keeping this section for the hens. At the time of mating the cocks soon follow the hens into their boxes and in a very short while all the inmates are settled. I must hasten to add that our nest boxes have no fronts but during the act of pairing up, fronts are clipped into position, the birds remaining therein allowing only one pair out at a time. The pairings are carefully watched, any that do not settle down quickly have an empty paint tin placed in the box in order that the hen will have means of retreat. Often is the case that when one visits the birds the next day they are quite happy and the tin can then be removed.

With the birds settled down I like to see the hens laying in about eight days, although I do not condemn the slow laying bird. One of our greatest racers 'Scruffy Girl' is always a slow layer. Once all the birds are sitting on their eggs exercising at regular times of the day begins starting with a forced fly of 20 minutes twice a day this building up to 30 minutes by the start of racing. By which time we have big squabs in the nest bowls with the cocks just beginning to call their hens again and it is from the cock birds that we expect to start the new season on a winning vein. Once the youngsters have been removed, this is generally at the 28-day stage, no more rearing is performed by the birds intended for racing, although I often send birds away to a race with a small chick which has been removed from a stock pair for this purpose. When the 150-mile stage is reached the exercising is increased to 40 minutes and by which time all the racers have had at least one race.

Many are the ways of winning these short events where extra keenness is required. To the observant inside the loft little things that are noticed can be exploited to the full. Our nest partitions are made of glass, this being shaded and nest bowls placed so that if one removes the shading the birds can be seen sitting side by side with just the clear glass separating them. Giving the keen young hen the feeling that another has encroached on her nesting ground, the cock birds can be treated likewise. Of course this sort of thing is only performed from time to time and would be of little use if repeated every week.

With the arrival of the 300 mile races the exercising is increased yet again this being 60 minutes twice daily. As can be seen this exercising has been gradually building up over a period of four months. I think this is essential, one cannot condition one's birds in a matter of weeks.

The distance events having now arrived our potential team is carefully scrutinised and nothing left to chance. All the months of planning and conditioning should now show itself and our team of birds should be bubbling with health. I never send any birds which I have any doubts about, this is not only a waste of £ s d but is most cruel. Another thing to be considered when selecting our team is one of experience plus the ability to stay the course. Here once again it is the keeping of past records that will give me this information.

With the completion of the OB programme the racing team are then removed from the breeding section, and in a very short while they fall into moult. At this period a little linseed is added to their daily diet. The less the birds are handled at this time the better and I like to see that peace and quiet are the order of the day while the moult is in progress. Our birds have always completed this well before the cold autumn days set in.

Young Birds

The YB come under our eyes more when their season is about to start. Previous to this they have had as many training tosses as we can give them. Our training methods are outlined later. The teaching of drinking from the basket is also most important. We achieve this by placing our youngsters in the basket and making them spend the night in it. This is repeated several times and in a very short time all their fears of the basket vanish, and they soon take to the water trough. I like to see our YB flying consistently and all have to fly the 200 mile race. The stopping of youngsters at the 100 mile stage to my mind is like putting on the brakes when a pigeon's learning is about to begin. Birds that we have stopped in the past are always the ones not to return when confronted with a difficult race as yearlings. Not that I advocate sending all our youngsters to every race. I like to split them into two teams racing them alternate weeks. YB that have flown the 300 mile YB open events for us have all turned out to be first class racers as OB.

On the completion of the YB season I separate the sexes. When the weather is suitable they are exercised in turn first one sex and then the other. This keeps our birds in fine fettle during the winter months and the question of fat is a thing we do not fear. Our feed during these long winter months remains the same and no cheap maintenance mixture is introduced.

Training

Very little of this is carried out for old birds, just three or four tosses from five to ten miles prior to birds' first race. This mainly for the benefit of the yearlings, birds two years of age and upwards receive only one toss. I'm of the opinion that you can only teach them the right way home once and this should have been done when the birds were young. Once taught I cannot see what good continually training one's birds over the same stretch of land can do, it's like sending a child back to school every year to learn its 2 x 2 table. Young birds have all the training that we can give them, starting at three miles working up to 20 miles before their first race and are given tosses through the season.

Feeding
Our feeding consists of a good mixture of maples, tic beans and tares. We prefer hopper to hand feeding, this way we know all the birds have their fill plus the fact that the corn stays clean and cannot be soiled. Coarse oatmeal that is steeped in Cod liver oil and mixed with bird seed is given to the birds each morning. They seem to relish this and none can resist coming off the nest to have their share.

Fresh water is given twice daily, not forgetting to wash and rinse the container each time. The water container if not cleaned will in a very short time become coated with slime and no one needs telling what germs and bacteria this can harbour. A tonic designed for toning up the liver is added to the drinking water once a week.

Green food in the way of spinach, cabbage and lettuce is given to the birds frequently, this being washed and a little salt sprinkled on it. Grit is always before the birds; this is given daily, any grit remaining from the day previous is thrown away ensuring that no stale or sour grit is consumed. A bath is put out once a week and when all the birds have performed this act it is then cleaned and turned upside down ready for the next. One thing they enjoy more than a bath is flying out in light rainfall. On these occasions their antics always amuse me, with wings outstretched they appear to be beckoning the rain on to themselves.

A Day to Remember
What of July 3rd, the day of the LNRC classic from Thurso 500 miles? To win this is the ambition of most fanciers within the London area competing on the North route. News came through that the birds were away at 8.30am, the weather at the home end was overcast with a chilly NE wind, not a pleasant day to be awaiting for arrivals. We were also competing in an open race from Selby 150 miles, this we hoped would be a guide when to expect the Thurso birds. Liberated at 10.10am the Selby birds covered the 150 miles in 3 hours and the Thurso birds we could assume would take roughly 10 hours. Our team of six birds were all experienced distance pigeons, all of them being well placed in previous Combine races. I must confess I was hoping for an all day fly, but the wind direction had dashed all hopes of this. After clocking our open race birds from Selby, we remained in the garden talking as always pigeon talk, every now and again scanning the sky for pigeons, but none was to be seen.

We were aroused from our sitting position, when a small batch of birds were approaching us direct from the north, the time 5.40pm, these passed the loft and we followed them out of sight, this batch of birds were quickly followed by another. The question we asked ourselves was, could these be the Thurso birds? At 5.47pm the sight of one single pigeon could be seen, coming direct from the north with winds folded and falling from the sky like a stone. Yes, it was 'Steptoe' and in a very short while he was clocked. A big improvement on our expected 10 hours, it was then naturally assumed the birds we had seen previously to his arrival, were birds of the LNR Combine and winning this event was never in our minds. 6.10pm saw the arrival of our 2nd bird, duly followed by our 3rd at 6.57pm, our 4th and 5th birds made home at 8pm, with the 6th entry having the night out, but was found on the loft early next morning.

The clock station revealed 'Steptoe' being the earliest, the next pigeon being some 7 minutes behind, this belonging to V & G Thorpe. The facts obtained from the clock station were that many members had failed to clock an arrival on the day and many only recorded one. Although fast in velocity, the race had proved difficult for the birds to negotiate and many must have been carried far and wide by the NE wind. It was Sunday when with many times to hand, that we realised that we had an outstanding pigeon and a chance of winning. This was confirmed a day or so later and what joy this brought us, more so because it was 'Steptoe' who had atoned for his narrow defeat in 1963, when he was 2nd LNRC Thurso, only 12 birds regaining their lofts day of toss out of nearly 3,900 birds. For this, he was awarded an Osman Memorial Trophy and RNHU gold medal for meritorious performance. Never before in the history of the LNRC has a pigeon come so close to winning this classic race twice and this must place him in the hall of fame alongside other great pigeons of our time.

I do not want to detail his many fine performances which might be regarded as plugging. But I would like to say that this great pigeon can be pitched into battle at any distance and leave the flock trailing behind him, whether it be 50 miles or 500 miles. Why should this be? In my opinion not that he possesses supernatural flying powers, but because he is an intelligent bird with an above-average brain. To the many who visited the Pigeon Olympiad at Alexandra Palace, he could be seen repreating Great Britain in the Long Distance sporting class section.

Novice and Newcomer
What can be said for the person about to enter this most fascinating sport of ours? First and foremost, he must have a love for our little feathered friends who possess this amazing ability of being able to return to their lofts from enormous distances. Patience is something else he must be blessed with, for the road to being a successful fancier is often a long and trying one with many disappointments. Finance is another thing to be considered, if found to be beyond one's means then one should try to find a partner who will share the expenses. With these three things in mind, the task of building or pur-

chasing a loft must be executed. Where possible I would advocate that the loft should face south thus enjoying to the full the sun which we know is something that we in this country are not always accustomed to.

It is often said, that one should start off right by purchasing only the best, when it comes to the point of stocking the loft. I do not recommend this, even the best of birds in the hands of the inexperienced are of little value to him, and I feel it is at this stage that the newcomer to the sport is most disappointed. After purchasing stock at considerable expense which he thinks will lead him to the hall of fame, he finds that he is still on the ground floor. I feel that the best time for the newcomer to purchase good pigeons is after he has been in the sport for a couple of years, by which time his knowledge of them will have increased tenfold, and he will be in a far better position to make his choice of stock. This annual will supply the reader with many of the country's most famous fliers, who advertise their performances, etc, for one to study.

The club room is another of the sport's places of learning, of all the good fanciers I know, all are good club men. This should be the aim of the beginner to take an interest in the welfare of his club; I'm afraid this is a thing far too many fanciers of today have little or no interest.

What of the fads and theories each and every one which, if followed, should take one to the top of the ladder, so we are told, yet in reality these lead to nowhere. I'm often asked my opinion on the most controversial of these, eyesign. I do not study it, therefore I'm not in a position to pass judgment. Naturally I have my views and facts come down against it. All fads and theories have to be put to the test and that means placing one's pigeons in the basket. This then is the time to make one's studies, yes, study the basket, this will tell you everything. From these one is left with commonsense facts, a thing no fad or theory can tell you.

The keeping of records is a must and nothing should be left to memory. From these one can follow the racing and breeding careers of all one's birds. An essential thing in the make-up of a successful fancier. These record books contain a power of knowledge and should always be studied. By looking at them you come to know your birds as individuals. Their favourite condition, and what is their best flying distance, can be checked in minutes.

WALK BEFORE YOU RUN is a very popular saying and how true this is when it comes to pigeon flying. Success cannot be built up in a week, month, year, but of several years' intensive study and hard work. Always look forward, when you have achieved one ambition strive to conquer the next.

In conclusion to this article I would like to convey my thanks to Squills for giving me the opportunity of putting pen to paper.

PIGEON RACING IS A SCIENCE

by FRED PRICE of Burscough

Since my previous article in SQUILLS I have had many letters and phone calls congratulating me. I have also had one or two fanciers complaining that my article will put a lot of fanciers off joining the sport and also stop the sale of stock from well-known breeders. This is a lot of 'poppycock'. My article will create greater interest and will help a lot of fanciers to look at pigeon racing as a scientific hobby. I have also had many phone calls asking advice on how to race birds and on how to treat many kinds of sickness and I hope my advice has been of some help to them.

BELGIUM
I have travelled in many countries and made many friends, but my staunchest friends were fanciers of world renown in Belgium. I was very friendly with Mons Stassart, probably the best fancier Belgium has produced, Mons Van Damme, Dr Bricoux and Mons Leroy and many more well-known fanciers. On the last occasion I visited Mons Stassart's loft he had specially caught all his birds and put them in show baskets for me to handle; he also had a short-hand writer making notes on what I had to say on each bird.

After I had finished I asked Mons Stassart if I could take a pair home with me. I told him I would pay him £20 for the pair (a lot of money in those days), although I knew they were worth more than this in the world market. He shook hands with me and told me to select my own pair and put them in a large pen on a table in the centre of the room while he went for a basket to put the two birds in. Fixed in the centre of the cage was a large pair of opera glasses. He had had the cage made specially for examining matings or observing any specific points of his many champions. I selected a yearling blue cock and a two-year-old black pied hen and put them in the pen. When Mons Stassart came back in the room he took one quick glance at the pen and put his hands in the air, exclaiming: "No, no, no, my two coming champions. No, no, no, my two coming champions". This went on for five minutes or more.

Being an old man I became concerned that he might get too excited and something happen to him. I went towards him and told him I would not take the pair of birds to England if he did not wish me to. Then happened one of

my most embarrassing moments. He leant over me and kissed me on both
sides of my cheeks and held on to me for a time. (He had a long thick beard
and nearly smothered me.) Then he shook hands with me. He said both birds
were big winners and his two coming champions. One had won five 1sts and
the other had won three 1sts. He kept on repeating I could have taken them
back with me; he could not have stopped me but I was a perfect English gentleman.

He then asked me to accept six squeakers next season with his compli-
ments and asked me to send on six rings, but unfortunately the war inter-
vened and I did not get them. Mons Stassart had one strain that he never parted with and evidently I had picked two of them. On another occasion he sent
me to buy some birds from a miner just outside Brussels who specialised in
the long-distance races only. He had been very successful and although I tried
very hard for five or six hours I had to come away empty handed. He wor-
shipped his birds and never sold any. They were a perfect type, a little over
medium, and very strong. Incidentally, this was the first loft I ever saw practising the deep litter system. He only cleaned his birds out twice a year but
the birds carried a tremendous amount of bloom and were in wonderful condition.

EARS NOT EYES

On another occasion I was with my friend, Mons Wery, and he asked me to
go to witness a demonstration. There was a big bet that two champion racers
would not home just a few miles after this man had had them for one or two
minutes. All he did was to smear a type of wax over their ears. They were put
on the top of the baskets but the birds never made any attempt to leave. This
gave me an idea which I have been studying ever since: the faculty of homing. I have proved beyond doubt that birds home with the aid of their ears.
Eyes do not enter into it at all; they are definitely only for avoiding objects
in the air and for landing, etc.

I carried out many tests with the late Dr Jennings, of Bracknell, and we
came to a definite conclusion with the help of eye specialists that pigeons can
only see a very short distance. I do not want to put the 'eye specialists' off
but you can get eyesign in many types of birds even though they do not fly.
I have had birds home in snow storms and heavy mists, visibility down to a
few yards; eyes could not have helped them to home. I am confident on this
point. Furthermore, when I moved my birds to my new loft a few years ago
I decided to race four yearling cocks that had raced well as youngsters. These
four birds came to win on most occasions but they came on a beam and tried
to drop where my old loft had been. They flew down to the level of the loft
and then struck up in fright. My new loft was only 30 yards to the north-east
and they could easily see it. They did not break off this trait until they were
two years old. This definitely confirmed my view that pigeons homed with
the aid of their ears.

Have we to start examining ears instead of eyes? No, but this brings me to
another point. It behoves every one of us to see that our birds are in perfect

health. Most complaints affect eyes and throat and this affects their ears as they are all connected, causing inflammation and possibly the loss of the birds. If in doubt leave birds out. They are better at home until they are fully recovered.

ADVICE TO 'NOVICES' AND NEW FANCIERS

Buy from a winning loft, not a 'has-been', a loft where the birds have to earn a perch, only sound pigeons can win and breed winners. It is my honest opinion a great many pigeons that are sent to races are not sound and it is therefore imperative that you acquire sound stock. This is more essential than pedigree. I personally like latebreds from birds that have scored that season from over the Channel. To prove my point I paired a latebed hen with five nest flights to my champion stock bird 'Bursco Jerk'. Quite a few fanciers thought I had gone mad, but the first youngster from this mating topped the Ormskirk West Amalgamation from Craven Arms and topped the Amalgamation from Rennes, first in two clubs, second Championship Club, 10th North West Combine, 3,832 birds competing. (He has also bred a winner.) So my judgement was not far out.

Pair latebreds up about 10th-15th March to your best racers just as the great Belgian fanciers have done for generations. Mons Stassart, Dr Bricoux, Mons Duray and Mons Sion exchanged each year six latebreds. They would not have continued doing this for many years unless it had been beneficial to all concerned.

EARLY MORNING TRAINING

Another point I must mention which is all against the present theories, that is regarding having your birds out for training early in the morning for exercise. Last season it was not practicable for me through building operations close to my loft, cats, and my loft man, Joe Halsall, having broken his leg the previous winter. He did not arrive at my loft until approximately 9.30am to turn out the birds. A lot of fanciers predicted a bad season for me but my birds flew just as well as ever. I had a wonderful season. My young birds were also treated in the same manner. It proves again that it is soundness and condition that win races, so it is not necessary any more to get up early in the morning. At least this is my belief and I have decided to give it another trial next season.

MEMORIES

I had a letter some time ago from a friend of mine, Alec McDougall, reminding me of a great performance of one of my birds before the war. She was liberated at Bordeaux, a strong east wind developed and only a handful of birds returned. My 'National Hen' was found in the loft on a Wednesday afternoon with a note on her leg saying that she was caught on board a boat in the Bay of Biscay on the Saturday night and the weather was too bad to liberate her on Sunday and Monday, but they liberated her on Tuesday off the coast of Portugal. I later thanked the gentleman for taking care of this bird and he wrote me that he had decided to take up the sport again as he was retiring and settling down in Newcastle.

In a weak moment I made him a present of the hen thinking it would encourage other people to help birds in similar circumstances. She must have covered the 900 miles on her own, over 600 miles of water. She was a glorious big hen, very tame. She would jump over my hand repeatedly and that is why I gave her the name of 'National Hen'. She would also drop on my shoulder in the garden and follow me everywhere. She was indeed the gamest and tamest hen I ever owned.

SPECIALS

As fanciers are no doubt aware, the sport of pigeon racing in Belgium is their national sport and all the large firms support pigeon racing in the same way as the big firms in England support horse racing. Consequently, they have motorcars, bedroom suites, carpets, china and practically every type of commodity presented to them to fly for as specials. These are usually displayed in some of the big shops in the centre of the town for everyone to see. If we can only get a quarter of the support from business concerns in this country it would put our sport on a very high plane.

WARNING TO ALL FANCIERS

Three years ago I had a very bad experience. I had liberated my birds when I heard the approach of a helicopter going over. It had been spraying crops in the vicinity. I tried to call all my birds in but seven of them were missing. Later that evening three returned and although I tried to doctor them, they were dead the following morning. The other four returned during the following morning and died that day. It is essential, therefore, that during any spraying in your vicinity at any time to keep your birds safety in the loft.

I also had an unusual experience in recent years. In July and August I had several birds that died suddenly with no apparent cause, but I knew they had been poisoned. I was watching my birds one afternoon when I noticed a silver hen. She had been my first bird from Manx and had also bred winners. I saw her picking at one patch of my lawn and to my surprise the following morning she was found dead on the nest. Upon examination I found a clover leaf in her mouth. I examined the patch where she had been on the lawn and found a small clover plant with a yellow flower (buttercup colour). On making further inquiries I found that this plant is a deadly poison. All fanciers who have lawns must keep a careful look out for this plant, otherwise they will have birds die suddenly and they will not know what has hit them.

MINISTER OF SPORT

Owing to automation it is anticipated that by 1970 the average working hours will be around 35 hours per week. This is the reason why a Minister of Sport has been appointed in this country. All other major sports may be supported, I understand, from the Government fund, but I have not heard whether our Royal National Homing Union or any other body of our sport have applied for enrolment and consideration as to receiving some of the grants for our sport. This is a point that must be taken up immediately if we are going to stand any chance of getting our sport put forward on the agenda for consideration at a later date.

WHAT OF THE FUTURE?
I predict lofts made of PVC (plastic) sheeting. They will never need painting and always be waterproof. Baskets made of plastic or fibreglass, cheap enough to fly out and leave after liberation. This will prevent the heavy costs of returning baskets. Clocks – I predict that in the near future clocks will be made of fibreglass or plastic, battery driven and will run to a few seconds a week. I would also like to see a Union rule made whereby all clocks over two minutes fast in 24 hours would be taken as a correct clock and all clocks over two minutes slow would have double the slow time added and in a short time I would like this brought down to one minute.

REAPING THE REWARDS

by EMRYS JONES, Newport

1st, 3rd, 14th Thurso Welsh Grand National

1967

When I was asked to write an article for Squills on "How I won the Welsh Grand National", I was very thrilled and gratified at the honour.

I don't think my formula or system differs greatly from most fanciers'. I can therefore but relate some of my experiences and tell of the wonderful companionship and friendliness of my club mates and the fanciers that I have met since I started flying young birds in 1965.

I owe my success to two old fanciers that have passed on – and I am only reaping the good work which they sowed. They both had lofts of a family of pigeons, renowned for their success in long distance flying. They are the late Ernie Durrant of Newport, and Jack Vaughan of Treharris.

In the summer of 1964 I was on holiday at my parents home in North Wales and my son was given two unrung youngsters and we brought them home with us and made a wall cote for them. They were fine strong pigeons of Swing Clear breeding and a fancier, the local policeman Ivor Jones, who came second to me in the National, greatly admired them and asked whether I was going to fly.

I knew nothing of pigeon flying but he got me interested, showed me his loft and gave me a pile of copies of *The Racing Pigeon* to read and the bug caught me. One day the two youngsters failed to return. They used to go fielding down at the Saltings for hours and I am certain they were shot.

I decided then to buy a loft and bought some latebreds from F Baker, Cardiff, but they turned out to be all cocks so in the spring I had to find hens for them, and bought two fine yearling hens at Durrant's clearance sale.

Soon afterwards, I heard that Jack Vaughan of Treharris was very ill in hospital and was not expected to recover, and that his pigeons were for sale. I called there and was lucky to have the first pick of his stock birds and I also bought eight squeakers which had just been parted. They were fine looking youngsters and have raced well for me.

The best of the stock birds was a blue pied cock Boneddwr NU59MDC242 who had sired many winners for Mr Vaughan and he is the sire of my National winner, and I kept him paired to the same hen that Jack had him paired to. He is a great producer and a yearling cock from him has flown every race on the card this year bar Lerwick and he has ben in the clock every

'Llygad Mawr', 1st Club, 1st Fed, 1st Open Welsh Grand National
Thurso 1967, 1,376 birds, velocity 989.

time, my first or second pigeon and I clocked him 7.15 in the morning from
Thurso.

Five of his youngsters were clocked in the Welsh Grand National from
Berwick with two clocking on the night and four of these are from different
hens. This cock was bought by the late Mr Vaughan as a youngster from the
late Mr Durrant of Newport and was kept prisoner for five years at Treharris.
When I tried to break him in he went back to his old loft in Newport. He now
enjoys an open loft during the breeding season and carries cut grass all day
long to his nest, and I think he is happier and fitter than he has ever been.

I did not race the National winner as a youngster as she was from a late
nest. When in the nest she was a fine looking youngster, jet black, wonderfully feathered, contented and well fed. Her sire is a fine breeder as well as a
producer, he loves his youngsters and always feeds two at a time. On the
floor he is a natural feeder and will feed as many as a dozen youngsters until
they are all satisfied, and never tires. To breed excellent top-class flying

pigeons, their period in the nest must be the most important and I tend to overfeed if that is possible.

In 1966 'Llygad Mawr' brought up a youngster on her own, as her cock failed to return from a training toss. I was astounded at her determination and care of this youngster. She would not leave the nest box except for water. Here I must say that as a novice I made many more mistakes than my birds. They have done all I have asked of them. Not knowing I put this hen to race after this strain in 1966 and she raced well always trying against all odds and really racing home.

Again in 1967 I made another mistake with her. I thought she was sitting but had become 'eggy' and she flew the hard race from Berwick to lay her second egg on returning. She held this egg all the way and Mr Jenkins who clocks for me had to massage her legs before she could stand up. She had crash-landed on the loft platform, her legs being cramped beneath her.

I had her sitting again for Elgin but again she fooled me and laid her second egg the day after, and I managed to get her down on this egg although the cock was away for three days late from Elgin. She recovered miraculously from Elgin showing extreme fitness and she was soon in excellent condition again.

I was at home in time to clock her from Thurso and what a thrill it was. It was a red letter day for me as I had just taken delivery of a new car and I had one eye on the car and one on the sky. It was a joy to see 'Llygad Mawr' coming flying high. She raced all the way bringing the other three birds over my loft and as she closed her wings she rolled two or three times like an aeroplane doing a victory roll and then dropped like a stone with her wings closed.

Three of my cocks hit a wire in a late training toss a week before Thurso. One never returned, another had his crop split open and I had to put 15 stitches in him. The third mated to the mealy hen was badly lacerated in the chest and wing. I managed to doctor him and he stuck to his eggs thus not upsetting his hen 'Gwenno'.

This was my second bird from Thurso WHU65H965, who was 5th Sect Welsh Grand National Elgin 1996. She also came well to be 3rd Open Welsh Grand National Thurso to win me the 2-Bird Average Cup. My club mates teased me no end, asking me for some Purple Hearts as I am a chemist and many of them think that I have some wonderful potion. There is far too much rubbish given to pigeons in the form of pills etc, many more successes would be achieved if fanciers concentrated on good wholesome grain and clean water.

I am only a novice and I can give no advice to the old fanciers. To any beginner I say – get your birds from established lofts that are winning long distance races. Be gentle and get to know your birds, caress and be kind to them. My birds are tame and I stroke each one on the perch, and they will eat out of my hand.

Why do the same lofts always score? They have the right birds for the job – they treat them well and condition them – in the words of Reg Bale (of Bale

'Boneddwr', Black Pied Cock.
Sire of 'Llygad Mawr'.

'Glaswen', Blue Pied Hen, 14th
Sect Welsh Grand National
Thurso 1967.

'Gwenno', Mealy Hen, 4th
Club, 3rd Welsh Grand
National Thurso 1967, winning
2-B Av with 'Llygad Mawr'.

'Coch-y-Bonddu', Strawberry
Mealy Cock. Sire of 'Gwenno'
and 'Glaswen'.

Bros and Griffiths, Newport), "Em – we loves them" and that about tops it all.

The fellowship of the Pill North Road fanciers is wonderful – it's a fine club in which I won Combined Av in 1967. I was told when I joined, that the club was the keenest in the country with names like Channing, Atwell, Bale Bros and Griffiths, Davy, T Samuel, etc, nationally known names in flying. They live 'pigeons' in Pill and in fact it is not only a sport with them, but a way of life.

In my first year flying all these grand fanciers offered to breed me youngsters, it shows the good sportsmanship of them all. Another fancier made me nest boxes but did not charge a penny and yet another, Charlie Pitman, gave me all the knowledge of his three score years and ten of pigeon flying.

Bill Jenkins has been a great help without which I would not have been able to race this year. His keenness, support and enthusiasm has inspired me to always try a little harder, and it is indeed gratifying to have this year beaten the giants of Pill and won the Grand National Victory Cup, the Grand National 2-Bird Average Cup, and the Combined Average Cup of 1967.

To these good fanciers and the help and challenge they gave me, to the life work of breeding put into my stock birds, I owe my early success in pigeon racing.

BE BOLD
BE DEDICATED
by J ADAMS jnr, Redditch
1st NFC Pau 1968

Grandad, father, son and now, I hope, my two sons, racing pigeons all our lives. It should be every son's ambition to emulate or even better a successful, good-living father. My father taught me to benefit by his own mistakes in life in general; as far back as I can remember he encouraged me to help with his pigeons. I must have been the youngest loft manager in the country. He was not only a kind and considerate father, he was my schoolmaster. Don't ever smoke, son, he would keep reminding me. Everything you eat, drink or breathe goes into the blood stream, don't pollute the blood that feeds the body, so I've never smoked in my life. I am 54 years old and feel 24 years. Father would keep on saying there is no substance for, and one cannot measure, experience, so read all you can, you don't learn if you don't read. So the pigeon books were always on the table. I call my father the memory man, for he can quote fanciers, their pigeons, sales etc, better than anyone I know – try him; he never fails.

No King's Cup winner ever had better tutors than I, no son a kinder father, it is that unquenchable thirst for the love of home and the nucleus that creates it, that leads to a happy family life. This applies to a pigeon loft. Once you have bred the pigeon for the job, give it confidence in itself created by scientific feeding and training, make life interesting, give it something to come home for, not just a nest, give it the atmosphere that all birds love, then it will punish itself more than any other living thing on this earth of ours.

Success or failure depends on the nucleus. You are the nucleus, you hold the reins. Before you condemn your pigeons remember this, you cannot buy success by buying the best bred birds in the land, if you could myself, Jimmy Warren and others before us would never have won a King's Cup. Some men would pay a fortune just for that honour, so you will see it is not all pigeon, although breeders would have you think so.

I race to a small backyard loft, 12ft by 7ft by 7ft, the ground is 18 yards long, 10 yards wide. Willow Way Road running one side, houses the other three. People walking past stop and admire not only the flowers but the pigeons, and the pigeons love it. 'As is the garden, so is the gardener', and it is the same with pigeons, 'As is the loft, so is the fancier'. My neighbours are interested in my pigeons, but I have been asked to scrape a little quietly at 5 or before in the mornings! I have very often been asked how many times do I feed and scrape out in a day, because they cannot see any droppings on the floor. The answer is ten years in the Territorial Army and six years all through the war in the RAF taught me to do things instinctively, as they put it in those days.

I go for the scraper every time I enter the loft, at the same time a pinch of seed for every bird. I have five tins all different mixtures I mix myself, though I use Red Band and Vigour food, how many times I enter the loft I've tried to keep count many times, but lose count each time. The dearest and the best peas available are always before them except at night, I put them in a drinking fountain. I never feed beans or maize, strength does not lie in the size of the grain, the bigger the grain the longer it it takes to get into the blood stream, only about six beans will pass through the crop to the gizzard per hour.

The largest living thing eats the smallest sea fungus, the whale, the smallest carries the biggest load, the ant. I do not handicap my pigeons when they are flying, and I want them in a position to do that at any time of the day. Peas are round, pigeons love all food round, it is easier digested, they know better than we do, try them. Maize is the last of the mixture to enter the gizzard, dust is the devil, and so is maize. I've coined this phrase and I stick to it, 'Give beans to the horse and maize to the fowl'.

When you are buying food do not consider money, it must be the best. Fanciers will pay big sums of money for pigeons, yet they tell you, I don't feed on peas, they are too dear. I make the Adams' cake, roll it into pills and pop one in their crop while they are sitting on the nest. They get one immediately they return from a race, this stops the green gall from the gall bladder getting into the stomach and on the loft floor. I do not force my pigeons to fly round home, I don't have to, they go out one at a time, the first bird out is the last one the next time. I aim to keep them going until I have no birds left in the pen. The hens are never taken off the nest to fly round home, they are taken for a short toss.

I paired my Pau birds up May 11, in fact Adam Boy had Frome and Templecombe races before he was paired up, he then went to Templecombe, Weymouth, twice Poole and the Nantes National. He had 13 hours on the wing, and it was a full wing. His hen laid her first egg of the second nest the day before he went to Nantes, he dropped his flight a few days after Nantes. My aim now was to keep him on eggs up to Pau, 26 days, the flight he had dropped would then be three-quarters grown a full wing, flights are very important. To assist buoyancy over 600 miles I fooled the pair with a youngster just hatched out. Adam Boy was not allowed to feed, only coddle it for a few minutes.

I usually let them rear one youngster in the first nest, but this year I tried no rearing before Pau. After Nantes a week's complete rest, plenty of open loft, make them happy. Then begins three weeks of hell for me, gradually get them used to the early morning light, 5 o'clock each morning, a minute earlier each day, open loft after night exercise. In 1968 I had my three Pau cocks

'Adam Boy', Blue Chequer Cock, 1st NFC. Owned and raced by J
Adams jnr, Redditch.

out together until it looked as if they wanted to get down, then I let up half
my youngsters with them, then the other half, then the remainder of old birds.
I can keep them flying for one hour without any flagging. If a clear morning
I sometimes took the Pau hens about three miles before I had the cocks out,
I do not take my hens off the nest to flag them round home.

Then, at night, I set off about 8 o'clock with all the Pau birds, I keep going
until 8.45, then the hens up first, followed by the cocks, twice a week. No
basket work a week before Pau. The day of the Pau marking I was up at
4.15am, I never go straight into the loft and roust them out, I just change the
water, let them see you pottering about first. They are ready to go when I say:
"Come on then, wakey, wakey!" Adam Boy was away with the youngsters
for 25 minutes, the three Pau cocks came back on their own. Out went the old

birds into them for another 20 minutes.

When I got back from the marking I said to my wife: "I have never sent pigeons to a race fitter in my life." "Maybe so", she said, "but have you looked in the mirror lately, you're a physical wreck". Never fly Pau candidates round home in the heat of the day, don't take off what you aim to put on. Hard condition of the body and feather bloom is important, feathers must resist the atmosphere if they are not to handicap the buoyancy and propelling motions. Dry feathers are not conducive to smooth flying. No one could have believed that Adam Boy had just flown 621 miles, only by opening the wing, which showed the pressure on the flights (upturned). Credit to Jack Donovan, he must be the best convoyer in the country, he looked even fitter than when I sent him.

Make no mistake about it, you cannot compare over 600-mile pigeons with 500-mile ones, intelligent pigeons may win up to 500 miles but, by golly, you need a little timber to go with it for that extra 100 miles. Look at the National winners and their velocities this year. I see no pretty ones in the Pau National, take the first two big chequer cocks and have you ever seen two better. The 4th Open belongs to my club friend, Jack Smith, a beautiful big blue hen, but look at the balance of these birds on the wing and about the loft, they are like Cassius Clays.

I have that happy knack of picking a good pigeon, so I am delighted with the Squire Ashton Logan first cross into the Royal Lofts Jurions, NURP57WSE327, a big blue chequer white flighted cock of Logan blood. 327 was a good consistent racer for Squire Ashton, of Cranhill, Bidford-on-Avon, right up to Rochefort, 432 miles. He flew Rochefort, 450 miles, for me as a broken bird. In 1963 I paired him to my big blue chequer hen, NU62R19215, pure Jurion. 19215 was the only bird home in race time in Redditch in the 1962 St Malo YB National. She was 30th Sect C. A big strong hen, she won 1st in Redditch Premier FC longest race, Niort, in 1964. She flew Pau in 1965 and was again clocked in from Pau 1966. She is a granddaughter of my noted RP49BM5, who won many prizes for me. He won the hardest Weymouth ever into Redditch, 4½ hours to do 120 miles; 2nd Marennes, 166th Open San Sebastian National, velocity 249, in 1953. He was bred by NU46V260, blue chequer hen x NU44V239, red chequer cock, both these sent to me by my 1942 wartime friend, the late Ernest Steel, Royal Loft manager. 260 and 239 came from those wonderful pigeons in the Royal Lofts under Mr Jones, GV919, 1st Banff; GV534, blue chequer, Lerwick five times; GV936, red cock, 1st Lerwick and four times Lerwick; GV155, dark chequer, 1st Lerwick, 1st Thurso; GV382, 1st Banff twice. There are no purer Jurions or Delmottes in the country than the Royal Lofts.

327, paired to 19215, bred the sire of Adam's Boy, his dam, NU63N10492, light chequer, is undoubtedly the best hen I have ever raced, four 1st prizes, including the terrible young bird smash from Frome in 1963. She was 2nd Worcester Fed, 7,595 birds, velocity 880, she was 30th Sect C St Malo 1963, 6th Sect C, 78th Open Pau, velocity 838, like mother, like son, Adam's Boy velocity 889. She was the first bird into Worcestershire and won the Grove

Trophy, bred from a son of No 5 and his half-sister, she is pure Jurion, so Adam's Boy is three-quarters Jurion, one-quarter Logan.

I have 16 old birds, ten youngsters, I shall race right to the end, Lamballe, and six beautiful late hatched, the eggs that my Pau entries left, so if I catch a bad race at Lamballe I have the late hatched to fall back on. It makes me laugh to hear fanciers say, "I've finished with far too many birds to winter, I must get them down". I tell them, send them to Pau, you'll need to buy some. My system is this, if they are still with me when they are six years old they are put to stock, this means they have flown Pau three times each, this way you will be a long time filling a stock loft. I have two such pairs of birds. I could not lose them racing and they breed youngsters I find hard to lose, usually they get to Pau before they are missing. But make no mistake about this, you can lose a champion pigeon in a 600-mile Pau race like this year.

I have, and now that the excitement of winning the race is over my heart is heavy each time I enter the loft, for the one real beauty that did not come home. Adam's Boy had an easy life for the first two years. I stopped him after he won 4th prize in the Templecombe YB race, he had a rough time when he broke his keel in October. I painted his nestbox and covered it with cardboard, one side fell down and his flights were smothered in paint, so I left him as a yearling.

At two years old he raced well, winning 2nd Weymouth in Birmingham Saturday Fed, 2nd Nantes, 14th Fed, 4,248 birds, I stopped him for the year. Three years old, raced up to Nantes, I thought he hadn't had enough experience for Pau until 1968. He is to my mind a perfect pigeon, strong powerful short legs, broad across the chest and skull, an eagle head with short thick neck, a wonderful long wing almost to the tip of his tail, the type of wing that performs well at the distance. Short wings handicap a pigeon, broad of vein of web, strong of quill, no cere behind the eye. All good long-distance birds are like this.

His eye is perfect in conformity, deep in colour, the red to brown green shade, it tells you that the blood is thick, so the bird is strong. If there is a weakness in the constitution of a pigeon it will show in the eye. All those that have seen Adam's Boy say he looks what he is a King's Cup winner.

What of his future? He has brought me the greatest honour in pigeon racing, what should his reward be – I don't want you, but the money I can get for you – though I am poor by some standards? No, a thousand times, no. There is not enough money in the banks of England to part me from Adam's Boy. You only have to look to those that have parted with their best, if we are to see their names again we shall have to have the results extended. There are so many brilliant fanciers and pigeons in this Pau race that 99 out of 100 think they have not got a chance of winning it. But I look at it this way, there isn't a fancier breathing who works harder with them or studies more for them, so if I can find the tools I can do the job.

I have timed three good birds in the last three Pau races and have yet to see one drop on the loft. This year I was eating ice cream sat by the door reading as usual about pigeons, when I looked up there he was on top trying to get in.

I heard nothing at all that day. Sunday I had to go to work at 6am to 10.30. Colin Mogg, president of the White Hart Invitation Club, came at 1.30, "all I know Jack is that Jack Smith down the road had one Saturday at 5.11 and my friend Percy Sambrook of Malvern, at 7pm."

Just as he was about to go Sid Davis arrived and said: "I've come to verify your Sect C winner and you are also the probable winner of the Open as things stand at the moment, but don't take this as final, I will let you know for sure tonight". I liked the way Sid softened the shock news, I kept thinking surely someone is in front of me. The final shock came around 8pm, going to the door a hand grabbed mine. "You've done it at last Jack, you've done it at last Jack, you've won the Pau National by a clear 13 yards". It was Mr Tucker, but even before he grabbed my hand my heart sank into my boots. One hour after he had left Sid Davis kept his word, he confirmed it and any doubts I had were now gone.

I have won everything there is to win in local club racing over and over again, but that proved nothing, position and mob flying are very misleading to the quality of your birds, though I was never a mob flier. My father will tell you I won the OB Average in Redditch Central FC with a pair of broken birds he gave me in 1947. The cock, a mealy, NURP44G5511, won 3rd Weymouth, 1st Weymouth, 2nd Worcester Fed, 3rd Templecombe, 1st Guernsey, 1st Sect G in the 'News of the World' race. His hen, a blue, won 2nd Templecombe, 3rd Weymouth, 1st Nantes. I completed the Combined Averages with only six youngsters, was top prize winner, £68 10s, and won all the trophies. As the years rolled by the Redditch fanciers coined this phrase: "If you can beat young Jack, you are in the money. On the nose, Young Jack, on the tail, Old Jack."

John McLaren wrote in 1968, five members out of 4,000 appear in all three National Flying Club race results in 1967, I wish he had mentioned their names, because I was one of them. I have one big regret (fanciers who say they have not got the birds good enough for National racing please note), I should have joined it and raced to win it in 30 years ago.

I met and worked with some brilliant Royal Air Force fanciers during the war. Three have won the King's Cup and another, Tommy Crotch, of Norwich, was 2nd Open NRCC. I often wonder why these outstanding fanciers are not asked to put pen to articles in the wonderful pigeon books of today, the stories they would have to tell would make the young fanciers of today sit up, for make no mistake about it, they are walking encyclopaedias. I thank God for this wonderful sport of ours; it has given me the happiest years of my life; the urge to work hard. If you knock on the door hard enough and long enough it will open, maybe next year it will be your turn, if you stand still you will not go forward. Be bold and, above all, be dedicated.

Upgrade Your Pigeons

by A H BENNETT, Church Stretton

1st NFC Nantes, winning record amount for one bird, and a motor car

Firstly may I thank the Editor for inviting me to write an article for the 'Squills' Year Book, and also I would like to thank all those fanciers who sent their congratulations on my record win from Nantes.

As most readers of these articles in 'Squills' Year book are usually looking for ideas on how to improve their own performances, I will try and give a few of my methods which I think have enabled me to have more than my share of luck with racing pigeons – I would like to stress that these are only my opinions, and that many fanciers may disagree.

I think that all pigeon fanciers when bitten by the pigeon bug cannot erect a loft and obtain their pigeons quickly enough. In my own case I was restrained from making this mistake by having the good fortune to make friends with a very successful fancier and I would suggest that all young novices try and get themselves adopted by a successful local flyer.

The first thing is to decide how many pigeons one can afford to keep and race properly as the biggest mistake for anyone to make is to have more pigeons than one has got time and money to look after, and remember quality always before quantity.

Selecting your first stock is probably the most important task for all fanciers; some buy birds from here, there and everywhere, in my own case my birds are bred down from only five pigeons which were obtained from successful local fanciers – direct children from actual Combine winners, with an outcross in 1958, a Doran hen which bred more winners than I can count, a Dutch cock in 1963 bred by A Nroeje, Gorincham, Holland, which is the sire of Champion Midas, and more recently a hen from Heber Fearnall of Chester which is also going to breed some excellent pigeons. So you will see I believe in inbreeding with an occasional fresh bird to invigorate my own pigeons, other birds have been tried but have been found lacking when it comes to the distance, so out they have had to go.

I consider a cross necessary when the hen birds begin to lose their size but a lot of care must be exercised when introducing fresh blood. I would like to stress that I don't think it's necessary to spend large sums on the purchase of pigeons to be certain of success, the best way is to buy about eight late bred youngsters from a successful racing loft and never put these pigeons on the

Midas, 1st NFC Nantes 1969, owned by A H Bennett

road, but to race all pigeons bred from them and form your own strain of pigeons. I can assure readers this would give you much more satisfaction than trying to buy success.

When I pair my own pigeons they are all confined to their own nest boxes until they lay their eggs, this is the only way to ensure the parentage of youngsters. Only 18 pairs are housed and about 24 to 28 youngsters are reared each season. I consider more pigeons in a loft of this size (27ft x 10ft) would be overcrowding. In addition I keep six pairs of stock birds, mostly retired racers and late breds from my best pigeons. These late breds I have found make the best breeders as the young from such pigeons usually have more vigour than birds that have done a lot of work, also the young from such pigeons are slower in the moult which is what I want for Channel racing. These are housed in a separate loft. Selecting the pairs to breed good pigeons is without doubt the most difficult task in our sport. I am convinced good pigeons are bred and not made by any amount of training. My method is to pair my best cock to my best hen providing they conform to the right shape that I think you need for pigeons to fly for long periods and I never breed more than one generation from a winner.

As to eyesign – after ten years of careful study of this subject I am not convinced. I wonder why most of the eyesign faddists keep so many pigeons?

Rearing youngsters is where so many fanciers fail, old and young alike. I like to enter the loft early in the breeding season and examine the egg shells before the old birds toss them out of the nest, any that have shown an excess of blood in the shell I make a note of and at seven days old when I ring the squeakers I turn it upside down, usually finding the cord has not come away from the squeaker cleanly. Such squeakers are quickly transferred to a bucket of water. I have also noticed when the youngsters are about 16 days old the parents, if allowed to have too much exercise, tend to neglect the youngsters, so my old birds are not allowed out very much at this period in the rearing. I am sure if more care and selection was taken in the rearing of youngsters there would be less lost in the YB programmes. Out of the two dozen or so that I breed, I usually finish up with 20 at least.

For feeding I use a mixture at all times of the year. I am rather a heavy feeder, believing you cannot overfeed racing pigeons. I like to send them away to races having enough flesh on them to last two or three days in case they get off their line of flight on their way home.

Training my youngsters, I like to allow myself at least a month to educate the young birds before their first race. I usually start young birds at about 20 miles on the south road, and then 20 miles in the opposite direction. After about a dozen training flights the youngsters strike straight for home. I consider then they are ready for racing.

Old birds have very little training compared with most fanciers I know. Perhaps a few details on the way my Champion blue cock was trained will give a little food for thought for those who think pigeons should be shown every telegraph pole. As a young bird he was raced on the north road as far as Ripon, and a week following the Ripon race was turned south without any training to win 5th North Shropshire Two-Bird Club from Weymouth. As a yearling he won 1st Nantes in the local club, 10th Fed, this being the only Channel race he has had in club racing. All his races since have been in Nationals and Open races, including Pau four times, 643 miles, winning prizes three times in the National FC. He has also won numerous other races. In 1969 he was paired on April 1. After he had been sitting seven days I separated him for a few days and then paired him to a different hen, transferring his eggs to a pair of yearlings (this is a method I use with all my best cocks and over the years I think has enabled me to up-grade my pigeons).

He was then allowed to rear 2 youngsters, given one training toss at 20 miles, a club race from Weymouth and into the Nantes National sitting 18 days on 17-day-old eggs, his hen laid one day too soon so I had to change the eggs so as not to have him chipping eggs.

The dam of Champion Midas was a good winner up to Nantes, being a half-sister to Champion Dual, being very inbred to all my best pigeons. The sire was a complete outcross, being the Dutch cock mentioned previously. Only two birds were bred from the Dutch pigeon and then I sold him. I believe if you keep a cross in your loft too long your birds soon begin to resemble the introduction more than your own family of pigeons.

The loft is cleaned as often as possible, once each day during the summer, and once or twice a week during the winter. One thing I don't believe in is so-called deep-litter. In my opinion this is the lazy man's way, and most of them end up with disease in their lofts.

To conclude, I would say to the fancier that has not yet won, set yourself a standard, and any pigeon that does not reach that standard suppress them without a further thought. If you have ten nest boxes, don't say you will keep ten pairs, only six pairs if the standard is not good enough.

I would like to wish all fanciers the best of luck in the coming season.

Methods, Notes and Records

by A G PLATT of A G Platt & son
1st London NR Combine Thurso

I am honoured to write for your great book 'Squills' and also *The Racing Pigeon*, which I have been reading for many years.

Now, to begin the year all my hens are taken out of the lofts and put in baskets. The same with the cocks. They are fed and watered for two or three days, that is in March. My lofts and nestboxes are sprayed out. When they are dry I put the birds back. A tablespoon of Glauber Salts is put in their water twice a week, this helps to get rid of any fat left over from the winter, and gets them right for pairing. Sawdust is put in all nestboxes. There are no nestbowls. I put all pairs of birds in their nestboxes with one small pot for grit, and another for food. I give each bird one ounce, two ounces to a pair, of a mixture of tics, maples, tares, wheat, maize, protein to value of 19 per cent. I let one pair out at a time to drink and mate, and use a white stick to shepherd them back again.

The birds are paired in the middle of March, stock pairs as well. The stock birds come in handy for hatching or feeding a baby when seven days old, or feeding a big baby when 16 to 18 days old, or for bringing up birds from my racing pairs, when I do not wish them to rear youngsters. I let each pair out on their own for four days to make sure that the youngsters I breed are from the correct pairs. After four or five days I put nestbowls in each nestbox and cut up straw about four to eight inches long putting it in the lofts. Then either my wife or myself feeds them in their nestboxes, and we let a pair out from alternate nestboxes until all pairs are out and the hens and cocks pick up the straw, take it to their nestboxes and make a nest. This straw carrying helps them to find and to know their nestbox quicker, and saves a lot of fighting. I believe they are happier because they have made their own nests.

The birds are let out in the loft when they have been sitting three days and are well settled when I let the cocks out in the morning at seven, leaving my wife to get them in. She gives the cocks just enough seed, linseed, rape, millet and dari to cover a sixpence, in a hopper, and also fresh water. Between one and two o'clock, my wife lets the hens out for a fly around. After their fly she gives them fresh water and the same seed as the cocks had.

At five pm, my wife gives them fresh water again. I do not think birds can have enough of this. I have water fountains outside all my lofts, with wire fronts on the shed, so the birds have to put their heads through the bars to drink out of the water pots, which are plastic and very easy to clean.

This way my birds are trained to drink from the basket by having to put their heads through the bars, the same as in a basket. I never have any worry with my birds learning to drink in the basket. Sparrows and other birds are not permitted to drink from these fountains. I also have half an inch wire mesh on all fronts of my loft to keep cats and birds out. The fronts of my sheds are all bars and wire, so are the doors and louvres.

I am a great believer in fresh air, in fact I do not think you can give birds enough of it. Rain, snow or dampness I take no notice of as long as the perches are dry. My lofts are sometimes full of water, when it rains heavily, but I just sweep it out, and the birds are always healthy. There are five small six feet wide sections just my own height, this is so that my birds do not become wild. I have all tame birds, they cannot fly past me in the loft, they have to walk past. Also, they cannot fly over my head. If any bird I lose comes home and enters my loft having picked up some disease, it is kept in a separate shed.

I also put a bird in a spare box with all wire round it if it comes home from a race two or three days late, to ensure that it is okay. I always give a bird one chance of making a mistake, after that it is put down, unless it is injured or has been shot. I train my old birds as soon as they have been sitting five days, ten miles, 20 miles, and upwards to 40 miles if possible, in any weather bar fog, two or three times a week until their eggs hatch, then I let them go out three times a day.

Once my youngsters are parted from their parents I race to pot eggs all the racing season. On the short races, I like a hen either with a chipping egg or if sitting twelve days, I give her a two-day-old baby for one hour, which I take from my stock loft. Cocks I give a fourteen-day-old baby, but this hen is taken away and he is left to bring it up. This again is where my stock loft comes in handy, for the stock bird can feed the baby, and the cocks I want for short races hardly feed and so nothing is taken out of them.

Also I like a cock on his own in a nestbox, and give him a rank hen on the day he is going away to a race, but only for short races up to 250 miles. For the long races I like a hen sitting five days and a cock the same. I nearly always send the pair. The reason for this is I have no trouble with cocks turning their eggs in while the hen is away, and vice versa. I believe a hen that is sitting five days is very fit or she would not have laid, and the cock if he has been driving, he is fit and there is no extra on him. Also sitting five days he gets nice and settled.

This condition is just right, that is how my bird was 3rd Open Combine Stonehaven in 1969, and this year when she was 1st Open Combine Thurso. Hens and cocks take good positions in the Combine races sitting the same as above. I like a bird to go to a long race with two new primaries. I think this is ideal. When training, hens go one day. Cocks and hens never go training together.

Any bird that comes home from training first is noted. If it is sitting and how long, as this gives me a guide in getting it ready for a race. You must study every bird if possible, and must have patience as pigeon racing is not learnt in a day or a year: I have been studying and racing pigeons on and off for forty years. In fact I was born into a pigeon family as my father and brother kept them. I have no mercy with any birds that do not work, and put them down, thanks to *Food for Novices* by Squills which I have read for years. Also, I like a bird to do well as a yearling and again as a two year old, as so many yearlings do well and when two year olds fall by the wayside. I like my birds to perform well in hard races of 800 to 1000 ypm. This to me is proof that I am on the right lines of racing and breeding, and shows the birds have plenty of courage.

Now to young birds. I take my youngsters away from their parents at eighteen to twenty-two days old when it is nearly dark, and put them in the young bird shed on the first long perch which is one foot above the floor. Some may say it is too early, but to me as soon as a baby gets on its nestbox front it is ready to be taken away. In the past I have come home from work and found them on the floor of the shed with heads and eyes pecked terribly and often have had to put them down.

To me, the baby is saying that it wants to come away, and it is crueller to be pecked nearly to death on the floor than to be taken away earlier and taught to fend for itself in the young bird shed.

I run the rule over my young birds, they must be plump, good looking with straight vent bones. I like a young bird that fights when you put your hand in its nestbox and blows itself out, and after three to five days of being in the young bird shed, to eat well and to drink. I pick them up and put them on the landing board and also put their heads through the bars into the water trough three times a day. I keep them in the young bird shed for three days. If there is any sign of a youngster going back instead of forward, it is put down, as I say to myself, "It's not got what it takes, and can't fend for itself". These I do not want, no matter how well they are bred. I take special note of the forward youngsters who learn to eat and drink well, and take a top perch in the shed and go there every time. This to me is proof that they have strength as well as brains.

Once my youngsters have learnt to eat and drink and take over a perch, I feed them at 6.30 am with a little linseed, canary seed, rape and millet, putting them in a big home-made hopper. There are six of these, all the same size and colour, two for young birds and one each for the stock sheds. When my youngsters are weaned, they know their hopper, and they have the same hopper for the rest of their life. A mixture of maples, tares, wheat, dari and groats is fed, no tic beans. I start to put their food on the hoppers and stand over them until they eat and start taking notice of me by talking to them while they are eating, pick them up and put them down, and push them gently away with my shoe or boot so that they get used to me, becoming tame, knowing who is their boss. When they have done this and learnt what I want of them I put them in a big crate with a drinker which I made myself, and put them on top of my young bird shed. It is left there all day, and when I come home from work the young bird shed is opened and cleaned out. If the weather is not too bad I put the food in the hoppers, open the crate door and call them into the shed. This I do three or four days running.

Then comes the big day, I let them out of the shed for the first time, always late when it is quiet. I open the shed doors and call them out into the garden. It is a pleasure to see them come running. They are given a little seed and I walk about the garden calling them for about half an hour. Then I put their food in the hoppers and call them in. I do this for about four or five nights, then open the doors and let them take off, or go on top of the shed roof. I live in a square with trees all around, having three tress in my own garden, with houses all around. As soon as a baby looks like going out of the square I call it and back it comes.

Some do not come back, they fly over the houses and I lose them. To me these have not got brains, although some do come back.

The reason why I take so much care over my youngsters is I have a neighbour to the left, who has five cats. These cats are always in my garden, though luckily they do not touch my birds. On the right of my garden my daughter has three Boxer dogs, they are in the garden nearly all day, and to see them run round the garden is like going to a greyhound track. Still, I have to get the youngsters used to it. As soon as my youngsters fly one hour, I basket, feed, and water them for three days. Then I put them back in the shed, cool them down for a day by talking to them and giving them a little seed. They are let out the next day for their fly.

Then I train them from ten miles up to fifty if possible. I like to train them when they have dropped two primary flights. I like to have more perches in the young bird shed than young birds, as this stops a lot of fighting for perches. Some youngsters I put straight into forty miles first toss. This is to prove they have the right blood and long distance in them, as there is always the chance of breeding or pairing wrongly.

Now for a quick summary of all my management during the year. From August until the birds moult out, I try and give them the best mixture, as much as they can eat. Also Cod Liver Oil on their food every ten days. When heavy in the moult I give them one or two baths a week in the shed with permanganate of potash in the water to keep the birds healthy and to keep vermin down. I give them a chopped-up cabbage with a little salt, all through the year, as I believe this helps to keep the blood and body in good condition. This I learnt from my mother, fifty years ago, as we had to drink the water from the greens which we had for dinner.

Once the moult is over, that to me is December 21, I cut their food down until the birds handle how I want them; then I start, or my wife does, to let them out, always around 12 o'clock, hens first, cocks second. Any bird that does not moult out properly I put down rather than keep it for another year.

To make sure my birds moult out properly I draw a blue pencil mark across their tail after racing, and then after the moult I handle them and you know the rest. I put down in my book everything I do in the sheds, the weather, the winds, the yearling management, the youngsters given away, eggs, and birds they have been bred from. I have never been mean, having given away over 200 birds and eggs to different fanciers in the last four years from my best birds. Fanciers have won with them. It has made me happy to know my birds can win in other fancier's hands as well.

I have always tried to help novices, and have put up three special prizes in my club every year for novices. Always trying to help if a fancier brings round an injured bird. I treat it for him with my own equipment, and tell the fancier when to go to the chemist. By this way they learn.

Any fancier who brings a bird with canker or a bad disease I tell to put it down. these are birds that can well be done without, no matter what they are bred from. I only breed from birds which have always been in robust health. My breeding secret is to pair robust health to robust health, guts and courage to guts and courage. This way I do not think a fancier can fail. I like a bird to have a good eye, good head, good silky feathering, good strong back, good strong straight vent bones, handle well and not to struggle. Width of flights primary or secondaries I do not take much notice of. I like to see the metal rings shine as if they have been polished, also no droppings on their feet, which should be red and warm, the wattle white or in some cases pink, eyes sparkling. Tight vents do not interest me as I have won with hens whose vent bones have been open from quarter to half an inch.

I like the glottis to be open and nice and round, the slit or cleft in the mouth to be well open and the mouth and throat to be pinkish white. I must mention here I think pigeon racing is a sport that is never conquered, you can always learn and experiment with pigeons.

The Individualist
by J & J G PALEY
1st Open Palamos (861 miles) 1971

Only in our most far off distance dreams did we ever imagine that we could beat all entries from Great Britain in a race from Spain. But by pursuing our dreams we have turned them into reality.

This article is devoted to our personal methods, or how we set out to fly 861 miles. It should therefore not be used as a guide for the many fanciers who are content to race their birds up to 500 miles. During the past few years we have sacrificed a great deal in order that we could succeed at the double long distance races. We have lost many pigeons which were capable of winning up to 500 miles by pushing them too hard, in an attempt to develop a pigeon to race the extreme distances.

Our system has been built to suit the time we have available for pigeons, which to many other fanciers would seem too little. For the past five or six years our aim has been to race successfully from the very long races, in our case 730 miles or 861 miles. Naturally to fly these distances in race time is very rarely accomplished therefore we did not have the help of other fanciers' methods, we simply had to study out how we could achieve this aim for ourselves.

Consider how many 300-mile winning pigeons there are in Great Britain today, we do not know, but there must be many thousands, again how many 400-mile winning pigeons are there and again there must be thousands, and the 500-mile winners again thousands of them, and no doubt many very good birds amongst them. Now how many 600-mile winners are there, well not so many, but you may say that there are not many 600-mile races, but this is simple because fanciers will not support a race over 500 miles.

No doubt the fanciers are correct that many 500-mile winners would not complete the course at 600 miles so why waste a good bird at 600 miles. But we were not satisfied at 500 miles so we pressed on. Now how many 700-mile winning pigeons are there and the answer is very few, it must be less than a score. Well we have gone so far we must go on to 800-mile winners. We know of one and that is 'Woodsider'. Make no mistake about this the pigeon was not sent to home from 861 miles, he was sent to race. At seven years old most pigeons are well past racing. Our opinion on Woodsider was that we pushed him so hard in his first five years he tired, and in 1970 it

**J & J G Paley, with their trophies received at the British Barcelona
Club dinner**

appeared he was over the hill. He went to Pau 730 miles when he was not 100 per cent ready, yet he homed in race time, then the training commenced for Palamos from the day he was home. He received full attention for 12 months and when sent to Palamos was as fit as any pigeon could possibly be.

It is quite simple to see that an 860-mile winner must be an individual. This was a point which we realised several years ago. Every pigeon is different from the next and we do not mean in size, shape or colour, but rather in character and temperament. To find these points it is a question of observation. The individual can be picked out by watching the birds in the loft, this is very simple to do just sit by the loft and watch the birds going about their daily activities. Your team of youngsters will have one or two individualists if you watch out for them, they are the ones which do not act as sheep and do as the rest of the pack. From these youngsters will come your champion, although watching for the individuals you will also pick out the one which is useless, but not to worry as this will soon be proved when it comes to racing. This theory of the individualism is the way we set out to race and it is the way in which we shall continue.

Our birds are housed in five lofts, all varying very much in design and construction, and we must add that they are not at all pleasing to the eye as some lofts go. Ventilation must be the main point in loft design in order to keep the pigeons healthy. The old bird racing loft in which Woodsider is kept is a wooden construction with overall dimensions of 13ft x 10ft. The rear of the loft has large apertures at loft floor level. A dummy back wall is situated 2ft from the main back wall and this dummy wall is constructed from peg board which allows a full flow of clean air to every nest box. The front of the loft is well ventilated by means of louvres and sliding windows. A section through this loft is shown on the sketch.

The loft contains 18 nest boxes each 2ft x 2ft x 2ft. The loft never houses more than ten pairs of birds and three weeks before the Palamos race the loft contained five pairs in order to ensure 'Woodsider' was happy. 'Woodsider' has two nest boxes and he has had the same two boxes since he was a yearling. The nest boxes have an internal perch 2ft x 1ft, which in effect makes the boxes two storey. One very important feature we have found is to position the nest bowl well away from the box side to prevent the bird from rubbing its flights when nesting.

The birds are fed on a mixture of beans, maples, tares and plate maize, in addition to the mixture they are given practically anything which we think they may desire. The birds are fed ample through the racing season with a hopper placed before them from morning to evening. The same mixture is continued through the moult and on until the mating season. Many fanciers

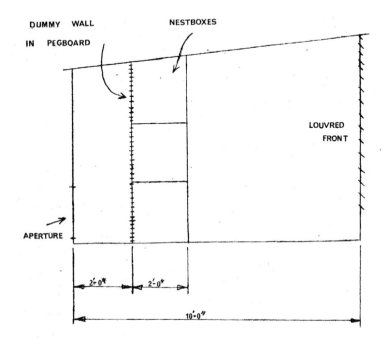

save pennies by cutting the quality of the feed after the moult but we believe it is far better to continue with a good feed in the cold months for the simple fact that it is far better to condition a well-stocked pigeon than to be faced with a team of under-weight birds, which are just a frame covered with feathers.

The drinking water is changed twice a day and on race days the water is never ice cold for them when they return from the races. Loft hygiene is very high although this does not mean the loft is cleaned out daily or even weekly, it is cleaned out when it is dirty. Droppings are not considered as being dirty provided they are dry and in a well ventilated loft, they are dry.

To cover the training of our birds it will be simplest to start with the youngsters and continue through to the old birds. When the youngsters are weaned they are put in the young bird loft which contains a basket for several days, the birds soon get used to being fed in the basket and they make this their home. As soon as the youngsters begin to pack and fly strongly they are basketed and fed and watered for two days followed by a liberation of two miles.

Then training tosses are carried out at approximately 4, 6, 8 and 15 miles, and then they are trained by the road transporter from 40 to 50 miles. The youngsters have a 40-mile training toss each week during the racing season. All the youngsters have to fly from the coast (220 miles) with very few exceptions. They do not compete in many races and often the youngsters' third race is 220 miles. Several of the youngsters are raced from across the Channel having to fly 307 miles or 374 miles.

Yearlings and old birds are trained from 8 and 15 miles prior to the transporter training of 40 and 50 mile tosses and then regular 40 to 50 mile tosses according to their planned race schedule. Yearlings are classed as old birds and have to cross the Channel. Woodsider raced from across the Channel three times as a yearling, he flew 364 miles being 10th, then 400 miles followed by 462 miles taking 6th place. The birds going to special races are given single up tosses late at night and early morning to induce them to use the full length of the day. Club races up to 220 miles are used solely for training purposes with very little importance being given to the club result but more importance given to the order in which our own birds arrive.

The birds are not forced to fly around the loft location by using flags etc. This type of exercise we consider to be detrimental to the individual type of bird, no doubt for the fancier who is keen to have all his birds racing each week it is one way to keep them exercised but we are of the opinion that this type of flying does not teach the birds anything, in fact it must degrade the larger percentage of them as it makes them fly at a speed which is much less than the bird would normally race at.

It also causes them to fly without having to use their eyes and brains, as they fly round the same small area for an hour they become so used to the obstacles that they could do the flying with their eyes shut. If you are the type of fancier who adopts this method of exercise just watch the birds next time you make them fly. And then ask yourself if this is the way you want your birds to fly when they are racing home from a long race. When a long-distance bird is racing it must use all its faculties on its own to speed it home.

Our birds are let out early morning for approximately one hour, the first ten to twenty minutes of this is spent in enjoyable flight and the rest of the period is more or less their free time. At lunch time they are allowed out similar to the morning in order that he hens can take exercise. In the evening the same procedure is adopted but the time is increased to about three hours, it is during these times that the birds are observed.

The subject of eyesign has been studied in great detail for many years with various breeding and racing experiments being carried out. We have also judged several eyesign shows with over 100 entrants per class. After the many experiments and observations which we have made our final conclusions are that the prime importance of eyesign is a guide to the bird's condition, by continual observation of the eye you can tell when the bird is reaching peak condition as the eyesign grows day by day.

The strain of our birds started in 1958 with birds of the Osman strain from F V Clarke of Paignton, Devon. One of this batch of birds which we purchased was Clifton Pride the winner of 1st Open News of the World St Malo in 1958 by a clear 70 ypm. Almost every year since that time new introductions have been made in a continual attempt to improve the strain but very few introductions have been allowed to make up much percentage of our originals.

Woodsider is a first cross with our old strain on the paternal side and the subject of the first cross on the maternal side, this being a Westcott hen which

WOODSIDER, 1st British Barcelona Palamos.
Owned by J & J G Paley.

had scored up to 400 miles prior to our purchase of her. The sire of Woodsider was one of our old favourites scoring many times up to 530 miles and winning over £200 during his career before finally going down at eight years old from a 578 mile race. The sire of Woodsider was an inbred grandson of Clifton Pride. The pigeon responsible for many winners for us and others.

Our loft holds the status of Master Breeders in the Thoroughbred Racing Pigeon Association and almost all the birds in the loft are pedigree registered to help guarantee the pedigree of our birds. The partnership was started in late 1958 after J G the son had raced a team of young birds in 1958, and from 1959 the partnership had been known as J & J G Paley (both Jacks). Jack had previous racing experience with the Silsden & District FC during the 1930s but this ceased at the outbreak of war.

We both put much of our lives into the pigeons, as we believe you must put in as much as you take out of the sport. Jack is president of the North East Lancs Federation, one of the oldest Federations in existence, and he is also clocksetter at Cowling HS. Jack Graham is secretary of the Thoroughbred Racing Pigeon Association, secretary of the Great Yorkshire Amalgamation, president of Cowling HS and vice-president of the Pennine 2-Bird Club.

We are grateful to 'Squills' for inviting us to contribute this article and we sincerely hope it will help to further British fanciers' efforts in attacking the true long-distance races.

The Art of Pigeon Racing - An Old Stager's Views

by G H DAVIES, Whitby

Every man if possible, and according to circumstances, should before he retires, make arrangements to have some sort of hobby or other. You would never regret having done so, especially if you took up the hobby of keeping racing pigeons. You will make friends with the most fascinating and wonderful creatures on earth. By their attraction and love of home, and you their provider, they will keep your mind occupied and away from many things, so if you know or meet anyone who keeps them, I am sure that any real fancier would do as much as possible to help you in this direction.

In the following paragraphs I have given a few hints or instructions, which you as the novice or tyro could use to your advantage. Most of these on your own would take a long time to accumulate, so you will be off to a good start, as advice like this has not, at least to my knowledge, been published in any pigeon racing or textbooks over the years.

The Loft

If the new fancier has a small plot of land, and is lucky enough to have a suitable brick building or outhouse so much the better, as this could provide the ideal loft. This could be built or made convenient for open-door trapping, flying straight into the nestboxes and well off the ground, where the birds seem safer and more settled. Otherwise, a loft could be purchased from those advertised in *The Racing Pigeon*. Then permission to erect one should be obtained from the local council, who now mostly allow tenants to do so. The dimensions allowed being approximately 12ft to 14ft long, by 7ft deep, and 7ft odd high, which should be on brick pillars of about 1½ft high. This would give head room of 6ft, which is quite enough and keep birds from being wild. After a while you will see and find all the necessary fittings and accessories such as baskets, drinking troughs, pot eggs etc that are needed.

All sorts of uses can be made if an aviary is attached, and no matter the size everyone should have one. My ideal would be to have half-inch mesh wire surround, with boarding about a yard high, so that the pigeons could not see cats etc, and keep out other predators, also sparrows etc. When this is possible, at least two small cotes or houses could be made, and put at ground level away from the main loft, just for one pair in each, or at both sides of the main loft, with bolting wires fitted to ensure captivity if and when required. Take care that no other bird gets in as the occupants may, and probably would, kill it. These cotes are small houses could be used mostly for inland races and various jealousy systems. Builder's sand is usually clean and dry, and should be used on all loft floors, nestboxes and perches. This should be raked over daily and renewed weekly or when necessary, cleaning when the birds are out. A Vapona bar should be hung up in each side of the loft. This bar lasts for between three and four months, and is sold by The Shell Oil Company. All lice will be dead in 36 hours!

Formation of the loft, and establishing a family
If YBs are bought, they should be bought from only one very successful fancier, and not from every Tom, Dick or Harry, as these are almost sure to be inbred, and usually all of one family, and all related in one way or another. I prefer latebreds, which should be mated, and their offspring trained up to at least 100 miles, and in the following season, give them the works, some for inland races, and the late arrivals over the Channel! You will soon discover the best at different distances if you want to specialise. Then line breed, mating your stud cock to his best daughters from say three different hens, which are closely related to him. Half sister to half brother etc, or best to best.

The survivors will produce of their kind and establish your own family. Many object to inbreeding on moral grounds, but this is not contrary to nature, as wild animals and birds mate as their instincts dictate, and is controlled by nature itself, by the survival of the fittest, and is not controlled by man. If OBs are bought, the stud cock especially should be super, and these birds should be from stock that neither they nor their ancestors have had any physical or respiratory troubles. Irrespective of age, these are the ones to breed from, especially if they have flown 500 miles two or three times, and are still hearty and agile, even if they are ten or even twelve years old. Many fanciers will entirely disagree with this, contending that only young and vigorous and virile stock should be bred from. Think this over, all the ten and twelve year olds are still with us and have not been lost, and the young and virile stock have still to prove themselves! Any eggs laid could be hatched under feeders. Longevity seems to be hereditary, and characteristics of strain are transmittable, so inbreeding, linebreeding, and intelligent selection, if provided with thoroughbred material should bring success.

Pairing up birds should be about the end of February or mid-March, according to the races being prepared for. The first egg after pairing will be laid in eight to ten days' time between 6.00 and 7.00 at night, and the second egg 46 hours later, between 4.00 and 5.00 in the afternoon of the next day but one. The incubation period will be about 19 to 20 days after the first egg. Both youngsters usually hatch within an hour or two of each other, and should be rung at six to seven days on the right leg. The hen will lay again when the youngsters are about 16 days old, and another nestbowl should be put in the nestbox at the opposite end. When picking up food at about 21 to 28 days old they should be weaned. Give the OBs plenty of food until the youngsters are well developed, especially during the first 12 days. Put the youngsters on top of the loft in their nestbowls for ten minutes if the weather is warm enough. At seven weeks old they should be nicely on the wing. Old birds and youngsters should, and must always be let out together, as this will prevent flyaways.

All birds, both young and old, should be kept in one loft, divided into two compartments by a lath partition with door to enable fancier to move from one part to the other, and to close at will, with nestboxes only on one side, and deep box perches on the other, with free access to either side, and all using the same exit and entrance. They will soon recognise, defend, and claim their own places. Individual study of each and every pigeon is absolutely necessary, and records kept, so don't rely on memory. All this takes time, but when the birds are in their usual places, talk to them, fondle them, and do everything to make them defend their selected places, and make them tame and contented under natural management, and be patient always.

Basket training
First take the waterbowl out of the loft, and teach your youngsters to drink in the basket. After exercise, feed them in with a little bit of hemp, wait awhile, then basket them, and let them settle down. Then place water troughs on the basket, and splash the water in. Do this until they know all about it.

Sexing youngsters
There is not really any definite way to do this. My way, though not guaranteed 100 per cent certain, though I have never found it wrong up to now, is to lay the three toes straight alongside each other, and if the outside toe is longer than the inside one, the youngster is a cock. If the two outside toes are the same length the youngster is a hen. This method can be used up to a month old, but afterwards walking about alters the foot of the youngster.

Training youngsters
When training youngsters, try to train in different teams, all reds or red chequers, all blue chequers, and all blues, etc. Get some old rubber race rings, and put these on the left legs of your youngsters, because on race marking nights when the rubbers are put on for their first race, this upsets them, and makes them nervous, and they are always trying to get them off, whilst yours will be used to them, and will be more settled. Do not mate youngsters unless you want to try the odd pair.

Smear the bottom of the bird's feet with vaseline before basketing for races, as this will stop any dirt or excreta sticking to them. Also dust the birds with a suitable insecticide. Be gentle and kind and do everything to make them comfortable when either training or racing. Treat them all alike, and barring accidents, they will fly to the one they love, and their home, and you they depend on. Try to train on sunny days if possible.

Youngsters should first be given a couple of tosses at about a mile in the direction of the first race point, morning and evening. Keep them at it, and the next tosses should be at three miles, then five miles. These tosses should be regular from eight, 12, 20 and 30 miles. Afterwards, if you have the time, do this as often as possible from the 30-mile point, especailly two up together. This is probably impossible for the majority of fanciers, but try to get as much of this training in as you can. This is good policy, and will teach your birds individuality, though they do fly better in pairs. There will be very many inexperienced youngsters amongst them, flying in races on different routes for the first time, and being gregarious, that is, that they like company and flying in flocks, so are easily diverted or deflected to fly in another direction, whilst your birds having been trained well at the last 30 miles will break away at this point, and come on their own. Too much emphasis should not be put on YB performances, as YBs which are home 10 or 15 minutes behind the winners or leaders often make really good OBs, as also do those that spend a night out, arriving back next morning on their own, so give them all a chance to prove themselves, and being well bred, this will surely pay off in the long run.

If any bird seems tired or run down after racing, give them some Glucose (grape sugar) in liquid form, and dilute with milk. Nothing does more to speed up recovery than this, and a warm bath afterwards. The day before basketing by the way, give the birds a teaspoonful of Rape (black) and add one tablespoonful of powdered glucose to one gallon of water, but only to those birds that are racing. This Glucose can be bought at almost any chemists in liquid or powdered form. During racing the birds should have one ounce of food per bird, per day, after exercise. Some fanciers give one and a half ounces, and others one and a quarter ounces, according to their particular system. Others stop feeding when the first bird goes for a drink. Anyhow, you will soon find your own system, but I find it best to give my birds one ounce per bird per day as previously stated, and they are well controlled, and always ready for meal times. This arrangement also gives every available opportunity to give training tosses without food being in their crops.

To control birds
Have a distinctive container, which they will gradually get to know containing only hemp and linseed, especially the hemp. They go 'nuts' about this, and will follow you everywhere. This is the only way to get complete control over your pigeons. Drop a bit and see!

Mated birds
I don't like the position of cocks driving, though there are exceptions which do well. But if you could manage for the hen to lay her first egg on the day before basketing the cock, this position would be ideal, for when he comes to take over sitting in the morning, he is aware of the first egg having been laid, and will race well enough to be pooled! Don't forget that he should be sitting between the hours of nine and five o'clock. Pigeons have a very fine physiological time sense. For instance, the regular hours at which the first and second eggs are laid, and the hours at which the sexes take turns at sitting on the

eggs. So it seems to me that the position of the sun must play a very important part in this timing. Cocks usually sit roughly between the hours of 9 am, and 4 to 5 pm, with the hen doing the rest of the 24 hours of the day. Therefore, the logical time to send the cocks is when they are due to be sitting and take their time on the nest. Time them to be hungry and due to be sitting when you send them for training tosses and see.

The racing season starts about the last Saturday in April up to the second or third week in September.

Different systems for jealousy etc
First of all, you must have feeders, or birds you do not breed from in your loft. Anyhow, you need a cock and a hen of distinguished colouring so that they can always be readily recognised, though they need not necessarily be a paired couple for these jealousy systems. Lots of fanciers have many tricks of their own, and perhaps some of the following (but would not admit it) nor would they let anyone see or hear about them. This one is very simple and I should imagine almost every fancier has tried or heard of the 'worm in the eggs' trick. It does no harm to tell it to novices, as older fanciers fade away, taking away their secrets with them, which is all wrong, and gives no encouragement to the younger generation of future fanciers.

However, this is done by puncturing the egg or eggs at both ends, and blowing out the contents, then getting worms to put in the now empty shells, and then sealing both ends with a bit of Sellotape. The eggs are now put under the cock after having taken the eggs he was sitting on away. So when the hen takes over to sit, both the cock and hen are under the impression that the eggs are about to hatch; here preferably race the cock. I think that both birds, cock and hen, should never be sent away at the same time, as the cock should always be expecting the hen to be waiting for him in their nestbox.

Mating two cocks to one hen
Get them all used to the same nestbox. You can do this gradually, so, when racing and getting ready for basketing, let the hen at liberty, and she will fly immediately to the nestbox. Then let the two cocks in the loft, when of course, they will both make for the same nestbox and where the hen is. There will be fighting, but don't let either of them think they have won if you are sending them both to race, though if you are sending only one bird, let the fight continue until the candidate you have selected seems to be winning, then pull him out and basket him. If you decide to race both candidates, keep them apart by using a compartment basket.

Racing a cock
Put a pied or distinguished coloured cock in the nestbox with the racing cock's hen, and lock the racing cock out of the closed nestbox, when he will try to get in, clinging to the front, then after a few minutes, take him to the marking station. Remember to always use the same cock or hen of distinguished colouring for these systems.

Here is another way to use the cock. Have a wire partition, and get some-one (not yourself) to put him in the house or nestbox where the cock who is going to race and his hen are. Keep repeating this from time to time and send for toss. The intruder will always be keeping the racing cock on edge, think-ing that the other cock is trying to take over, and steal his hen. You may get good results with this. Anyhow, try different birds with different systems, especially with inland races. Then if you are successful, you can keep any bird for any particular race or distance. I know all these things take time, and all cannot be done at the same time. Nevertheless, try to study each individ-ual and keep your eyes open!

Racing a hen
Remove the cock from the nestbox, and place a distinctive coloured hen which is used for these systems in the nestbox where the hen which is going to race is alone. After a few minutes, take her to the marking station. This can be used and repeated by always keeping the racing hen on tenterhooks against further visits.

Racing two hens
If you can make, or have a wire or lath partition between two nestboxes so that the occupants can see each other you can gradually bring the two nest-bowls together until they are close to each other. Do this about an hour before basketing for the race.

Any of these little tricks could sharpen up certain candidates and surprise and bring you success, so don't be afraid to try any with your OBs. There are numerous theories, systems, and beliefs sworn to by certain fanciers.

Some put aniseed in the nestboxes, while others put mirrors. Some with plenty of time on their hands practically rear some youngsters themselves by hand feeding to weaning. Many will not keep any bird with a white streak in the quills of the tail feathers as this denotes that the bird is of unsound con-stitution.

Now with youngsters
You can make your own system and keep to it, as once they get into certain ways, these are very hard to break, so select one of the following if you wish. Try feeding once daily at evening after exercise, but don't overfeed, and give one ounce of food per bird per day. Or try with no exercise around the loft at all, with twice daily tosses, two up together, gradually increasing the distance as far as you possibly can. This teaches them to break away without too much circling. If you can get no help with this, then let them trap on their own, and expect some hemp in their nestboxes upon their return. Feed up afterwards.

Or you could try letting them out for exercise, then get them in with a lit-tle hemp. Basket and take them for a few miles, then feed in with the evening meal, of one ounce per bird per day. Repeat this on successive evenings, and train in the direction of the first race point, gradually increasing the distance up to 30 miles. Afterwards, if you have the time, try tossing three up together, then two up, and then solo. They should do about a mile in two minutes.

Young bird races usually start about the third week or Saturday in July. All birds when basketed on Friday evening for Saturday's racing should not be let out. They should be fed early, either in their nestboxes or deep perches with titbits and half of their daily ration, viz half an ounce. (One half pound of mixture is enough for 16 birds.)

Setting leg fracture

Birds are often injured through hitting television aerials and telephone wires etc, returning with injuries so bad that it makes one wonder how they ever got back. In this case, take the bird to the vet, especially if it takes two to do it. One may have to hold the bird, whilst the other deals with it. In the case of a leg fracture on the leg without the metal ring, you could mix some Plaster of Paris with water, then soak a narrow piece of gauze bandage. Fix the leg gently, and wind this round the leg. Don't wind it too tight, or this could stop the circulation. Then hold the leg until the Plaster of Paris is dry, which should be in about 10 to 15 minutes. Most bird's bones set in about 15 to 16 days. Put the bird in a nestbox with plenty of hay or straw. After this time the Plaster of Paris will soften in vinegar.

Feeding etc

I germinate or sprout all peas of my mixture by buying these separately. I steep them in boiling water overnight, then dry them off in the morning, always keeping a supply handy for alternate days, no more. Rock salt or an iodised salt block must be in the lofts with grit and minerals. In any aviary or garden under slight cover, put some peat or virgin soil, which can be got from the tops of mole hills out in the country. Then scatter over it some fish or haddock bone dried in the oven and crushed or bone meal. This is calcium! Your birds need vegetables, and the very best of all is garlic! Other vegetables which are very good are parsley, mustard and cress, and shredded raw carrots. All these things will stop your birds from fielding. Rain water is also good for pigeons if caught in a retainer. Also a lemon squeezed in the drinking water once a week is a good refresher.

When birds return from the race they should be given no hard corn, just a titbit with tepid water, and brown bread steeped in milk and covered with brown sugar. Wait as long as possible until every bird is home, and treat the last one as well as the first one. Remember to always time the first bird in if two come together. Take particular note of the position each bird returns on your own home-made league chart each week, then if you wish, you can retain any pigeon for the race he or she prefers. Note the condition in which they were sent, such as sitting on eggs from four, six, ten days, etc, or on youngsters. Then you can pool any one in the following season for that particular race again. In other words, little fish are sweet as they say. Later on, you can if you wish, go in for the long distance races, for which, by the way, your pigeons are bred, having come from originally long distance stock, you have every chance of success. The best long distance birds usually have small coverts with short wings and strong thick keels. Sixteen to 17 per cent protein will do for any distance when racing.

After August until about the end of January some fanciers feed only on wheat, gradually changing over to one ounce per bird per day. This change may cause loose droppings for a few days, but they soon get used to it. I think that the best all the year round diet is one made up of four parts germinated peas, two parts of Cinquantina maize, one part tares, one part wheat, and one part lentils. In the winter I sometimes give my birds a toss of a few miles if the weather is not too bad, but those at the bottom of my chart go to ten miles in any weather if the visibility is OK.

Before the coming new season starts, all necessary alterations should be made, as pigeons do not like changes, and should as far as possible retain the same nestboxes. It would be better if there were two nestboxes for each pair. Always keep the sexes together after racing as they are much happier this way. See that you don't overcrowd, and before the end of September, close all the nestboxes, and mount plenty of perches. About a month before remating keep the birds in, and separate the sexes, putting the cocks on one side and the hens on the other. Now, many of the older fanicers may ridicule many of my suggestions, but what suits one does not suit the other, and according to environment etc, etc.

Everyone has some system of management, or should have, and stick to it. After all, if you are not successful, it may not be the pigeons, it could be you!

A Full Year's Job

by G MITCHELL of Mitchell Bros, Seaham
1st Up North Combine LIllers

We feel very highly honoured having been asked to write an article for Squills 1974, having to confess, we are more at home with the scraper and corn tin than with the pen, nevertheless we will try and put our methods and management into writing.

First I would like to start by saying we are the third generation of Mitchells to be flying, so you can see we were born into the sport and take our flying very seriously and find pigeon flying a 365 day a year job. You must have the love of the birds at heart and be willing to make numerous sacrifices for them, in fact you must be dedicated to them.

South Road stock

At the present we have two families of birds, the Vandeveldes and the Krauths, the Vandeveldes bred through the best of the old Lucas and Askew families have been worth their weight in gold to us over the years, keeping us at the top or near the top year after year in both Club, Fed and Combine. One bird I must single out, our good red cock 58R1904, this cock had nine 1sts in Club with many other positions, two 1sts Fed and seven other Fed positions in the NE Counties Fed, approx 2,500 birds a race, he also won every inland race from Selby to Ashford 275 miles.

This cock proved himself just as capable at breeding, being the sire and grandsire of numerous winners in Fed, Club and Combine, a son from him, mealy cock bred for stock, paired to our Vandy blue hen, bred F Swan's good mealy hen. This hen up to being a two-year-old has won approx £1,000, as a YB she had a 1st Up North Combine Earls Court, and 2nd Up North Combine Ashford, 11,099 birds. Up to being a two-year-old she has eight 1sts to her credit. Mealy cock also bred 4th Combine Angerville, 6,080 birds, also grandsire of 7th Combine Tours, 4,856 birds, 520 miles.

Our other family of birds are the Krauths we introduced from two racing pigeon auction sales, the first in 1963, the first draft of Krauths into this country, the second lot from F Haylock's sale in 1966, plus introductions since, these are a grand team of birds and a ready made family and have proved themselves excellent racers, our 1st Up North Combine Lillers 20,380 birds, was a chequer cock from this family. As a yearling this cock also had a 1st

Club, 2nd Fed Hatfield. We were on the Combine sheet five times in 1973 in Channel racing. In the last seven years alone we have won over £1,000 from Lillers, our first Channel race, with other good wins up to Bourges, 520 miles.

What I am pointing out, is, you have got to start with the best of stock, nothing second-rate, it must be the best, go to any established fancier who has flown well over the years and don't be afraid to offer a good price for a good start to your foundation stock, if he has no OBs, he may be able to breed you a couple of pairs of youngsters or even latebreds. Failing this, through the medium of *The Racing Pigeon*, a must for all fanciers young and old, some of the best birds in the country are put at your disposal, breed from these birds then make the basket prove their worth and what comes up to your standard make a family around. As a rule we always believe in going to see what we are buying to see that the type, size, etc, is up to our standard.

The Loft
We are firm believers in plenty of ventilation, the top half of our loft is open dowels from end to end, with six sets of louvres bottom half, six sets of ventilators in the back and side, the roof open 2in all round, but we also believe in dryness with ventilation and have special perspex windows along the whole front of the loft, which are made to act as vents, these are used during the winter months and can be put up or down in a few minutes.

Our loft consists of racing end, trapping end open door, this is also a must in the NE, for example where we are flying within a radius of approx 144 square yards there are 24 flying members alone and in fast races where you have a number of birds together every second counts. Next we have our YB end and stock end with a good size aviary, this being used for stock birds and YBs as needed. We clean every day summer and winter, and the loft floor is covered with beach sand as we only live a couple of miles from the sea front.

We are also firm believers in cleanliness, we always keep at our disposal a majority of Harkers Remedies and the loft is sprayed with Duramitex at least three times a year, it is no good waiting until the bug gets in, keep it out, whatever we think is wrong we give as Harkers' directed. I know this is not cheap as working men ourselves, coal miners, we have to work damned hard for our coppers, but as a safeguard to the birds it is worth it. A book of records is always kept at hand and everything recorded, everything is put into writing, as memory very often fails us.

Breeding
Breeding is taken very seriously and is a full year's job. Our birds often pair on paper three or four times, before going together and it takes very careful selection. We believe in the keeping of a winning family of birds together, which involves quite a lot of inbreeding in fact, if we introduce we always go for birds that have originated from the same fountainhead as our own family, but, of course, only if the performance is there. Once our birds are paired, February 14, they are not allowed out of their boxes until they have laid,

drinkers are put on each box and they are only let out for cleaning purposes.

Wherever possible we only allow our racers to rear a single youngster. Once they start feeding, water and corn with a little Hormoform is always at their service. Once we see a youngster starting to pick up it is put into the YB end, when our youngsters go over we always dip their heads into the drinker for the first two or three days, whenever we have any time to spare, we sit among them, fondle them and gain their confidence and affection as soon as possible. When this is done they are put into the basket as they accept it better the younger they are. Only the healthiest youngsters are kept, any doubt at all irrespective of pairing, they are destroyed, we never allow sentiment to creep in.

Don't forget to keep the spray handy at all times against vermin, every other day youngsters are given a little Hormoform. We always breed ten or twelve youngsters more than we need in case of losses, so at the end of the year we can go and select what we are going to keep, rather than be left with maybe only seven or eight and more or less, having to hang on to them whether we want them or not.

Racing

Our birds are always raced on the Natural System sitting ten to 14 days being our favourite position, hens occasionally race to small youngsters, or cock bird just looking at the hen. Our cock birds never seem to give their best with youngsters up to a week or so old. Our Combine cock as a yearling won 1st Club, 2nd Fed 12 days sitting, and in 1973 he topped the Combine 12 days sitting. When breeding in our racing loft is finished, we start racing, excepting for the odd pair or two. Regular outings every day, regular feeding times, and observation at all times keep an eye on their wattles for a sign of good health, white to pinkish colour. A fancier with a few years' experience should be able to go into the loft and within a few seconds note any bird out of sorts, once again eyes should be at the ready for any bird showing any extra keenness on eggs or that extra enjoyment in exercise. In fact there are times they tell us when they are going to win by their actions; only fit birds are sent, any doubt at all and they are kept back, we are never afraid to lift a fit bird up to 200 miles or over.

Helping to keep our birds in racing trim we believe in two 30 mile training tosses a week right until we reach the Channel. Our birds are very good exercisers and very rarely do we have to make them fly. Our YBs are given as many training tosses as possible as this is a start to their racing education, starting at five miles twice or so, then in stages up to 40 miles, at least three times a week weather permitting. As we live near the coast we have always got to be wary of east winds, any doubt at all we keep them in the loft, in fact any weather doubts at all, leave them in the loft. Once they start racing we still like to give them a couple of tosses a week, our YBs are flown naturally and never paired, still getting good results up to Ashford National 271 miles. We are not wing faddists but youngsters coming up to their two ends flights are stopped, half our YBs are raced up to 217 miles Hatfield, the other half going on to Ashford 271 miles.

Feeding

One time we used to be faddists on feeding, it had to be the dearest and the best, but one time through not being able to get our delivery, we had to take a lot cheaper grade in which it was still sound feeding and our results never altered. Since then we have not been afraid to buy cheaper corn providing it is sound feeding. As a rule we always buy our needs for a full racing season, so there is no change during racing, this consists of beans, maize, peas and a little wheat, each bought separately and mixed by us. We are good feeders, a reasonable amount after their first outing then their fill at night. We give them their fill till they are through the moult, then put on a cheaper quality, for winter feeding.

The Eye

It's remarkable, I can remember going back 44 years ago, I was then a boy of ten years old, my father and grandfather were at the loft with a big magnifying glass in hand looking into the bird's eye. I can also remember asking what are you looking for father, and his saying to me, "Well there is supposed to be something in the colour of the eye which denotes its quality". Yes, today we are still looking. To me the only proof of the bird is the basket, but I am still not saying there are not qualities in the eye by all means. My own idea being that if you start with a colour pattern with an inbred family you can more or less expect the same colours of that pattern, the same thing applying with the eye if you breed down from a family of birds you can expect the eyes of your foundation stock in the majority, so why not follow your own family pattern as the eye is a family characteristic?

Note the wins by different birds, note the colour of eye plus distance etc, also colour of stock bird's eyes that breed winners etc, this at least may give you a lead to different qualities of different eyes in your own family, but as I have said no guarantees. We have had eyesign men come to our loft and say this bird should fly 500 miles and win and we have known by its basket experience it could not fly the mile, others say this bird might fly up to Ashford 270 miles and the same bird has scored at least twice over the 400 miles stage. What I am trying to point out, is that the owner himself should know best, the quality of his own family of birds, all the way round.

In conclusion

We would like to thank all those people on the administrative side of the sport, in the Up North Combine who have helped to make our racing what it is today, our Combine secretary Bill Towers and the rest of the Combine staff, who have transformed pigeon racing and training in many ways, in the last few years. Also good wishes to many of our friends in many parts of the country who sent us congratulations on our Combine win this year and hope they in their turn may find the same reward.

Luck Plays its Part

by Mrs BARKER, Boosbeck

1st Up North Combine Bourges

When I first started to fly pigeons in 1949, a brother-in-law sent me a pair of reds. There was no pigeon corn available and I fed them on rolled oats I borrowed from my husband's pony's ration. The men in the club christened my first red pigeon I raced 'Quaker Oat King'. Unfortunately I sent him to Guernsey and lost him.

After this the club members took pity on me and gave me a chequer hen and a dark chequer cock and a lovely blue Logan hen. This latter bird topped the East Cleveland Fed by three quarters of an hour from Dover and I felt I was really in the big time then. A descendant of the two chequers also won me the YB Average in the Fed and after this the men decided I was on my own and no more gifts.

They even did me the honour of making me secretary of the club for a while, but keeping pigeon fanciers in order is a man's job. Secretaries are born and not made and are the salt of the earth.

Pigeon racing is a fascinating hobby and I never cease to marvel at the feats a pigeon is capable of. I keep 10 pairs of actual racing pigeons in a racing loft and also two or three stock pairs for breeding purposes in a small loft beside it.

My racing loft has two compartments, one with twelve pens in to race my old birds to. I like a spare pen for any bird that may temporarily lose its mate and pair to something else, this way when the mate comes home, eventually, the pen is still there to recover in. I also keep a pair of Fantails in the racing loft and use them to give my birds confidence when dropping from the races.

The second compartment is to fly my young birds from, I like them to be able to see the parent birds, they settle down to looking after themselves better. I feed on good quality corn all the year round, a mixture of wheat, peas, beans, tares and Indian corn. I mate my birds in the last week in February and, if possible, take all my young birds from the first round of eggs.

This way when the youngsters are strong on wing they have plenty of time to get the habit of climbing into the sky when taking exercise. They circle for quite a few miles round the loft on clear days and acquire a sound knowledge of the country round home before their training begins. Pigeons need to feel secure in the loft and need to like being prisoners, a wild, uneasy pigeon unsettles a whole team of birds and in confined space this is damaging to flight and body.

I spend a lot of time moving about among my birds and occasionally handling them, they get used to a voice a footstep and respond to a familiar sound. My old birds I race as naturally as possible, getting my best results when they are fourteen days sitting on eggs. I have had one or two birds that preferred a four-day-old youngster in the nest and have put down a good performance when flying in this condition, but feeding puts a strain on a pigeon and I do not really like this method.

My pigeons do not exercise much at home and I do not like to force them to fly so I give them an open loft for three or four hours each day, but remain close by to see they are not disturbed by cats or anything else. I give them two midweek tosses all the flying season and prefer them to be sent to the same training point each time, about 20 miles distant.

Young birds require a more extensive training than from one fixed point and I give mine a good spell all round the southern part of the country up to 40 miles. Immediately I start racing I go back to the routine of my old birds and give them two midweek tosses from the same liberation point each time. I find youngsters exercise more freely than old birds but I still like them to have their freedom for a few hours each day.

I do not like them to mate and do my utmost to discourage it, I have found that when a young hen lays it is very little use after the young bird season and seems to have lost its stamina. I have a small grass plot outside my loft and a bit of building waste scattered about but I also keep small containers full of grit and minerals available inside the loft. A pigeon bath is a necessity and

Mrs Barker with her winner of 1st Up North Combine Bourges. This is the lady's second time to top this massive organisation.

Derek Towers, secretary Up North Combine, verifying the winner and congratulating Mrs Barker on her second Combine win.

mine is just placed in front of the loft for two or three days a week.

I am not a fanatic about strain and pedigree, my birds are a blend of Hansenne, Barker and Bricoux and I do not pay big prices for pigeons. My best results have been from gift pigeons, but I have bought pigeons that have proved to be good breeders from fellow club members. A good way to buy pigeons is to buy from successful fanciers in your own locality, who have proved their birds by being club average winners nearly every season for years.

These fanciers usually keep about twenty pairs of pigeons and build up a good family strain. When the season ends he will have a few birds surplus and will be glad to sell them rather than kill. I do not keep plenty of birds to feel that I can give value for money if I sell, I keep pigeons because I like them and find pleasure seeing them fly freely round the loft. They are just a hobby and relaxation to me. A man is more single-minded and will dedicate himself to the Fancy and is much more likely to reach the top of the ladder.

I have had one or two setbacks since I first started flying pigeons and quite a few lean periods where the birds just did not respond however hard I tried. Fanciers nearby have quite often inquired if I have left the sport or changed my name, but this is part of the game. Small children broke into my loft and damaged a good breeding hen beyond repair and almost broke my heart, but I am still flying the birds, it looks as if I have got the pigeon fever in my blood. The good wins more than make up for the bad seasons.

My loft is situated in the best position in the Up North Combine, I have the shortest distance to fly in this large organisation. I am two miles inland from the coast and as west winds prevail most of the summer this is a great advantage in the Championship races from France. It was a vital factor in my latest success from Bourges as my winning pigeon was clocked 38 minutes before darkness.

Birds flying 60 miles further had no chance to reach home on the day. A good fit pigeon will look for water as soon as the light is good enough next morning and be on his way, but seconds are ticking away and a tired, hungry pigeon has lost that vital spurt for home and has to be content with a position down the list.

Condition in these races is the deciding factor, its feather must be sleek and close and the eye clear and bright. I find the early inland races help to prepare a bird for a Channel fly, the sustained flying for five or six hours each week is of great benefit, more so than a lot of midweek tossing when the pigeon has to be disturbed on the nest.

It was a great feeling when I won my first Up North Combine race from Hatfield in 1965. That confirmation, after hours of suspense, when the late great Mr Bill Towers shook my hand in congratulations was quite something. This time it was his son, Derek Towers, who came to give me the great news that I had managed to top it from Bourges, but I had never really convinced myself that I had achieved what so many great fanciers are striving to do, and I have not quite taken it in even yet!

It is a big feat to top the East Cleveland Federation from any race. We have had so many Combine wins to our credit and positions in the first 50 in the Combine too numerous to mention. Even in my own club we had a runner-up to the Vaux Tankard winner in 1974 and an outright winner two seasons ago. My two Combine wins stemmed directly from the club, a fellow member gave me a lovely red hen that besides breeding the young bird winner, also bred young birds for quite a few seasons that won Up North Championship Club places and money, and my Bourges winner was bred from a blue Barker cock purchased for a nominal fee from a Boosbeck Club member.

I had a very good position in the Vaux race from Beauvais with a gift bird from the late John Hall who himself had four Combine wins to his credit in his racing career. As you can see if you look in the breed books of the Up North Combine, there are some fanciers of the top grade flying in the East Cleveland Federation.

I have found in pigeon racing that it is the attention to small details that counts. The blue hen, winner from Bourges, I bred in 1971 and gave to a new starter flying in a nearby club. He flew her for three seasons but although she was what he called a good honest pigeon and he flew her down the line to Beauvais, she never got to any better than the sixth position in the club. When he gave up flying in 1973 I asked to have her back again and brought her to my flying loft.

It took me a long while to get her accustomed to the new home and I had to watch her for hours until I found a mate that she was prepared to accept. When she finally paired, after just one night out she settled straight away and I've had no further trouble with her, but I have given her a little special attention all summer, such as giving her a gallipot in her pen for corn and a little titbit most days.

This was more to ensure that she did not home to her old loft position and sit about on the houses than anything else. Some good fanciers will not enitrely agree with me but as in this instance, luck has to play a small part in the sport. I might never have had the hen in my loft if my friend had not given up the sport.

I have been fortunate to top the Up North Combine twice but it isn't the winners that keep the sport going. I think we pay too much attention to them. The backbone of this game are the fanciers who fly their pigeons to their best ability year after year and have to be content with club positions because fortune has not smiled on them. This is enough for a man with the sport in his blood and I hope a few more of them can have the luck that I have had with me these last ten years.

Many Roads to Rome

by W MATHER

1st Section, 4th Open NFC Pau

Winner of Oliver Dix Memorial Trophy NFC

It is with great pleasure that I respond to the editor's invitation to contribute an article to Squills.

I had my first pigeons when I was seven years old and I am writing this on my sixty-third birthday. So you see I did not get into print very quickly although I suppose it would be fair to say that I have had a few near misses over the years.

When I read an article I like to hear how the fancier concerned manages his birds. His methods may not be possible for another person under different circumstances but they make more interesting reading than just being treated to a list of wins.

I have flown pigeons under many different methods and situations. On looking back some of them seem quite mad but they still produced winners at the time. I can't write about all the things I have tried in one article. The last ten years have been very successful ones for me. During that time my management has hardly varied at all. I will try to explain it as briefly as possible.

My loft is the upper story of a large brick building. It is well ventilated, dry and vermin proof etc. It is situated in open fields and the only real drawback is trapping. The birds fly through an open door but if they decide to land on top or fly round there isn't much one can do about it. It faces south and I often get bad traps in tail winds.

I clean out every time I get really ashamed of the state of things. This is usually about every two months. After scraping out I put a little ground limestone on the perches and the floor and then add a liberal amount of grit or river sand. The loft is big inside but I have various partitions and sliding doors so that birds can be temporarily confined in small places for basketing etc. I have a couple of places which are kept a bit cleaner with plenty of sand and the birds are fed in same. I use Kilpatrick's grit and minerals. I don't like medicated grit. The birds eat too much just to get salt.

I separate the sexes as soon as the last young bird race is over and they never go together again until they are paired. They are given liberty on alternate days with a clean bath always before them. If the weather is bad they are kept in until it improves but still given baths in the loft. From the time they

Bill Mather holding 'Captain's Pride', winner of 1st Sect, 4th Open National FC Pau.

are separated I feed once per day as much as they can eat. They get about half an ounce of mixed seed which has had 50% of wheat mixed into it. Then they are fed with a mixture of beans and maize until they are leaving a bit on the floor. The time of this feed varies between 11am and 2 pm according to the weather and other personal circumstances. On days when there is a club show at night I would feed them very early so that they would not eat too much and have little or nothing in the crop for the show. The cocks are never kept short of food; sometimes it becomes necessary to cut the hens down a bit if the weather is mild in January and early February.

I pair all my birds some time between February 14th and 21st according to the weather. I have tried holding some back for a month or more but don't think it makes any difference. In fact I have a friend who pairs all his birds in January, lets them lead a free natural life and rear a youngster in nearly every nest; he's very hard to beat in any race.

I go to a lot of trouble to prevent fights when the birds are first mated. Any that are stupid are confined to their box with jam jars of food and water. I put the jars in a corner with a brick against them so they can't be knocked over.

As soon as they are settled down I give them the open loft and let them build their own nests. I used to continue the open hole but in recent years I have had a lot of youngsters ruined by birds fielding and picking up artificial manure and treated grain etc. So now I stop the open hole just before they are due to hatch. All the time they are feeding I only let them out late on in the evening when I am sure it is too late for them to go off to the fields. While the birds are feeding youngsters I hand feed twice a day. The first feed is never before lunch time and the second one is at night. The feed is now two parts peas and one part maize with a little mixed seed. They are always fed until some is left over.

The reason I don't feed in the morning requires some explanation. Pigeons are creatures of habit. When they pick corn up they are not really eating. They are storing it away in their crops to be eaten later. If they are used to being fed early in the morning they will have empty crops in the morning. When fed they will come off their young when it is cold. They will eat very little because the days are still short and they themselves are still fat and not getting much exercise. Then they fill the youngsters up with water and a bit of food. If they are never fed until lunch time they become used to this system and store a lot of food in the crop the night before. In the morning they have a lot of softened grain left over to give the youngsters. On this system you will find the youngsters are always full of food and the old ones always ready to grab some more for them.

When the youngsters are twenty days old I put some peas in the nest bowl each morning. This may sound dirty but it really isn't. If the nest bowl is not clean and dry the youngsters are no good anyway. When they are twenty-four days old they are removed and never fail to eat. When the youngsters are first removed I put them in a compartment on their own. They have a bath on the floor which is kept very clean. I put a nest bowl full of peas in twice a day and remove it when all have eaten. Later the bath is turned upside down and a drinker placed on it.

As soon as the youngsters start perching they are allowed to mix up with the old ones and the whole lot get open hole for the rest of the season. I lose very few young ones off the loft.

I start training my old birds five days before the first race. They are sent by road about twenty-two miles every morning including Friday and then into the race to win. During the short races I only train yearlings and one or two old birds which are still fast at short races. The birds are now fed after the morning toss at 8.30am. After this they have open loft and are fed four or five times more during the day right up to last thing at night with no rationing. The feed is two parts peas and one of maize. I am also fairly liberal with seed. The birds going to short races are not fed after 1pm on Friday and are not let out after this time.

As the season progresses the Channel birds are jumped in at about the fourth and fifth races. They are then added to the training baskets for at least ten days before the required race. I give the youngsters one toss at about three miles just to let them learn how to get out of the basket and look for home. When the Channel races start there is more room in the training baskets and I start to add a few youngsters each day until all are trained to the twenty-two mile stage.

The young birds are sent to this stage every day from a week before the first race. The youngsters have to be kept a little bit short of food for the first three races. Nothing very drastic. They are still fed morning noon and night but any that fail to come in quickly miss out. At night I go into the loft with a tin of peans and any that will eat out of my hand can have as many peas as they like. This only applies to young bird racing for three weeks until they learn the ropes. Soon the young bird races are over and another year has gone by and we are separating the sexes and back to where the article started.

I realise that one can't put everything into a short article. So I will try to anticipate a few questions that some of the new fanciers would no doubt like to ask. I don't use any pills or potions nor do I believe in them. I have tried Widowhood with some success but don't like it. I hate the look in the eyes of the hens when you are taking them away. I hate wasting good hens that can sometimes walk on cocks. I think Widowhood gives you a good trap and can without doubt give good results on easy days if done properly. In a climate like ours I much prefer the Natural System. I never fuss about the state of the wing. The late F W Marriott always said that he would have missed two of his five King's Cup winners if he had bothered about wings. I certainly don't think that soft food goes sour and stops pigeons winning. In fact one of my favourite conditions for a hen is feeding a youngster three to six days, but no older.

Now a few words about my 4th Open Pau winner flying 730 miles because without him I would not be writing this article. A certain racing pigeon correspondent christened him 'Captain's Pride'. I told him to change it to 'West Coaster' but he would have none of it.

He is a pencil blue and was 3rd in the Channel class at the Doncaster show January 1975. This seems to show two things, namely that he's a good looker and that sending a bird to a show doesn't spoil it for racing. He was bred from a Dordin cock and a hen of mixed long distance blood of L Massarella.

Although I have my race youngsters in with my old birds I also have another team of young birds in another loft about fifty feet away from the racing loft. These are usually only trained and are 'rods in pickle' for a rainy day. 'Captain's Pride' was a second round youngster and was in the spare loft as a young bird. But he was a crafty 'Captain' and found out that there was seed and peas as opposed to beans in the big loft so he moved over and got trained. In his first race he was home to top the Amalgamation easily but sat on the other loft and refused to move, so was not raced again as a youngster. As a yearling he was paired and established in the main loft. He came to win two inland races but sat out on the other loft. At Avranches he was home to top the Combine and was out 35 minutes and was still 4th Club. I then sent him to Nantes as a yearling. There was only one bird in the Combine on the day in a north wind. 'Captain's Pride' made it the next morning looking sadder but wiser. In 1974 he won one prize from 120 miles and flew two Channel races but not to win.

In 1975 I decided to make him trap by taking his hen away. A sort of mild Widowhood. Just left on open hole with the other birds but finding his hen always there from training tosses and races. In the first race of 60 miles he was 6th Club, 10th Amalgamation and pools coming with the winner but caught second bird. He won all pools in another short race from 120 miles. Then in spite of the semi-Widowhood he took to landing on the young bird loft before coming in. He won the club nom pool from Avranches and was then paired up and sent to Pau sitting nine days.

Somebody phoned me and told me what time birds were in on the South Coast. I worked it out that one had to time in at 3pm to be with them. At this distance I thought 4.15 would be nearer the mark. So it was a big thrill when the 'Captain' dropped in at two minues past three and I knew I had a good one. He had green droppings on his feet from the basket and was dehydrated. I feel sure he had not had a drink since the liberation on the Friday. His throat was 'made up' and it was some time before he could even swallow water. The next day he was almost his old self again.

Oh, I nearly forgot to say that he threw his third flight the morning I basketed for Pau.

Thanks to the old maestro Marriott I still sent him with £5 on for luck.

Of Many Things

by ALF JARVIS of A T Jarvis & son,
Bridgwater
4th Nantes National 1969, 1st & 2nd Pau
National 1971, 1st Nantes National 1976

In 1971 after winning the Pau National I was asked to write an article for Squills, an honour which many men strive for in their hearts, but which few achieve. This I did and wrote a quiet article, hoping against hope that Alf Jarvis would not show through too much.

The little man that has been on my shoulder all my life thought otherwise, which is where I begin this article. You will find as you read my account that that little man has been very busy.

I have reached my early sixties with a lot more luck than judgement. My wife many years ago gave up any hope and as long as I stay teetotal and non smoker I may, with a bit of luck, see my pension. My children of which I have six when speaking of their sire say "my father is mad!" or when speaking to me say, "Pop, you are mad". My wife just shakes her head in despair.

If any of you who have kept pigeons any length of time think you are going to learn anything new in this article forget it. This article is for newcomers only. Quite a lot of older fanciers have tried to impress me with their knowledge, many of them on the strength of a couple of club wins. It's the quiet ones I fear.

Have any of you heard the story of the wife who went fishing with her husband and on speaking to her friend said, "I did everything wrong again today, I talked too much, used the wrong bait, reeled in too quickly (and caught more fish than he did)". Well that is the story of my pigeon life. So novices, when you are listening ask yourself how many prizes has he won or what's his excuse for not winning. If he is a 'dud' his excuses will be many, amusing and in some cases hair raising.

Now, according to all the rules, I should have been content with my modest win of 4th Nantes National in 1969, and some club prizes. The little man on my shoulder said "No, Alfie boy, you can do better than that", and I decided to ease off my club racing (by the way I won every old bird cup in my club in 1969 and a Combine prize or two). In those days my club averaged between fifty to sixty members. It has over forty now.

In 1970 with a little nudge, and an east wind from Nantes I won 19th, 22nd, 26th Open and another at 73rd. I also had six out of six home from Pau. In 1971 I had a bigger nudge, with 1st & 2nd Pau National, winning the Two

Alf Jarvis holds 'The Maverick', 1st Open Nantes NFC 1976. Mrs Jarvis holds 'Pretty Lady', 1st Open Pau NFC 1971.

Bird Average and Oliver Dix Trophy for best average in the two races. In 1972 I had a quiet year – 6th Sect Nantes.

In 1973/74/75 I had a public works contract within three feet of my loft, a job that was not finished until the end of March in 1976, when the galvanised fence within a foot of my loft was replaced. However I still endeavoured to compete, winning 6th Sect, 62nd Open Pau in 1973, and 14th Sect, 113th Open Nantes 1975. A friend told me to wait at least three years and rebuild, but the whisper in my ear said "No, faint heart will get you nowhere fast!" After my 1976 Nantes win I had a prize or two in my club from Rennes and Nantes. I had eighteen birds which had flown Nantes to pick from for Pau, winning 7th Sect, 26th Open and another at 25th Sect, 105th Open. Having five out of six on the winning day.

I have for a long time had a very well-known saying in my mind at National racing. It is; "If you can't stand the heat, stay out of the kitchen". My brother in Wiltshire thinks it is that same heat that keeps me away from Palamos. He may be right. I have a very great admiration for the men who fly and win from Palamos.

As this article will seem to be all about me I will try to make it as interesting as possible. A fancier may win one National and if he wishes sink back into being an ordinary quiet sort of fancier. But I am not naive enough to think he may win two. Someone somewhere will spot him.

Now about my loft and my likes and dislikes. If you are looking for long pedigrees in my birds you are going to be very unlucky. The only pures I am interested in are pure triers and pure winners – in that order. I think I have written six pedigrees in my life. This will come as no surprise to those who know me. This is not because my pigeons have not got any. There are men in our hobby that can make a pedigree out of thin air and little else.

I winter between twenty to twenty-five pairs. My stock birds are my pensioners that have finished their stint on the road. I keep the hens until they are nine years old and the cocks a year older. I then have to harden myself.

I have not purchased any birds for many years, the last time must have been in the late sixties. The latest introduction was a gift from Captain W Mather

'The Maverick'

of Blackpool, in 1975, which was the first one for a very long time. Of course not every bird bred here is a champion, and like everyone else I breed my share of good, fair and indifferent. Sometimes I have had the feeling that the indifferent ones have outnumbered the others by a large margin. It is ony when you are perfectly honest with yourself on that subject that you start to be a fancier. If it were not so why is it that on looking through your loft your find yourself with a greater number of pigeons of a certain age than others? Please do not count your yearlings. The good breeding years will be few and far between. Some years it will be nearly all duds, another you hit the jack-pot.

I like a bird to look intelligent or wide awake. I have been told that my birds have a little eyesign (whatever that may mean), one fancier said that one of them even had a violet. I did not worry much because that particular bird's performances over the years told me all I wanted to know. The rest the basket tells me. I am very much in favour of the Belgian method of studying their feet. The first feet over the loft floor on race days are the best. The more often they are the better they are.

As this has been written for novices I will end by giving a résumé of my methods. First the loft. It must be as dry as possible with a very good flow of air. It must also be as quiet as possible, this is very important, more impor-tant than it is given credit for. The food is a very controversial question in these times and it is a subject I am not going to get involved in. I wrote about this in 1972 'Squills' saying I used beans as a base, I will leave it at that.

Exercise is also a thing I think deeply about. I am a 'flag man'. I put these along the front from April until the Pau National week, working up their exercise rate to three hours per day. I start this, or my wife does when I am at work, as early as possible, increasing it to one and a half hours in the morn-ing and one and a half hours in the evening. By the first week in June they are really swinging and are a pleasure to watch. They soon get used to the idea, if you can keep them off the tiles, or out of the fields. I have only one place for tile pitchers myself, they don't live very long.

The youngsters are only let out in the evening, after the old birds have exercised, until the week of the Pau race. By then I have finished all flagging, in that way they are not frightened in any way, and being very tame will drop out of the sky without any fear whatever.

What you sow, you reap. And you must make up your mind what you want to be. You can set out to be a long distance man or the club champion or get a lot of fun just pottering along with the odd prize or two every season. Or finish up like me – a pigeon-mad fool.

Take great care of your young birds, they are your old birds of the future. A young bird prize or two is very nice but an old bird prize or two is a lot better. Leave the young bird specialist races to the experts. They breed young birds for the job and don't let them kid you they don't.

I race my birds for the pleasure of winning and I take a great deal of care to make sure that when they go to a National marking they are as fit as I can get them. It does not worry me that they carry little money in pools. But don't worry on that score, I have no illusions about the quality of my birds, or my own ability as a fancier.

Keep Bouncing Back

by **J NICHOLSON & SON**, Hull

1st Vaux Lorimer International Falaise

Ever since my youth I have had pigeons of some sort. Shortly after the war I purchased a couple of pairs as pets and with a small number of local fanciers formed a short distance club, flying no more than two or three miles on Sunday mornings. My father raced this type of pigeon and I treasure a large, ornate clock he won with his birds some 50 years or more ago.

After one season our short distance blue disbanded, I think owing to the fact that one man won all the races (his father was a 'spiv' and knew all the wrinkles). The few birds I had were stolen and that was my lot until 1963 when a change of employment found me working with two keen fanciers. The same year my youngest son Geoff asked if he could have a pet, and as like most fathers I would end up looking after it. In the end I gave way and ended up with an unrung youngster from one of my workmates who was a good fancier and had a bird that flew Faroes, 620 miles twice.

After a lot of barracking from my workmates I knocked a window out of my garden shed and kept the bird in it, but one night I forgot to close the bob wire and our only bird finished up as supper for a cat. Three pairs of old birds were given to me by a local fancier and my garden shed ended up as a pigeon loft.

We decided to join the Hull Newland HS and at this time it was one of the larger clubs in Hull with forty-five members. For two or three years we were content to take part in the races and get the birds home, also just to go to the prize table and receive our prizes, although often small it gave us a great thrill. One race day all the club had clocked in early and we took an empty clock in at 6pm. At 7pm my young son walked into the club HQ clutching a rubber ring, to the good natured cheers of all the lads. My son who was still at school did get a wisecrack from one of the members; he did us a great favour although he did not realise it. We went home and got our heads together to beat this chap and my other son who also had the bug was 100% for it. We became better fanciers and improved our breeding and management.

We became runners-up and were top prizewinners for the next three years with old birds only. After some time we turned South to have a go at the Channel joining the Endyke HS, a club with fifty-five members, one of many clubs in this real hot bed. We were third from top in 1971. That year was our

Jim and Geoff Nicholson with their International winner.

first young bird race from Lillers, it turned out to be a bad race and a bird bred the same way as our Vaux-Lorimer winner won it when only two were clocked in this club. In 1972 we were fifth from the top.

In 1973 my youngest son and I purchased a hen to go with the sire of this winner. We concentrated on breeding from this hen, and made a big mistake. Instead of going forward we went backward. On to 1975, a pal of ours returned a hen which we had given him, and our Lillers winner's sire was seriously injured twice so we retired him from racing. We were looking forward to getting our old blood back when in January 1976, whilst I was down with the flu, once again we were wiped out by a cat and only a few survived.

My son Geoffrey was married and living away, my other son Les married and now lives in Liverpool where he races his birds. I had been off work for some time through illness and was down in the dumps and felt like giving up the sport. As my health improved my lad and I decided to give it another go.

We were offered young birds from several fanciers, some of these lads had only been in the sport a couple of seasons and were not established themselves, but came forward like true sportsmen. I should like to take this opportunity to thank them all. A fancier who works at the same firm as myself heard of my loss, he was a complete stanger to me at the time, but nonetheless gave me some young birds; they flew consistently for me and one of them won Lewes. These type of people make the Fancy great.

Our loft is 18 feet long, 6 feet 6 inches at the front down to 6 feet at the back. It has a 2 feet 6 inches corridor with all the front completely louvred. The compartments are 6 feet deep, two for the birds and one for equipment. The nest boxes are 2 feet wide of square mesh on all sides, with a sand tray under the nest box floor which is easily raked out or slid out and tipped. We do have sand on the loft floor which is removed when its gets too dusty. A 9-inch cavity wall is at the rear of the loft, this forms a kind of suction duct. There is PVC on the corridor top for extra light, but I had to put wire mesh under it as some birds tried to fly straight through it. At the moment we use the 'drop in and catch' method, but hope in the future to try the straight fly-in system, as we have had trouble with trapping.

With regard to feeding, we hand feed a good sound mixture, thus keeping the birds tame and easy for catching. We give a tit-bit of linseed and hemp. The young birds get as much as they want to eat except on basket night, then they have a reduced ration. Clean water is there all the time, and the drinkers are washed often. Training the old birds is twice a week if possible, otherwise they are raced when fit on the Natural method.

Young birds are trained as often as possible, and are sent to all or most of the programme. We do like to give the yearlings at least one single up of 70 to 100 miles. We insist that the birds must be fit before going to a race and like to see a good clear eye. My son is interested in the eyesign, but I cannot pass comment as I know little about it.

Our winner from Falaise, 'Jubilee Express', was very well bred and I

An interior view of the Nicholson loft showing nest box design.

should tell you about its grandam, 'Newland Queen' and sire, 'Red Baron'. The grandsire was bred for stock and was unraced. The grandam took about twenty prizes from the shortest to the longest race. One year she won with the fastest velocity of 2018 ypm and was 3rd Lerwick, velocity 1093. Six days later she won Dunbar with a velocity of 971, all pools etc. In one year this hen bred seven youngsters (by floating the eggs) and all of them took cards as yearlings, some as many as five, with velocities ranging from 800 to 1400. Some were turned south and took prize cards. The best of these birds was the sire of the Vaux winner, he took cards as a young bird and as a yearling, being 2nd Whitby, 2nd Thurso, 3rd Lerwick. The following year he took Federation cards being 5th Redcar, 5th Berwick and then like the dam won the shortest and fastest race of the year, velocity 1998, and then went on to win Lerwick, only bird on the day with a velocity of 772. There were only six birds home in the Federation. Returning home so late at night this Lerwick winner was almost locked out for the night. He was once again injured and retired to stock, bred two latebreds and was one of the birds killed by the cat. The latebreds were in another compartment.

The Vaux winner was bred too late to be trained or raced in 1975, but was given hard training in 1976. In 1977 he was raced to Lewes then trained and held back for Falaise and we are over the moon that he won for us. He was driving to the nest before the race, then I slipped him an egg that he took to straight away. He did have a Nottingham single up just beforehand. Now in semi-retirement we hope to breed some more of the old blood that did so well for us.

As to new starters, keep our ears open and listen to everyone, then sort out the best from it. You can never stop learning as there is always something new to learn. Pick a system that suits you and your pocket, don't be shy to try new methods if the old way won't work. A good fancier is like a champion boxer, if he takes a knock or has a set-back he bounces back for more. Take part in your club activities, then you will get to know what's going on. Have a good sense of humour, you will need it, because this is the only sport where you and your birds are thrown in at the deep end amongst the champions but don't forget, they are only there to be knocked off the top perch.

Keep it Simple

By S G BISS

1st British International Championship Club

Marseilles

I have been asked to write about my methods but of course, this is equivalent to being asked to write enough for a book which is certainly much more than could be done in the short time between the end of racing and the printer's deadline for 'Squills'. I must therefore, confine myself to what I consider are a few simple basic requirements for making a good start. I mention the word simple to emphasise that is most respects successful pigeon management is a matter of plain commonsense and that where so many would-be fanicers go wrong is that they allow themselves to make it too complicated.

As we are all aware, lofts come in many sizes and styles but the function, whatever the style, is to provide shelter and an abundance of fresh air without draughts. I suggest that no one can hope to reach the top unless his stock and racing lofts are built to meet these essential requirements.

As far as possible the construction should also be directed towards keeping the loft as dust-free as possible for the elimination of dust should be a major concern in everyday management. Dust getting into a pigeon's system can in itself cause respiratory or other troubles. Much of the dust can be germ-laden which makes it increasingly a health menace.

There must be plenty of room. I regard overcrowding as gross mismanagement. I am one of what seems to be a comparative few who take the view that the biggest enemy of young birds and the major cause of young bird losses racing is disease. Only if the loft is built on the right lines will the birds enjoy the good health they need for success. Always guard against dust and overcrowding.

The breeding of healthy first-class stock calls for the finest ventilation and I consider there is nothing better for the loft roof than pantiles – the old pantils usually associated with old farm buildings. They are still obtainable and they look most attractive. They never fit tightly yet they keep out rain and they facilitate a gentle movement of air within the loft. This excellent ventilation is achieved without opening up the front too much which of course avoids draughts within the loft which pigeons certainly do not like and which are bad for them.

It is well known that the better the ventilation the less the likelihood of respiratory troubles. If a bird brings respiratory trouble home to a loft that is

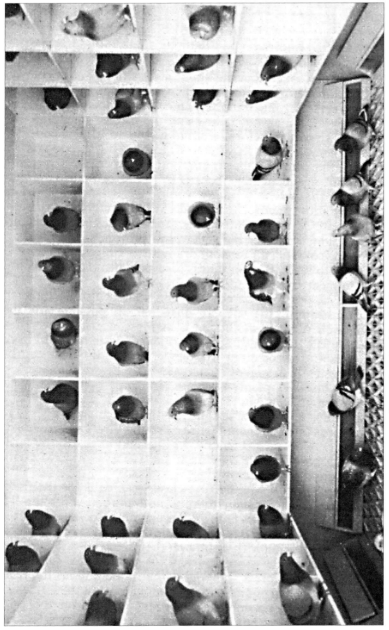

A section inside the lofts of S Biss.

properly ventilated and the birds are not overcrowded the risk of the infected bird setting up an epidemic is much less because the physical condition of
the other birds is such that they can resist infection. Compare this with an
overcrowded and ill-ventilated loft and it will be realised how quickly one
affected pigeon can contaminate the rest. Compare it also with the advice we
usually get from our health authorities when health hazards such as flu are
about – avoid stuffy, crowded premises such as cinemas, public gatherings,
etc.

   Dust carries the majority of pigeon ailments and it cannot but harm birds
that are continually exposed to drawing dust and germs into their systems

'Hillside Cuff Link'
2nd British International CC Perpignan owned by S G Biss.

especially their respiratory tract. A pigeon that is required to fly 500 miles must be able to breathe with complete freedom and take the full benefit of the air it breathes. If not, what chance has it of success? Very little, and no human athlete would have much chance in an endurance test if his breathing system was less than perfect.

The drinker is the No 1 source of infection, especially canker. The drinker, even if you are able to let each pigeon have its own individual vessel should be sterilised with frequent regularity. And remember to sterilise that damp patch on the floor around or beneath the drinker.

A word here about illness. Nowadays there appear to be so many problems compared with what there seemed to be when I first started racing pigeons. Fanciers are seeking help from their local veterinary surgeons in ever-

'Hillside Fortune'
1st British International CC owned by S G Biss.

increasing numbers but I am afraid that although they understand poultry and the treatments they need they are not familiar with pigeons apart, of course, from a few, a comparative few. They cannot always make the correct diagnosis without which the correct treatment cannot be prescribed. Here is a potentially dangerous situation. The wrong drug, or even the right drug in the wrong dosage, might produce a slowing down of an illness of temporary relief but it will not be a cure.

For instance a course of drugs for poultry might require to be given over a longer period for pigeons because the illness in the pigeon could be more virile. Another important point is that after the use of certain drugs pigeons should be given a course of vitamins because the drug, in burning out the bad things have also burned out some of the good things which require vitamins to replace them. I am against the use of vitamins in any other way. I think putting vitamins in the drinking water is playing with fire because of the dangers of imbalance.

Regarding our friends the vets; we cannot expect every vet in every town to be a pigeon specialist. We could not give them sufficient practical experience. I would however, like to see things so arranged that in each town we had two or three vets whom we could get sufficiently interested in pigeons to the extent that they could be regarded as the ones from whom pigeon men would expect the most assistance with their problems.

To build up a good family of pigeons, start by getting the best foundation birds you can afford. My advice is to get birds from one successful family, and by a successful family I mean one which has been producing and reproducing champions in each generation and has been winning within the immediate past five years. There is no value in going for a family that has not been winning for any length of time no matter how good its performances might have been over an earlier period. If a loft has not won anything of note for ten years I would have to assume that the owner has not been doing the job properly and that the winning genes have gone. Once this comes about nothing short of a clear-out and a fresh start can bring back success.

In recommending the beginner to go for one family I am not advocating inbreeding for this can easily be taken to a stage when it reduces the good health and vitality essential for success. So a cross becomes necessary and this also must be introduced with great care, for too much out-crossing breeds a mixed lot of mongrels.

In advocating having the best that can be afforded, I suggest that two good pairs are a better proposition than four inferior pairs costing the same.

Apart from physique and balance, aim also for beauty and a quiet temperament. These qualities add so much to your own enjoyment; for the true fancier should find it a joy to be among such pigeons. And see that their progeny have these same qualities. If you keep ugly, wild or otherwise ill-tempered pigeons they will breed pigeons with the same deficiencies. In my 40 years in the sport I have had no success with birds of this type. All my good racers and breeders have been quiet, good-looking and have handled well. I am of course, talking about winning in the long races. Anything can win a short race. You can take a young bird from the nest off the church roof and win a short race with it. I know it has been done; but you need quality to win the long ones.

As for my own methods: I fly my birds entirely on Widowhood and have done so for the past seven years and from the shortest race through to 750 miles but I will not go into more detail on this here. I have no fads and fancies with perhaps one exception – if that can be called a fad – I like quiet pigeons with personality – trusting pigeons that are pleased to see you and not rush away from you.

There are too many fads in the sport that cannot stand challenge and I am afraid many young fanciers are sadly misled by them. I relate the eye to the state of the pigeon's health and general well-being but I regard the eyesign theory as utter nonsense. Nor do I take any notice of the so-called wing step. There are no short cuts to what will breed or what will race. The old saying "The proof of the pudding is in the eating" can certainly be applied to this sport of ours. The proof of a reproducer is its ability to breeding winners and the best fancier I know and the only one with 100 per cent ability to select correctly is Mr Basket who tests and provides us with the true answer.

Good Pigeons, Management and Patience will win Through

by FEAR BROS
1st Open Palamos, 1st Open Pau NFC, 1st British Section Barcelona

We are honoured to be asked to contribute to the 1980 Year Book, and will try and give our thoughts on the management of our team of long distance racing pigeons, although we fly our club programme right through from beginning to end, 46 miles to 528 miles.

To start with a well ventilated loft to suit your environment is a must. Then your choice of pigeons for your foundation is vital, for if they have not got what you are looking for you will not be able to put the distance there. We were extremely lucky at the start to know such a great fancier in Bert Thatcher of Radford, Camerton, near Bath who was flying 500 miles plus.

He bred us six youngsters and left us to it. One was 4th Section, 6th Open Nantes NFC and also became a great stock bird. Also 700 milers and another in the early sixties to fly Barcelona with the BBC to be verified 37th Open.

A few years later we were introduced to Bob Legg of Seaborough, Dorset, by our neighbour E Sheppard. Bob was flying Thurso to Dorset 536 miles and winning. He had a Shearing Logan from Mr Sheppard to mate with an unrung Logan cock, of which he said he would never have had if he had been rung. Being a farmer he said he did not have the time to fly in young bird races so every year after we had got to know him, we took down a couple of Bert Thatcher's older birds to mate with his own and he bred us a dozen youngsters.

One of these (NU73P66663) took 9th Section, 27th Open, 2nd Section 5th Open Palamos 700 miles, then 1st Section British International Championship Club Barcelona 718 miles.

These two gentlemen of our sport did everything to put us right, if they said a pigeon was right for us, we never worried about the pedigree, wing, eye etc. We knew the birds had the class for what we wanted for distance racing. Bob Legg is now out of the sport.

At the beginning of the year in January our birds are fed on beans and a little Hormoform and also a touch of linseed. We continue this until about the middle of February, then we use a mixture of beans, peas and tares as a base which we use throughout the season with Hormoform. Around the beginning of March we mate most of the birds up at the same time. Then when they have laid and are sitting we have one hopper with the base mixture and one

Roly and Sam Fear outside their lofts holding their 1st & 3rd Open Central Southern Classic FC Pau 1978.

for Homon pellets which we think helps with the development of the young birds, we also have a gallipot in their nest for any tit bits such as a carbohydrate mix, linseed and hemp Hormoform, but very little of each as they will not eat their main food. We allow them to bring up their young in their first nest then we can see how they have come through the winter.

We are a little late for the early races in the beginning of May but fly these races with our cock birds. When the hens have laid again we start training them. We like to give them the basic training, but the yearlings we try and start them like our younger birds with plenty of short tosses and like to wait, if possible, until it gets a little warmer.

We like to give our distance birds a couple of races as early as possible to see if their constitution is alright. Our blue hen 1st British Section BICC Barcelona made a mess of a Plymouth race 97 miles, taking 3-4 days, so we left her alone for a whole season (1978) and with patience it paid off in 1979. We like to get the feel of the general fitness of the loft early in the season, then work it from there.

We have, in our area, a few open or what we call Sweep races around early season and holiday time, and we find it is the best training to give our birds because it is not a great distance and it saves a lot of training, but if we think a bird needs an extra toss we send or take them to Weymouth, 40-50 miles.

Red chequer cock, 2nd British International Club Barcelona 1979, previous winner of 3rd Sect, 7th Open Palamos 1975, 2nd Sect, 8th Open Palamos 1976, 36th Sect, 127th Open Palamos 1977 and 2nd Sect, 10th Open Palamos 1978.

'Clandown' mealy cock, winner 1st Sect F, 1st Open NFC Pau 1979, 536 miles.

We try to do the simple commonsense thing properly, no fads, and the birds will show you when they are fit. So whatever the condition of the nest, sitting, driving, hatching etc, they go to the race that week. It is surprising how some will fly, but the thing we try to see is that the bird which goes into the race that week is the one that we consider to be the fittest in the loft.

With our young birds we feed them a varied diet. A mixture of beans, peas and maize as a main feed, but also a small amount of Homon pellets plus a little Hormoform. We use a carbohydrate seed mix to trap them with.

Their training starts as soon as they begin to roam, then short tosses of anything from 3-5-10-20 miles. We stop around 20 miles until nearly the first race then give them one or two tosses to 30-40 miles. Then they go to the first race. We train most days when the weather is alright, anything from 15-25 miles all through the season.

Blue hen, 1st British International Club Barcelona 1979. Previous winner 9th Sect, 27th Open Palamos, 2nd Sect, 5th Open Palamos.

To find the best birds we fly them to their perch right through the season to find the genuine racers, but there is always an exception to this rule, but we only send the birds that are in good enough feather and condition to complete the race.

Hoping these few lines help someone somewhere, and remember good pigeons, management and patience will win through in the end.

'Merci Pierre'

by BRIAN MINTON of Minton & son
1st London NR Combine Berwick and 1st Section NRCC Lerwick

For the past few weeks since our editor's kind invitation to contribute to Squills, my thoughts have centered around the golden opportunity I now had to pen a few pages that could be looked on as something rather different from the usual Year book article. However, when thinking of the possible alternatives, I asked myself just what it was that we all looked forward to and expected from the fancier in question. The answer is, of course, a bit more of the same.

Our family has been associated with pigeons from the days following the second war when both my grandfathers kept pigeons at Enfield Lock within earshot of the Royal Small Arms factory. Mum's father, Bill Ford's particular passion being for tipplers and rollers, and Dad's father, Vic Minton housing the racers.

The pigeons provided me with many happy hours, spent in their company as a nipper in the early fifties, and I often think what a gap there is in the lives of today's generation of 'tele' kids, so many of whom have little or no contact with birds and animals as we did. I feel sure that to grow up with a love and respect for animals is one of the first steps to becoming a responsible adult.

A couple of moves for our family during my school days saw dad and I dabbling with the hobby, but it wasn't till 1970 that we took up pigeon racing again in earnest from our home in Brimsdown. Unfortunately, for my long suffering wife Carol, this more or less coincided with our getting married in 1971 and moving about 6 miles away to Cheshunt. So the last 9 years have seen me continually making the journey back and forth between home and the birds. I'm sure she thinks that setting up the loft was some grand plot on the part of mum, dad and my sisters to stop me straying too far! If so, it certainly did the trick.

Our first team of youngsters consisted of birds from Mr Moore of Hatfield Peverel, Bill Wills of Ealing and Mr Gorton of Hayes. All these were successful for us, winning prizes right from the start in the North London 5 Bird Club. I still think that one of our biggest thrills was winning 3rd Club and pools in our first race with one of Bill Willis's youngsters.

1972 saw our first real setback. From the start we had decided that for us the only way to fly pigeons was to allow no 'roofing' and to train the birds by exercise round the loft rather than by continual roadwork in the car. Things began to go wrong however when a couple of yearlings started pitching the thick railway cables that run along the rear of the gardens.

All attempts to break the habit failed and a matter of weeks saw the whole team strung along the wires. Now, it is impossible to maintain race fitness in a team of pigeons without at least 1½-2 hours per day on the wing, and at the height of the season when things are starting to get tough they cannot be made to do this if they are in the habit of landing anywhere except on the loft. For this reason we reluctantly decided to dispose of all the old birds except a couple of yearlings of Mr Gortons which had definitely not been on the wires. This being early summer of 1972 we had already bred our team of young birds and we were able to retain these as they were not flown out with the old birds.

This new start had coincided with some different blood being introduced into the loft in the form of a pair of yearling half-brother and sister Dordins from Chris Hollingworth, from 51624, a son of 'Ramses' and 'Louvre II'. A '67 son and a '68 daughter of Wally Hollis's famous 614/2570 (brother to West Bros 'Hollis Cock'). Also, our pride and joy, a yearling grandson of the 'Hollis Cock' presented to us by Mrs Nellie Went in July 1972.

From the Dordins we bred but two young cocks, then promptly lost the parents through the left open door of the stock section. These two went on to become our 'Champion Flysign' and his brother, both superb breeders and racers. From the Went Bros/Hollis birds, winner after winner has been produced through to the present day. Indeed the Went Bros cock, 'One Eye' as he came to be known because of a damaged eye, proved to be a superb producer, breeding 1st prizewinners with six different hens up until late 1977 when he had to be put down.

So these pigeons were to set the scene up to the present day with just about the only other introduction up to 1980 being three eggs presented to us by Les Went in September 1974 (a son of 'Startoe', a daughter of 'Skytoe' and a son of 'Champion Singtoe'). Then two Dordin hens, 'Ecaillee D'Aristo' and a Jim Biss daughter of 'Remuant' purchased in 1975, and 30640, last son of the 'Hollis Cock' and sire of 'One Eye' purchased from Went Bros in 1977.

Having 'pinched' the Went Bros pigeons we were now to go on and pinch their methods. I well remember talking to dad back in 1972 about the general subject of ambition with pigeons and how well we hoped to do with them, and we agreed there and then that it was going to be a game of patience. We had chosen our birds and our system and we would stick to them both for an absolute minimum of five years, and our sole ambition would be to do a little better each season than we had the one before. The year is now 1980 and looking back I have to admit that the birds have exceeded all expectations. They have indeed improved each year and the thing that gives me greatest pleasure of all is to be able to say without contradiction that the main quality that has become associated with our team, is that of consistency at the distance.

Of course it has all been done before, but I think the best method of illustrating our way with pigeons is to briefly run through the pigeon year. At this point however I feel I must state that I would not like anyone to think I am trying to say that their way is wrong or their birds are wrong, etc. We all know, that there are as many winning ways with pigeons as there are fanciers, thank goodness! I am acutely aware that there will be fanciers reading this edition of Squills who have more knowledge of the pigeon game in their little finger than I will ever learn. However I come to the defence of all the contributors when I say that it is extremely difficult to avoid giving the wrong impression when writing about your own birds and I am sure that every writer does his best to put it across in a modest way.

Many scribes have given their opinion as to the start of the year in a racing loft, and I must say I've yet to read one to agree with the way that I think of it. Given that the climax of 90% of serious pigeon fancier's year is the longest race, then it seems quite proper to me that the beginning of the year in the loft must be the day after, for this is the day when work must begin with the goal in mind of winning it next year. Of course, this period of time also heralds the beginning of activity with the babies, so I think the best thing I can do is to briefly run through our young bird method from time of weaning to end of racing, so that I can then concentrate on the old birds.

It's not that I think the training and racing of the babies unimportant, of course, it's very important, but I do believe that it is a very straight forward and simple part of the hobby and I don't think it requires a 'pigeon fancier' to race babies successfully.

After weaning into the young bird section the babies are hand fed twice a day until the time at about 7 weeks when they start to go outside. From then they are fed once per day in the evening. At all times from the very beginning they are encouraged to respond to the corn tin and for the rest of it's life any bird that doesn't trap when required will miss it's meal. Our babies are only flown out in the early evening, the routine being, out, approx 1 hour fly, then called down onto the loft and in to feed. Except for the odd few hours once a week when they are allowed to scratch around the garden at will. Good habits learned at this age will last forever. This is reflected in the race-trapping of our birds. The worst we have had over the past ten years was one of about two minutes in 1978.

Our ideal training schedule for babies is about 6 tosses at 15-20 miles then one toss every two weeks from 25 miles during racing. But I must say that some years I haven't trained them at all and they've flown just as well with no noticeable increase in losses. Our continual aim is to create good Combine performers, not good baby racers. There's no such thing as a champion young bird. Looking back I find that some of our best birds flew very well as babies, but some of our very best were not trained or raced at all till they were yearlings or two year olds. During the races our babies are split into two teams up to 200 miles then usually 4 or 5 are pushed on to 250 miles with the Combine. I suppose at the back of my mind with the babies is the feeling I have, that the season is much too long. I feel that many fanciers and of course the pigeons would be a lot happier, if the season were about 3 weeks shorter.

With Thurso over, the battle scarred and late comers are tended to and the old birds are given a couple of weeks 'holiday' with an open loft and little or no forced exercise. However any birds that arrive home really exhausted are got into the air again as quickly as possible; for we find that the surest way to get the strength back into the wings and the confidence back into the bird, is for it to be made to fly again. Even if for only 10 or 15 minutes at first, recovery is far quicker than if the bird is allowed to mope around the loft for days on end. As an example, our 'Stingstep' and 'Maisie', after their fine effort with NRCC from Lerwick this year were out exercising that same evening, as I was afraid of wing-lock after 8 days in the basket. By the evening of the second day they were almost back to normal. Indeed such fitness was shown by 'Stingstep' during those next few days, that Wednesday night saw him back in the basket, once again to do the Thurso trip with flying colours. A great tribute I think to Mr Lawson.

The old birds rear another chick each after Thurso, then remain together usually sitting dummy eggs till the end of young bird racing. From early August the old birds are flown 1 hour per day each, cocks and hens. The moult is ignored except for the day or two when a bird will let you know that it really doesn't want to fly, when the 10th flight is pipping through, then the bird is allowed to stay in the loft. From October onwards when the sexes are separated the daily exercising is continued. The only weather that is avoided being heavy mist or fog. It is during these cold winter months that the birds really do look a picture. I get goosebumps just looking at them, and not because of the cold!

The diet is never varied, they get beans, peas and maize throughout the year, with a little Red Band as and when we decide to give them some. I find that as the birds are worked hard throughout the year they can be given as much food as they care to eat without fear of putting on excess weight. I imagine this must help a great deal with latebreds with which we have been very successful. The bird will be exercising well, developing it's mind and it's muscles during it's growing months, instead of like many of it's contemporaries just sitting in the loft from October to February, or at best 10 minutes round the garden then up on the roof once a week or so. Although under this treatment old birds seem to manage. I think it would be a sheer waste of time rearing latebreds and then expecting them to mature into 15 hour on the wing performers.

In many ways the off season is used as a time of schooling and discipline for the race team and come February they are on their toes in every way and as fit and muscled-up as any July race candidate.

Many hours are spent during the winter months arranging the pairings, a job which I find absolutely fascinating and enjoyable. The birds are always paired on pedigree first and type and performance second, with eyesign and any other petty fads ignored. Usually only about 5 or 6 pairs are seriously bred from, but even the non-breeders are paired the way we want them so that if it is decided at any time during the season that something is wanted from

'Merci Pierre', 1st London NR Combine Berwick.

a particular bird then he is already paired the way he should be. One thing we have always made a point of doing is to split even our best breeders after a season or two together and re-paired them with a different cock or hen the following year.

The reason for this is that breeding-wise you have far more potential for future years with half-brothers and sisters than with a loft full of performers all the same way bred.

All birds, racers and stock are paired March 7. No special preparation is necessary for this event as the birds are as fit and keen as can be. First eggs appear on the 8th, 9th and 10th days and each pair goes on to rear a single chic. This routine goes on for the rest of the season except for sometimes one round when a pair or pairs will be raced 'dry'.

I've found this to be the ideal date to kick off and things work out quite nicely for our three Combine races. For Berwick, a cock's race, the race team will be feeding big chics and just looking at the hens, a good condition for this ultra keenly contested event which has produced for us a 1st, 2nd and 4th Open in the last 3 years with 8,000 to 10,000 plus birds competing. For Stonehaven a fortnight later the team will be sitting approx 10 days and race candidates can be selected at will. For Thurso 3 weeks later, pairs can be either on approx 12 day chics, or, if eggs are removed at Stonehaven, sitting 10 day eggs again. Invariably this timetable will be upset by a pair or two but by and large the routine will not vary one year from the next. Lerwick is not promoted particularly seriously by our Combine, it being lumped together with Stonehaven day. This does seem a terrible waste and it would appear doomed to remain a 100 or 200 bird race until it is given it's own day and made part of the Combine programme.

From the March pairing date all pairs are normally broken to their boxes within 3 or 4 days and once this is done they are given open loft for the entire season. Although the birds spend a lot of time on the wing, this is not taken into account when the time comes for their exercise stint. In reality if it could be totalled up, at least 3 hours per day early in the season and 4 hours later on (when forced exercise is stepped up to 1½ hours per session) for each bird would be about right. To give pigeons this amount of work on the road would entail 100 miles every day in a head wind or 200 miles in a tail wind. This is where I feel the clue lies to what I call the 'tossers trap'.

Many thousands of birds are trained day in and day out throughout the season on feeble 20 mile spins. The fancier is sure in his mind, and rightly so, that he is doing all he can, but in reality of course the team just isn't getting enough exercise. So what can be done, go further and more often? Ok this would seem the answer but the problem is of course that once May comes and the proper races start to arrive the majority of these birds are beginning to become stale. They are rarely lost and returns are good "All home in an hour", but the first one was 20 minutes behind the leaders. The racing 'edge' is being lost. Then later on when 8, 10 and 12 hours on the wing are asked of the birds the problem often becomes worse.

There are always exceptions with pigeons, of course, but I do believe that this is the root of the problem for many fanciers who fly year in and year out, always thereabouts but never really winning anything except a few early races. We know that pigeons don't have to be shown the racing line, the reason for training is the same for any athletic performer, to gain fitness. So surely if a way can be found to achieve this without resorting to John Basket it must be worth a try. With this system, as practiced so successfully by Went Bros, and so many other excellent distance fanciers up and down the country, you have all the advantages. All your time (and your pigeon's) is spent at home, hardly anything is spent on petrol. Your birds can have say 8 races with perhaps a total of 9 or 10 times in the basket during the whole year, allowing for a couple of well-timed tosses, and they will be fresh right through to the last race.

A word, before my space runs out, about this thing that has happened to me in 1980, it's called 'winning the Combine'. A couple of years ago after winning the consolation prize I thought what a fantastic feeling to be right up there with our old favourite 'Flysign' and did not mind one little bit being beaten by that fine Busschaert 'Loughton Star'. Until this year, when, with the company round mum's house that phone call came through that we'd won it. I just sat back, then jumped around a bit. It was the supreme moment. I only wish that one by one the same feeling will come to one and all of my fancier friends. Not until then did I have any idea of the enormous difference in the feeling between being the winner and the runner-up. The one thing of course that casts a grey cloud over the whole affair is that dad should have died in January 1978 just too soon to see the fruits of his labours, but we are sure he knows just what has been happening.

I will take the opportunity to thank sincerely my friends Doug, Les and Mrs Nellie Went for advice, encouragement and pigeons all freely given.

One hope for the future of the sport that I have thought a lot about this past year is that one of our knowledgeable scribes, I will mention no names, should sit down, press-gang some help with the mountain of research and write for us a big fat volume, 'The life and pigeons of Pierre Dordin of France', and send me a free copy. Once again; "Merci Pierre".

My Secret for Success

by LEEN BOERS

The racing ace from Holland, in collaboration with Wilhelm Wulfmeyer, Rinteln, West Germany

Reports of Leen Boers' fantastic competition results have appeared in most of major pigeon racing magazines in countries in which the sport enjoys a measure of importance. As a result, Leen Boers has received innumerable letters requesting information about his best racers and details of his formula for success. By now revealing his success formula, Leen Boers is redeeming a promise he made at the Pigeon Olympiad in Tokyo.

Introduction to successful flying:
On September 1 every year I separate the cocks from the hens and put them in a special loft. On the same day I select the racers I wish to keep. I then put the selected breeder pairs together for the following season. From September 1 onwards, if it is possible, I like the birds' loft to be in the sun. They also receive a special diet.

From September 10 until September 24 they are given daily Colombine Tea and a Colombine Moulting Feed. Every three days a teaspoonful of glucose is added to their drinking trough. On Mondays and Thursdays the birds are given a bath with a measure of bath salts added to the water.

I feel I should also mention, in passing, that the breeders are given the following daily diet for the whole year – one spadeful of earth from a mole-hill, two handfuls of chopped lettuce, one handful of calcium, redstone and sandstone, a little domestic salt and one handful of English peas. These ingredients should be thoroughly mixed and served in a large bowl. From November 15 onwards all the birds should be fed half each of Moulting Feed and Breeding Mix.

Mating takes place on December 1 in the late evening. In each nest box I place a nest pan containing some tobacco stalks, a dish of water and an additional dish with sandstone, calcium, redstone and vitamins. When the pairs are put together they are fed the breeding mix. At the same time the day is artificially extended by having a light burning from 8am till 8pm.

If the weather is not too bad, there should be two eggs in each box after a period of ten days. When the YBs are ten to twelve days old my feed plan comprises, per twelve pairs, initially two handfuls of English peas mornings and evenings. The bowls in the nesting compartments are then filled with breeding mix. On Sundays the birds are given a tablespoonful of vitamin complex and on Wednesdays a soluble tablet of vitamin C in the drinking water. As soon as the YBs are 12 days old, I place a second nest pan in the compartment which has already been filled with nesting materials.

Separation:
Before racing commences, the cock and the hen rear one YB. They then incubate the second egg for a period of twelve days. When the cocks have their free flying time on this evening, the hens and the nest pans containing the eggs are removed. The used, dirty nest pans are to be replaced by clean wooden nesting bowls. Placed upside down. When the cocks return on the Saturday from their first flight, they are allowed a period of approx 20 minutes with their respective hens in the nesting coop. However, before the hens are put in with their respective cocks, they should have full crops. This is important.

After the 20 minutes are up, the hens must be removed . . . very carefully. The nest pan should be turned the right way up again. the cock should never be allowed to see where the hen has gone – ie where she has been taken. He must believe that she is under the upturned nest pan. One must never enter the loft carrying a travel-basket. On Saturdays, ie on the day they return from a flight, the birds are not allowed free flying time. They will have done enough flying during the race. They are not released on Sunday morning either.

When they are let out on Sunday afternoon they are given a bath, should the weather be clement. However, the birds must be dry before they re-enter the loft; if this is not the case, the loft must be heated slightly. My birds have the following free flight schedule for the working week: Monday morning 15 minutes; Monday evening 20 minutes; Tuesday morning 20 minutes; Tuesday evening 20 minutes; Wednesday morning 25 minutes; Wednesday evening 25 minutes; Thursday morning 30 minutes; Thursday evening 30 minutes; Friday morning 45 minutes; Friday evening 30 minutes. They are then ready for racing flights.

In addition they receive the following for the subsequent four day period. Sundays – vitamin complex. Mondays – a bath. Wednesdays – a soluble tablet of vitamin C. Thursdays – the aforementioned bowl of each and chopped lettuce etc.

If the racing flights are scheduled for a longer distance, the time allowed for free flying can be extended accordingly. But it should never exceed 45 minutes. The following point is to be noted on the feeding. When the birds return from a flight, the feeding trough is to be completely filled and left for one whole hour. This applies to each meal the pigeons are given. The trough must, as I have stated, remain in position for one whole hour. What has not been eaten can then be removed. (cf of feeding of the widow-hens). I am of the considered opinion that the pigeon knows best what it needs and what does it good.

Before the cocks are taken off for racing flights their respective hen is put

FRIENDS
left: Wilhelm Wulfmeyer, right: Leen Boers. Co-authors of this article.

back into the next box. This has to be done as the nesting box is turned over, so that the cock thinks that his hen was under the box the whole time. The cock and hen have 3-5 minutes together in the nest box. Whether the cock treads his hen, or not, is quite immaterial. Sometimes it can even be a good thing if he does.

When the cock is lying in the nest pan and the hen is rubbing against his head and neck – the cock usually calls a little – the moment has come to remove the cock; but, very carefully I repeat the word 'carefully' to stress that the removal of the cock from the nesting box must be carried out gently and slowly. You should then cover the cock's head with the hollow of the palm of your hand and keep it covered until you are able to place the cock in the carrying basket. This latter must not be done inside the loft, but outside in the fresh air. In this manner the bird will not be subject to any kind of distraction and will arrive for the race flight raring to go. If you travel to the marking by car, leave the basket in the boot until your turn is called.

If you are taking part in a particularly important event, you should place a nest pan filled with nesting material in the cock's nesting compartment one hour before he is given his hen. The day before the last flight the birds can be allowed free flying time from 9am onwards, if the weather is fine. Once they are out, clean the lofts and nest boxes quickly and place a nest pan in each compartment, filled with nesting material. Then release the hens. Choose the most propitious moment, ie when the cocks are circling high about the loft. You will be surprised how quickly the cocks notice that the hens have been released. Cocks and hens can enjoy their freedom for the whole day. This trick often produces fantastic results.

Finally, two bits of advice. If it is at all possible, one should always basket all the birds in the loft. Routine and habit are of paramount importance during the race season. The birds must always be basketed every week, or for longer periods every other week.

Maintaining and Feeding the Widow-hens:

When the separation period begins for the cocks, the hens are placed in individual boxes. These should be constructed in such a way that the side-walls project forward. The grille is hence recessed and the hens cannot see each other, even if they stick their heads through the grille. If one did not have this arrangement, some of the hens would pair-off visually. This must never be allowed to happen. If it does, you can forget your racing prizes. So you must not erect the widow-boxes facing each other, either. For food, the widow-hens should be given the left-overs from the trough allocated to the widow-cocks and YBs. These will have had a full trough to select their food from. The left-overs should be liberally mixed with barley. The hens should be limited to one meal per day and it should be frugally measured. The calorie requirement of the hens is very low, as they spend five days of the week sitting in their boxes with practically no movement and no exertion.

The Young Birds:

There are some fanciers who take their YBs out of the nest before they are 22-23 days old. I consider that far too early. I only transfer them to the YBs' loft when they are 22-23 days old. For the first four days they only get English peas: for the next four days 50% peas and 50% breeding mix. For a further period of four days they have 25% peas and 75% breeding mix. After that food is as normal, ie only breeding mix. If the weather is fine the YBs should be put out from 10.00 hrs to 16.00 hrs. They should be completely free. Then a system should be introduced whereby they are allowed free flight from 08.00 hrs to 09.00 hrs, 12.00 hrs to 12.30 hrs and 17.00 hrs to 18.00 hrs. Food should be rationed to a minimum.

Once the YBs are three months old they should start private training. In the first week all the YBs should be released every morning at 08.00 hrs from a distance of 3-5 kms. In the second week they should be basketed three times a week and be released again at 08.00 hrs from a distance of 10 kms. The same arrangement should apply for week three, only the distance should be increased to 20 kms. They should also be released 20 kms from the loft on Mondays, Wednesdays and Fridays during the racing season.

All YBs must be basketed every week. Firstly they should fly a distance of 80 kms, followed by flights over 120, 250, 380, 500 kms, and another flight over 400 kms. These flights should be interspersed with shorter ones, but some of the YBs should fly over 600 kms. Daily feed should comprise the standard breeding mix. That also applies to the racers returning on Saturdays from races. On the following four days they receive the following additives, to their diet and welfare. Sundays: vitamin complex; Mondays: a bath;

Wednesdays: vitamin C, soluble tablet; Thursdays: bowl of earth, chopped lettuce etc – as previously mentioned.

Special Tricks of the Trade:
The following situations are applicable to older birds as well as young ones. If older racers, or young ones, have been alone for a period of up to about four days with an eight to 12 day old YB and an egg, one can place a small two day old young bird in the box with them, but this latter must have a crop full of soft food. At midday the racer should be placed in the carrying basket for 15 minutes and then released in front of the loft. The racer will fly immediately into the loft and should be allowed 3 minutes with its two youngsters. It should then be re-basketed for 15 minutes and taken 3 kms away before being released. This process must be repeated until the racer, be it young or old, flashes into the loft like the proverbial greased lightning after its 3 kms flight home. The bird should then be re-basketed – alone – and placed in a warm part of the house until the time comes for it to leave for the race in question. This method of race preparation can also be used for hens.

The following method is also very successful. Let's assume that the cock (young or old) has been running for four days. On the 5th day he is allowed to carry on for the morning period. Then he is to be locked out until the late afternoon, ie from 12.00-15.00 hrs. On day 6 he will be locked out for three separate one hour periods. On day 7 – the day for basketing prior to a flight – he should be allowed to mate for a few minutes, and then he should be placed in the carrying basket for 15 minutes before being taken 3 kms away and released. He can then mate for a period of 5 minutes on his return to the loft. Then he is to be basketed again for 15 minutes and taken away for a 3 kms training flight. This process should be repeated until he almost breaks down the door of the loft in his eagerness to get in! He should then be placed somewhere warm in his carrying basket until the time for the event. Be careful, however, that he is not disturbed. At this point he must see nothing and hear nothing.

The success of these methods is clearly demonstrated by Leen Boers fantastic track record.

Why Lose So Many Young Birds?

by A H BENNETT

1st National FC Pau 1982

Having been invited by the editor to write something for this years Squills on my method, I suppose when you have had the luck to win the premier race of the year the novice readers are looking to see if they can learn anything new from your writings. Well I am afraid I can only repeat things I wrote about the last time I was invited to contribute an article for Squills, and perhaps express my views as to some of the reasons for the ever increasing losses with young birds.

I am more convinced than ever, that without the right pigeons, pigeon racing can become very hard work and that the foundation of any loft is the most important thing to remember. It is very much like building a house if you don't get the foundation right it is very difficult to correct things afterwards. Being a reader of *The Racing Pigeon* for the last thirty-seven years, the first thing I used to read (and still do) was 'Food for Novices' 'old hands barred' it was always stressed to go to a good racing loft for your foundation stock, or when you needed a bird for a cross.

I have always followed this advice, and have chosen, where possible, fanciers that have had pigeons longer than myself this way the sorting out has been done for you, especially if its long distance pigeons you are looking for. I would never buy pigeons from prisoner stock as stock; birds deteriorate very quickly if not allowed their liberty. I know there are fanciers who will disagree with this statement, but I can assure the beginner he will have many disappointments if he obtains his pigeons from prisoners. Two things I cannot stand are the sight of pigeons kept prisoner and a dog on a chain, just study the expression on the face of these poor creatures!

Which I suppose brings me to the next thing, don't get too bogged down with too many theories; as my friend Andre van Bruaene says, "The man who can pick the best pigeon on looks is not yet born". So you can say I'm not an eyesign faddist but I think you can tell the relationship of one pigeon with another in your own family of pigeons by studying the eye and it is a sure indicator of the pigeons health. When a pigeon is really fit you can hardly see it blink its eyes, but when not on top form this is quite noticeable. I have my own ideas about the wing and I think there might be more in wing theory than eyesign, but there are always exceptions, and the ultimate judge as far as I am

LONG MYND HERMES
1st Grand National Pau 1982, 1st Federation Exmouth 1982.

concerned is how they breed or race.

As to my own breeding methods I try to retain my very best hens for stock. This is probably why I have had much more success racing cocks than hens, I have found if you overwork hens they never breed very robust youngsters. I have also found that latebreds have bred the majority of my best racers. My best stock hen at the present time was hatched on October 8. An advantage with pigeons bred from latebreds is they tend to moult more slowly and are therefore suitable for the important races in June-July.

Most of the cock birds in my stock loft are birds that have won from Pau several times mated to younger hens that have been unraced. All the pigeons are paired at the same period the second weekend in March, this has occurred every year for the last thirty odd years. I try to keep my management as near nature as possible at about three weeks before pairing up, I deworm my pigeons then feed on the very best grain I can obtain. Pigeons when pairing up must be as fit as possible, you only get out of the egg what you put into it. I prefer to have two young to be reared in each nest, my records show that all my champions have been double reared, the only advantage of single rearing is to save the energy of your racers.

I feed a good mixture of grain all through the racing and moulting season not for me the so called depurative mixtures, the grain one can obtain these days is poor enough, only in the winter do I change the mixture in about the middle of December. I then add 70% barley to the mixture, during the racing and moulting period birds are fed twice a day as much as they can eat. I do not believe you can over feed pigeons, "I have never seen a fat stray".

My training methods or the lack of them are probably where I differ from many. I like to give my young birds a good training with as many as eight to ten training tosses up to 80 or 100 miles after which they are allowed to develop and moult, as I have said before if your original stock came from a good racing loft you don't have to race the young to prove them. The yearlings have one or two training flights before racing starts and as many inland races as I have time to send them to but no training flights occur after the racing starts.

The old birds, ie two years and upwards get no training whatever, in fact my 1982 Saintes winner had her first toss of the season at Rennes which is a distance of 311 miles. I not only believe training old birds is unnecessary but I sincerely believe it is detrimental to the development of long distance pigeons. My own pigeons have an open loft and fly freely at home, stock birds and racers alike. Perhaps if I lived in a town I would have to revise my methods. My own pigeons are raced both Natural and Widowhood, up to and including Pau.

I first flew pigeons on Widowhood in 1952, and I can assure those that have not practised it long that they are in for a few shocks it creates as many problems as it solves "interfere with nature and nature will take her own back" is a saying my father used to use, which is very true when it comes to flying pigeons on the Widowhood, just note in your own area a few of the fanciers that have been practising it for a few years I'll bet a lot of them are not winning as much as when they started! Also just observe the tremendous losses with young birds which have occurred these last few years.

Some writers talk about sun spots and such theories as the reason for the losses, all that I can say is that we've always had the sun but not some of the present methods of racing such as Widowhood which if you are not careful brings about stress especially in widow hens which is then transmitted to the young pigeons bred from widow hens. Also the import of thousands of Belgian pigeons, not all of which are as good as their pedigrees might suggest has not helped the situation, as most of these are bred from Widowhood pigeons.

A.H.BENNETT.
1982 NATIONAL LONG DISTANCE CHAMPION
1st GRAND NATIONAL PAU
644 MILES 6928 PIGEONS
1st TWO BIRD AVERAGE
1st THREE BIRD AVERAGE

HURST WOOD
ALL STRETTON
CHURCH STRETTON
SHROPSHIRE SY66LA
CHURCH STRETTON T2274E

THE 1982 PAU WINNERS

   Another cause of stress is the nutritious value of the grain we are able to obtain, which is certainly not as good as it was thirty years ago. This has all been brought about by the commercial necessity which has required farmers throughout the world to grow new varieties of grain producing greater yields per acre regardless of the nutritional value, all these things have helped to bring on stress in many lofts which in turn brings on illness although the average fancier is not aware of this, it takes on a mild form of respiratory disease this to my mind is the biggest cause of losses in young birds.

   Now just a few details about my Grand National winner, he is a medium sized bird unraced as a young bird flown to Rennes as a yearling and in 1982 won 1st Federation from Exmouth clocked from Nantes Nat, and then 1st National Pau being the first pigeon for fifteen years to win the National at over 600 miles. I also clocked five on the winning day which gave me the two and three bird averages. The sire of my winner which we have named 'Hermes' also sired my first National birds in 1980 and 1981 and is bred direct from a five times prizewinner from Barcelona. The dam is bred down from Andre van Bruaene's 'Electric' which was the ace pigeon long distance of all Belgium in 1972 so you might say 'Hermes' was bred to fly the distance.

   In closing I hope the newcomer to long distance pigeon racing gets as much pleasure out of it as I have had for the last thirty five years and can assure him one good position in the National will give him more satisfaction then winning fifty 1st prizes in his local club.

A Few Views on Long Distance Racing

by COLIN & KEITH BUSH

1st & 2nd Midlands Championship Club Lerwick

Winner from Lerwick Colin writes:
As you will see from the title, I will give a few views on the management of long distance racing pigeons. I underline long distance as I have no interest in the sprint scene.

The Loft

The loft is of wooden construction with asbestos roof. It is fitted with traps to allow the pigeons to be on the open hole system. Ventilation is through wire mesh, this being 3 feet high along the front top half and a 3in mesh ventilation along the back of the loft at floor level. The front was left uncovered until end of 1982 when it was covered with Norplex sheeting which is adjustable for ventilation.

You may wonder why the loft was closed in after the 1982 season and I will try and explain. The 1982 summer was a rather cool damp affair and I found I could not get the pigeons into condition. Looking back through my records, I found that my pigeons had always flown at their best during warm summers. It therefore seemed reasonable to assume that the temperature in the loft was not high enough, hence the reason for closing in the loft. It seems to have worked as the pigeons looked far better and the racing results from the distance races in 1983 were very satisfactory.

Feeding

Feeding is done with the hopper system for the old birds, grain being before them all the time. The young birds are hand fed twice a day with all the food they can eat. I emphasise the point all they can eat. The grains I have fed have varied almost every season, with the protein content always kept high. The last 2 seasons peanuts have been added as the long races approach. I don't know if it did them any good, but the pigeons loved them and the results are encouraging.

I think for distance racing, if you feed a balanced diet of approximately 60% protein type grains to 40% carbohydrate type grains and the pigeons are good enough, success should follow. The only seed that is used is linseed, a small pinch of this is fed to the old birds on their nest box fronts every evening with the young birds receiving a handful between 30 young birds. Grit and minerals are before old and young birds constantly. Oyster shell being the grit of my choice.

Race preparation and general management

All the pigeons are paired during March. When they are settled on eggs about 10 days, the eggs are taken away from some of the best pigeons and floated under the yearlings. They then go to nest again and rear the second round. This allows two pairs of youngsters old enough to race and at the same time retards the moult in the distance pigeons.

When the Lerwick candidates have reared this round of youngsters and laid again they get 2 or 3 training tosses and into the 4th race at about 100 miles. They get the next 2 races at 140 miles and 200 miles which takes them up to 4 weeks before Lerwick, when they are stopped. Their eggs are taken away on the Monday, by the time 2 eggs are laid it is usually about 18 days before the race date. This leaves the week prior to basketing for training. They are trained intensively all week then given 4 or 5 days rest before basketing day. During this period of rest they can be seen to come into condition. Most of the pigeons therefore go to Lerwick sitting up to time or even a little oversitting on race day.

The races up to 200 miles are regarded as training for both old birds and yearlings with no training being done midweek at this stage. With all the old pigeons going to Lerwick it is left to the yearlings to fly Thurso, nearly 400 miles. The only time the old birds are doubled back is if they get an easy Lerwick race. It will be seen that from this type of preparation the old pigeons are only in condition for a short period each year, at the time of the longer races. This I feel to be important, to get the very best from them. If your ambition is to win distance races concentrate on them, to try and win all the way through the programme will usually leave you struggling at the far end.

The young birds get very little training before their first race or in-between races. Young bird racing is not taken seriously and is regarded as education. Most of them go to the end of the programme but some are put on one side and only raced lightly.

Another change I made at the beginning olf 1983 season, was to put all the pigeons, old and young, on straw deep litter. The contentment of the pigeons on this system has to be seen to be believed. I think that is the end of the cleaning out ritual for me. The lofts are sprayed out 4 times a year with sheep dip, a very potent insecticide, no problems with insects of any kind have been experienced since this was used.

Stock selection and breeding

The selection of stock is the most important step of all. To know where to buy these pigeons one must make a study of the long distance race results at the classic and National races. Study these results carefully, they are the most important item that appears in the pigeon press. Look back through these

'Lyndale Double'
Chequer pied cock GB81X11243. 1983 1st Sect, 1st Open MCC Lerwick.
Bred and raced by C Bush, South Normanton.

results over a period of not less than 10 years, longer is preferable, and take note of any names which consistently appear well placed in the result. Another thing to look for in these results is to take special notice of the person who consistently wins 'out of position', the hardest thing of all to do. You have only to look at the results over a few years to see where the bulk of the prizes are won. Go to the person who can overcome these odds because he is the fancier with real class pigeons. When you have sorted out the fanciers contact them and try to buy pigeons as closely bred to the winners as possible. Don't haggle over the price, good pigeons don't usually come cheaply and these fanciers are not normally keen to sell anyway.

My method of breeding is to inbreed to good pigeons and bring in a cross occasionally (selected as described) from these crosses good pigeons will often occur. First you have to establish your good inbred family before you can put a cross into it, otherwise you will be just cross-breeding and the odds of getting success will be lengthened.

Both my 1981 and 1983 Midlands Championship Club Lerwick winners were inbred pigeons and provided ruthless selection has been carried out, my opinion is that this system is the safest way to proceed. Don't inbreed for the sake of inbreeding. Inbreed to good pigeons and then practice ruthless selection with the basket test. This is inbreeding with a purpose, the only reason to inbreed, but first you must obtain your good pigeons. It is pointless to inbreed until you have.

Having established a winning family of distance pigeons there is always room for improvement and I can see in my own team a need to do just this.

Don't sit back and let the others overtake you, always be on the look-out for something good. Not all the crosses will be successful, therefore, be on the look-out most seasons. The ones that turn out to be good will be infrequent.

Finally thank you to a dedicated and helpful family, Mary, Mark and Joe.

Brother Keith 2nd from Lerwick writes:
I have been involved with pigeons since early childhood, and along with my four brothers, all of whom now have their own lofts, spent many happy hours helping my father and learning from him. Although there are benefits to be had from having a successful father, twice 1st and once 2nd King's Cup winner, it does make it difficult to get recognition for ones own performances but this in all probability makes you more determined to succeed.

After university and marriage I set up my own loft in 1975 with about twelve youngsters bred by my father or from birds obtained from him. These flew the programme through to 250 miles but without any success. However, out of these came two good distance hens so I have never bothered too much about young bird performances being quite happy to get them home in good condition.

Being a retail pharmacist my time is somewhat limited as I work every Saturday and frequent rotas until 6.30pm. I therefore decided to concentrate my lofts towards the thing I enjoyed the most – distance racing. My 20-25 pairs of racers are kept in a 16ft x 6ft loft and raced on the Natural system. They have a deep litter of straw and once the youngsters are weaned have an open hole, but no forced exercise. They receive no training whatsoever before the first race and they have approximately 600 miles of club racing before their big event. Most are sent sitting from 14 day eggs to small young-sters and get 3 or 4 training tosses the week prior to basketing.

The young birds have about ten tosses to 30 miles and most of them fly the programme through to 200 or 250 miles. I never force the youngsters and try to dissuade them from any nesting activities. They are fed the same general mixture as the old birds but are fed twice a day with more beans added at night. Both have grit and minerals always before them. A sprinkle of linseed is given in the moulting period and up to 50% barley in the winter depend-ing on the weather. Water is changed three times daily, my wife doing the midday change and retopping the hoppers if necessary.

With breeding I try to pair two good birds together or pigeons bred from two good ones. I believe that if you cover all four corners (ie all four grand-parents are good birds) then you have a chance of breeding something worth-while. You never know where the good ones will come from. I like youth in my matings and have found quite a few of my best birds have come from latebreds, so each year I breed a few for my own use after racing is done. Both my winner of 2nd Lerwick and 1st Thurso were from latebreds.

My Lerwick hen has scored well from every race point from Berwick through to Lerwick and as she is breeding the goods may well now be left at stock. She was sent sitting fourteen day eggs and handled really well after her 12 hour fly having lost hardly any weight at all. She flew through to 200

'Lyndale Naevus'
Mealy pied hen GB79X08932. 1979 1st Club, 1st Fed Morpeth 142
miles. 1980 Thurso on the day 388 miles. 1981 1st Sect, 173 birds, 1st
Open, 653 birds Midlands Championship Club Lerwick, vel 1410. Bred
and raced by C Bush, South Normanton.

miles before going down on eggs for Lerwick. She then had three tosses the
week before basketing.

My Thurso winner flew Lerwick on the day three weeks before Thurso.
She was spare at Lerwick but took to a cock, was very keen on her eggs. She
had one toss the week before basketing and went straight to her nest on
arrival. With both my birds I got most enjoyment from seeing their arrival
and knowing they had flown a good race and the fact that I just got beaten
from Lerwick didn't really disappoint me, perhaps because it was my brother Colin who beat me.

Looking to the future I would like to fly consistently from the distance
events but providing I continue to get the same enjoyment and relaxation
from my birds I will be well rewarded. To all fanciers, whatever your goal
may I wish you every success and my thanks to the editor for inviting me to
write this article.

Barren Plans for Olympus Triumph

by ALEC KEEVIL

1st Welsh Grand National FC Lerwick

I would like to take this opportunity of thanking the editor for allowing me the honour of contributing an article to Squills 1985 Year Book.

I first kept pigeons in 1938 and I can still vividly recall the trauma I experienced at the commencement of the war, when, owing to the restrictions imposed, they had to be disposed of. I can recall all these years later the grief I felt on seeing a small pile of crumpled bodies on the garden path. At the end of the war, I obtained some more pigeons and I have never been without them from that day to this.

In 1955 I commenced racing in the Caerphilly HS. My wife has never forgotten that we returned from our honeymoon one day early, in order that I could compete in my first race. I sent four birds and won 1st prize at my first attempt, against an entry of 440 birds. I flew from this loft at the rear of my parents' home at East View for twenty years.

In 1975 I erected a loft in the garden of my present home. My lofts are of wooden structure and my main racing loft measures 24ft x 10ft. It is many years since I completely closed the front of my loft with PVC sheeting. Ventilation is by means of louvres. In my main racing loft there are no perches, only nest boxes of the Widowhood type. The loft is perfectly dry and houses 24 pairs of birds which are raced on the Natural system.

Since that magical day in June, many fanciers have asked me how I won what I consider the greatest test for any pigeon flying from the North into Wales – Lerwick. If I knew the answer I would kick myself for not doing it before. What I can tell the readers is the planning and preparation which eventually brought about my achievement. All my life I have derived tremendous pleasure in conditioning pigeons to fly successfully over long and arduous journeys. Winning is an added bonus.

To this end, many years ago I obtained birds from the late Edwin Payne of Cwmgwrach, West Glamorgan. These birds had for many years flown hard Lerwick races and figured prominently at National level. Edwin had obtained these birds from the late Jack Felton of Merthyr Tydfil which were of Westcott strain. I also obtained progeny from Mr Massarella's famous 'Grizelda'. My National winner in 1984 was a 4-year-old grizzle cock which I have named 'Olympus'. The grizzle colouring comes from his famous ancestor.

This was my number one priority, to obtain birds of a 600-mile winning family. There are many birds capable of flying 500 miles, but the numbers drop dramatically when asked to fly the extra 100 miles. Needless to say, birds which successfully fly 600 miles year after year, by their sheer endeavour prove that they have the necessary physical attributes for prolonged flight.

This is the philosophy that I have adopted. 'Olympus', dam flew Lerwick seven times and on six occasions was placed in the prizes, winning many positions at the highest level. Having obtained birds of proven background, it is now up to the fancier to bring them to peak performance – this is how I prepared my National winner.

I paired my champion 'Olympus' up on 21 March 1984. Prior to this all birds had been treated for worms, canker and coccidiosis. Owing to work commitments, I was unable to spend much time with my birds and they were left for long periods on their own; never, I hasten to add, without water, minerals, grit etc, and a hopper full of high protein corn, ie beans, maples and tares. Although the birds picked what they liked first, no corn was soiled or

'Olympus'
1st Caerphilly HS, 1st Welsh Grand National FC, 1st Welsh North Road Fed, 1st Welsh Combine Lerwick 578 miles vel 1171. Bred and raced by Alec Keevil.

wasted. I also fed by hand maize, barley, wheat, oats, dari, hemp and linseed. I have always believed in a wide and varied diet.

As the season progressed, gradual exercise at home was encouraged. This was done by simply closing the loft doors whilst at exercise. They soon learned that they must fly whilst the loft doors are closed. This method of exercise is indoctrinated into the team each year, ie the older birds who have been on the system for many years teach each year's yearlings.

In mid-May, he had his first basketing of the year, two tosses from Hereford, which is about 40 miles. On May 19, he flew his first competitive race of the season from Lancaster (170 miles). In this particular race I won 1st & 2nd Club with two leading Fed positions. 'Olympus' came about one hour behind the winner, but as always showed no sign of stress or effort. Exercise at home was increased, followed by two more 40-mile tosses. On May 23, I removed his hen from the loft.

At this juncture, I should add that he had only ever been paired with one hen. This hen had always been barren and because of this I was able to create circumstances which perhaps would be difficult in a normal pairing. One thing I had noted in previous years with this pairing was that the drive to nest as associated with normal pairings was not present. 'Olympus' just looked at his hen. With this in mind I continued his preparation.

On June 2, he flew his second race of the season from Roslin Park (295 miles). He returned about half an hour behind the winner to find his hen awaiting him in the nest box. Exercise was again increased. One week later he had a 90-mile toss. I now became aware that both cock and hen were becoming anxious to sit. On June 9, I placed the first egg in their nest pan, and they both immediately began sitting. Two days later I placed a second egg. On June 15, three days before basketing for Lerwick, he had his final training flight, another 90-mile toss. He homed in perfect condition and rushed to his eggs.

On Monday June 18, he was basketed for his first attempt at Lerwick (598 miles). As the week progressed, one's thoughts centred on the birds en route to Lerwick. The weather forecasts were watched with added interest and analysed. On Friday June 22, news came through that the birds were away at 8.15am in a WNW wind. The race pessimists immediately predicted a smash, stating that the wind was far too strong from the west and would take the birds across onto the Continent. The optimists predicted a fast race with plenty of birds on the day. I just queried in my own mind why the liberation was 2¼ hours after the proposed liberation time. As the day progressed, it looked as if the optimists were going to be right. The day was clear with good visibility and the wind NW.

Our liberation at Lerwick had been simultaneous with the North Yorks & Humberside Fed and the South West of England Lerwick Club. With some members of the Welsh Grand National located in North Wales and flying some 100 miles shorter, it was expected that news would filter down the pigeon grapevine of arrivals in North Yorks and Humberside, followed by the North Wales birds.

At about 8.00pm I spoke with that superb fancier and gentleman of the sport, Albert Bennett of Church Stretton, who is situated about 80 miles north of me and on a direct line of flight. He informed me that the visibility was 'one hundred miles' and the wind NW and predicted day birds. At home the visibility was perfect and the wind NW; one could not have picked a better evening. As the evening began slipping away towards night, a little anxiety came over me. Had the pessimists been right? As dusk fell, I put the lights on in the loft and waited – no luck.

Darkness had now fallen and the phone began to ring. No birds in North Wales. Another call, one bird into Yorkshire at about 10.00pm flying about 456 miles. Out with the calculator, to be on for birds should be arriving in South Wales at about 8.30 in the morning. Check clock, loft lights out, a few more phone calls, everyone seems agreed anything before 9.00am would be good.

The race was due to open at 4.30am. Was there need to get up. I kept thinking of what Albert Bennett had said: "Visibility a hundred miles, wind NW". I decided to rise early. At 4.20am I rose. It was dark. A cup of tea and I was at the loft at 4.30am. I doubted the wisdom of my early morning vigil.

Situated two miles to the NE of my loft is a range of mountains rising to about 1,300 feet. The sun had not yet appeared above the mountain. The morning was clear, nothing was moving. At 5.10am I was gazing hopefully towards the north, wondering hard for the birds who were away when my attention was drawn to something white moving against the background of the dark mountains. Was it a seagull? No, there was something familiar about the wing-beat. Yes, it was a pigeon coming low and fast from the north. My heart missed a beat. It's one for me. With a swish of wings he passed in front of me on to the trapping platform in front of my loft. I quickly clocked him in. It was 'Olympus' my grizzle cock.

I was transfixed. Soon I came to my senses and allowed him into the loft. He immediately went and sat his eggs – the time 5.13am – he never ate or drank. My first thought was that he should get on the National sheet. As per National rule, I rang through to record his arrival. At about 5.30am I had a call from Tal George, the secretary of the WGNFC, informing me that he was the first bird recorded. It never dawned on me that I had clocked the eventual winner. I began to look for a second arrival. At 6.30am, another call from Tal George: no more birds, but early days, fanciers have two hours in which to record arrivals. A call from Larry Phillips, secretary of the giant Welsh North Road Fed congratulating me on my arrival. More calls from Tal George, no more birds recorded. I began to think, was this time going to be good enough. At 7.45am, another call from Tal George: a second bird recorded to Lance Boswell at Merthyr Tydfil at 7.10am. At 7.50am – Radio Wales – Dan Driscoll. Dan Dan, the pigeon man, with liberation times for the Welsh Feds – congratulations to Alec Keevil, probably winner of the WGNFC Lerwick race. I nearly died.

Back to the loft for another look at 'Olympus'. He still looked in superb condition and had retained good body weight and as if to prove a point he took off and made a few circles of the loft as if to say, "Yes, I could have gone further". So ended a few of the most traumatic hours I have ever spent in connection with this wonderful sport of ours.

In this one historic flight 'Olympus' won 1st Caerphilly HS vel 1171, only bird in race time, 1st Welsh Grand National FC 894 birds, 1st Welsh NR Fed 878 birds, 1st Welsh Combine 1,096 birds, recording the highest velocity of the 1,350 birds liberated and beating the second bird by over two hours.

On arrival home, 'Olympus' had dropped his third flight. I have noted over the years that birds in this wing condition often perform well. Last year I won 1st & 2nd Club Elgin (420 miles) flying 15 hrs 24 mins and 15 hrs 38 mins respectively. Both these birds carried the same wing condition. In the same week that I won the National, 'Food for Novices' by Squills in the RP was making the same observation in relation to this wing condition. As I pen these notes, I have been informed that I have won the coveted Dan Watkins Memorial Trophy awarded to the fancier obtaining the best average in the five Scottish races Roslin Park (295), Perth (334), Elgin (420), Thurso (485) and Lerwick (598) competing against over 3,500 fellow fanciers.

How perceptive of the late Dan Watkins in this age of sprint pigeons to remind fanciers of the object of forming the Welsh Homing Pigeon Union – to encourage the sport of Long Distance flying.

Long may it be.

For the record, 'Olympus' flew the programme as a YB through to Roslin Park (296 miles), Thurso as a yearling winning 3rd Club, 37th Centre Sect 2,770 birds vel 1045, 67th National 4,358 birds, 1983 3rd Club Elgin clocked at 5.20am second morning. He was vaccinated together with all my birds against paramyxovirus in the autumn of 1983.

Loft of Alec Keevil. The greenhouse in the foreground and vegetable garden show that Alec has barely enough time for other interests.

Even as I write these notes, South Wales, and my locality in particular, has once again been ravaged by the spread of paramyxovirus. To those who question the effects of vaccination: on August 18 I vaccinated some young birds; on August 25 one of these birds won 1st Club, 1st Centre East Sect, Welsh NR Fed against 4,553 birds – need I say more.

I hope what I have written has been of some interest to my fellow fanciers.

Almost a Hat-Trick
by GEOFF KIRKLAND of Coalport
1st & 3rd Sect, 2nd & 8th Open Rennes MNFC; 1st Open Nantes NFC; 5th Sect, 45th Open Pau NFC; 1st & 7th Sect, 1st & 11th Open Angouleme MNFC

I can never remember a time when pigeons were not kept by my family, my father had them when he was a young man and also two of my uncles. I raced for many years in partnership with my father and after I got married and had my own place I started racing on my own.

During these early years the old Natural methods were practised with a fair amount of success, then in 1963 I realised that these methods would not be able to live with the Widowhood system. I have flown Widowhood ever since. The one thing I did not like about Widowhood racing was the waste of so many good hens, so I devised a system for racing my best hens, which has paid off very well.

My present Widowhood loft was designed by me after a lot of reading and visiting a lot of Belgian racing men. It is 32 foot by 8 foot with three equal sections, each section has sixteen boxes. Outside of the loft is exterior ply and the inside is lined with white hardboard, there is a false pegboard ceiling under the apex roof. The front of the loft has three sets of open doors and is half glazed with frosted glass. For ventilation I have two sets of adjustable glass louvres in front of each section and a louvre in each gable end.

The bottom boxes are set up about two feet from the floor as I have found that the cocks do not like the bottom boxes. There are no perches at all but I find that some cocks like to perch on the partition doors.

The loft faces south east right down the Severn Valley and is situated half way up the slope of the garden in the orchard, it is actually higher up than the roof of the house. Some people say that a lot of trees around a loft is no good for the pigeons, but my birds do not seem to know that because they are surrounded by old apple trees and love to sit in them for hours when they are on open hole early in the year.

Over the last few seasons I have concentrated more on National racing than club and Federation racing and this season, 1985, everything was geared to the National racing programme from the word go. I paired my widowers up later than usual with the idea that I would get them to peak when the good races came round. They went to the first race with a big youngster, the second race driving or having laid one egg and it was only the third race that they were put on Widowhood. Two things which I have done differently in

Geoff Kirkland inside his loft. *Picture by Albert Dean.*

1985 was first I kept a thick layer of wheat straw all over the loft floors and secondly I let all the cocks pick their own hens and boxes. This has worked very well and they have never raced better.

During the time that the widowers are paired they are on open hole all day unless the weather is very bad. I will sometimes give them open hole for a day when they are on Widowhood.

The first few races on Widowhood in 1985 I showed the hens for a few minutes before they went and also gave a few short tosses to teach the yearlings the system. After this I did not show the hens very much at all, in fact all the club Channel races which were basketed Wednesday or Thursday I did not show the hens so as not to disturb the National birds which were staying at home.

Feeding for the Widowhood cocks is standard depurative early in the week and then racing mix for the last few days. To the depurative I add a little Hormoform and to the racing mix I add Hormoform, Homon pellets and plenty of peanuts as the races get longer. I do not use much small seed mixture at all, if you use too much seed the widowers will not last the full season.

The few hens that I race are in a small loft next to the widower's loft. They have one section with 12 Widowhood boxes and one section 12 'V' perches. They are paired in the usual way and rear one pair of youngsters. When they are on the second nest they are put on Widowhood and the hens are placed in the small section with 12 'V' perches. They are flown like this through the inland programme, seeing the cocks before they go to the races. When the longer races come round they are paired again and left on open hole all day. They are very keen on their nests after being separated for so long and I have had some very good performances from them in the Channel races. Feed for the racing hens is the same as for the widowers, ie depurative early in the week and then built up for the last few days with the strong mixture.

First classic race of the season was Rennes MNFC and I decided to use these hens in that race and leave the main team of widowers for the Nantes NFC the week after. The hens were re-paired so that they would be sitting 10 to 12 days. I entered ten hens in this race and my first bird was a yearling having her first race across the Channel, she was 1st Sect, 2nd Open, just beaten by a decimal. Second hen was a two year old, she was 3rd Sect, 8th Open. I had a perfect trap with the first one and could not have saved any time, so I must congratulate the winner, Mr Chapman.

The next week was NFC Nantes, I entered 15 Widowhood cocks. They all had one Channel race and then rested for two weeks prior to this race. My first bird in this race was a two year old dark cock, which was well clear of my next bird and I thought he would be well up in the result. On the Sunday it looked as if he could be the probable winner and it was later confirmed as 1st Sect, 1st Open NFC Nantes, an ambition had been realised.

The next race was two weeks later, Nantes with MNFC. This turned out to be a difficult race with a lot of birds missing. I timed three birds in, the first being 12th Sect, 33rd Open, a Widowhood cock, and the second 16th Sect,

40th Open, a paired hen.

I now had to decide what I was going to send to Pau NFC, and decided on three widowers and two paired hens. this again turned out to be a hard race for this area. My first bird was a three year old widower which had been in the Angouleme smash of the year before, he finished 5th Sect, 45th Open and was well clear in the clock station.

Last classic race of the season was Angouleme with MNFC, by now the widowers have flown a few miles in Channel races as I think nothing of giving some of them five or six races at the distance, including the yearlings. After a lot of thought I decided to use the oldest trick in the book for sending widowers to the last race of the season. The day before the race about 6 o'clock I let the cocks out for exercise and while they were out I opened all the boxes and covered the bowls with a thick layer of straw. I then let all the hens out with the cocks and left them open hole until dark, this can only be worked once in a season. The cocks were basketed next morning with some of them starting to build nests and others calling in the bowls.

I had my first bird in 10 hours 17 minutes for 495 miles, he won 1st Sect, 1st Open MNFC, the second bird, a yearling, was 7th Sect, 11th Open, so you see it really does work, if you catch them just right. I also timed the nestmate to the Angouleme winner from Saintes, 481 miles, on the same day which finished 2nd in the club. What was even more satisfying was that the sire of these two cocks was 1st Open MNFC Nantes in 1981, so it is a case of a classic winner breeding a classic winner.

A few things I did not mention at the start of this article are that I use the usual preventive medicines before racing and always give multivitamins midweek when racing. I do not train the widowers after the first few races but they do fly very well at home.

My young birds have a separate loft the same size as the widowers but with Sputnik traps instead of open doors, I find they are easier to control with these. I like to have youngsters all the same age if possible and from the time they are weaned until I start to train them they are on open hole all day, the

Geoff Kirkland's Widowhood loft, housing 48 cocks in all.

only time they are called in is when the widowers are exercising. Early in the year when they are just starting to fly we sometimes have trouble with female sparrowhawks, which will take them off the loft top in a second. Later on when the female sparrowhawks are sitting and the smaller males are hunting we never have any trouble. Towards the end of the old bird racing I start to train them and will train them every day, weather permitting. All of them have to race and I like to get them all down to the coast, the only time they stop at home is if they are rough in the feather, especially cover flights, or when the middle of the tail has gone.

I have tried nursing the cocks and giving them a couple of races as recommended in a lot of Belgian books but I have found that these are the first to go down, the first time they hit a hard race as yearlings. All my best birds have been through the mill as young birds even if they have not won any prizes. I feed them a light feed in the morning and as much as they want at night of good racing mixture. I do not believe in flying them to the corn tin for the sake of a few prizes.

In conclusion although I have won everything there is to win in club and Fed, racing for many years, National racing gives more satisfaction in one race than all the others put together. If you are thinking about it, have a go, you won't win if you do not send.

Selection – The Dutch Way

by PETER VAN DE EIJNDEN
European Marathon Champion 1986

What is the best type of pigeon? The best type is always the one which produces good racing results. It is extremely difficult to give a definition of 'the best type of pigeon', so selection becomes the key.

When our pigeons are a year old we transfer them from the young bird loft to the racing loft. We then carry out selection in the hand. After that the basket becomes very important, in a way it selects for us.

When we select pigeons by hand, we observe the pigeons on the following characteristics, or lack of them. We prefer a slender pigeon, with a fairly deep breast-bone, long enough though to add to the breast muscles. The breast

Peter van de Eijnden outside his racing lofts, on this occasion he was waiting for his entries in the Barcelona International. His good results in this race helped make him European Marathon Champion for 1986.
Picture by Rick Osman.

itself should neither be too broad, nor too slim. It is essential that a pigeon has a good, hard coccyx (upper-part as well as lower). It must not be too heavy, but has to be strong. I strongly believe that when one holds a pigeon in one's hands and gently presses the upper part of the coccyx with one's thumb, the bottom part should not move, and the separate tail bones should not move from one another.

The whole body must have a harmonious build. The forearm should be short and strong, and may not show any process from underlying joints. The forearm should be close to the body when the wings are at rest. The wing should feel strong over its complete length. Wing feathers' plumage should be large and even. The last three or four primary flights should be a little shorter than the rest. We don't want pigeons with wide and flat feathers. Many Janssens of Arendonk pigeons showed wide and/or flat feathers fortunately not all of them. It is just a matter of careful selecting. Wide, flat feathers have hard and coarse ribs which are a disadvantage because it makes the wings less flexible. We have seen few successful pigeons with wide, flat feathers.

What we especially like in a wing is a somewhat smaller feather which, if one looks over the feather-tips, seems to be placed a little tilted in the wing. This decreases the air pressure (backwash) on the bottom of the wing when a pigeon is flying. The pressure is much greater when the pigeon's feathers are wide and have coarse ribs. Because slightly tilted feathers let through only a small percentage of air, a pigeon with feathers like this will be spared when flying long distances. These pigeons will have some reserves at the end of a long distance race, which is always an advantage. All the top pigeons I ever owned had tilted feathers.

Many pigeons are condemned for having too long a backwing. Even more are condemned for too long secondaries than for flat feathers. The secondaries should not be excessively long, but may be a little longer than the standard many fanciers recognise. We have our own ideas. The 'De 37' and the 'De Perpignan' had what some people might have called long backwings. But they had such an enormous muscular forearm and such beautifully tilted feathers that it was not a problem. That so-called long backwing was part of a complete wing which was built very harmoniously.

Heads of the pigeons should be strong and nicely marked. Its expression must show, willpower, we really hate those with mere beauty in the head, the big round ones judges want to see during shows. Maybe these pigeons are able to win prizes in short distance races but they surely will not during tough long distance races.

It is very important that a pigeon has an eye cere which is a complete circle. If our yearlings do not have a complete circle we de-select them. These pigeons will never do well in long distance racing. it is not so important for a pigeon to have a complete circle eye cere in its first year of life but it is a must for pigeons that are older than a year. When pigeons are young they can suffer from so many growth problems that cause eye ceres which are not perfect, these we do not reject at that age.

Peter van de Eijnden, European Marathon Champion for 1986. *Picture by Rick Osman.*

We never give our young pigeons cures or vaccinate them. We select them on their health and their way of recovery. A pigeon with an incomplete eye cere will never be a topper, according to us, why do we think this?

A pigeon that is flying a race experiences a certain amount of air friction on its eyes, caused by the wind. When a pigeon has got partly incomplete eye ceres that wind can cause tears, because the tear-ducts which are behind the ceres start to produce moisture. This is a disadvantage because the pigeon will lose a certain amount of concentration. There is not a fancier in the world who could change our minds about this notion.

This problem is greatest in long races. it is not so serious when a pigeon has to fly about 125 miles with a head-wind or 190 miles with a tail-wind. It is a serious problem though when a pigeon has to fly 12 hours one day, and six to 10 hours more the next day.

We say once more a yearling which has eye ceres that are not a closed circle is not allowed into the racing loft.

Personally we prefer green-brown eyes. It is not a necessity, however, for a pigeon to have eyes of that colour. There are good and bad pigeons in every group of eye colours. What we think is bad, is when a pigeon has lacklustre eyes. This is often the case with the colours such as yellow, pink, red (without depth), light green or light brown.

Lacklustreness is often seen when the iris is exactly in the middle of the eye. I believe it should be placed slightly towards the pigeon's nose. Eye colour is not important for pigeons under a year. From the time they are a year old there should be a clear colour correlation though.

When selecting it is very important to check whether a pigeon has hard bones or not. The bones should be hard, but not coarse. Keen bones often mean insufficient attachment of the muscles. As a result pigeons with such bones and muscle-attachment will be easily injured and will not recover from injuries very fast. The disadvantage of coarse bones is that they put too great a demand for energy on smaller muscles. Hard bones which are at the same time fine are best.

Very bad is a pigeon which puts on weight easily. For us this means that such a pigeon is unable to function on a normal level.

Most important for long distance racing is that they stay fit and healthy when given clean water and a tray of food only. They should create their own naturally balanced food intake. Pigeons which don't succeed in doing so are removed from the racing lofts. A pigeon should have a thick glossy plumage. When one holds a pigeon in one's hands it should feel as soft as silk.

Our selection by hand is very rigorous. We carry it out year after year, though. We breed with pigeons which belong to the group that is good but not fit for racing. Each year we see an improvement. Not only do the offspring become better and better each year, the results of the racing pigeons improve too. Our way of selection is based on our own experience. We think it is a very good way to separate pigeons of high quality from pigeons of mediocre quality. It is a procedure which anyone can follow but it certainly is not the only one. Maybe there is a fancier somewhere who has an even better method. One thing we do know though, nobody knows everything where pigeon racing is concerned.

Burning the Candle at Both Ends

by G MacALONEY & SONS, Coatbridge
Winners of 1st SNFC Sartilly (2)

Being a winner is the ambition of all no matter the sport, to achieve this at the highest level, is the ultimate, such was our good fortune in the SNFC Sartilly (2) race in July. Of course there is always a price to pay and our attempt at putting pen to paper for this article is part of that price.

'Burning the candle at both ends'! Well, it has long been preached in pigeon circles that a loft must concentrate on either sprint, middle or long distance, that's not for us, in fact if there was a race from the set of traffic lights a mile up the road we would send pigeons. Naturally 500 miles is our goal but I don't understand how fanciers can sit out 90% of the racing season waiting for the Channel events.

I was born into a pigeon family, my father and uncle and later my brother and myself all involved in the Chryston club racing in Glasgow Fed. Like all young lads I was keen as mustard and did my share of the loft chores. It was very much a pigeon family and is much the same today with pigeons being the number one topic in our household, indeed the enthusiasm and energy of two of my sons William and Kenneth ensures there is no slacking in our efforts for success. After getting married in 1956 I moved to the Coatbridge area (a few miles south of Glasgow), it was then I started racing on my own competing in the local club which was the biggest in Scotland at that time, we raced in the Midland Fed (but have recently joined the big Lanarkshire Fed). I was a regular visitor to the winner's table with much of my success coming from birds I received from the late Willie Warwick of Symington, birds also came from another close friend Bob McLaughlin of Salsburgh. Bob of course is the well known SNFC secretary, but I prefer to think of him as an outstanding pigeon breeder and racer. A chequer hen 'Caroline' was the result of a first cross Warwick x McLaughlin, she was a Fed winner inland then 1st Scottish Cental Combine Avranches 1974. Other notable distance winners include dark pied hen 1st West Sect, 11th Open SNFC Nantes 1980, she was a grand-daughter of Bob's famous stock pair 'Mac' and 'Mary'. Since that race we have competed in 24 Nationals failing to time once and winning positions from 20. One of our biggest disappointments was the loss of our mealy cock from a 130-mile flight, he had been a winner in Open competition half a dozen times from the coast 360 miles, and 3rd West Sect, 14th

Open Rennes, he was down from Venner of Street x Mariens.

The 1987 SNFC Channel programme opened in mid June with the Sartilly race approximately 505 miles to our loft, our timer was a mealy Busschaert on Widowhood winning 17th West Sect. He is a club winner plus minor positions, this was his first Channel crossing.

Rennes was next and it was a bit of a disaster probably the worst Rennes race since the early sixties, we timed a blue cock to win 41st West Sect, he had won twice before from the Channel, he is down from Bobby Carruthers of Bonnyrigg bloodlines.

Nantes is 606 miles to our loft, it was an excellent race, we took 12th West Sect with a mealy cock (McGilwary x Carruthers), he is the start of the present team, having flown the Channel six times, winning three times from Rennes including 17th Open, once in Fed Sartilly and now at Nantes. His only failure was at Sartilly when he was timed two minutes out of the prizes.

Sartilly (2) was a late liberation 11.30 and like Rennes was a bit erratic, we timed our black cock at 3pm the second day to win 1st West Sect, 1st Open. There is a bit of a story to the breeding of this cock. I visited clubmate Andy Durning's loft and was particularly impressed with two cocks these being bred by the late Jim Nash of Croy (1st Open Rennes with 'Daphne Delight'). Andy agreed that I would give him two hens to pair with the Nash cocks, with both of us to get a youngster from each nest. All four pigeons have since won, Andy scoring from Rennes with one of his. The sire of the black cock is a mealy, he is nestmate to Jim Nash's 4th Open Rennes, the hen we sent to mate to him was a dark chequer, twice a winner from 500 miles and 1st Club from 285 and 365 miles. She is from a son of 'Caroline' our SCC winner x a daughter of Bob McLaughlin's 'Red Cockade'. The National winner raced the YB programme. As a yearling twice 285 miles, twice 365 miles and once from 606 miles, yes he flew Nantes as a yearling. In 1986 he was 1st Club 160 miles and second bird to the loft from Rennes. This year he flew four races to 265 miles then timed at 1.15pm from Sartilly (1), our second bird, but just missing the prizes. In the three weeks prior to his win he was a dozen times at 50 miles plus he was being flagged one hour each evening with the rest of the team and was sent sitting 14 days on eggs at basketing, the same nesting condition from which he won earlier in the season. At this point on behalf of my sons and myself I would like to thank all who have phoned and offered congratulations on our win.

Our club at present sends an average of 350 birds per race, competition is very keen with the members taking many prominent Fed prizes. We ourselves haven't been out of the first three in the club in the past two seasons, winning the Fed OB, YB and Combine Averages in 1986 (12,000 birds in the early races). I don't know how many Fed positions we have taken this year, but in the Hastings smash 400 miles we sent eight Widowhood cocks, and had the only three birds in the club in race time, all three taking good Fed positions and winning the best 2-Bird Average in Fed. Our first bird winning 7th Open is a blue chequer Busschaert, he is an outstanding racer, his wins include three times 1st Fed, twice 2nd Fed plus other Fed prizes, he was 1st

West Sect, 52nd Open SNFC Appledore 400 miles in 1985, this cock was also 14th Open Fed Sartilly in July (not bad for a sprinter). A half-brother of this cock also racing Widowhood has among his wins four 1sts Fed, ten 2nds Fed beaten by loftmates.

Our 1987 Channel performances have enabled us to win the coveted Lanarkshire Social Circle Gold Cup for Best Average from five Channel races with four nominated birds per race.

General Management

Stock birds and widowers are paired on January 19, all rearing a pair of young. Hens are taken away after sitting ten days on the second round approx three weeks before the first race. The widowers are given four 30-mile tosses in this period, with the yearlings having their mates on their return, the cocks go to the first five races then on alternate weeks, the yearlings seeing their mates for a couple of minutes before basketing, and all are with their mates for half an hour after the race.

We do not break the birds down, firstly because of the colder climate in Scotland, and secondly because we want our widowers to fly to at least 350 miles, therefore, we keep them on 1oz of Widowhood mixture through to Wednesday then increase to 1¼oz. We have also installed heating in the Widowhood section, this is kept low all week and put up to 65°F on Thursday, this extra warmth we feel is essential particularly early in the season. Recently skylights have been put on the roof to further help heat the loft. Widowhood will begin to dominate the Channel races in the years ahead, it is also worth noting that the chequer cock when he won Hastings mid June had just dropped his first flight and was on his second at the Fed Sartilly in mid July, this after being paired in January.

We use yearling hens that have flown the young bird programme as mates for the widowers, they are then put back on the road as two year olds, in fact it would appear they develop quicker with the year off the road. The hens are kept in a cold aviary all week and fed only Barley with Widowhood mixtures Thursday and Friday.

Readers may be interested to note that we have tried seven different Busschaert lines, with three of these responding to our management, ie Jeff Horn, Eddie Lauder, Andy Pullman and a grand-daughter of 'Domingo', the latter performing better over 200 miles. We recommend Busschaert lines that were introduced prior to 1975. I should add that no matter how well a bird may be bred, if it doesn't work to our system of management, it is out, you cannot chop and change your routine to suit individual birds.

The distance families are paired in mid March, they have four club races to 270 miles, and sent in varying nest conditions depending on the individual bird. We pay particular attention during the yearling and two year old stage, to what motivates each bird, often it happens by chance, ie a mate being lost and the pigeon sits the nest the rest of the week etc. Having found what makes them tick we set about getting them fit. We flag from 4.30 to 5.30 each evening from the 150-mile stage, with a dozen tosses from at least 50 miles, three or four of these flights maybe 75 miles our favourite spot being east of Longtown which we consider the line of flight for many National winners. These birds are fed a mixture of 45% Maples, 5% Beans, 30% Maize, 20% Wheat with a few Peanuts added.

Young Birds
Education is the secret, I am convinced that a large percentage of losses can be attributed to dehydration caused by careless fanciers who don't basket train their youngsters. After ours are eating and drinking out of the basket we start tossing in mid June, they go every day until the first race, and then three to four quarter times 30 miles between races. Where we differ from most is that every time we train YBs we send old birds with them and I stress not old birds that have been sitting about a loft doing nothing, but actual winners with sometimes as many as 30 liberated with the YBs. The reason for this is we want to teach them early the true line of flight and racing home from A to B as quicky as possible. Pigeons that take detours won't win. Most of our youngsters race the programme, but a few likely looking distance birds are stopped after three races.

Loft Set Up
The loft is big by average standards for our nine pairs of stock birds and 45 pairs of racers. The main loft is two tiered built with grey facing brick, no birds are kept downstairs, this we use as a store etc. The upstairs area measures 30ft x 8ft in four sections, since the photo was taken we have put on an apex roof with perspex skylights, with a false ceiling made of pegboard for ventilation. We also installed steel gratings 9ins above the floor with granules

The main loft of George MacAlloney & sons.

on the floor itself, this means that we only have to lift one grating in each section once a month and rake off the droppings, the birds and loft remain perfectly clean and of course it's less work for ourselves. Two timber lofts one 19ft and the other 10ft with aviary are used for Widowhood hens and young birds.

The Future
It should always be on the mind of the good fancier, planning matings, race programmes etc, on the lookout for a cross perhaps, on the latter I would advise tread warily if you have a winning line of your own, don't change your bloodlines because you buy the son or daughter of a National winner, by all means mate it to the best but test the offspring thoroughly before they go through your loft. We are always on the lookout for new stock to improve our own, at the moment we are working on some Bricoux x Sion and the Van Bruaene from A H Bennett.

Advice to Novices
To put it simply there are five points: 1. Dedication on the part of the fancier; 2. Good stock; 3. Environment (that doesn't mean a big fancy loft); 4. Motivation; 5. Fitness. Finally I would thank the editor for recognising our win by asking us to write this article, I hope it may help some novice on the road to success.

Patience is a Virtue

by LES VARNEY

1st Open London NR Combine Stonehaven

Once again I have the privilege and the honour of writing for Squills International Year Book about my Cattrysse family. During the years I have learned a lot, mostly by trial and error. One thing I did get right was my single flight breeding pens, a set of five pens to house five pairs, it does guarantee parentage of youngsters. The pens are 6 foot deep by 6 foot high, 3 foot wide, with a perspex roof and one single nest box in each, with wooden slatted floors so that droppings go through and small mesh wire so sparrows and vermin cannot enter.

My breeding aviaries cover an area of 35 foot long by 16 foot deep. This was once an aviary of scaffold tubing all covered in half-inch mesh, three parts of the roof being convered in perspex. There is a corridor 3 foot wide that goes through the length of the aviary with four doors going off into four compartments. The corridor does cut down the possibility of the birds escaping. Eight inches above the concrete floor there are wire grilles so that birds do not come into contact with their droppings. I had the nest boxes made 2 foot 6 inches long and 15 inches deep with wire grille bottoms, leaving 15 inches in height in the nest boxes. Again birds do not come into contact with their droppings. All wire floors and bottoms of nest boxes can be taken out for cleaning.

I have another aviary built in the same way. This is 15 foot long, 6 foot wide and 7 foot 6 inches in height. All stock birds get plenty of air and sunlight. We have the breeding in the early part of the year. Time switches for lighting which comes on at 4am, off again at 8pm comes on again at 6pm and off at 9pm. This gives the old birds plenty of time to feed as early or as late as possible. All stock birds can get, when they want it, a varied assortment of minerals and grit. They get multivitamins once a week. They are fed on a no-Bean mix and Barley is increased by 50% at the end of the month. Drinkers are washed out every day and water changed three times a day. Any food that goes through the wire grilles is thrown away. Underneath the grilles the nest boxes are cleaned out every other day. Grilles over the floor are cleaned out once a fortnight. Aviaries are disinfected out three times a year and nest bowls are cleaned and changed after every round. One reason behind the wire grilles besides cleanliness is that we can observe the droppings which are

One of the stock lofts, allowing the inmates plenty of fresh air and adequate protection from the elements.

checked by the local vet before breeding. All stock birds as well as the race team are treated for worms, coccidiosis and canker before the breeding season starts.

My old loft and new loft design
My original loft was 20 foot long, 6 foot 6 inches high at the front, 6 foot high at the back and 6 foot deep. Sixteen Widowhood boxes of my own design in an 8 foot compartment at one end and young bird compartments, 8 foot wide at the other end and leaving a 4 foot section in the middle for my Widowhood hens. This loft was a flat sloping wood roof with half-board and half-mesh front. Ventilation at the bottom and back and also at the front came from the bottom of the loft and also holes drilled at the top. These air vents went the length of the loft.

I had two very good seasons with this loft; 1984 when 'Blue Metro' won London North Road Combine Berwick and the Louella car and up in the Combine results in 1985. Well, they say you should not change a winning loft but I had great ideas for the loft to make my birds more successful. This was a drastic mistake. I enclosed more of the loft and introduced heating. At the same time as this I had an accident at work and was told that I would never work again and as months passed by I was helped by the Manpower Services, took a gamble and turned my hobby into a business. I bought out my good friend Les Davenport; he now takes a keen interest in golf. All Les's birds were put to stock with what I already had and most of my Widowhood team were pulled out, some of which were 'Blue Bullet', 'Blue Shadow', 'Blue Jet', 'Blue Strada'.

One of the louvres in the Les Varney loft.

Louvre removed showing the bob wire trap for late arrivals.

Getting back to the loft and big mistake, shutting out the air, introducing the heating and taking out good birds. My racing performances decreased and rumours coming down the grapevine were that his blood had gone and he is finished. But Rowley & Irons who I was helping at the time with my methods of racing and feeding, plus they had my birds, were doing the talking for me. They won in 1987, 1st London North Road Combine Berwick and the Louella car with 'Fighting Mick'. 'Fighting Mick' being the son of 'Blue Roan', sister to 'Blue Metro'. Then my keenness came back to race and win. I would like to thank members who said my blood had gone because they made me more determined to make a success of what I had started. A phone call to Roger Carr, a trip was soon arranged to Belgium, Holland and Germany where I would be looking for good Cattrysse pigeons but mainly looking at the lofts. After visiting a few lofts I soon saw my mistakes. Thanks Roger for your help.

I came home by plane which for me was not the best of journeys, a bit too near headquarters for me. In future my birds do the flying!

That same night the plans were drawn up for the new loft. My new loft is now 27 foot long, 6 foot deep and 7 foot 6 inches high and 9 foot to the apex. I have three compartments which are one 10 foot section each end and a 7 foot compartment in the middle section with Vee perches. This section is where the young birds are raised as well. The roof overhangs the front, making a 3 foot canopy. It is a peg tiled roof, there are three skylights in the foor slanted over each section, these can be opened when necessary. There is an extractor fan in each gable end of the roof, these are thermostatically controlled. Inside there is a false ceiling with an 18 inch wide air vent that goes the length of the loft. I can control the air flow with sliding shutters. On the front of the loft at the bottom I have four louvred air vents which can be pulled out to reveal bolt wires for late-comers as I use open doors for trapping. There are four Georgian bay windows, they are glazed in blue perspex. The birds cannot see out but it lets the sunlight in. The two end sections come to the centre of the two middle bay windows, this was done for a purpose. Small doors are fixed inside the two middle bay windows so that I do not have to keep handling the hens. I just open the doors and in they go to their appropriate sections to the cocks. There is another air vent which goes the length of the loft at the back, 6 inches from the bottom. I have no doubt in my mind that the air flow is the most important in the loft. The Widowhood nest boxes are approximately 2 foot 6 inches long, 15 inches high and 15 inches in depth. I have used one Widowhood front and the other half of the box is solid to darken down the box. To darken down the box in this way saves darkening down the loft itself. The birds do need natural light in the loft. The loft is lined and clad with shiplap.

With 'True Grit' winning 1st Open London North Road Combine Stonehaven this year, a colour television and over £1,000 to his credit plus my other Combine results, I now feel my methods and loft are completely right and I will not alter a winning loft again.

I pair my Widowhood race team up on or as near as February 14. They are left to rear two youngsters each and are hopper fed. They have open doors. If it is not too cold the cocks are trained while on eggs. I take youngsters away at 21-24 days old. By now the hens will have laid again, youngsters and hens are taken away together. The cocks are left to sit the eggs for one or two days, after this nest bowls and loft are cleaned, bowls are put back, turned upside

The Widowhood boxes showing the hinged door which darkens half the box.

Blue cock, 'True Grit', winner of 1st Open London North Road Combine Stonehaven, bred and raced by Les Varney.

down and half the nest boxes shut off. While it is only club and Fed races the cocks are given 25-mile training tosses, two to three times a week. They come home from training to their nest boxes open and bowls turned upright and left for about half an hour, there will be a couple of peanuts in their nest boxes as well. The nest bowls substitute for the hens and I do not like to keep putting the hens in. The only time I like to show the hens is for the classic races, as birds are basketed at least two days before the race. Sometimes we basket for an ordinary race on a Thursday night, there is food and water in the basket, they are kept in the basket and taken down to the club on Friday and off to the race. They then come home to the hens, they do not come home to the hens on a one night of basketing. On any holdover the hens are shown, the cocks eventually learn that the longer they are away in the basket the hens will be waiting. Whenever the hens are shown it is for the weekend, this helps maintain their form. On short races sometimes another cock is used in their boxes to 'wind' them up. This can be as good as using a hen to some cocks. This was the only way we got 'Blue Bullet' to race and he won £1,700 in two seasons.

Birds are exercised one hour in the morning and one hour in the evening. During this time doors are closed and the loft is cleaned out, water changed and food put in. Birds are fed in their nest boxes, they are not teaspoon fed and are fed three times a day. Leading up to the classic races the birds are trained 60 to 70 miles if possible, taken up to Les Davenport and he singles them up for me. We have found that the Cattrysse on this method must not be kept hungry.

Basketing for the classic races, the cocks are put in an ordinary race pannier, they are taken out of the loft, the hens are let in and the cocks are put into the loft on the floor in their baskets so as to see the hens in the nest boxes for about two minutes. This is done between 3 and 4 o'clock in the afternoon. They are then put in their basket on the lawn and covered up to settle down before going to the club. This method has proved very effective.

My advice to a new starter to the sport is get your loft and ventilation right first. Choose your birds from a racing loft. Don't be afraid to put your hand in your pocket. You want quality not quantity. Feed a good corn. Keep one family of birds and don't be afraid to ask for advice. One thing we all learn at some time in our life is 'patience is a virtue'.

The Two of Each Theory
by DAVID A DELEA
Winner of 1st NFC Pau

It was indeed a great honour to receive a letter from the Editor congratulating me on my above performance and inviting me to contribute an article for Squills.

I race my pigeons to a 16ft x 6ft loft, my birds number anything from between 36-42 during the winter months.

I started racing pigeons in 1959 with birds from two local fanciers, G MacAllister and J Montgomery both Combine winners at the distance, stars in their own right. I crossed their birds, and they went on to win races from 65 miles up to winning the Combine at 599 miles on the day, flying nearly 16 hours from Lerwick. Although I might add they did not adapt too well to SR racing.

In the early 1970s I introduced to my loft ten pigeons from B Spencer, again a Combine winning family. From his clearance sale, these being Osman and Kirkpatricks, added to these came birds from R Wilson, yet again a Combine winner, these birds came from his 'Champion Major' bloodlines, a cock from this family I paired to a hen in my brother Paul's loft from his G King Kenyons ('Fairlight King' – 1st NFC Pau 1952) from these fanciers my family evolved. In 1973 I moved to my present home then started to race south in 1975.

All my birds I pair in late February or early March. The first week in May the OBs start to train, I've always trained my birds with my brother Paul directly to the coast about 60 miles. Once at the coast it's then double and single up from there, direction is of no importance to us, the birds are taken where the mood suits us. This I find makes the birds independent and confident for their tasks.

As I don't club race, my YBs first race is the YB National or BBC Rennes 250 miles. All my YBs go to one or both races. The same goes for OBs and yearlings, Nantes NFC or BBC 306 miles, should they miss these, they are then sent into one of the Classics all over 565 miles. Over the years I have been given youngsters and exchanged them for my own, with the exception of one from G Hunt they never adapted to my methods or they fell by the wayside.

The food I use is locally grown – beans, maize, barley plus a little seed.

'Greenacres Florence', blue hen GB88T65758. Bred and raced by D A Delea, 1st Open Colombovac Pau NFC National 1989, flying a distance of 565 miles, beating 5,378 birds, as a yearling in only her second race.

Water twice a day in the summer, also black minerals and grit. Anything else the birds require the fields next to my loft supply. With my open loft birds are out at the crack of dawn in all elements collecting nest material or roaming the farm lands. To the birds, the field next to my house is an extension to their loft.

I believe the essential requirement of National racing is the ability to fly single up for very long distances, the only way to ensure that this ability is carried forward is to test every generation. That is what I have done and I have produced a family which I know works for me.

My advice to anyone attempting to fly the Nationals is to test every individual to the distance and fly them single up. Remember the winner of the National always has the right wing, the right eyesign, the right coloration and goes in the right nesting and mental condition. In other words take no novice of fads. Fly your birds and the results will sort out the eyesign. I do not know what kind of eyesign my National winner has and frankly I do not care.

In passing I would like to say thanks Paul, without him it would not have been possible. Also Helen for looking after the birds and taming them down while I'm away. Charlie & Micky Harvey for the many nights you kept me up talking. Also John Ryan.

My answer to eye theory and wing theory is that it is preferable for a bird to have two of each.

Let's Keep it Simple the Widowhood Way

by DEAN PALLATT

This article is aimed at the young novices of our sport, I am going to try and explain the ways for them to gain success in this highly competitive hobby of ours. First we are going to look at the housing of our pigeons. Lofts can now cost anything up to £10,000 and over but believe me this does not make pigeons win. My own race loft cost £420 brand new 24ft by 6ft in 1988. As long as it is a dry loft you will have no problems. The best and most cost-effective way is to go to your nearest garden centre or shed manufacturer, remove the glass windows and fix a Sputnik trap to the front. This will cost you a tenth of the price of going to a pigeon loft manufacturer. You must on all accounts stay away from secondhand lofts as these lofts may carry unknown diseases that you will never cure. Regarding ventilation this must be controlled by the amount of pigeons you keep. You must always remember never to overcrowd your birds. If you fix a 4in soil pipe with a cowl on at every 6ft on the roof this will be sufficient to keep the air circulating, the air will come in through the Sputnik traps and out the soil pipe chimneys on the roof.

Behind the Sputnik you must have a shutter to close at night and when you put the cocks on Widowhood. The nest boxes should be as large as possible, the Widowhood type that can be partitioned in half are best. These measure around 30in by 18 in and you must have no more than 8 boxes per 6ft section. You do not need heat of any kind, this is a falacy, I have been to many famous winning lofts where they have had heating installed but no longer use it. In my opinion it can only bring on a false condition and moult; your cocks must reach a natural form that they will keep for a season. The floor of your loft must be of a sheet material to make it easy to clean. You do not want dust and dirt, that you cannot get at, in cracks on the floor as this will only create a disease problem.

Your most important decision is going to be your choice of pigeons. This is where you will succeed or fail. You must decide what you would like to achieve. It is no use trying to fly a Fed programme of races between 70 and 400 miles with 500- and 600-mile birds. You must obtain a short to middle distance family. There is no way these 600-miles will obtain success week after week at club and Fed level, they just do not fly fast enough. If you want

Dean Pallatt in his Widowhood loft with 'Georgey Boy'.

to win National races and 500-mile races it is a different ball game altogether. Then you are specialising in just two or three races a year and I am not the
person to give advice on this type of racing. So we will get back to club and
Fed racing of up to 300 miles which 75% of all races are. You must be looking for a winning loft at that time; it is no use looking at strains that were at
the top ten years ago.

Ourselves we went to Staf van Reet who at that time was right at the top at
sprint races in Antwerp, Belgium. These were the type of races we wanted to
win and it has paid dividends we have just not looked back. What you do not
want is a one pigeon strain but a family of pigeons that breed first prizewinners across the board. You should be just as confident to breed winners from
your yearlings as your stock birds. In fact yearling hens are prone to breeding champions in their first year, so it is a must to breed from your yearling
hens. You should obtain your stock as close to the champions of your chosen
strain as your pocket can afford. Take the advice of your chosen fancier of
which birds you should obtain. One golden rule look at the birds first and
pedigree second. If you are going to fly Widowhood at 300 miles then you
need pigeons that have done the same with success for generations.

There are many systems of racing pigeons but I believe Widowhood is the
only way to achieve continued success week after week. I am going to try and
explain my own system of flying Widowhood that works with great success
for us with our own family of Staf van Reets. Some families adapt to different methods but the basics are the same and you must always try to keep it

simple.

At the end of the old bird season all birds go on open hole. The cocks will stay on open hole to come in and out of the loft at their own liberty all winter. At the last race all cocks will be paired, and will rear a round of youngsters whether it is one or two youngsters is immaterial but they must rear to give them that love of home. They will go back down on eggs, these will be changed for dummy eggs and when they leave these eggs all hens will be taken away, the nest box closed all nest bowls taken out. The cock now sits in the open part of the box, but still has his liberty, on open hole. At this time the complete loft must be disinfected, myself I use household bleach. From now until the end of November they should be fed a good breeding type mix with approximately 40% maize once a day as much as they will eat. I feed my cocks communally on the floor by hand, in this way I feel I have good control over the cocks.

Around the end of November or when the cocks are three-quarters up on their last flight I change the feed mix to 50% depurative, 50% breeding mix fed once a day the same time every day. I always pair the Widowhood on February 1 all birds must stay in the loft until the hens have laid. Then they will go back on open hole, they are fed the same 50% depurative, 50% breeding mix until the youngsters are about six days old, then they are fed a full breeding mix. When the youngsters are about 12 days old or just before the hens re-lay all hens will be taken away and placed in their Widowhood hen boxes so they will not pair together. The cocks must now finish rearing their youngsters on their own. On every fine days from now on the cocks are trained from 12 miles, but they are still on open hole. I take the youngsters away at about 24 to 28 days old and all youngsters are taken away at the same time. On the same day the loft will be completely disinfected, the shutters closed down behind the Sputnik; the cocks are now on Widowhood. I will now treat the cocks for whatever disease we feel they may have, with today's technology you should take your droppings to a veterinary surgeon to be analysed. He will tell you if you have any problems and prescribe a treatment accordingly. It is always best to treat the whole team if there is a problem.

The cocks are now on a strict routine. They are let out in the morning at 8.30am and locked out, but are allowed to land on top of the loft at there own will. They will clap up and down for five to ten minutes at a time, at 9.30am they are let in and fed by hand communually on the floor 50% depurative, 50% race mix, until they start to leave the barley, then no more. The shutters are then closed and on no account must you go to the loft during the day. At this time the cocks must learn to rest. At 4pm I basket the cocks and give them a training toss from 12 miles back to their evening feed, that is the same as in the morning. It is very important to go to the loft at the same time every day, this is the only way you will get your cocks to rest. Also it is as important to feed at the same time every day. This will help bring on a lasting condition of the cocks.

This system will be carried on every day with the cocks just coming home to their evening feed until the night before the first race. Usually it is a

'As Good', 2nd MCC Perth 1990, second to car winner.

Saturday morning basketing open race so I train the cocks on Friday as normal, but on their return to the loft their hens will be waiting for them in the closed side of the nest box. This will be the first time the cocks have seen their hens since their youngsters were twelve days old, approximately five or six weeks ago, and as you can imagine the cocks get very excited but must not be allowed to make contact with their hens. After about three to five minutes take the hens away and place back in their own separate boxes, then feed the cocks as normal on full race mix as much as they will eat. On race day place a nest bowl in the closed side of the cock's box and open the door of the box to allow the cock to his nest bowl, then come out of the loft and leave him with his bowl for about 30 minutes, then go and basket all the cocks for the race. The cocks will have remembered the night before when the hen was waiting for them, on the return from the race they will find their hen waiting for them. You must now let them go together for about two hours, then take all hens away at the same time. The Widowhoods have now had their first race and a strict routine must be obtained.

FEED SYSTEM SATURDAY TO FRIDAY

Morning feed 9.30am		Evening feed 6pm
Saturday		100% Depurative
Sunday	100% Depurative	50% Depurative 50% Race Mix
Monday	50% Depurative 50% Race Mix	50% Depurative 50% Race Mix
Tuesday	50% Depurative 50% Race Mix	50% Depurative 50% Race Mix
Wednesday	50% Depurative 50% Race Mix	50% Depurative 50% Race Mix
Thursday	100% Race Mix	100% Race Mix
Friday	100% Race Mix	100% Race Mix

Always feed as much as they want by hand until they start to leave the barley, on Friday evening feed bring feed forward so they will have two clear hours before basketing.

DRINKING WATER SATURDAY TO FRIDAY

Morning drink after feed		Evening drink after feed
Saturday	Clear Water	Salts Tonic
Sunday	Clear Water	Salts Tonic
Monday	Clear Water	Salts Tonic
Tuesday	Clear Water	Salts Tonic
Wednesday	Clear Water	Multivitamin
Thursday	Clear Water	Citacon B12 liquid
Friday	Clear Water	Clear Water

As you can see the cocks receive clear water at least once a day, this is because they will not always take enough water when other products have been added. The salt tonic we use is a combination of salts mixed together that I mix myself, but the nearest commercial product I have found is a product called Antec extra high potency soluble powder, this product is used in the poultry industry, especially for rearing day old chicks and also for times of stress.

EXERCISE AND TRAINING SYSTEM

Morning		Evening
Saturday		Stay in loft
Sunday	Stay in loft	5pm fly out 6pm let in
Monday	8.30 fly out 9.30 let in	5pm fly out 6pm let in
Tuesday	8.30 fly out 9.30 let in	Train from 12 miles on line of flight, no hen
Wednesday	8.30 fly out 9.30 let in	Train grom 12 miles on line of flight, no hen
Thursday	8.30 fly out 9.30 let in	Train from 12 miles see hen on return
Friday	8.30 fly out 9.30 let in	Stay in loft basket for race

Mr & Mrs D Pallatt holding 1st & 2nd Open, 1st & 2nd Leicester Section MCC Perth 1990.

On night of basketing for the race there is no need to show the Widowhood cocks their hens, give them their nest bowl for approximately 30 minutes and basket them, they will soon learn that their hen will be waiting for them on their return from the race. Once you have established the right team of pigeons, the most important factor in winning pigeon races is the conditon of the pigeons, you will only learn this from experience.

As you can see from my results this system has worked for me, and I hope it can help you to achieve success with your racing in the future. In closing now I would like to thank the editor for giving me the opportunity to write this article for Squills.

Going Natural for the Long One

by RON STRONG, Hexham
1st & 2nd West Durham Amal
Tours (540 miles)

My name is Ron Strong. I have kept and raced pigeons since I was a small boy of five years old. Being the youngest of four I followed my father to the loft as soon as I could walk. He gave me birds to race as soon as I could manage them. I then joined the club and our partnership of Strong & son, Hexham HS, was formed. That was 50 years ago. My first recollection of pigeons was walking up the path and helping to basket birds for a race. My job then was to look after the basket lid and also to put the snib on the garden gate each time we went through. I soon graduated to – yes you've guessed it – scraping out with a hoe and brush. When I got to be ten years old I took over the running of our lofts and started to put all my ideas into practice. I had been spending a lot of time at the loft of the club champion W P Davison. He was a wizard at racing pigeons and he passed on all his secrets and taught me all about breeding and racing that I had not been able to glean from my father. I had all these new ideas and wanted to put them into practice.

Being young and very keen on the birds my father said – go ahead, try out your ideas – as he had always admired Pratt Davison. My father was a very competitive man and from young had competed in target shooting and at World Championship level in Cumberland and Westmorland wrestling. I must have inherited this trait from him; it is just like breeding good pigeons, the will to win and determination. If you have not got that in yourself and your birds you are wasting your time and your money. To be the champion you must eat, sleep, think and dream about pigeons, and always think positive. If you go to the loft listless and down your birds will not win that week I can tell you that. The pigeons pick up confidence from you, it rubs off and they do well. Pratt also told me you cannot both be the champion and be liked by everyone. I soon knew what he meant by that.

My father and I had always liked the Channel and long distance racing so I set about improving my stock. I have always had this fascination about winning the longest race, I called it 'the impossible dream' not just locally, but the Open race. I was looking for birds to fly 500-600 miles. We had Vandeveldes at the time, they were nice pigeons and I had a great respect for them for distance racing. By now I had been on National Service duty for two years in the Army and when I returned our birds had been neglected in terms of training. My father did not have any transport and was some distance from the railway station. I went to a good friend of my father, Will Oswald, Purdhoe, and got eggs from all his best birds; I got a beautiful blue cock from Baston Bros, Alnwick, from their Combine winner. I blended them in with our own Vandies and set about breeding a super bird. I trained and raced them hard and always kept detailed records of everything, which I still do to this day. Pairing, laying, hatching, training and everything to do with the racing. I raced the young birds hard. They went every week to the bitter end. I then took stock and everything that was returning late I culled. As old birds I did the same. It took patience but slowly we developed a formidable team of racing pigeons which won from 93 to 540 miles.

Looking back through my records we have won every race, every cup and every average many times. I don't want to bore you with a lot of wins, but from 1958 to 1991, we have won 109 Channel 1sts and many hundreds of inland races. In the last few years alone we have topped the Tyne and Derwent Valley Fed eight times from the Channel, five of these from the longest race point, Tours, 540 miles. When one thinks where Hexham is situated miles and miles away to the north west of Northumberland I am very proud of our achievement. As the race birds come out of the south east many people have said to me that it would be impossible to top the West Durham Amal and very difficult to win anyway. You have to get into the first few in the Amal to win our Fed from the Channel. Anyway, we are now up to date.

So, having had this obsession for many years with the Tours race which is the Blue Riband of racing in the WDA, on 26 June 1991 I realised my lifelong ambition by winning 1st & 2nd Open WDA Tours, clocking a three-year-old dark chequer hen at 7.39pm on the day. It was a Wednesday evening, the wind had been blowing all day from the north west and the birds had been away since the previous Tuesday, having spent eight nights in the baskets due to bad weather conditions both sides of the Channel. They were liberated at 6.30am on the following Wednesday into a strong north west wind. The convoyers told me afterwards that they did not expect any in on the night the wind was so strong and after the long holdover. But, there were 21 game birds timed in from our Fed which is the farthest flying Fed in the WDA. We got another five-year-old hen at 7.46pm but did not get her in to the loft until 7.49pm. She caught us and a fellow clubmate coming out of the loft and veered away on to the chimney top. She still won 2nd Open, another miracle I would say. Two sisters, Busschaerts, flying to the farthest position in the Amal. The Busschaerts now can win from all distances!

We got the first Busschaerts in about 1980 and have crossed them with our Vandies very successfully ever since. The first we got from Eddie Lauder, Sacriston, were Latcham Busschaerts which go back to the old Busschaerts of the 60s. The Barcelona cock, The Crack, The Little One, The Tempera Cock and De Schoon. I blended them with another line from the Louella Stud and they were The Joker, Broken Toe, The Little Black, Ladybird, The Long One, The Crayonne, Soontjen, The Claren, De Wittentik and on the other side, Kanibal, Capito, Taco, Rapido and sister to the Little Chequer etc. In fact, all the best Busschaerts. A lot of other people are winning with our Vandy/Busschaerts as well. They got better and better as they go used to our management. This is important, when you get in new birds or a new breed, give them time to get used to your methods so you must be patient with them.

I have been asked what we did differently this year to other years. I think we concentrated on the health of the birds more. I improved the ventilation of the loft by digging out the far side which was built into a bankside. I cleaned out all the vents and I could feel the air being pulled through the loft, but not a draught. That is not good for pigeons. We never, ever hunger our birds, they are fed as much as they want all the year round. In the winter they get beans with a little barley and wheat and a little Widow mix fed by hand, also linseed, as they have always done. The birds are on open hole all day until November when I part the birds, then it is cocks one day, hens the next. We changed the feed for the racing.

In the past we always hopper fed two-year-old beans and hand fed Widowhood mix night and morning. This year after I had reared the youngsters on beans, a little barley and winter mix, I put the race birds on Wilkins Turbo Widowhood mix, a very good mix with a few old beans in it. I added some dogtooth maize near the end of the week for the races and fed a little seed and a few peanuts for trapping. They also get linseed every day and clean grit, clean water twice a day, except Wednesday when multivitamins are added. I also altered the water system. I made some wood boxes with doors and fitted them to the front of each compartment and put the water fountains in. This kept the drinkers cleaner and made it easy to change the water from the outside, not disturbing the birds. It also prevented freezing up in all but the most severe weather.

Throughout the winter I only see the birds at weekends because of the dark nights. My wife, Molly, looks after the birds, feeding and watering each evening. Since my father died she has taken a keen interest and has been a great part of our success. I could not have had the success without her support. Always keen and interested in wild birds, she enjoys the pigeons and spends a lot of time with them. There is a message here for all pigeon men; families can play a really important part in our sport. You may not feel like leaving precious birds to youngsters or novices, but start in the winter months when nothing much can go wrong and encourage all you can. I can tell you after raising a few children, pigeons are childsplay for most wives!

I remember in 1975 when we were in the process of moving our loft we raced to both lofts for a season or two. I must have been like an old broody hen. My father, then aged 90 at one loft and Molly, clocking in for the first time on her own, at the other loft. Just as I backed out of the gate to go to my father's loft I saw two fantail trappers disappearing head first into the water bucket. The top had not quite been put on properly and in they went. However, we soon became proficient and we had some fun sending messages between the two lofts on Channel races! Now, Molly lets out the Widow cocks and baskets the young birds ready for me to take away as soon as I come home from work.

We train the youngsters to go through a little trap door straight into the basket. The first time or two you do need a tin hat, but they soon learn and the birds do not get knocked about, losing bloom etc. This also goes for the old birds, and, taking out the birds we don't want to go that particular time, we are away within minutes. We can basket 60 youngsters in two or three minutes without any fuss.

Having selected my pairs for the next season's breeding – some will be old pairs that have bred well before – I will change a few. Line breeding is what I like to do and often a first cross is lucky and you get a good one. It is quite a skill selecting breeding pairs. My father used to laugh at me, I put two birds together and look at them, I concentrate and I can see the two youngsters they will breed standing alongside them, all ready to race. If I don't fancy the youngsters, I change the pair! Once I have got the pairs sorted out I then spend the next few weeks up to pairing-up time letting them get to know each other. They are then all paired up ready for the nest box. It saves a lot of trouble. Anything that doesn't toe the line must go. In 1991 the birds were paired up altogether on January 26 and all went down on to eggs well. I check all the eggs after laying and they were a good lot. Anything not perfect is eliminated. I selected all the best pairs and these included the two winning hens, and shifted their eggs after they had been sitting ten days. I use yearlings mostly for this job. Ours rear very well, have plenty of soft food and are vigorous. Remember, your young birds are your future. Breed poor youngsters and you will nose-dive from the top very quickly. At this time the old birds are on open hole, all running out together, with the bath out every day, weather permitting. The birds are left to sit in peace, Everything must hatch by the 19th day or is not kept. The youngsters are run at six days, anything small, same applies.

I keep a close watch on the youngsters every day through the box fronts until they are 26 days old. By then they are all over the place and the bottom boxes are full of youngsters from the top boxes. Having been fed in the boxes for the last fortnight they are all eating well. When I shift them usually about 26-28 days, I inspect them noting the ring numbers and colours for my records, then move them on to the young bird end. I dip their beaks in the water for the first few nights, after that it is up to them. I bed them on shavings for the first few weeks and keep the drinker up on a box to deep the water clean. They are nice and cosy and fed on wheat, barley, winter mix and beans. When they are eating well I feed the beans first, then the rest, as much as they can eat. They are on to the perches and flying in no time.

Training starts about five weeks before racing. This gives me plenty of time to pick good days and give them about 15 tosses at three, five, ten and 20 miles, before they go on to the transporter training. They then get about six of these transporter tosses up to 50 miles then straight into the racing. They go every week after that until the Stevenage National which is 227 miles. After this I select the best and stop them. The rest go on to 227 miles again, and then Brands Hatch 268 miles. Most of what are left I keep.

Old birds. Once the youngsters had been moved I took away the Widow hens. The cocks sat for a few days then I cleaned out the boxes and they were on Widowhood. I like Widowhood, you get the birds fitter and they hold their form for about two weeks. I have only raced Widowhood for seven years so am only a novice really. I put down 6 inches of thick shavings (having first sprayed the floor with Jeyes Fluid) and covered the front of their compartments with black plastic. These 25 Widow cocks are trained five times up to 50 miles and then go in to the races. They were raced every week until being paired up at the start of Channel racing. Then they are trained twice a week with the Natural birds. I changed the floor dressing to sand as I did with the Naturals as the shavings were blowing about. Other years I have tried straw, ordinary deep litter and sawdust, but I like heavy sand the best. It also keeps the loft very dry and soon settles down. Droppings and feathers can be carefully raked off the top from time to time.

Our Natural birds were on open hole all day. When Molly gets home from work at 4.30pm she puts in the Naturals and lets out the Widowers and they fly about until I come home at 5.54pm. I put them in, either let the youngsters out or the Naturals for the rest of the night while I do the odd jobs around the loft. I try to get finished in the loft by 8pm to give the birds some time to settle down. I let out the Widow cocks for a second fly before I leave. Then they are shut in and fed for the night. When it is dark, I slip up and drop the lets so the Widowers can get out at first light. They are flying from about 4am in the morning. I get them in and close them up at 8.15am and feed as I leave for work. The Naturals are out then for the day.

The two hens which were 1st & 2nd Open Tours were treated the same as the rest. The winner, 'Windmill Supreme', has been a good one right from a youngster. She won as a YB and as a yearling, and was also 2nd Abbeville as a yearling. Our yearlings go to the 375-mile and 422-mile Channel races if they are on form. Last year, as a two year old 'Windmill Supreme' won Tours at her first attempt. She was 28th WDA and of course this year she won both clubs, Fed and Open WDA. Both the winning hens should have been sitting 16-day eggs for the big race, but the long holdover changed all that.

Our second Open winner is a lovely light chequer called 'The Merry Widow'. She is a five year old and a sister to the winning hen. Both are pure Busschaert. She was a fourth nest youngster, and raced all stages as a young bird. As a yearling and two year old she was kept for a Widow hen until I paired the Widow cocks up for the Channel. She was then raced to the coast. Last year I paired her up Naturally. She flew all stages to Tours winning seven positions. She was our fourth bird from the eight we clocked on the day from Tours, 74th Open WDA. Four weeks after Tours she won Abbeville 375. This year she again flew through to Tours, winning six positions and was 2nd both clubs, 2nd Fed, 1st Championship Club with all pools to £5 and 2nd WDA Tours. She is a great pigeon. Many brothers and sisters of these two hens have been winning both inland and Channel races for us. In 1988 we were 4th Open Tours with a sister to the two hens, so the Busschaerts can certainly get the distance if you get the right ones.

Preparation of the big race. Our big race candidates for Tours are raced every week, except when driving or laying. I like hens best for the 540-mile race. They are better for the long hard races. They are not as big, they are very game, and will fly long hours to get home at all costs. I like quiet, intelligent birds for long distance. A good shape with a feather of silk, strong, with the heart of a lion to be able to fly 16 hours on the day into a head wind and still have that love of home to kept them going. All this rolled into one makes a champion 500-miler. That is why they are so difficult to produce. Flying Natural pigeons for the long distance is entirely different to flying Widowers. All good long distance birds are individuals, all are completely different. You have to train, race and handle them slightly differently. It is a slow build up and they fly themselves into condition with good, careful management.

I selected 25 Tours' candidates during the first weeks of the season. As the race programme progressed some went for shorter Channel races and the 20 of these went to the first Abbeville a month before Tours. I studied their performances and how they finished after that race, and then selected 13 which I thought could win and fly in on the day. After Abbbeville they idle about for a while, then I get them back down on to eggs for the race the week before Tours. They would be sitting about six/seven days on eggs for Stevenage, 227 miles. Sending them the week before the big race is a risk, but I do keep one or two back in reserve. This year both of the winning hens went the week before and got a good workout. They were held over until the Sunday which doesn't do your digestion any good. They were liberated at 8am in to a WNW wind. 'Windmill Supreme' was our fourth bird at 1.42pm, fiver hour and 32 minutes flying. Our second 'The Merry Widow', was our 29th pigeon, at 3.44pm, seven hours 44 minutes workout.

I inspected them on the Monday evening as they were going away the next night, Tuesday, As I handled the birds they were perfect, I thought. I left the two good hens until last and 'Windmill Supreme' was the last one I handled. I opened her wing and to my horror she had thrown two flights from both wings. She already had one half-grown flight she now had almost two and a half flights from each wing. I said to Molly that I was going to send her. I have won with them like that before. They are super-fit when they throw flights like that. When birds have a full wing they are definitely not right and should not be sent. So, we sent them off the following night and it was the end of a year's work for us.

Holdover races on work days often present problems for us, but not Tours as we have time to prepare the loft for the returning birds. We like a nice, clean quiet loft for the birds to come back to, everything in order. With a light feed and water at the ready. Their partner, eggs or young should be ready for them to see as soon as they enter the loft. That is what they have raced to get back to and most important to them. A spoonful of glucose in the water, some seed in the hopper and on the box front after clocking welcomes them home. We make a fuss of the birds as well, they know your voice and that you are pleased to see them. Handle them gently when clocking, go back and talk to them an give them extra seed.

When the birds are away at the race their boxes are closed. This prevents a lot of problems for the home-coming bird. Their domain is safe. All other birds are shut in their boxes giving the race birds peace and quiet and time to recover. The next day we concentrate on helping them to forget the race. They get extra food to help replace the energy lost. Even on fast racing days the winners will have used up some of their reserves. Check the birds for injury – bumps and scraps never noticed on return often show up next day when stiffness sets in. Any latecomers need special treatment as they could have flown many miles more than the winners, having been off course. They may be the winner next week. A bath is most important for the next few days, very soothing after a race.

This year we got the two birds on the day from Tours, last year we got eight, but we were still up at 4.30am on the second morning to welcome others. This year we clocked five pigeons very early the next morning starting from 6am onwards so those had not been very far away the night before. Three of the five flying Tours for the first time at two years old. A new experience for them. Another year they could be the winner on the day.

Advice to new starters and novices

We often read articles advising new starters to go to the loft which is flying the best and get late breds. It is good advice, but the champion fancier is not always waiting for Tom, Dick or Harry going along for his late breds. It is often best to get eggs if you can. They are cheaper and you often get some for nothing, and a lot more than you expected. You must go armed with the laying dates of your hens. The eggs you are putting under them must not be more than three days either way. So, if your hen lays on the 10th, the eggs you get to put under her would be OK laid on the 7th-13th; better as near to hatching together as possible. Soft food would be available at the correct time.

Give the chap some warning and he may select you eggs and make sure they have been sat on while his birds have been away racing, training etc. They are always better reared in your own loft, for if not reared well, they will not be any use for anything. He may have a few newly hatched youngsters which you could put under your own birds that have just hatched. Remember they must get a regular supply of soft food. Always say whatever will be convenient for you will do fine for me. Try to build up a relationship with the champion fancier, it could be to your advantage. When you visit him don't stay too long. If he is the champion he will be busy.

When you first start out keeping pigeons don't expect to win, just be content to get them back safely and work from there. Try to get a team built up. Lots of patience is what you need to keep and race pigeons. Treat your birds like pets. I talk away to mine all the time. Slowly, you build up a love for home and their confidence in you and they will race for you. Your loft must be safe, cosy and cat-proof. It is their home. I remember at the old loft being bothered by crows, so I made a very realistic scarecrow holding a long stick like a shotgun. It gave my father and me a heart attack for weeks every time we came out of the loft.

The loft must be free of disease and pests. Always keep a mouse and rat trap beneath the loft where the pigeons and wild birds cannot get into them. Spray out regularly and watch the health of your birds. I read all the time about this but no-one ever says go out and buy a book about pigeon diseases. Any good pigeon man will tell you 'droppings are everything'. My wife gives me stick about my obsession with droppings, but it is true, get the droppings right and you are on the right lines. Nowadays, with all the stuff fanciers put into the drinking water, different remedies every day, no wonder the birds have loose droppings. You must recognise the difference between greenish droppings after a race or when you are dosing them for worms, coccidiosis, canker etc and the so-called 'watery droppings'.

My advice for novices with birds with loose droppings. First of all, get rid of all the fancy stuff in the drinker, take out the minerals and medicated grit, and put the birds on a staple diet of corn and clean water for a few days. Put something on the floor to dry up the area, kill youngsters in the nest that have loose droppings. If they do not improve within a day not two, ask the pigeon men at the club, then go to the vet and dose them. A lot of the medicine advertised will be of no use if you don't know what is wrong with the birds. In many cases, I suspect over-dosing with various remedies and multivitamins.

I have thought a lot since the Tours race. When we were standing to win the race many, many people rang and called to say congratulations, and you realise just how many friends you have made over the years. Some previous WDA winners said, 'How does it feel?'. They knew what we were going through, you keep shaking yourself thinking you are in a dream. It is a feeling of sheer disbelief, after all the years your dream has come true. I look back at all the years of hard work and it has been work and pigeons, and just when you think you will never get your name on the famous trophy, it comes out of the blue and you win it.

Nowadays, with studs all over the country, fanciers can buy good quality birds of all breeds, because there are good birds in all breeds. The studs have an important part to play in our sport. The champions at stud allow the ordinary fancier access to birds which would never have been available before. As with life, there are ups and downs. Cloud 9 one week, the bottom the next, like the time I found the best pigeon we had sitting the nest, eyes open, perfect – stone dead. On another occasion, we won the race on the Saturday from about 400, our winner went out for a fly on the Sunday morning, straight into the wires – gone!

Amusing incidents too, I noticed some suspicious characters hanging about our road and, thinking they were after our birds, rigged up a trip wire at the loft. My father wandered up after dark for a quiet smoke, tripped over the wire and ended up in the muck heap. Only his dignity hurt fortunately for me. It was the businessman's safe up the road that they were after! There have been many, many more stories and memorable races, and now we have grandson Matthew named as his great grandfather. At two years old he won't stay out of the loft, 'helps' me and remembers everything that we do. His job is to watch the lid of the basket, and he never forgets to shut the snib on the loft gate . . .

London's Double Combine Winner

by TERRY ROBINSON of Cheshunt
1st London NR Combine Stonehaven and 1st Thurso,
winner of Tommy Long Trophy Best Average
in LNRC races

After receiving a phone call from the editor about writing an article for Squills I was very pleased, it was something I had always thought I would like to do, but when you actually sit down to start, you realise it's not so easy. I suppose the best and only place to start is the beginning. I have always had an interest in pigeons and looking back my mum put up with quite a lot as we only had a small backyard. I packed up pigeons during my teenage years, and only went back to racing after I had got married and moved to Southbury Road, Enfield in 1970. I started with Heinz Variety and racing in the Hyde. It was through being a novice and having a flyaway one Sunday that I met up with Sid Jones, we struck up a friendship straightaway. As Sid wasn't racing the following year, he generously gave me some old pigeons to break in and race. I did really well with them and it started me off winning.

   After a couple of years I started to look around and buy pigeons in. I went

Terry Robinson with the two year old cock who won 5th in the LNRC from Berwick this year.

The main loft of Terry Robinson showing the top section of the Widowhood cocks.

to Alf Baker, and bought a daughter of 'The Laird' which bred a good cock called 'The Tryer' which had four Open London NR Combine positions, his best being 6th Open Combine Stonehaven. I also bought pigeons from the London Auctions, one being a blue Dordin cock from Ron Aldridge of Luton which Frank Hall kindly picked out for me. It couldn't have been a better choice as when paired to a hen I purchased from Alfie Wiseman this pair filled my lofts with winners. They included my good cock 'The Machine' which won fifteen 1sts for me, including 4th Open Combine Berwick over 11,000 pigeons in 1982, he sired my 1992 Berwick winner 'Sparrow Hawk' being 5th Open Combine with over 10,000 birds. Sire of 'The Machine' is also grandsire of my 1992 Thurso Combine winner 'Blue Hawk' and also grandsire of my good pigeon called 'The Combine Cock', because of his five Open positions from Berwick to Thurso. I also had pigeons from Les Massey and Pat Newell which had a hand in the breeding of 'Blue Hawk'.

In 1982 I went with Bob Taylor to visit Arthur Beardsmore. We purchased a pair of pigeons which Arthur said would breed winners straightaway. How right he was, because these must have been the finest pair of breeders in London. They have bred winners racing from 50 to 500 miles. The list is endless. One such pigeon sold to Peter Pedder is sire and grandsire of at least 12 Fed winners, and also 2nd and 6th Open Combine.

In 1984 I moved to Melba in Goffs Oak, which was an upheaval in more ways than one. Les Massey had packed up pigeons and gave me his old loft, in fact, he put it up for me, it was just a question of keeping the birds fed and watered, as I hadn't decided where to put the lofts permanently. It was about this time that John Leisi packed up, and I bought his complete team, including a squeaker, which turned out to be my 1992 Stonehaven Combine winner called 'The Kestrel'.

This pigeon has caused much sorrow as well as joy. In 1988 from Stonehaven, he sat out for 23 minutes, causing Les Massey and myself near heart attacks, but ended up 45th Combine. It's a day I am not going to forget. But he's made up for it, he was even runner-up for the car in 1991, in which he ended up 10th Open Combine Berwick. He is a very versatile pigeon he flies short and long races being 14th Open Combine Thurso (500 miles) in 1990, 13 hours on the wing.

It was in 1987 that I started to get my team of pigeons together. I had by this time relocated my lofts to the bottom of a big garage-cum-shed, but I felt that I was not getting the best out of my pigeons and never seemed to get the right condition on them. I carried on in this way for a couple of years, and then had a good re-think about the loft and pigeons. As I was racing in the strongest club in the Fed, the Cheshunt, and being one of the most westerly fliers, I decided to concentrate solely on Combines, and hopefully have a better chance of scoring. It was then that Les Massey came in again to help. We sat down and started to work out a plan to put the lofts in the roof of the garage; it was quite a big undertaking. We started in the autumn 1990 and it took a month to complete. The most important thing in the loft to me was the ventilation. I wanted a loft with a good clean air environment, and not a stuffy pigeon smell. We did not go for the traditional apex roof, but a flat roof with square chimneys in it, which cost next to nothing to make but has the desired effect.

The lofts contain three compartments 6ft x 8ft, there are 12 boxes in each, but I only use at the most ten boxes, as one contains my nest bowls when I am racing. As I race all Widowhood I find having three separate compartments makes it easier to manage. I try to keep things as simple as possible, as I don't get a lot of time with my birds. I normally start off racing pigeons I want for the Combines at Doncaster, jumping most of my old pigeons straight in without training, but I do train the yearlings and two year olds. I race the pigeons I want for the Combines every race until the first Combine, then I stop them every other week. Perhaps I feed unorthodoxly but I only break my pigeons down on a Saturday or Sunday depending when they race, as I feel too much heavy food after a hard race doesn't do the pigeons any good at all. I feed all my pigeons on the floor, I do not feed in the boxes as it takes up too much time, but I also feel that certain pigeons like different grains and they pick out what they want when you feed on the floor. I buy all my pigeon food from the farm and mix my own; after all that's where the manufacturers buy theirs from.

In 1991 with everything in the loft finished I was looking forward to racing, the birds looked a hundred times better, their condition was marvellous and I had the feeling I could do well. It certainly proved to be a good year, the highlights being 2nd Open Combine Berwick, I topped the East Herts marking station three times (over 1,000 birds), I was runner-up for the car, and took ten Open Combine cards and won eight 1sts in Cheshunt club.

Little did I know 1992 was to prove even better, it has been a wonderful year. Bryan Cowan has always been around watching me time in when I've

Having a look at his own wing 'The Kestrel', the LNRC Thurso National winner.

scored well so he was invited back as my lucky mascot (by the way lads I'm his agent). I scored well in the Berwick Combine, being 5th and I won Stonehaven Combine. With these two good positions behind me, people were saying you must win the Tommy Long Trophy but, as I said to Bryan, you have won no trophy until you time in from Thurso, because to me that's the daddy of them all. I was remembering back in 1982 when I won 4th and 6th Combine, and the same things were being said, and I never timed in. In fact, I did not see a Thurso pigeon for a week. So it was a good feeling when I timed in my Thurso bird, about the time I had predicted, it was only then that I said to Bryan, "I think we are in with a chance of winning the trophy". But once we had got up to the marking station and I found I was in with a chance of winning the Thurso Combine, that really put the icing on the cake. I did win the Tommy Long Trophy, but I feel I must say a word for the runner-up Steve White, he put up a very good performance which would have normally won the trophy, he was 9th, 3rd and 2nd Combine and then won the YB Combine race.

   In finishing I would like to think this article gives novices something to think about, especially when purchasing pigeons, never turn a good pigeon down because it is a cross and not a pure bred. If I had done that I wouldn't have had the success that I have had, go by your instinct and not always the pedigree. Plus my two Combine winners are aged six and seven years, many fanciers would not have kept them that long.

"You Need a Bit of Luck"

by JIM RICHARDS, Basildon

1st London NR Combine Berwick (Thurso)

When Colin Osman phoned and asked me to write an article for Squills, I said to him that I had achieved my ambition of winning a London NR Combine race. By being asked to write an article makes this a season to remember.

I am not going to bore readers with my results, because as far as I am concerned last year's results are now history. My aim is to do better than last year. I shall try and give some insight on my way of racing pigeons. I race total Widowhood and have done so for the last six years. But you must have a bit of luck.

Six weeks prior to pairing my widowers, all my birds are fed barley only. I pair my Widowhood team up on February 13 & 14. The older cocks usually have the same hen from previous years. With my yearlings I like the birds to have a choice rather than forced pairings. I call these 'Love Pairings'.

All birds have previously been wormed and treated for canker etc. I try and give the birds an open loft, whilst sitting on eggs, with plenty of baths. Also plenty of nesting material outside the lofts. All this keeps the birds in fine fettle.

I use Versele-Laga various food mixtures all through the year plus breeding pellets whilst rearing the youngsters. When the cocks start to take an interest in going to nest again I remove the nest bowl along with the youngsters and hens to a loft with a good layer of straw and plenty of wood chips. So into this loft goes all the young birds (aged about 14-15 days) with all the hens to finish rearing them.

I do not allow my cocks of sit a second round of eggs. The hens will rear the young birds alright, in fact, better than otherwise. Also I find that I can wean my youngsters a lot earlier. After removing all the hens and youngsters, the Widowhood lofts are thoroughly cleaned and disinfected out and I even use a gas burner. All of my race lofts have a good layer of straw on the floor. I clean the widowers' boxes out twice a day and use the gas burner every time. This also helps to keep the lofts warm earlier in the season. I close all vents shut and make sure that none of my cocks can see out of the loft. The front of my lofts are covered with Norplex for warmth.

Each cock is locked in his box and his only freedom is when at exercise,

which is one hour am and one hour pm. But no exercise on the day of basketing. They are fed and watered in the boxes from galleypots. The yearlings
usually take to the system alright, but sometimes one or two misbehave. I
find by locking them out of the loft all day they soon toe the line, but sometimes a little patience is needed. During this time my cocks are given an
eight-day pigeon tea treatment and are fed depurative mixture.

On the Tuesday and Wednesday before the first race I give my cocks a toss
from about eight miles single-up. The hens are waiting in the boxes for them.
No contact is allowed and after a few minutes the hens are removed (the
youngsters by this time have been weaned). The hens are fed barley until the
day before the cocks are basketed. They then are fed the Widowhood mix for
you must look after your hens. I keep mine in an aviary which I can close up
when the cocks are exercising. I also put my hens in separate boxes from
Wednesday evening to keep them rank for the cock birds when they come
home from the race. If I find that the cocks are not flying too well I flag them
as I like to have them flying for an hour every exercise period, more so as the
longer races get nearer. I sit and watch my birds at exercise and they tell you
when they are on form. I watch them through my binoculars and find that
when certain birds mess me about not wanting to go in or fighting all the
other birds, that is when they are on super form. It's hard to explain but you
know when they are right, 'they tell you'! No exercise if raining. I have no
set time on how long the hens stay with the cocks on race days, I usually
remove them all about 4.30pm.

I use the basic Widowhood system, my birds are broken down until
Tuesday evening when they are fed Widowhood mixture. Gradually feeding
to appetite by Friday. On the longer races when the birds are in the basket two
days, my birds are on Widowhood mix for five days previous. I use Versele-
Laga mixtures and tonics etc. I treat my birds with Ridsol after all Combine
races, also I hand bathe my cocks in warm water every Sunday after a race.

I try and keep my system very simple, there are no secrets and no doping.
The only dopes are the people who think I use it. I do add extra maize and
peanuts to the feed as the races get longer also I use Red Band and hemp. But
these must be used with care if you require your cocks to race from the first
race to the longest as I do. My birds have to go every week except Thurso but
get a 40-mile single-up on the Sunday before Tuesday basketing. My aim is
to have my birds on song for the longer races in June and July, with Thurso
being my main race of the year. Any averages won are a bonus, but I don't
worry about averages at all. 'If you win 'em, you win 'em'. I class YB racing as training for the future and hold no esteem in YB winners. I would like
to mention the London NR Combine Thurso/Berwick race, when I read in
The Racing Pigeon an article by press officer Jean Andrews, that LPW
would be presenting a first class pigeon, I decided to send an extra four yearlings to the race, with the hope of winning this bonus prize. One of these four
yearlings, now called James won 2nd Open.

The weather at Thurso was awful and the birds were brought back to
Berwick. So one of my four yearlings that I intended to keep at home only

flew Berwick 305 miles. LPW decided to give a pigeon to the 2nd & 3rd winners as well so I won a first class pigeon after all. I said you have to have a bit of luck!

To end this article I would like to explain to the Squills readers how my 1st, 2nd, 3rd & 4th London NR Combine winners were bred. The base of my family of birds are Busschaert which I crossed with LPW Lefebre-Dhaenens, hopefully to add some speed. I paired a good 1988 Busschaert cock to a Lefebre-Dhaenens hen, which was a sister to The Rascal, 3rd LNRC Stonehaven. This pair bred Tommy Three, 3rd and 8th LNRC Thurso and 5th Essex Combine Berwick. Tommy Three paired to a daughter of 7th LNRC Berwick winner bred Eddy the Eagle, 1st LNRC Stonehaven. Eddy the Eagle paired to another sister of the Rascal, bred James, 2nd LNRC Berwick/Thurso winner. A sister of the 1988 Busschaert bred Tommy Two, 2nd LNRC Stonehaven and the Rascal is sire of Nellies Boy, 4th LNRC Thurso.

As I said before you need a bit of luck. I hope you can understand the above and thank you for taking the time to read this article. Keep your birds lean and mean and have confidence in your birds and yourself.

Roundabout Triumph in the Mighty Up North Combine

by IAN JAMIESON, Morpeth

1st Up North Combine Provins (511 miles)

The Beginning

I first became involved with pigeons at the age of 12 when I comman- deered an empty hen-hut on my grandmother's small-holding in the then mining village of Shillbottle, Northumberland. I lived some four miles away in Alnwick and travelled by bus every day after school to tend the few strays I had acquired from the farm buildings. A number of my schoolmates in Alnwick had small backyard lofts and we would organise our own makeshift races, sending the birds by rail to be liberated by Stationmasters at the various stations up to about 100 miles. This was our apprenticeship before eventually joining the local homing society in Alnwick which is in the Coquetdale Federation and part of the mighty Up North Combine. The distance between the front and back lofts in this organisation is about 100 miles. Alnwick lies about 70 miles into the Combine and some of my old schoolmates still fly in that club to this day. Names like Mr & Mrs Mallaburn, Murphy & Shepherd. Thompson & Hope, D Hayes, J Brown and others have all scored well over the years in the Combine races against upward of twenty thousand birds. I myself moved from the area whilst in my teens to study and work in architecture which involved much travelling and pigeons quickly became a thing of the past. My one young bird season in the senior club is however well remembered as my novice status was broken in the Young Bird National race from 200 miles. That was all back in the 1950s.

The Restart

It was not until 1987 when, having moved back to Northumberland with my family, I met an old schoolmate Dennis Fife. It so happened that Dennis was living only some five miles from us and kept a fine team of Widowhood racers. It wasn't long, after a few visits to his place that the urge to get involved in the sport took over. Dennis has a huge collection of books and videos on the 'Modern' sport and I had use of these whenever I wished. Over the next six months Dennis and his wife Eileen spent many, many evenings with Yvonne and myself just talking pigeons and we must have studied every article and loft report available in that time. I have no doubt whatsoever that we owe a great deal of our current success to those friends who did everything possible to get us restarted on the right track.

The Birds
I took the advice of Dennis with regards to stock birds and indeed jointly purchased with him a top Janssen hen 'Superdam' at an entire clearance sale. This bird had already bred nine Open, six Fed and two Combine winners. Her mate 'Superstud' was purchased by Dan O'Donohue of Southern Ireland and I persuaded him to leave the two birds together. The intention being that Dan, Dennis and myself would have them for one year each in their breeding prime. In actual fact the pair have never left my loft and we simply shared the young birds for as long as they remained fertile. Dan has never seen the bird he purchased at a high cost and I am grateful to him and Dennis for the trust they placed in me. The two birds are still together sharing a well-earned retirement as they are now responsible for many more winners.

In addition to these I purchased two proven Busschaert stock birds. The cock 'Pied Prince' originated from G Busschaert's clearance sale and is from 'Jeroom' x 'Sara'. The hen purchased locally is from a brother/sister pairing, both from the famous 'Box Pair'. In general our current race teams are based around these four birds and some of their direct children are retained for stock also.

The Lofts
After a short spell with a small loft at Morpeth we flew young birds and immediately won 1st Fed, 2nd Amal in the Young Bird National. This fuelled the fire and with Yvonne taking a keen and active interest also we decided to find a location with space to plan a suitable loft. Within a year we had moved to our current home on the outskirts of Amble and successfully applied to join both the Saturday and Wednesday clubs which have about 27 members each. This done the lofts were planned and construction immediately started. There are six sections in the 52-feet length. Two central sections for cocks, two adjacent sections for hens and two further extreme sections with aviaries for young birds.

The central sections have large dome rooflights which allow plenty of sun to stream down on to the boxes in high summer. All sections have plenty of light and adjustable ventilation and are bone dry with walls, floor and roof having built-in insulation. The birds are cleaned out twice a day 365 days a year and no floor dressing is used except on the rare occasion when rain may have leaked in through open doors. Emulsion paint with Duramitex mixed in is used internally each year and all boxes and perches are removed and scrubbed. Preservative stain is used externally.

The System
When considering the system to use at our new location two main factors governed our decision. First we wished to fly both cocks and hens, and second we could not afford the time to train the birds on the road during the racing season. We eventually decided on a Roundabout system where cocks and

hens could be treated identically. Yvonne plays an equal part in managing the birds where all chores are done together before and after normal working hours. We manage two old bird teams – one for inland racing up to 312 miles and the other for Channel racing up to 600 miles. Our season begins with the pairing of the stock birds and the inland team in mid-February with stock eggs being floated across. Three weeks later the Channel team are paired and the second round of stock eggs are floated. The stock are allowed to rear their third round and are then parted.

When young under the respective race teams are about ten days old the hens are moved to the adjacent sections and the Roundabout system started. Cocks with the young during the day and hens at night. During this period between three and six training tosses are given and then exercise around the loft only. Hens 6am till 7am and cocks 7am till 8am. At 5pm a similar pattern takes place. When racing starts the cocks go on Saturdays and the hens on Wednesday flying to the young initially. Cocks never see the hens during this period until the young are weaned at about 24 days. Our inland team go every week while the Channel birds go every two weeks. This gets the Channel team into a two-week routine which corresponds with the Up North Combine Channel programme. When Channel racing begins, about eight weeks into the full programme, cocks and hens in that team go together and have performed equally well. All yearlings in that team go to 511 miles with their older loftmates. Our longest race is 600 miles. As soon as racing is finished all racers are allowed to rear latebreds before being parted for the winter.

The Feeding

For no special reason we have always used De Scheemaeker corn and found it to be very satisfactory. During the racing season the standard breakdown method is used using Depurative early in the week and building up to full Widowhood mixture prior to basketing. The adjustment from Depurative to strong mix during inland racing depends on a number of factors such as how tough the previous race was, the weather conditions during the current week, and the weather forecast for the coming race. When in doubt we bring the stronger mix in sooner rather than later.

The birds are fed after each exercise, on the cleaned floor, by hand and are always given as much as they want without corn being left over. A handful of fresh grit is given at each feed and minerals are before the birds at all times. During Channel racing that team receive 50% Widowhood mix and 50% Young Bird mix with a few peanuts added for a full week before each long race. Garlic is added to the water every Tuesday and Multi-vits the day before basketing. After the racing season the strong mix and linseed is continued until the moult is complete and then taken down to 50% good mix and 50% Depurative for the winter months.

The Results

The ambition of every flier in the Up North Combine must be to win a

Classic race in that organisation. As I said before there is about 100 miles between the front and back lofts and we now lie about 60 miles into the pack. Clearly in such a huge range, weather conditions may play a part in where the leading birds are likely to be. Nevertheless leading birds into all 21 Federations have to be in tip top condition in such fierce competition.

Our first Channel race this year was from Lillers, 369 miles, with 19,867 birds competing and we took 8th position with a yearling cock flying to our fantail trapper. He was followed by six loftmates to take 48th, 59th, 61st, 107th, 151st and 170th Up North Combine, to make a great start to the programme for us. Two weeks later at Beauvais, 435 miles, 11,500 birds were competing and we took 17th position with a two-year-old hen who flew for a full two minutes before entering the loft. She was followed by three loftmates to take 73rd, 93rd and 111th Up North combine – one of which again was the 'Lillers Cock'

At Provins, 511 miles, we hit the jackpot with 'Speckles' the nestmate to the 'Lillers Cock' 5,430 birds were liberated at 6.25am and we timed in with a good trap at 5.51pm for 'Speckles' to make a velocity of 1039 ypm. His aunt followed him to take 76th position securing the two birds average trophy also and again the 'Lillers Cock' came to take 85th position and more valuable points. To win the Up North Combine race was more that we dared expect – the last time in Amble was 1939. The 24 hours or so to confirmation coming through seemed much longer.

Along with news of this race I was contacted by two local scribes, Wiley Flight and Barry Pearce to say that we were leading the Up North Combine Channel Averages with one race to go. Up till then we had never even thought of Channel Averages and I must admit that had it not been for this news most of our Channel birds would have been stopped after performing so well. However, this was probably a once-in-a-lifetime opportunity so we again began a further ten-day preparation for the Channel birds – first with complete rest and then a gentle build up to basketing for the last event.

On 23 July at 7.30am, 15,422 birds were liberated at Lillers into a light head wind turning on route. During the day Yvonne and I could think of little else. There was the usual visits by clubmates to discuss the ETA etc. At 3pm we were past ourselves with anticipation. At 3.57pm we timed in the three-year-old sire of 'Speckles' and 'Lillers Cock' to take 4th Up North Combine at 1309 ypm. He was followed by four closely related loftmates including – who else but his son, 'Lillers Cock' – again to take further Up North Combine positions. These Up North Combine results, 14 in total in the first 150 birds from four Channel races, will at the time of writing this article make us probable winners of the coveted Up North Combine Channel Average Trophy and if that is so we wouldn't mind if we never win another race. I have not mentioned our inland team but they too performed well enough to win the club Inland Average for the last three years running.

Almost There

by ARNOLD DAVIES of Davies, son &
Cockcroft

2nd Welsh Grand National Flying Club five
times – 1986, 1988, 1990, twice in 1994

I would like to thank the editor on behalf of the partnership for his invitation to write this article for Squills 1996. The Davies, son & Cockroft partnership was formed in January 1980 and consists of myself, Arnold, my son Keith and our friend, Doug Cockcroft, I would like to say at this point that the three of us take a very active role in the management and care of our pigeons, and it is a partnership in the true sense of the word.

A few words now about the partnership: I am employed by the Ford Motor Co Ltd at Bridgend, as a security guard, working 7 days continental shifts. Not the best arrangement to fly pigeons, as I have to work Saturdays and Sundays and have to rely heavily on Keith and Doug to keep our system going. Keith is employed as a wood machinist and is responsible for making the nest boxes, fittings and alterations to the lofts, as well as doing his bit with the birds, cleaning, exercising and feeding. Doug worked for British Rail for 28 years and has now retired, so is able to spend a lot of time at the lofts, which is an asset to any partnership, enabling us to have a regular system with the birds.

Since his retirement Doug has taken up gardening in a serious way, and has a garden to be proud of, which we all benefit from!

Before the partnership was formed we all flew pigeons in our own right. Doug has been involved for over 33 years, and his performances from Lerwick (600 miles) are known throughout Wales. Keith and I started flying in 1975 with little success at first. I can remember our very first race in Trehafod HS which was one of the strongest clubs in Wales at the time; when our first bird arrived, and knowing nothing at all about receiving pigeons, I caught the young bird, wrote it's ring number down for future reference, took off the rubber, put it in the thimble and clocked it. When the clock was read in the evening we were 3rd Club, beaten by a decimal for 2nd place! Needless to say it was our very first lesson of many in pigeon racing and we soon got our priorities right.

The 1977 season saw our first 1st prizewinner from Perth (330 miles) winning 1st Club, 1st Section, 4th Open Welsh NR Fed. We were to top the Centre Section again that year. In 1978 we were 4th Club Lerwick (597 miles) and other positions. 1979 saw us with 1st and 4th Club Lerwick with

Left to right: Doug Cockcroft, Arnold Davies, Keith Davies.

twice 1st Section – improving and learning all the time.

So as can be seen there was a good partnership in the making, which has proved to be the case over the last 16 years. It was decided at the outset that we would fly the Widowhood system. At the time we were very friendly with the late George Fear, of the Fear Bros partnership of Pontypridd, the performances of this partnership were phenomenal. One day we were in conversation with George, and the topic of Widowhood flying arose. George mentioned that he was friendly with a fancier in the Midlands with whom he had exchanged pigeons. The fancier was so pleased with the birds, and the system to fly them, that he in return sent George the system to fly Widowhood, which he had obtained on a visit to Belgium.

It was all written down in great detail and covered all aspects of the system, George said he was 'too long in the tooth' to start flying this system, and if we would like to have it we could. Of course we readily accepted his offer, and the rest is history. We still use the system today with a few additions of our own, and it has enabled us to win from 60 to 597 miles. Thank you deal old friend we are indebted to you.

The first season for the partnership 1980 we raced 12 yearling cocks on Widowhood The results gained from these cocks was something you dream about: 1st, 2nd and 3rd Section with up to 5,000 birds competing, this was

'The Tic Hen', 1st Club, 1st Section Fed (756 birds), 1st Centre Section WGNFC (1,480 birds); 1st Centre Section Welsh Combine (3,154 birds); 2nd Open WGNFC (2,653 birds); 5th Open Welsh Combine (6,618 birds) Newtongrange YB.

repeated again plus 3rd Open (15,000 birds). We finished at the end of the 1980 season top prizewinners in our club, Trehafod HS, which was at the time one of the strongest clubs in WNR Fed. This position we held on to for 15 of the last 16 years, being runners-up on one occasion, beaten by a loft based on our birds and flown by good fanciers.

The base of our birds came from a pair of pigeons I bought as young birds in a sale in 1980, at the Hen & Chickens public house in the Midlands. They were direct G & M Vanhee Janssen and I paid £250 for the pair. With me on this occasion was the late Walt Thomas, the 1969 Lerwick WGNFC winner. I can recall his words to me: "Boy you must be made paying that amount for a pair of young birds". I did begin to doubt whether I had done the right

'The Nom Hen', 2nd Open WGNFC (2,825 birds); 2nd Open Welsh NR Fed (3,176 birds); 3rd Open Welsh Combine (5,577 birds) all from Roslin Park.

thing! However time has proved that the right thing had been done, as this pair of birds are the base of our loft today, and are responsible for over a hundred 1st prizewinners. Their children, grandchildren and great-grandchildren are still breeding 1st prizewinners today. So successful was this pair that I wrote to G & M Vanhee requesting to buy another pair of birds bred the same way as the original pair, but was regrettably informed the birds had been sold to Taiwan for a large sum.

Later we introduced Janssen Van den Bosche from Herman Beverdam, Mr & Mrs George Litherland and John Kirk & son with great success.

In 1987 the great partnership of Lee & Cooper was dissolved, and the birds came up for sale. I considered this partnership to be one of, if not the best in Wales, so we didn't hesitate and purchased two pairs of birds, which have been very successful for us. The strain of these birds are Lefebre-Dhaenens

'Lerwick Cock', 1st Club, 1st Section, 2nd Open WGNFC Lerwick; 1st Section, 2nd Open Welsh Combine Lerwick.

and Verheye. We still make seasonal introductions of Janssen, Janssen Van den Bosche, Lefebre-Dhaenens and Vereheye, as we believe you must always be on the lookout to improve your stock, and should leave no stone unturned to remain successful. The above strains have enabled us to win from 60 to 597 miles at all levels of competition.

We have two racing lofts, one for old birds and for one for young birds, plus a stock loft. The three lofts are former British Rail cabins which we were able to purchase through Doug for a nominal fee. They are all 32ft x 8ft, and have flat roofs, and are sectional. They were assembled by Keith assisted by Doug, myself and a few friends and are double lined which enables them to maintain a reasonably constant temperature. They are ventilated at floor and roof levels, as we believe you must have good ventilation, but it must be a

'Double Top', 1st Fed Beeston (9,600 birds); 1st Fed Thurso (4,228 birds); 2nd Open WGNFC Thurso (3,780 birds); 2nd Open Welsh Combine Thurso (9,691 birds).

controlled ventilation, as this plays an important part in bringing your birds into form and maintaining their health and fitness.

The cocks are paired up in the second or third week in January, and are allowed to rear a pair of young birds. When the birds are 12 days old we take the hen away with one young bird allowing the cock and hen to each rear one young bird. A week after we wean the young birds we re-pair the birds. The hens normally lay within 8 to 10 days, and they are then allowed to sit these eggs for 10 to 12 days. All the hens are then removed, even if they have only been sitting for four days. We feel this is very important. The cocks are allowed to sit the eggs on their own, which usually lasts for two to three days, then all pans, eggs etc, are removed from the box, and the boxes are then given a good clean and are sprayed with a good disinfectant. The cocks are

'Windbeater', 1st Club, 1st Section Fed (1,358 birds); 2nd Section WGNFC (2,382 birds); 2nd Open WGNFC (2,892 birds); 2nd Section Welsh Combine (5,072 birds); 9th Open Combine (10,399 birds) Crieff as a yearling.

then on Widowhood. For the first two or three days we give them an open loft for one hour in the morning and evening. During this time they do very little flying as they are still upset after the hen has been removed. On the fourth day we close the loft door for 20 minutes and the cocks are made to fly. This is repeated in the evening, increasing the time the cocks are made to fly every day for about five minutes, until the cocks are flying one hour morning and evening. This is the only time the cocks are forced to fly, thereafter they are shut out one hour morning and evening and allowed to do as they like for that time. The only time the cocks are not allowed out to exercise is if there are high winds and heavy rain as there is a danger of them being blown into elec-

'The No 1 Cock', cock of the No 1 pair. Bred by G & M Vanhee.

tric cables that run across the front of our loft. If the weather permits, we give the birds a few 12 mile tosses, making sure the hen is always in his box when he returns. We have found over the years that this tossing is not essential and we have won many races and positions putting them into the race straight from the loft, providing of course that they are exercising well around the loft which they should be by this time.

When we started Widowhood we mixed our own feed from the system given to us by George Fear. After a few years we found that there were many feeds available on the market that were similar to our own mix, so we decided to buy a few top mixes of what we fancied and then mixed them with the same results, and the same applies to the depurative. We use the breakdown system when we feed our cocks on returning from a race on Saturday they are fed depurative until Wednesday morning. They are then put back on the

'04 Cock', son of the No 1 Cock. Responsible for scores of winners up to National level.

racing mix, until they are basketed for the next race on Friday. The problem we have found with this system is that some cocks react differently. For example some cocks have to be put back on the racing mix on Monday after the race, or they do not come into form for the Saturday, and only cocks must be kept on depurative until Wednesday evening or they will come into form too soon. This is something you have to find out for yourself by observing your birds. The cocks are fed all they want to eat and are fed individually in their boxes. This we have found the best way to feed as we can observe what each bird is consuming. Fifteen minutes after each feed any food that is left in the pots is removed as we find this very important in maintaining the birds appetite.

All our young birds are treated for cocci and canker when they are parted over, which is normally between 21 and 23 days old. They are given a multivitamin once a week usually on Wednesday, and are fed all they need as we feel it is very important that they have sufficient food to allow them to mature properly as they are our future. We feed twice a day, morning and evening. The YBs are trained up to 30 miles before the first race which usually takes about 12 training tosses. They are then exercised twice a day, one hour morning and evening, and after they have had their first race they are not given any more basket work, only exercise around the loft. This system works for us as we have won many times from 60 to 300 miles with young birds, at all levels of competition.

We blanket treat all our birds for canker, cocci and worms before racing and breeding, after this treatment they are put on a multivitamin for two to three days. We do not treat during races unless we have to. All birds are injected against paramyxovirus every 12 months, and we have found that this does not affect them in any way. For the first ten years of our partnership we have never needed to visit a vet, but things have changed so much over the last few years with so many different viruses and diseases that affect the birds, it has now become unavoidable and inadvisable not to pay him a visit. We have a very good vet in Frank Harper, one of the best, if not the best in his field. The quality I like about him the most is that he will not treat your birds for the sake of treating, and the advice he gives is second to none. To win races out of turn it is essential to have good pigeons. These are first selected on conformation, feather and constitution, the final selection is made by the basket, without doubt the best and truest selector of them all.

The environment that the birds are kept in is most important, a dry well-controlled ventilated loft is a must. This will ensure that the birds remain fit and healthy, which we all know is essential for success. You must be dedicated and observant and pay attention on detail, for it is no good knowing what to do if you are not prepared to do it. Over the last 15 years we have won many 1st prizes at all levels. We were champions of Welsh North Road Federation winning the silver pigeon on two occasions, and runners-up once. The WNR Fed is reputed to be the biggest Federation in the UK, with up to 1,500 members, and 21,000 plus birds a race. We have topped our section with up to 5,000 plus birds from every racepoint from 60 to 597 miles, and have been highest prizewinners in our club, Trehafod HS, 15 out of the last 16 years. We were runners-up in 1991, beaten by our own bird and a good partnership. When this partnership had a sale in 1993 we bought the principal bird back that had been bred by us, and what an investment he has turned out be.

We have had many thrills and much enjoyment from our birds over the years and we hope this will continue for some time to come, however two of the biggest disappointments we have ever experienced was when our stock loft was broken into in 1988 and 1995, when 24 and 18 birds respectively were stolen. This is something I would not wish on anyone, and I thank God the person or persons concerned are in a minority in our wonderful sport.

Finally our sincere thanks to our wives, Diane, Clare and Maureen for supporting and putting up with us for the last 20 years.

Winning at all Distances

by R O JONES, Resolven

1st Ludlow (60 miles) 1st Fed &

1st Lerwick (600 miles) 2nd Fed

I thank the editor of Squills for asking me to write an account of my racing pigeons based on their National performances through the 1980s and 1990s. I had the honour of writing in the 1983 Squills when the team was responsible for successes mainly in the 400-600 mile races. They were, you might say, the Jones strain, based around Krauth and Vandevelde bloodlines with a later influx of Kenyon bloodlines. They were hard day pigeons, not great lookers, but they had fantastic stamina and would fly all day if need be.

It became apparent, in this day and age, that one needed to be able to compete throughout the programme if possible, especially with the establishment of the Widowhood System. However my love in pigeon racing is to win the distance races and in trying to establish a team of all rounders I was determined not to detract from this aim. We easily won the Federation Scottish Averages this year to prove it is working.

Two pieces of luck came into play during the mid 80s. Firstly I started exchanging birds with R Griffiths & sons of Flint. Their team was very similar to mine being a good honest consistent family especially at the distance based on Delbar and Channing bloodlines. Their annual introductions up to the early 1990s reinforced my own family of birds. Then secondly, the continued National success came to the attention of Michael Johnson of Clwyd Lofts. Basically he wanted to promote his Jan Aarden pigeons and wanted me to test them out for him with the obvious mutual benefits. In 1986 he sent down three pairs of YBS for me to try out on the road. The following is briefly what happened:

Pair 1 – bred from 'Kipo' bloodlines – Jan Aarden, smokey chequer hen, WHU86B9541 – raced as a YB only winning 1st Fed at 80 miles. Dam of 1st Lerwick, 10th National 1990. Chequer cock, WHU86B9542 – raced young and old winning 2nd Thurso, 11th Section National. Sire of 1st Thurso.

Pair 2 – inbred to 'Super Stud', Jan Aarden, light chequer hen, 86B9543 – raced as YB only winning 3rd Welsh Combine (253 miles). Grand-dam of 1st Welsh National Carlisle 1992. Light chequer hen, 86B9544 – raced as YB and old, won four 1sts including 1st All Wales Combine Roslin (289 miles) beating 16,087 birds. Dam of 1st Thurso.

Pair 3 – bred from direct Ko Nipius birds, dark pied hen, 86B9545 – lost

**Blue w/f hen WHU94S, 2nd
Thurso, 4th Fed, 10th Section
1996.**

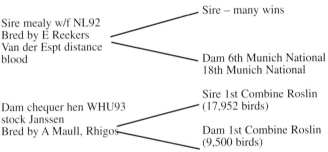

Sire mealy w/f NL92
Bred by E Reekers
Van der Espt distance
blood

Sire – many wins

Dam 6th Munich National
18th Munich National

Dam chequer hen WHU93
stock Janssen
Bred by A Maull, Rhigos

Sire 1st Combine Roslin
(17,952 birds)

Dam 1st Combine Roslin
(9,500 birds)

training as YB. Dark chequer cock, 86B9546 – raced young and old, won twice 1st and was sire of 3rd Welsh National Lerwick 1991.

What luck, five of the originals when paired to my own family, bred winners from the start, but more importantly the offspring started winning at the shorter distances also. Over the ten years since their introduction they have produced an outstanding type of pigeon, well-built apple-bodied frames with excellent feathering and rich coloured red/yellow eyes and extremely tame. Perhaps the most important factor is that the birds have still maintained that vigorous stamina so evident in the early days. They always look well after races, never tired out. For example, last year's Lerwick winner, winning 1st Section, 10th National in a hard race, on returning early the following morning went straight to her water drinker and started pumping it into her two week old youngster. You would never have thought she had just flown 600 miles in poor weather conditions.

The final chapter in the team's development has been the introduction of Janssens from top racing lofts. I've had two or three from each loft asking the owners to select what they think to be their best. They were Tracey & Dean Peart of Gloucester, Alan Maull, Rhigos; Gareth Watkins, Tonypandy and Brian Clayburn. These birds are very different to my own being large robust types and with a shorter wing length ratio to body size. I have blended the best of them into four pairs that breed winners every year since their introduction in 1992.

Again Lady Luck has played a part in the success of the Janssens. A Dutch YB came into my loft in August 1993. I reported the bird and found that it was bred by a fancier by the name of Cor Kouwenhoven of South Holland. He is the current leading flier in his area. He sent me the pedigree which is full of direct Janssen birds. This cock has bred me winners and prizewinners in every nest.

Since the introduction of the Janssens I have only experiemented on two occasions with pairing my own family with a Janssen as a cross. The first bred me 1st Lerwick, 4th Section Welsh National this year and the second pairing bred me 2nd Thurso, 10th Section National. Is this the start of the Dream Team? Only time will tell. At present though these Janssens are kept pure and are winning YB races through the programme and OB races up to 400 miles.

Now you know the background to the birds, what about my methods of flying them? Well I'm lecturing at a local College of Further Education by day, run a local youth club on two evenings a week and I have a family with two sons that are very keen on sport. The amount of time I can spend with the birds is confined to early morning and early evening. I must point out that my father lives next to me with the loft strategically placed between each bungalow. He's a retired pharmacist and although not a pigeon fancier as such,

'Thurso Cock', 1st Thurso, 2nd Section 1995 and 1st Thurso, 7th Section. Came after dark, timed in first thing next morning 1996.

Sire bred by Griffiths & son chequer cock 87N98191 2nd Rennes, 4th Fed 3rd Rennes 10th Fed

Chequer cock 4th Rennes – only four on night

daughter of 6th Open NFC Nantes into N Wales

WHU86B9546 bred by Clwyd Lofts

Dam blue WHU87 ring removed sister to 3rd National Lerwick

Chequer hen 86B9547

Chequer pied hen WHU93M, 1st Lerwick, 4th Section, 37th Open 1996; 11th Section Thurso National 1995.

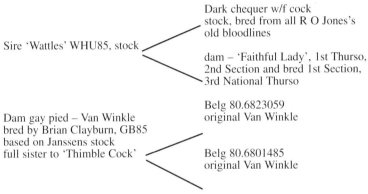

Sire 'Wattles' WHU85, stock

Dark chequer w/f cock stock, bred from all R O Jones's old bloodlines

dam – 'Faithful Lady', 1st Thurso, 2nd Section and bred 1st Section, 3rd National Thurso

Dam gay pied – Van Winkle bred by Brian Clayburn, GB85 based on Janssens stock full sister to 'Thimble Cock'

Belg 80.6823059 original Van Winkle

Belg 80.6801485 original Van Winkle

he's always on hand to carry out any odd jobs that have to be done. The tame and trusting nature of the birds is enhanced by his easygoing quiet nature. Perhaps more importantly, it's my father who does any training necessary – a tremendous help because I can watch and observe the condition of each bird on its return.

With my sons being born during the mid 80s I had toyed with the idea of flying Widowhood now that my energies would be needed increasingly with the family. But what about my hens? I was having so much success with them that I decided to keep flying Natural but only using the hens seriously in races from 300 to 600 miles. They would be targeted at certain races and individually trained for these specific events.

I pair all birds up in late February into Widowhood boxes, always having more boxes than pairs of birds. This gives the birds much more room and stops overcrowding which is one of the main causes of failure in racing pigeons.

As stated, I concentrate my racing with the cocks. No bird is raced until May, no matter when the Federation starts its programme. I always get the YBs weaned off the racers before training commences. The cocks are left out

Blue hen 'Lerwick Gem', 1st
Section, 10th Open 1995; 10th
Section, 53rd Open 1996.

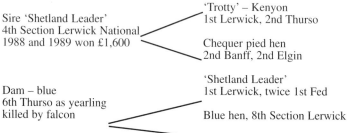

Sire 'Shetland Leader'
4th Section Lerwick National
1988 and 1989 won £1,600

'Trotty' – Kenyon
1st Lerwick, 2nd Thurso

Chequer pied hen
2nd Banff, 2nd Elgin

Dam – blue
6th Thurso as yearling
killed by falcon

'Shetland Leader'
1st Lerwick, twice 1st Fed

Blue hen, 8th Section Lerwick

for approximately an hour morning and evening. They fly for about 20 minutes and clap about for the rest of the time. If they exercise well, there's no training, and vice-versa. The hens are largely ignored during May and are brought into fitness with a series of training spins for the later races. I only spend about 20 minutes in the loft morning and evening during the week. While the birds are out I change the water, top up the hopper and clean any excess droppings. With only a few pairs in each section and plenty of room available it takes no time at all to do the cleaning. The floor has a deep layer of straw which is changed regularly. The loft is dry and cosy, birds happy and contented.

The birds get a light feed on race days and Sundays but get the main mixture of 35% maize, 35% maple peas, 15% beans, 10% tares, 5% dari the rest of the week. They have grit and mineral powder in front of them at all times and get vitamins once a fortnight. They are wormed and treated for canker twice a year. It's a simple system with very little time involved and it works for me.

Now the incentive to get the birds racing hard to the loft can be achieved in a variety of ways. With my cocks I like two year olds and older driving their hens back to lay. I've won most Thurso and Lerwick races with cocks in this condition. For the short and middle distance races I've had a lot of success with racing two cocks to one hen. Again very simple but effective.

With hens I prefer sending them either on pipping eggs or with a YB about

Chequer w/f cock, WHU96S, 1st Fed Ludlow as YB this year and 1st Amal (4,500 birds). Sire and grandsire won 1st Lerwick.

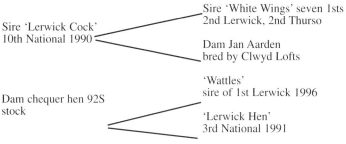

Sire 'Lerwick Cock' 10th National 1990

Sire 'White Wings' seven 1sts 2nd Lerwick, 2nd Thurso

Dam Jan Aarden bred by Clwyd Lofts

Dam chequer hen 92S stock

'Wattles' sire of 1st Lerwick 1996

'Lerwick Hen' 3rd National 1991

a week old. Conditioning pigeons to win the Lerwick race just means that I work on them for about three weeks before basketing with regular exercise and a few training spins. Birds selected for Lerwick are given an easy racing programme before the event. Good quality rest is as important as good quality training in getting birds to win from 500 and 600 miles. The two hens that won 1st & 2nd Lerwick this year only flew 160 miles before the race. The winner was sent on pipping eggs and the six year old hen sent on a week old YB. She won Lerwick last year in the same condition.

My team of YBs usually starts with 40 to 45. They are raced throughout the programme to Newtongrange (289 miles). So far this year they have been successful winning 1st & 3rd Fed and Amalgamation Ludlow (60 miles). The winner is inbred to two recent Lerwick winners. Basically the YBs are treated the same as my old bird cocks, out morning and evening. They are flown to the perch and fed the same as the OBs. They are trained every day for two weeks before the first race and then kept ticking over for the rest of the season. YB racing is not taken seriously, no Darkness Systems etc here. These YBs are not thought of with short-term success in mind. They, hopefully will win Thurso and Lerwick as two, three, four year olds.

I hope somebody gains something out of my article. There's no secrets to pigeon racing – be patient, don't overcrowd and let the birds have peace and quiet would be general rules. Try and get three pairs of latebreds off a con-

'Deano' – Janssen, chequer cock, WHU94S. 1st Fed Hexham as YB, 1st Amal; 1st Fed Hexham, 2nd Amal this year, won six prizes this year so far.

sistent winning loft to start, preferably off their principal breeding pairs. Don't forget the people responsible for all the hard work behind the scenes at Federation, National and Combine level. They tend to take a lot of unnecessary criticism when they are just doing their best to help us enjoy the sport we love.

Finally a mention to my fellow clubmates at Glynneath club. All I can say is that it's a pleasure to know you all and compete against you. The weekend when we get together are times to look forward to and cherish, thanks for your company lads.

The Ones To Do The Job

by RONNIE NEVIN of Nevin Bros of
Coleraine
1st NIPA Dinard Derby

It gives me great pleasure and is a great honour to be asked to contribute an article for the *Squills International Year Book* on its 100th anniversary edition. We started racing pigeons in 1996 under the name of Nevin Bros, but prior to that, while still at school, we would have had a few pigeons flying around home which was out in the country. These had all disappeared except for two red chequers, a cock and a hen. One day these two birds went into an outhouse and my brother Hugh crept up, closed the door, caught the two pigeons, put them in a cardboard box and set off up the road a few miles where they were liberated. They returned home all right and it was then that we decided to acquire a few more pigeons with the idea of racing, we had been bitten by the bug.

Application was made to the strong Coleraine Premier Club with a membership of 40 plus fanciers competing in NIPA which takes in most of Northern Ireland. We were duly accepted and the fun then began. We bought birds from different fanciers also our good friend, H A Montgomery gave us a few pairs which made their mark in the early years of racing. At that time Mr Montgomery held high office in the NIPA and in our local club was made an honorary life member although he retired from racing some years ago.

We had reasonable success over the years winning odd races and even timing in from the distance which gave us a great thrill. We continued racing for some years, when both of us got married and the birds were moved to my address. Circumstances changed and it was in 1975 by mutual agreement that the decision was taken to race on our own which made racing even more interesting with myself retaining the name Nevin Bros.

I moved house again in 1988 where I now live and race but it was not until 1991, after the sudden death of my very close friend Billy Smyth, that I really decided to devote more time and energy to the longer races. That year I purchased a few latebreds from Joe Shore of Comberbach in Cheshire which proved to be a real asset when crossed in with my own bloodlines, one of these being the sire of my 9th Open King's Cup hen and another being a sire of a meritorious award winner. A gift hen from R Hunter & son of Limavady in 1989 has also proved her worth in the stock loft, her dam being a good racer winning from 350 miles and also 1st Sect, 9th Open Avranches, approx-

Ronnie Nevin of Nevin Bros with his NIPA Dinard Derby winner.

imately 490 miles, in a hard race.

The birds are housed in an ordinary garden-type loft made up of five sections plus a small loft which is used for young birds, stock birds or for whatever is required. I would normally pair up in mid-February but this year with extra yearlings I mated up six pairs on January 27 with the idea that the pick of these would go to the Yearling National (350 miles) which was the case. I sent nine, had four on the day of toss and three next morning.

The longer distance candidates were paired on February 25 then after sitting ten days I would swap a few of the best pairs' eggs and let them go back to nest again. This I find helps to curtail the moult and improves form which I think does not really start until they have fed a youngster.

Training would begin about two weeks before the first race with most of the birds. When I see they are flying freely and exercising well around the loft they get a 15-mile toss, then two or three more up to 30 miles. This is to get the mental aspect right for racing as I believe this to be an important factor with the racing pigeon, especially leading up to the distance events. The same training would apply to the longer distance candidates but this year,

Ronnie Nevin's loft at Coleraine.

after rearing their young, they were separated and flown in this condition until the time arrived to re-mate them for the races intended.

The Derby winner was earmarked again for the King's Cup, this year from the new racepoint of St Nazaire (574 miles), but circumstances changed as the time approached for these events. Coming into fine fettle earlier than expected, she was set for the Derby. This turning out to be a very hard race, no day birds and only 37 in race time from a convoy of 2,503. She excelled herself being 1st Club, 1st Sect A and 1st Open on a velocity of 688ypm, this for me was a dream result.

As a matter of interest I am the first fancier to have won the Derby from France in this part of the province although some years back my good club friend A McDonnell won the King's Cup from Les Sables (633 miles). A fine performance.

The following week one of my two candidates for the King's Cup went missing from a 20-mile toss so the two-year-old hen was my only entry, again this turned out quite a stiff race where she finished 8th North Sect, 9th Open on a velocity of 764ypm, of which I am very proud as these two results won the French Diploma which is awarded for the Best Average of these two races.

Dam of the King's Cup Hen won from Bude (308 miles) and was also 1st Club, 2nd Sect A, 31st Open Dinard (489 miles), again a hard race. The grandsire was also in the club prizes three years running from France, best position being 2nd Club, 6th Sect A, 43rd Open, another stiff race.

For the young birds training would be more varied and would start about six weeks prior to the first race. But to start with, when weaned at 20 to 24 days old I make a point of sitting with them while feeding to get their confidence built up so that when I enter their section they do not scamper off into

First Open NIPA Dinard Derby 1997 for Nevin Bros.
Picture by Sid Collins.

a corner but do the opposite, looking to be fed. I also make sure that each bird has a drink for at least a couple of days after which they are on their own. After being basket-trained I would put them on the road starting at three miles and working them along gradually to about 15 miles tossing them in three or four batches after the initial shorter tosses. The idea is to try to avoid the birds of prey which are getting to be a real problem for the Pigeon Fancy in general. Training would continue on to the 30-mile stage when they would then be ready for the first race.

As I am more for the distance, I am always on the look out for the one which I think will do the job, this being the case with the Derby winner. The past few years the young birds, and even the yearlings, do not win too many prizes against the present-day systems, but as long as they are racing consistently this does not bother me. After two or three races I would select a few of those I think the best and stop them. The same would apply to the yearlings, paying particular attention to the condition on their return from the

Another view of Ronnie Nevin's loft.

harder Channel races, as this could be an important indicator as to how they might perform in later life from the distance.

Going back to the Derby winner she won her first prize as a yearling from 308 miles, as a two year old 2nd Club, 4th Section A, 38th Open Dinard beaten by a loftmate, as three year old only entry in the King's Cup race from Rennes (527 miles) she was 174th Open, and then her performance this year.

For feeding I use a good mixture plus depurative when they return from the races, otherwise I have no fads regarding any particular brands of feeding.

For the novice or younger fanciers starting I would suggest they go to an honest fancier with a good working family, whether it be sprint or long distance, purchase a few late breds take his advice and I think they would be on the right track. To finish I thank all those who called personally or phoned with messages of congratulations and I wish you all enjoyable and good racing in 1998.

'Squills'

The original Squills, Alfred Henry Osman
born 12 July 1864, died 30 March 1930